Grundlehren der mathematischen Wissenschaften 289

A Series of Comprehensive Studies in Mathematics

Springer-Verlag Berlin Heidelberg GmbH

Yuri I. Manin

Gauge Field Theory and Complex Geometry

Translated from the Russian by
N. Koblitz and J. R. King

With an Appendix by Sergei Merkulov

Second Edition

Springer

Yuri Ivanovich Manin
Max-Planck-Institut für Mathematik
Gottfried-Claren-Straße 26
D-53225 Bonn

Title of the original Russian edition:
Kalibrovochnye polya i komplesksnaya geometriya
Publisher Nauka, Moscow 1984

Library of Congress Cataloging-in-Publication Data

Manin, ΙΟ. I.
[Kalibrovochnye poli͡a i kompleksnai͡a geometrii͡a. English]
Gauge field theory and complex geometry / Yuri I. Manin ; translated from the Russian by N. Koblitz and J.R. King. -- 2nd ed.
p. cm. -- (Grundlehren der mathematischen Wissenschaften, ISSN 0072-7830 ; 289)
Includes bibliographical references (p. -) and index.
ISBN 978-3-642-08256-6 ISBN 978-3-662-07386-5 (eBook)
DOI 10.1007/978-3-662-07386-5
1. Geometry, Differential. 2. Geometric quantization. 3. Quantum field theory. I. Series.
QA649.M3613 1997
516.3'6--dc20 96-35228
CIP

First Reprint 2002

Mathematics Subject Classification (1991): 81EXX, 14-XX, 18-XX, 53-XX, 35-XX

ISSN 0072-7830

ISBN 978-3-642-08256-6

SPIN 10868052 44/3111 – 5 4 3 2 1 – Printed on acid-free paper

PREFACE TO THE SECOND EDITION

In the seven years since this book was published in English, the exciting interaction between complex geometry and theoretical physics continued to bring new results and ideas to both domains. For this edition, Prof. S. Merkulov has written an Addendum and compiled a list of additional references. Thanks to him, several dozens of misprints in the first edition were corrected as well. I would like to express my thanks to him for this help.

Bonn, January 1997 *Yuri I. Manin*

So here I am, in the middle way, having had twenty years —
Twenty years largely wasted, the years of *l'entre deux guerres* —
Trying to learn to use words, and every attempt
Is a wholly new start, and a different kind of failure,
Because one has only learnt to get the better of words
For the thing one no longer has to say, or the way in which
One is no longer disposed to say it.
— T. S. Eliot, *Four Quartets*

FOREWORD

The 1970s were a transitional decade in elementary particle physics. At the 1978 Tokyo conference the standard Weinberg–Salam model, which combined the weak and the electromagnetic interactions in the framework of spontaneously violated gauge $SU(2)_l \times U(1)$-symmetry, was finally acknowledged to have the support of experimental evidence. The quantum chromodynamics of quarks and gluons, based upon strict gauge $SU(3)_c$-symmetry, was also able gradually to acquire the status of an accepted theory of strong interactions, despite the lack of a theoretical explanation of confinement. The work of Pollitzer and Gross–Wilczek in 1974, which showed that quarks are asymptotically free at small distances, contributed to this acceptance of the theory.

At the same time, the success of these two theories was always regarded as a provisional situation until a unified theory could incorporate all interactions, including gravity. A totally unexpected step in this direction was taken in the last ten years with the discovery of supersymmetry, which intermingles bosons and fermions, and the observation that the localization of supersymmetry inevitably leads to curved space-time and gravity.

It is natural that cooperation between physicists and mathematicians, which is as firmly rooted in tradition as is their difficulty understanding one another, received a fresh impetus during these years.

What was probably of the greatest importance from a technical point of view was the discovery of new methods of solving nonlinear partial differential equations. The famous inverse scattering problem method is effective in one- and two-dimensional models. But in realistic quantum field theory, which works with more complicated Lagrangians (the Lagrangian of the unified $SU(5)$-model contains more than five hundred vertices), the role of nonperturbational effects became clearer. The inclusion of these effects in the quasiclassical approximation is connected with the existence of localized solutions to dynamical equations of monopole type, solitons, and instantons. The solutions are studied by means of topological and algebra-geometric devices.

From a philosophical point of view, one can speak of the geometrization of physical thought; more precisely, of a new wave of geometrization which for the first time is sweeping far beyond the boundaries of general relativity. Tables of the homotopy groups of spheres and Čech cocycles have started to appear in physics journals, and nilpotents in the structure sheaf of a scheme or analytic space, which in the 1950s might have seemed little more than a caprice of Alexander Grothendieck's genius, have acquired a physical interpretation as the supports of the external degrees of freedom of the fundamental fields in supersymmetric models: the statistics of Fermi induces the anti-commuting coordinates of superspace.

This book is intended for mathematicians. It is a modest attempt to introduce the reader to certain types of problems which are motivated by quantum field theory. But we shall keep to the level of classical fields and dynamical equations, without going into secondary quantization. The reader can learn about quantization from the classical monograph *Introduction to Gauge Field Theory* by N. N. Bogolyubov and D. V. Shirkov and the excellent book *Introduction to the Quantum Theory of Gauge Fields* by A. A. Slavnov and L. D. Faddeev.

Following this foreword we shall devote a few pages to helping the mathematician reader translate the physicist's terminology into the geometrical language of this book, which is the standard jargon of the theory of complex manifolds and sheaf cohomology.

Part of the material presented here was taken from lectures given by the author at the Mechanico-Mathematics Faculty of Moscow State University and at various mathematics and physics meetings. I am deeply grateful to many people whose influence is reflected in one way or another in the pages of this book: my teacher I. R. Shafarevich; I. M. Gel'fand; L. D. Faddeev; M. F. Atiyah; and my friends, colleagues and coauthors A. A. Beilinson, A. A. Belavin, V. G. Drinfel'd, S. I. Gel'fand, S. G. Gindikin, G. M. Henkin, I. Yu. Kobzarev, D. A. Leites, V. I. Ogievetskii, I. B. Penkov, A. M. Polyakov, M. V. Savel'ev, A. S. Švarc, Ya. A. Smorodinskii, I. T. Todorov, V. E. Zakharov.

Yu. I. Manin

FOREWORD TO THE ENGLISH EDITION

Only three years have gone by since the publication of the Russian edition of this book. But in this short time complex-analytic methods have taken center stage in quantum field theory. This is connected with a series of papers by E. Witten, J. Schwarz, M. Green, A. Polyakov, A. Belavin and their collaborators, in which one begins to see an amazing picture of the world at high (Planck) energies:

(a) Space-time is ten-dimensional; six of these dimensions are compactified, and perhaps form a complex Calabi–Yau manifold.

(b) The elementary constituents of matter are one-dimensional objects called strings (or superstrings). The mathematical theory of these objects is based upon the classical Riemann moduli spaces of algebraic curves; the fundamental quantities in the theory are complex-analytic.

(c) Supergeometry replaces ordinary geometry wherever fundamental interactions are described in a Grand Unified fashion.

I would like to express my sincere thanks to Profs. N. Koblitz and J. King for their long and difficult labor in translating this book.

Yu. I. Manin

TABLE OF CONTENTS

GEOMETRICAL STRUCTURES IN FIELD THEORY

1. The Feynman path integral. The mathematical underpinning of modern elementary particle physics is quantum field theory (QFT). The fundamental quantities in QFT are expressed in terms of Feynman's path integrals $\langle A\rangle = \int A(\Psi)e^{iS(\Psi)}\,D(\Psi)$. This is a form of symbolic notation which one strives to make mathematically precise; much of the formalism of QFT is concerned with various ways of giving a precise meaning to this notation.

In this integral Ψ denotes a set of fields in the theory; A is an operator constructed from these fields; $\langle A\rangle$ is the average value of this operator; $S(\Psi)$ is the action functional in Planck units, which usually has the form $\int \mathcal{L}(\Psi)\,d^4x$, where $\mathcal{L}(\Psi)$ is the theory's Lagrangian density, which is integrated with respect to spacetime; and $e^{iS(\Psi)}\,D(\Psi)$ is a symbol for a certain measure on the function space of fields Ψ which satisfy some initial, boundary, or asymptotic conditions. This measure is rarely defined in the mathematical sense.

In most realistic models, the path integral is given a precise meaning and is computed using the methodology of perturbation theory. This leads to a formal series of divergent integrals. The integrals can be evaluated in various ways; and then the series can be approximated by a finite sum whose terms have been regularized.

2. The dynamic equations. The Euler–Lagrange equations $\delta \int \mathcal{L}(\Psi)\,d^4x = 0$ are called the (classical) dynamic equations or equations of motion. These equations can have various physical interpretations.

(a) If Ψ is an electromagnetic or gravitational field, then the solutions away from the sources can be interpreted as the "classical force fields." It was in this way that Maxwell's and Einstein's equations arose historically. In the quantum theory of electromagnetic fields, these forces appeared as the result of averaging many-photon interactions, i.e., as the mean values of certain quantum field operators. One expects that it will some day be possible to give an analogous description of the classical gravitational field.

(b) If Ψ is the field of a Dirac spinor particle, then the solutions of the equations of motion (perhaps in a classical external field) can be interpreted quantum mechanically as the wave functions of the particle. In quantum field theory these solutions become the vectors of one-particle states. With some qualifications one can interpret the analogous solutions of Maxwell's equations ("photon wave functions") in a similar way.

(c) The classical solutions in "imaginary time," i.e., analytically continued to a suitable region of complexified Minkowski space, can be used heuristically to compute the principal part of the path integral by the stationary phase method.

(d) Finally, localized classical solutions of the field theory equations of soliton or monopole type can indicate the presence of special quasi-particle excitations of the quantum field.

3. The Lagrangian of electrodynamics. The Lagrangian of an electrodynamic field A_μ interacting with a Dirac field ψ has the form

$$\begin{aligned}\mathcal{L}(A_\mu,\psi) &= -\frac{1}{4}F_{\mu\nu}F^{\mu\nu} + i\bar{\psi}\gamma^\mu d_\mu\psi - m\bar{\psi}\psi - e\bar{\psi}\gamma^\mu\psi A_\mu,\\ F_{\mu\nu} &= \partial_\mu A_\nu - \partial_\nu A_\mu.\end{aligned} \tag{1}$$

Here m is the mass of ψ and e is the charge.

As a classical system, the free field A_μ is the superposition of infinitely many oscillators, whose standard quantization leads to the photon picture. The interaction of the fields A_μ and ψ is described by the "vertex" $-e\bar{\psi}\gamma^\mu\psi A_\mu$. The charge e plays the role of a constant of interaction between the fermionic current $\bar{\psi}\gamma^\mu\psi$ and the field A_μ. The interaction term is assumed to be small in the perturbation theory of quantum electrodynamics (QED), and so one can use series expansions in powers of that term. Using various combinatorial rules, one indexes the terms in the power series by Feynman diagrams; for this reason $-e\bar{\psi}\gamma^\mu\psi A_\mu$ is called a "vertex."

The full Lagrangian (1) in QFT may be regarded as an abbreviated notation for the list of fundamental fields and interactions of the theory. From the quadratic part one obtains the free field propagators. The interaction terms give the vertices of the diagram, in terms of which the amplitudes of the processes are expressed.

4. Fermions–83. As in QED, the fundamental fields in the basic models of modern QFT are divided into two classes: matter fields (ψ in QED) and gauge fields (A_μ in QED). The first type of field obeys Fermi–Dirac statistics; the second type obeys Bose–Einstein statistics.

Matter fields occur in the Lagrangian as sections of vector bundles over space-time, while gauge fields occur as connections on these vector bundles. The coordinates along a fibre correspond to polarization (spin) and internal degrees of freedom, e.g., the color of a quark. This is a simplified picture, which does not take into account such terms as the Faddeev–Popov ghosts (an artifact of quantization); but we shall not use such effects.

Matter fields correspond to particles which at the given level of resolution are assumed to be without structure. In the paradigm which took root in the 1970s, these are: the six leptons (ν_e, e), (ν_μ, μ), (ν_τ, τ); the six quarks (u, d), (c, s), (t, b); and their antiparticles. In addition, the electroweak model of Weinberg–Salam postulates the existence of Higgs bosons, which have not been observed, supposedly because of their large mass. The fermions (i.e., the leptons and quarks) are divided into three generations, of which only the first generation, consisting of (ν_e, e, u, d), is important for understanding the basic phenomena of the world around us. The connected systems of u, d-quarks give the neutron and proton, and the residual forces bind them into the nucleus. The electromagnetic interaction of nuclei and electrons leads to the formation of atoms and molecules.

The vector bundles on Minkowski space-time whose sections correspond to these fermions are trivial, and their structure group is reduced. It is convenient to describe these vector bundles in terms of the tensor algebra generated by the following vector bundles:

$\mathcal{G}_l$ and $\mathcal{G}_r$, the left and right two-component Weyl spinors, which describe a particle's polarization;

the rank three $SU(3)_c$-bundle $\mathcal{E}_c$ corresponding to the colored degrees of freedom;

the rank two $SU(2)_w$-bundle $\mathcal{E}_w$ corresponding to so-called "weak isospin";

the $U(1)_{\text{em}}$-bundle $\mathcal{E}_{\text{em}}$ corresponding to electric charge.

For example, the first generation left particles $(\nu_e, e^-)_l$, $(u, d)_l$, e_l^+, $\tilde{u}_l$, $\tilde{d}_l$ correspond, respectively, to the vector bundles

$$\underbrace{\mathcal{E}_w \otimes S_l, \qquad \mathcal{E}_w \otimes \mathcal{E}_c \otimes S_l}_{\text{without including charge}}$$

$$\mathcal{E}_{\text{em}} \otimes S_l, \qquad \mathcal{E}_{\text{em}}^{-2/3} \otimes \bar{\mathcal{E}}_c \otimes S_l, \qquad \mathcal{E}_{\text{em}}^{1/3} \otimes \bar{\mathcal{E}}_c \otimes S_l.$$

In the $SU(5)$-unified scheme, where the electroweak and strong interactions are different components of a single interaction at large energy, until the $SU(5)$-symmetry is violated the matter field is a section of the direct sum of all these vector bundles.

In the language of physics, the structure group G of the matter bundle (more precisely, the structure group of the internal degrees of freedom, as opposed to the polarization degrees of freedom) is called the symmetry group of the theory. The group representation corresponding to the vector bundle carries information about the quantum mechanics.

Until recently, the associated principal G-bundle was always assumed to be trivial. This state of affairs changed as a consequence of studying monopoles and instantons; the Pontryagin and Chern numbers of the resulting vector bundles came to be called "topological charges" and to be recognized as a special sort of quantum number.

5. Gluons, photons and intermediate bosons. As mentioned before, the particles which carry the interactions are the quanta of connection fields on the vector bundles of the fermions' internal degrees of freedom. The covariant derivative of a connection field is written in the form $\nabla_\mu = \partial_\mu - igA_\mu$, where g is the constant of interaction and the A_μ are potentials of the connection field. Of course, here one must assume that a choice of trivialization has been made (otherwise ∂_μ would not make sense). Changing the choice of trivialization changes A_μ according to standard formulas. If the contribution of A_μ to the Lagrangian does not depend on the trivialization, then the field is called a *gauge* field. In other words, the Lagrangian of a gauge field is invariant under the action of the group of sections of a principal G-bundle.

In the standard models, the connections A_μ act on the following vector bundles (and then extend to the tensor algebra generated by them):

$\mathcal{E}_{\mathrm{em}}$	(the photons);
$\mathcal{E}_w \otimes 1_B$	(the intermediate bosons and the B-field; in the Weinberg-Salam model a photon is a linear combination of the B-field and one of the components of the connection on $\mathcal{E}_w$);
$\mathcal{E}_c$	(the gluons).

The Lagrangians of these models are constructed along the lines of the Lagrangian of QED:

(a) Every connection field makes a contribution proportional to the square of the modulus of its curvature.

(b) The fermions contribute kinetic terms. The terms for the interaction with the connection field are obtained using covariant derivatives.

(c) The Higgs field in the electroweak model contributes the potential term; its minimum is interpreted as a nonzero vacuum average which spontaneously violates the symmetry and gives mass to the fermions and intermediate bosons. These considerations can be understood in terms of "quantization without quantization," insofar as information about the particle and vertices can be obtained directly from the Lagrangian.

This naive picture, when supplemented with radiation corrections and renormalization considerations, often turns out to be remarkably effective. Predictions that are made based on unified models, such as the $SU(5)$-model, generally use these simple ideas. In recent years, experiments have been conducted to detect the decay of the proton that is predicted by these models. This decay is mediated by the X-boson, which is an $SU(5)$-connection quantum or, more generally, a G-connection quantum, where G is the group of the unified theory.

The role of gauge invariance of the Lagrangian in models with nonabelian symmetry group became clear gradually, starting with the pioneering work of Yang and Mills [119]. Since that time, connection fields on external bundles (in distinction from Levi–Civita connections, for example) have been called Yang–Mills fields.

Within the framework of a more or less naive perturbation theory, the prototype for which is quantum electrodynamics, the geometrical language does not really give anything new. But perturbation theory is not even sufficient for the basic problem in quantum chromodynamics (the theory of strong interactions). This is the problem of explaining the binding of quarks — more precisely, the fact that all observable states are "colorless," i.e., correspond to the identity representation of $SU(3)_c$. Chromodynamic forces keep two- and three-quark systems tightly bound within nuclear distances. These forces, which are somehow made possible because $SU(3)_c$ (unlike the electromagnetic group $U(1)_{em}$) is nonabelian, have not yet been incorporated into a consistent theoretical analysis.

A complicated statistical-geometrical picture of strong interactions was developed to understand the confinement of color. One can point to several highly nontrivial geometrical features which have come to the forefront in recent papers on the subject. Polyakov and 't Hooft have proposed certain tunnel processes which change the topology of a vacuum. In the simplest approximation, these processes are described by the instanton solutions of the Yang–Mills equations, i.e., by fields with finite motion in imaginary time. These fields have turned out to be exceptionally beautiful mathematical objects, and one would think that, as some physicists have said [105], "such beauty cannot have been created for nothing." Nevertheless, it must be said that the role of instantons in the fluctuations of gluon fields remains unclear.

Another quasiclassical picture is that of gluon strings, which join the quarks together and form tubes for the force lines of the gluon field. The effective classical fields in this theory are two-dimensional surfaces in space-time, the world surfaces of a string with Nambu action or some variant of it.

Finally, the geometrical characterization of a connection as a set of parallel translation operators is a central aspect of the dynamics of loops and the Wilson contour integral, which were developed in part to derive a picture of the gluon strings from the Yang–Mills picture in the binding phase.

From the point of view of particle physics, all of this is still conjectural. But it is already clear that quantum field averages over the complicated nonlinear function space of a gauge field can be computed only if one studies structural questions in some depth.

6. Space-time and gravity. We have not yet said anything about gravitational interaction, which has been the object of an essentially geometrical theory ever since the creation of general relativity. At the energy levels attainable by all current (and foreseeable) accelerators, this force is negligible in elementary particle interactions. For this reason, QFT can work in flat space-time. On the other hand, the unification of general relativity and QFT has turned out to be a very difficult problem, and a quantum theory of gravity has not yet been developed.

According to Einstein, the classical theory of gravity is at the same time a theory of space-time. Space-time makes a contribution to the Lagrangian like any gauge field; this contribution is the scalar curvature of the Levi–Civita connection. It seems that the space-time degrees of freedom, along with the internal degrees of freedom, are different aspects of a single geometrical picture. Thus, in any cosmogenic scenario a description of the origin of matter must be accompanied by a description of the origin of four-dimensional space-time.

7. Penrose twistors. Penrose's twistor program, some mathematical aspects of which are presented in this book, is one of the nontraditional attempts to construct a quantum field theory by rejecting the space-time M as the background for physical processes. Instead, the space L of light rays is proposed as the support of the quantum fields. Whereas M is the (extended) configuration space of the classical system "a massive point particle," L can be interpreted as the configuration space of the classical (relativistic) system "a particle of zero mass." A point x of space-time is represented in L by its "heavenly sphere" $L(x)$, and the field transformations from M to L and back are highly nonlocal. A large part of the present book (including almost the entire second chapter) is devoted to the properties of this transformation, which we call the Radon–Penrose transform. Penrose also developed the groundwork for a diagram technique on L, but we shall not treat such questions here.

Both space-time and its space of null-geodesics are treated complex analytically in this book. The role of complex analyticity in the technical side of QFT has been known for some time: the theory of dispersion relations and cross-symmetry makes very serious use of analyticity. Since the quantum degrees of freedom are in essence complex, perhaps the genesis of space-time degrees of freedom as a collective "graviton condensation" effect passes through intermediate stages which lend themselves to mathematical description within the framework of complex-analytic methods.

The geometry of simple supergravity also leads to complex-analytic worlds (the left and right worlds). Although we are probably nowhere near having a consistent theory, many different considerations suggest that holomorphic geometry will take the traditional place of differential geometry in such a theory.

In any case, it is this point of view which is adopted as our dogma for the main body of the book and which is developed using the tools that are familiar to mathematicians.

CHAPTER 1

GRASSMANNIANS, CONNECTIONS, AND INTEGRABILITY

This chapter is introductory in nature. In it we first encounter several classical themes which will later recur in various forms throughout the book.

The first theme is the realization of Minkowski space-time as the manifold of real points of a big cell in the grassmannian of complex planes in the twistor space. This is described in § 3, after we first give the basic facts about grassmannians and flag spaces in § 1 in a form suitable for later use.

The second theme is the cohomology of natural sheaves on homogeneous spaces and its connection with representation theory. In § 2 we describe a typical — and for us the most important — special case of this vast theory: the Borel–Weil–Bott theorem on the cohomology of invertible sheaves on a full flag space.

The third theme is integrability conditions as a mechanism both for generating nonlinear differential equations and for finding their solutions. After giving the classical definitions in §§ 4–5, in § 6 we introduce the basic concepts in this theory: the conical structure and the conical connection. These concepts arose when the geometrical data in the self-duality theory of the Yang–Mills and Einstein equations (and also the Yang–Mills supersymmetry equations) were axiomatized.

These three themes come together in § 7, where generalized self-duality equations are defined and investigated. The genesis of these equations in field theory will be discussed in the next chapter.

§ 1. Grassmannians and Flag Spaces

1. The grassmannian as a topological space. Let T be a finite dimensional complex vector space. The "grassmannian" of d-dimensional subspaces of T is the set $G(d;T)$ of such subspaces with the following topology.

Let H be the set of d-tuples $(t_1, \ldots, t_d)$, $t_i \in T$, such that the t_i are linearly independent. Let $\varphi: H \longrightarrow G(d;T)$ be the map which takes each such d-tuple to its linear span. We say that a subset $U \subset G(d;T)$ is open, by definition, if and only if $\varphi^{-1}(U)$ is open.

The grassmannian $G(1;T)$ is called projective space; we shall usually denote it by $\mathsf{P}(T)$.

Suppose that $\iota: T \longrightarrow T'$ is a linear imbedding. Then the ι-image of any d-dimensional subspace of T is a d-dimensional subspace of T'; in this way ι induces a continuous map $G(\iota): G(d;T) \longrightarrow G(d;T')$. In particular, if ι is an isomorphism, then so is $G(\iota)$. Thus, $GL(T)$ acts on $G(d;T)$; and the corresponding map $GL(T) \times G(d;T) \longrightarrow G(d;T)$ is continuous.

Let T^* denote the dual space of T, and set $c = \dim T - d$. If we let a subspace of T correspond to the subspace of T^* which is its orthogonal complement, we obtain a canonical isomorphism $G(d;T) \xrightarrow{\sim} G(c;T^*)$.

Later we shall introduce an algebraic variety structure on $G(d;T)$. It will be left to the reader to check that all of the maps in this subsection are morphisms of this structure, and a *fortiori* are morphisms of the underlying complex and smooth structures.

2. Tautological sheaves and vector bundles. Let $x \in G(d;T)$, and let $\mathcal{S}(x)$ be the d-dimensional subspace corresponding to x. We denote $S = \{(x,t) \mid x \in G(d;T), t \in \mathcal{S}(x)\} \subset G(d;T) \times T$, and we let $\varphi: S \longrightarrow G(d;T)$ be the projection onto the first component. S is called the "tautological vector bundle" on $G(d;T)$; the fibres of S are d-dimensional. Along with S, it is useful to consider a second tautological vector bundle: $\tilde{S} = \{(x,t') \mid x \in G(d;T), t' \in \mathcal{S}(x)^{\perp} \subset T^*\}$; its fibres are c-dimensional, where $c + d = \dim T$. One should think of $G(d;T)$ and $G(c;T^*)$ with tautological vector bundles $\mathcal{S}$ and $\tilde{\mathcal{S}}$, respectively, as two different but equivalent realizations of the same grassmannian. Choosing one of them means choosing a particular realization.

In § 3 we shall see that in the case $c = d = 2$, when the vector bundles S and $\tilde{S}$ are restricted to the Minkowski space, they become the two vector bundles of two-component spinors.

Instead of S and $\tilde{S}$, we shall usually work with the corresponding sheaves $\mathcal{S}$ and $\tilde{\mathcal{S}}$ of analytic sections of the vector bundles in the analytic structure which will now be described.

3. The standard covering, and coordinates. When it is necessary to indicate the dimension of a space, we shall use superscripts for this purpose. We fix a subspace $S = S^c \subset T^{d+c}$, and set

$$U(S) = \{x \in G(d;T) \mid \mathcal{S}(x) \cap S = \{0\}\}.$$

Clearly, $U(S)$ is an open subset of $G(d;T)$, and $G(d;T) = \bigcup_S U(S)$. The grassmannian's "big cells" $U(S)$ have a canonical structure of cd-dimensional affine space. More precisely, $U(S)$ is the affine space associated with the vector space $\mathrm{Hom}(T/S, S) = S \otimes (T/S)^*$, where the difference map

$$U(S) \times U(S) \longrightarrow \mathrm{Hom}(T/S, S), \qquad (x, y) \mapsto x - y$$

is defined as follows. Let $\tilde{t} \in T/S$, and suppose that $t(x)$ and $t(y)$ are representatives of $\tilde{t}$ in $S(x)$ and $S(y)$, respectively. Then we define $(x - y)(\tilde{t}) = t(x) - t(y)$.

We now choose a system of coordinates in $T = \mathbf{C}^{c+d}$ and show that $G(d;T) = \bigcup U(S_I)$, where I runs through the d-element subsets of $\{1, \ldots, c+d\}$ and S_I is the coordinate subspace spanned by the standard basis vectors whose indices are not in I. Namely, any element of T^d (which here denotes $\overbrace{T \times \cdots \times T}^{d \text{ times}}$) can be written as a $d \times (d+c)$-matrix Z whose rows are vectors of $T = \mathbf{C}^{c+d}$ in our chosen coordinate system. The subset $H \subset T^d$ in § 1.1 consists of the matrices of maximal rank d. Any such matrix has a nonzero minor. Let $I \subset \{1, \ldots, c+d\}$ be the indices of the columns of this minor. Then the subspace $S(Z)$ spanned by the rows of Z lies in $U(S_I)$.

Moreover, if $x \in U(S_I)$, then there is a unique matrix $Z_I(x)$ whose columns with indices in I form the identity matrix and whose rows span $S(x)$. The entries of this matrix are natural coordinate functions on $U(S_I)$. They are affine-linear functions with respect to the affine structure defined above.

We summarize what has been said.

(a) Starting with a coordinate system on $T = \mathbf{C}^{d+c}$, one constructs the affine covering $G(d;T) = \bigcup U(S_I), I \subset \{1, \ldots, d+c\}, \operatorname{card} I = d$.

(b) $U(S_I)$ has canonical coordinates $x_I^{\alpha\beta}, \alpha = 1, \ldots, d; \beta \in \{1, \ldots, d+c\} \setminus I$.

(c) Construct a $d \times (d+c)$-matrix Z_I by putting $x_I^{\alpha\beta}$ in the $\alpha\beta$-place and filling in the remaining columns with the identity matrix. Let B_{IJ} denote the submatrix of Z_I formed by the columns with indices in J, where $\operatorname{card} J = d$. Let $U_{IJ} \subset U_I$ be the open subspace where B_{IJ} is invertible. Then $U_{IJ} = U_I \cap U_J$, and the two coordinate systems on U_{IJ} are connected by the relation $B_{IJ} Z_J = Z_I$. Thus, the transition functions are algebraic, and so are *a fortiori* analytic and smooth. The grassmannian is an irreducible algebraic variety of complex dimension dc.

In particular, when $d = 1$, the one-row matrices Z of rank 1 are the homogeneous coordinates on $G(1;T) = \mathbf{P}(T)$, and for $i = 1, \ldots, c+1$, the matrices Z_i are the nonhomogeneous coordinates with 1 in the i-th place. For this reason, in the general case one sometimes calls the entries of a rank-d matrix Z, which is a point of $H \subset T^d$, the "homogeneous coordinates" of the corresponding point of $G(d;T)$. But one should not forget that the transition to the nonhomogeneous coordinates $x_I^{\alpha\beta}$ involves dividing by a matrix, not just by a scalar. In Chapter 4 we shall see how this construction generalizes to supergrassmannians.

(d) We represent $GL(T)$ by the group $GL(d+c;\mathbb{C})$ of matrices acting on row-vectors in T. Then in the homogeneous coordinates on $G(d;T)$ this action becomes right multiplication of Z by $GL(d+c;\mathbb{C})$.

(e) The tautological vector bundle is canonically trivialized over U_I: the rows of the matrix Z_I form a basis of sections of $\mathcal{S}$. If we write the sections as coordinate rows in this basis, then the transition functions are the matrices B_{IJ}^{-1} in the notation of (c) above.

The second tautological vector bundle is also canonically trivialized over U_I, since the map $S_I \longrightarrow T/S(x)$ is an isomorphism for all $x \in U_I$.

4. The Plücker mapping. To each d-dimensional subspace $S \subset T$ we associate the one-dimensional subspace $\wedge^d(S)$ in $\wedge^d(T)$. We obtain a mapping $p: G(d;T) \longrightarrow G(1, \wedge^d(T)) = \mathsf{P}(\wedge^d(T))$, which is called the Plücker mapping. A basis (t_i) of T determines a basis (v_I) of $\wedge^d(T)$, where $v_I = \bigwedge_{i\in I} t_i$. In the corresponding homogeneous coordinates the mapping p has the form: $p(Z) = (\det_I Z)$, where $\det_I Z$ is the minor formed by the columns of Z whose indices are in I. This set of minors is called the Plücker coordinates.

One can show that p is a closed immersion, and the image of p is a closed algebraic subspace in $\mathsf{P}(\wedge^d(T))$. To do this, one writes out the full system of (quadratic) identities that must be satisfied by a set of numbers (d_I) in order for it to be the set of minors of a single matrix. In § 2 we shall carry this out in the case $d = c = 2$.

In Chapter 4 we shall construct the supergrassmannian and see that, except in degenerate cases, there is no analog of the Plücker mapping.

The sheaf of sections of the (first) tautological vector bundle on the projective space $\mathsf{P}(T)$ is customarily denoted $\mathcal{O}_{\mathsf{P}}(-1)$. From the definition of the Plücker mapping we see that on $G(d;T)$

$$\wedge^d \mathcal{S} = p^*(\mathcal{O}_{\mathsf{P}(\wedge^d T)}(-1)).$$

Similarly, $\wedge^c \tilde{\mathcal{S}} = \tilde{p}^*(\mathcal{O}_{\mathsf{P}(\wedge^c T^*)}(-1))$, where $\tilde{p}$ is the Plücker mapping for the second realization of the grassmannian.

From the exact sequence $0 \longrightarrow \mathcal{S} \longrightarrow T \otimes \mathcal{O}_G \longrightarrow \tilde{\mathcal{S}}^* \longrightarrow 0$ we obtain $\wedge^d \mathcal{S} \simeq \wedge^c(\tilde{\mathcal{S}}) \otimes \wedge^{c+d} T^*$, i.e., there is no essential difference between the two sheaves.

5. The tangent sheaf. Let M be a complex manifold. We let $\mathcal{T}M$ denote the sheaf of holomorphic vector fields on M, i.e., derivations of the structure sheaf $\mathcal{O}_M$. Our first goal is to express $\mathcal{T}G(d;T)$ in terms of the tautological vector bundles.

Let $\mathcal{T}$ be the sheaf of sections of the trivial vector bundle with fibre T on $G = G(d; T)$. By definition, we have the exact sequence

$$0 \longrightarrow \mathcal{S} \longrightarrow \mathcal{T} \longrightarrow \tilde{\mathcal{S}}^* \longrightarrow 0.$$

Let X be a local vector field on G, and let s be a local section of $\mathcal{S}$ defined on the same region. We define X_s as a section of $\mathcal{T}$ by the formula

$$X(\sum f_i t_i) = \sum (X f_i) t_i,$$

where $t_i \in T$ and the f_i are local functions on G. One easily checks that this is well-defined (the definition does not depend on the choice of representation). This action of $\mathcal{T}G$ on $\mathcal{S}$ is $\mathcal{O}_G$-linear in X, and it satisfies Leibniz' rule in s: $X(fs) = Xf \cdot s + fXs$. Hence, the map $\overline{X}\colon \mathcal{S} \longrightarrow \mathcal{T}/\mathcal{S}$ given by $\overline{X}(s) = Xs \bmod \mathcal{S}$ is linear, and the map

$$\mathcal{T}G \longrightarrow \mathcal{H}om(\mathcal{S}, \mathcal{T}/\mathcal{S})\colon \quad X \mapsto \overline{X} \tag{1}$$

is also $\mathcal{O}_G$-linear.

6. Theorem. *The map (1) gives a canonical isomorphism*

$$\mathcal{T}G = \mathcal{H}om(\mathcal{S}, \mathcal{T}/\mathcal{S}) = \tilde{\mathcal{S}}^* \otimes \mathcal{S}^*.$$

Proof. Let (t_α), $\alpha = 1, \ldots, c+d$, be a basis of T, and let $I \subset \{1, \ldots, c+d\}$, $\operatorname{card} I = d$. Over U_I a basis of sections of $\mathcal{T}G$ is formed by the $\partial/\partial x_I^{\alpha\beta}$ with $\alpha = 1, \ldots, d$, $\beta \notin I$. The rows of the matrix Z_I form a basis of sections of $\mathcal{S}$, also over U_I, i.e., the basis has the form

$$s^\alpha = \sum_{\beta \notin I} x_I^{\alpha\beta} t_\beta + t_{I(\alpha)},$$

where $I(\alpha)$ denotes the α-th element occurring in I. Hence,

$$\frac{\partial}{\partial x_I^{\alpha\beta}} s^\gamma \equiv \delta_\alpha^\gamma t_\beta \bmod \mathcal{S}.$$

Finally, the t_β mod $\mathcal{S}$ form a basis of sections of $\mathcal{T}/\mathcal{S}$. This completes the proof. □

Thus, the structure group of the tangent bundle on the grassmannian $G(d; T^{d+c})$ reduces canonically to $GL(d) \times GL(c)$. The entire tensor algebra of grassmannians therefore has additional structures coming from this reduction. We now mention some of them; they will be given an additional "physical" interpretation in § 2.

7. Null directions. We say that a tangent vector $\tau(x) \in \mathcal{T}G(x)$ at the point x is "null" if it decomposes as $\tau(x) = \tilde{s}(x) \otimes s(x)$ for suitable $\tilde{s}(x) \in \tilde{S}^*(x)$, $s(x) \in S^*(x)$. The null directions in $\mathcal{T}G(x)$ form a cone with base $\mathbf{P}(\tilde{S}^*(x)) \times \mathbf{P}(S^*(x))$. In Penrose's model, the null directions turn out to be the complex directions of light rays, i.e., the null directions in the (conformal) Minkowski metric.

When $d = 1$ or $c = 1$ — and only in those cases — all directions are null. The grassmannian then is a projective space.

8. Grassmannian spinors. Theorem 6 implies that there is a canonical identification $\Omega^1 G = S \otimes \tilde{S}$, where $\Omega^1 G = \mathcal{H}om(\mathcal{T}G, \mathcal{O}_G)$ is the cotangent sheaf. If we combine this identification with the contraction, we obtain two linear maps:

$$\sigma: \Omega^1(G) \otimes S^* \longrightarrow \tilde{S},$$
$$\tilde{\sigma}: \Omega^1(G) \otimes \tilde{S}^* \longrightarrow S.$$

In particular, $\Omega^1(G)$ acts on $(S \oplus \tilde{S})^*$, and takes this sheaf to $S \oplus \tilde{S}$. The matrices of this map in suitable coordinate systems are the analog of the classical γ-matrices, which are usually defined in terms of the metric's Clifford algebra. Thus, we shall sometimes refer to the bundles S, $\tilde{S}$ and S^*, $\tilde{S}^*$ as "grassmannian spinors."

9. Decomposition of 2-forms. Let $\Omega^i G = \Lambda^i(\Omega^1 G)$. Using the decomposition in the previous subsection and a simple lemma from tensor algebra, we obtain:

$$\Omega^2(G) = S^2(S) \otimes \Lambda^2(\tilde{S}) \oplus \Lambda^2(S) \otimes S^2(\tilde{S}) = \Omega^2_+(G) \oplus \Omega^2_-(G).$$

In the special case $d = c = 2$, this is called the decomposition into "self-dual" and "anti-self-dual" components. Since the curvature forms for connections are 2-forms, we can write down analogs of the Yang–Mills self-duality equations for any grassmannian manifold (they become trivial only if $d = 1$ or $c = 1$).

10. The sheaf of volume forms. As above, we have

$$\Omega^{cd}(G) = [\Lambda^d(S)]^c \otimes [\Lambda^c(\tilde{S})]^d = p^*(\mathcal{O}_{\mathbf{P}}(-c)) \otimes \tilde{p}^*(\mathcal{O}_{\tilde{\mathbf{P}}}(-d)),$$

where p and $\tilde{p}$ are the two Plücker mappings in § 1.4.

11. The functor of points on a grassmannian. Let M be an analytic space, and let $\mathcal{F} \subset \mathcal{G}$ be two quasicoherent sheaves on M. $\mathcal{F}$ is said to be a local direct summand of $\mathcal{G}$ if every point $x \in M$ has an open neighborhood over which $\mathcal{F}$ has a complement in $\mathcal{G}$. A local direct summand of a locally free sheaf is locally free, and the quotient sheaf is also locally free. For example, on $G(d;T)$ the tautological sheaf $\mathcal{S}$ is a local direct summand of $\mathcal{O}_G \otimes T$.

Let T be a finite dimensional complex space. To every analytic space M and every positive integer d we associate the following set:

$$G(M) \;=\; \text{the set of rank } d \text{ direct summands of } \mathcal{T}_M = \mathcal{O}_M \otimes T.$$

This correspondence extends in the obvious way to a functor $\mathrm{An}^0 \longrightarrow \mathrm{Sets}$. Namely, to a morphism $\varphi\colon N \longrightarrow M$ one associates the map $G(\varphi)\colon G(M) \longrightarrow G(N)$ which takes a subsheaf $\mathcal{F} \subset \mathcal{T}_M$ to the subsheaf $\varphi^*(\mathcal{F}) \subset \varphi^*(\mathcal{T}_M) \subseteq \mathcal{T}_N$.

12. Theorem. *The functor $G(M)$ is represented by the grassmannian $G(d;T)$. More precisely, for any analytic manifold M and any morphism $\psi\colon M \longrightarrow G(d;T)$ one constructs the subsheaf $\psi^*(\mathcal{S}) \subset \psi^*(\mathcal{O}_G \otimes T) \subseteq \mathcal{T}_M$. This determines a map*

$$R(M)\colon \mathrm{Hom}(M,\, G(d;T)) \longrightarrow G(M),$$

which is a functorial morphism giving an equivalence of functors.

Sketch of a proof. We shall omit the formal verifications, and shall only indicate how the inverse map $R(M)^{-1}$ is constructed, i.e., how, given a rank d local direct summand $\mathcal{S}_M \subset \mathcal{T}_M$, one constructs a morphism $\psi\colon M \longrightarrow G(d;T)$ such that $\mathcal{S}_M = \psi^*(\mathcal{S})$. We first choose a coordinate system in T and construct an open covering $M = \bigcup U_i$ of M by Stein manifolds such that the sheaf $\mathcal{S}_M$ is free of rank d on each U_i. We choose a set of generators consisting of d free sections s_i^α of the sheaf $\mathcal{S}_M$ on U_i, and we write these sections in matrix form with respect to the basis of T:

$$s_i^\alpha = \sum_\beta f_i^{\alpha\beta} t_\beta, \qquad f_i^{\alpha\beta} \in T(U_i, \mathcal{O}_M).$$

Every point of U_i has a neighborhood where one of the $d \times d$-minors of $(f_i^{\alpha\beta})$ is invertible (this follows by Nakayama's lemma from the fact that $\mathcal{S}_M$ is a rank d local direct summand). Replacing the covering by a finer one, if necessary, we may assume that for each i we can choose an $I(i)$ such that the I-th minor is invertible on U_i (here I has the same meaning as in § 1.3). Thus, by changing our basis s_i^α, we can choose $f_i^{\alpha\beta}$ so that $(f_i^{\alpha\beta})_{\beta \in I(i)}$ is the identity matrix. We now define morphisms $\psi_i\colon U_i \longrightarrow U_{I(i)} \subset G(d;T)$ by setting $\psi_i^*(x_{I(i)}^{\alpha\beta}) = f_i^{\alpha\beta}, \quad \alpha = 1, \ldots, d, \quad \beta \notin I(i)$.

One can check that the morphisms ψ_i are compatible on the intersections $U_i \cap U_j$, and so can be patched together to obtain a global morphism $\psi\colon M \longrightarrow$

$G(d;T)$. It is obvious from the construction that $\psi^*(S) = S_M$. To check that the construction does not depend upon the various choices, the simplest approach is to use the observation that if ψ exists, then it is unique, since the local set of generators of S_M with identity matrix in the I-places is unique. □

13. Relative grassmannians. We shall now relativize the definition of a grassmannian as follows.

We consider an analytic base space N and a locally free sheaf $\mathcal{T}$ of finite rank on N. To any space over N, i.e., to any morphism $\varphi: M \longrightarrow N$, we associate the following set:

$$G_N(d;\mathcal{T})(M,\varphi) = \{\text{rank } d \text{ local direct summands of } \mathcal{T}_M = \varphi^*(\mathcal{T})\}.$$

This correspondence extends to a functor $\mathrm{An}_N^0 \longrightarrow \mathrm{Sets}$ (where An_N is the category of analytic spaces over N). The special case when N is a point and T is the sections of $\mathcal{T}$ over N gives the functor in § 1.11. Thus, we can visualize $\mathcal{T}$ as a set of vector spaces parametrized by the analytic space N, and we are interested in the family of grassmannians formed by subspaces of these vector spaces (this family is also parametrized by N). The existence of such families of grassmannians is guaranteed by the representability of the functor $G_N(d,\mathcal{T})$, which we now prove.

14. Theorem. *The functor $G_N(d,\mathcal{T})$ is representable by an analytic space over N, which will also be denoted $G_N(d,\mathcal{T})$.*

More precisely, there exists a morphism $\pi: G_N(d,\mathcal{T}) \longrightarrow N$ and a rank d local direct summand $S \subset \pi^*(\mathcal{T})$ such that the maps

$$R(M,\varphi): \mathrm{Hom}_N(M, G_N(d,\mathcal{T})) \longrightarrow G_N(d;\mathcal{T})(M,\varphi)$$

which take a morphism $\psi: M \longrightarrow G_N(d;\mathcal{T})$, where M is an analytic space over N, to the subsheaf

$$\psi^*(S) \subset \psi^*(\pi^*(\mathcal{T})) = \varphi^*(\mathcal{T}),$$

give an isomorphism of functors. The sheaf $S \subset \pi^*(\mathcal{T})$ thus corresponds to the identity morphism of $G_N(d;\mathcal{T})$.

Sketch of a proof. We first cover N with open Stein subsets over which T is free of constant rank, i.e., has a trivialization $\mathcal{T} = \mathcal{O}_{N_j} \otimes T_j$. It is not hard to show that the functor $G_{N_j}(d;\mathcal{T}|_{N_j})$ is represented by the analytic space $N_j \times G(d;T_j)$. Next, we note that any isomorphism from $(N,\mathcal{T})$ to $(N',\mathcal{T}')$ determines an isomorphism of functors from $G_N(d;\mathcal{T})$ to $G_{N'}(d;\mathcal{T}')$ and an isomorphism of the spaces representing these functors if they exist. Using this observation, we can define a map which glues together the relative grassmannians $G_{N_i\cap N_j}(d;\mathcal{T}|_{N_i\cap N_j})$

considered as open subsets of the i-th and j-th larger grassmannians. Patching in this way, we obtain $G_N(d; \mathcal{T})$ along with a structure morphism $\pi: G_N(d; \mathcal{T}) \longrightarrow N$ and a tautological sheaf $S \subset \pi^*(\mathcal{T})$. The verification of all of the necessary properties is lengthy but completely mechanical. □

15. The relative tensor algebra. Given any morphism $\pi: G \longrightarrow N$ of analytic spaces, one can define the sheaf $\Omega^1 G/N$ of relative differentials and the sheaf $\mathcal{T}G/N = \mathrm{Ker}(d\pi: \mathcal{T}G \longrightarrow \pi^*(\mathcal{T}N))$ of vertical vector fields.

Let $G = G_N(d; \mathcal{T})$ be the relative grassmannian, and let S and $\tilde{S}$ be the two tautological sheaves on G. Then the construction in § 1.5 applied to the vertical vector fields gives isomorphisms

$$\mathcal{T}G/N = \tilde{S}^* \otimes S^*, \qquad \Omega^1 G/N = S \otimes \tilde{S}.$$

This follows because, as noted above, locally on the base space N the grassmannian G is the product of the base and the ordinary (absolute) grassmannian.

16. Flag functors and flag spaces. Again let N be an analytic base space, and let $\mathcal{T}$ be a locally free sheaf of finite rank on N. Let $0 < d_1 < \cdots < d_k < \mathrm{rank}\,\mathcal{T}$ be a sequence of integers. By a flag of length k and type $(d_1, \ldots, d_k)$ in $\mathcal{T}$ we mean a sequence of subsheaves $S_1 \subset S_2 \subset \cdots \subset S_k \subset \mathcal{T}$ such that each $S_i \subset S_j$ is an imbedding as a local direct summand, and rank $S_i = d_i$.

To any space $M \xrightarrow{\varphi} N$ over N we associate the following set:

$$F_N(d_1, \ldots, d_k; \mathcal{T}) = \{\text{all flags of type } (d_1, \ldots, d_k) \text{ in the sheaf } \mathcal{T}_M = \varphi^*(\mathcal{T})\}.$$

This correspondence extends to a functor $\mathrm{An}_N^0 \longrightarrow$ Sets. In the special case when N is a point and $\mathcal{T}$ is a vector space, one obtains the absolute flag functor.

Let $d = (d_1, \ldots, d_k)$ be a type, and let $d' = (d'_1, \ldots, d'_l)$ be another type. If d includes d' (as a set), then there is a natural morphism of functors "projection onto the subflag of smaller type" $\pi(d, d')$: $F_N(d; \mathcal{T}) \longrightarrow F_N(d'; \mathcal{T})$ given by

$$S_1 \subset \cdots \subset S_k \mapsto S'_1 \subset \cdots \subset S'_l, \qquad S'_j = S'_{\iota(j)}.$$

17. Theorem. *The functor $F_N(d_1, \ldots, d_k; \mathcal{T})$ is representable by an analytic space over N, which will also be denoted $F_N(d_1, \ldots, d_k; \mathcal{T})$.*

The space $F_N(d_1, \ldots, d_k; \mathcal{T})$ has a structure morphism $\pi: F_N \longrightarrow N$ and a structure flag (also called the "tautological" flag) $S_1 \subset S_2 \subset \cdots \subset S_k \subset \pi^*(\mathcal{T})$, which corresponds to the identity morphism of F_N. The description of the isomorphism of functors is similar to the one in Theorem 14.

Sketch of a proof. We use induction on k. For $k = 1$ the functor is represented by the relative grassmannian. To prove the induction step from k to $k+1$, we consider two types $d = (d_0, d_1, \ldots, d_k)$ and $d' = (d_1, \ldots, d_k)$. By the induction assumption, $F' = F_N(d'; \mathcal{T})$ is representable. We consider the smallest sheaf S_1 in the tautological flag $S_1 \subset \cdots \subset S_k$ on F', and we construct the relative grassmannian

$$G_{F'}(d_0; S_1) \xrightarrow{\pi_0} F' \xrightarrow{\pi'} N$$

(where π_0 is the structure morphism of $G_{F'}$ and π' is the structure morphism of F'). We set $\pi = \pi'\pi_0$ and consider the following flag on $G_{F'}$:

$$S_0 \subset \pi_0^*(S_1) \subset \cdots \subset \pi_0^*(S_k) \subset \pi_0^*(\pi'^*\mathcal{T}) = \pi^*(\mathcal{T}),$$

where $S_0 \subset \pi_0^*(S_1)$ is the tautological sheaf on $G_{F'}$.

We claim that as an N-space $G_{F'}$ is $F_N(d; \mathcal{T})$ with the above tautological flag, and the morphism $G_{F'} \xrightarrow{\pi_0} F'$ represents the functorial morphism $\pi(d, d')$. The proof of this makes use of Theorem 14 and is completely formal. The key step is, given $M \xrightarrow{\varphi} N$ and a flag of type d on M, to construct the corresponding morphism $M \longrightarrow G_{F'}(d_0; S_1)$. One first constructs the morphism $M \longrightarrow F'$ corresponding to the shortened flag (it exists and is unique by the induction assumption), and then uses the smallest flag component to lift this morphism to $G_{F'}$. □

18. Notational principles. Many of the spaces we shall later study will be flag manifolds; moreover, they will have various representations as relative flags over different base spaces or as fibre products of such flag manifolds. One thus needs an appropriate notational system which conveys the information necessary to describe such representations.

The notation $F_N(d_1, \ldots, d_k; \mathcal{T})$ will usually be detailed enough; the structure morphism $\pi: F_N \longrightarrow N$ will also be indicated if necessary. The base space (the target of the arrow, or the space on which our sheaf is defined) will be indicated by a subscript, and the rank of a sheaf will be indicated by a superscript. For example, the complete notation for the tautological flag on $F = F_N(d_1, \ldots, d_k; \mathcal{T})$ is as follows: $S_F^{d_1} \subset S_F^{d_2} \subset \cdots \subset \pi^*(\mathcal{T}^d)$. The orthogonal flag will be denoted by a tilde: $\tilde{S}_F^{d-d_k} \subset \tilde{S}_F^{d-d_{k-1}} \subset \cdots \subset \pi^*(\mathcal{T}^*)$.

Suppose we are interested in one of the projection morphisms onto a grassmannian: $\pi(d, d_a)$: $F_N(d_1, \ldots, d_k; \mathcal{T}) \longrightarrow G_N(d_a; \mathcal{T})$, where $1 < a < k$ and $d = (d_1, \ldots, d_k)$. It can be represented in the following way as the fibre product of two structure morphisms over the base $G = G_N(d_a; \mathcal{T}) \xrightarrow{\pi} N$:

$$F_N(d; \mathcal{T}) = F_G(d_1, \ldots, d_{a-1}; S_G^{d_a}) \times_G F_G(d_{a+1} - d_a, \ldots, d_k - d_a; \pi^*(\mathcal{T})/S_G^{d_a}).$$

This generalizes the device used to prove Theorem 17.

Later we shall extend this notation to isotropy flag spaces and to superspaces. We conclude this section with an example.

19. "Light rays" on grassmannians. Let $2 \leq d \leq \dim T - 2$. We consider the diagram

$$L = F(d-1,\, d+1; T) \xleftarrow{\pi_1} F = F(d-1,\, d,\, d+1; T) \xrightarrow{\pi_2} M = G(d; T).$$

Here π_1 and π_2 are the projection morphisms onto the subflags.

In the notation of the preceding subsection, we can represent π_1 as the structure morphism of a relative grassmannian: $F = G_L(1; S_L^{d+1}/S_L^{d-1}) \longrightarrow L$. From this one sees that F is a relative projective line. In particular, this means that the fibres of π_1 are projective lines. The tautological sheaf on F relative to this representation can be expressed in terms of the tautological flag on F relative to the original representation, namely, it is the sheaf S_F^d/S_F^{d-1}. It will be left to the reader to compute the second tautological sheaf and $\mathcal{T}F/L$; we shall return to this later. From this calculation one will observe that the π_2-images of all of the fibres of π_1 have null direction at any point. See the discussion in § 6 below.

The above diagram $L \xleftarrow{\pi_1} F \xrightarrow{\pi_2} M$ is the first of many examples of "double fibrations" which will occur in this book.

§ 2. Cohomology of Flag Spaces

1. Relative projective spaces. Suppose that M is a manifold and $\mathcal{T}$ is a locally free sheaf on M of rank $m+1 > 1$, so that we have $\mathbf{P} = \mathbf{P}_M(\mathcal{T}) \xrightarrow{\pi} M$. Recall that we write $\mathcal{O}(1)$ for $S^* = (S_{\mathbf{P}}^1)^*$ on $\mathbf{P}$. We also denote $\mathcal{O}(n) = \mathcal{O}(1)^{\otimes n}$, and $\mathcal{E}(n) = \mathcal{E} \otimes \mathcal{O}(n)$ for any sheaf $\mathcal{E}$. According to Theorem 1.6, we have $\tilde{S} = \Omega^1(1)$, where $\Omega^1 = \Omega^1\mathbf{P}/M$, and this gives us the following short exact sequence for the second tautological sheaf: $0 \longrightarrow \Omega^1(1) \xrightarrow{a} \pi^*(\mathcal{T}^*) \xrightarrow{b} \mathcal{O}(1) \longrightarrow 0$. We then obtain two exact sequences for the kernel of the morphism $S^j(b)$ and the cokernel of the morphism $\wedge^j(a), j \geq 1$:

$$0 \longrightarrow \pi^*(S^{j-1}\mathcal{T}^*) \cdot \Omega^1(1) \longrightarrow \pi^*(S^j\mathcal{T}^*) \xrightarrow{S^j(b)} \mathcal{O}(j) \longrightarrow 0, \tag{1}$$

$$0 \longrightarrow \Omega^j(j) \xrightarrow{\wedge^j(a)} \pi^*(\wedge^j\mathcal{T}^*) \longrightarrow \Omega^{j-1}(j) \longrightarrow 0. \tag{2}$$

Setting $j = m+1$ in (2), we obtain the identification

$$\Omega^m(\mathbf{P}/M) = \pi^*(\wedge^{m+1}\mathcal{T}^*)(-m-1). \tag{3}$$

2. Theorem. (a) *$R^k\pi_*\mathcal{O}(j)$ is nonzero only for the following values of k and j:*

$$k = 0,\ j \geq 0: \qquad \pi_*\mathcal{O}(j) = S^j(\mathcal{T}*),$$
$$k = m,\ j \leq -(m+1): \qquad R^m\pi_*\mathcal{O}(j) = S^{|j|-m-1}(\mathcal{T}) \otimes \wedge^{m+1}(\mathcal{T}).$$

(b) *$R^k\pi_*\Omega^j$ is nonzero only for $0 \leq k = j \leq m$, and in that case $R^k\pi_*\Omega^k = \mathcal{O}_M$.*

(c) *If $i \neq 0$, then $R^k\pi_*\Omega^j(i) = 0$ except for the cases $k = 0$, $i \geq j+1$ and $k = m$, $i \leq j - m - 1$.*

Proof. The first part is (a relative version of) a classical theorem of Serre. We note that the isomorphisms $\pi_*\mathcal{O}(j) = S^j(\mathcal{T}^*)$ are given by $\pi_*(S^j(b))$ (see (1)), and the identification of $R^m\pi_*\mathcal{O}(j)$ comes from the exact pairing $\pi_*\mathcal{O}(j) \times R^m\pi_*\Omega^m(-j) \longrightarrow R^m\pi_*\Omega^m = \mathcal{O}_M$ (Serre duality) and the equality in (3). In order to obtain the isomorphisms in part (b), we multiply (2) by $\mathcal{O}(-j)$ and apply π_* to the resulting exact sequence. We have $R^j\pi_*(\pi^* \wedge^j (\mathcal{T}^*)(-j)) = 0$ for $1 \leq j \leq m$, by part (a), and this implies that the boundary homomorphisms give us successively

$$\mathcal{O}_M = \pi_*\mathcal{O}_{\mathbf{P}} \xrightarrow{\sim} R^1\pi_*\Omega^1 \xrightarrow{\sim} \cdots \xrightarrow{\sim} R^m\pi_*\Omega^m.$$

A similar argument, where we tensor (2) with $\mathcal{O}(i-j)$, leads to part (c): one uses ascending induction on k for $i < 0$ and descending induction on k for $i > 0$. □

3. Demazure duality. Now suppose that $\operatorname{rk} \mathcal{T} = 2$, i.e., $\mathbf{P} = \mathbf{P}_M(\mathcal{T} \xrightarrow{\pi} M)$ is a relative projective line. We set $\mathbf{P}^* = \mathbf{P}_M(\mathcal{T}^*)$. A special feature of projective lines is the existence of a canonical isomorphism $\epsilon\colon \mathbf{P} \xrightarrow{\sim} \mathbf{P}^*$ over M: this isomorphism is induced by the convolution isomorphism $\mathcal{T} \xrightarrow{\sim} \mathcal{T}^* \otimes \wedge^2\mathcal{T}$ if one takes into account that $\operatorname{rk} \wedge^2 \mathcal{T} = 1$ and $\mathbf{P}_M(\mathcal{T} \otimes \mathcal{L}) = \mathbf{P}_M(\mathcal{T})$ for any invertible sheaf $\mathcal{L}$ (and $\mathcal{T}$ of any rank). (Here the isomorphism on the functor of points reduces to tensor multiplication of any direct subsheaf by $\mathcal{L}$.)

The isomorphism ϵ interchanges the two sheaves $\mathcal{S}$ and $\tilde{\mathcal{S}}$: $\epsilon^*(\mathcal{S}_{\mathbf{P}^*}) = \tilde{\mathcal{S}}_{\mathbf{P}}$, $\epsilon^*(\tilde{\mathcal{S}}_{\mathbf{P}^*}) = \mathcal{S}_{\mathbf{P}}$. It also has very good functorial properties: it is compatible with change of the base M, it is functorial relative to isomorphisms of pairs $(\mathcal{T}, M) \xrightarrow{\sim} (\mathcal{T}', M')$, and it "does not change" when $\mathcal{T}$ is replaced by $\mathcal{T} \otimes \mathcal{L}$, where $\mathcal{L}$ is invertible.

This isomorphism is connected with the following construction of Demazure. Let $\mathcal{L}$ be an invertible sheaf on $\mathbf{P}$. Its degree $d(\mathcal{L})$ is uniquely characterized by the properties: $d(\mathcal{O}(a)) = a$; $d(\pi^*\mathcal{K}) = 0$ for any invertible sheaf $\mathcal{K}$ on the base

M (which we are assuming to be connected); and $d(\mathcal{L}_1) = d(\mathcal{L}_2)$ if $\mathcal{L}_1$ and $\mathcal{L}_2$ are isomorphic. We have: $d(\Omega^1\mathsf{P}/M) = -2$; the map $\pi_2^*\pi_{2*}\mathcal{L} \longrightarrow \mathcal{L}$ is an isomorphism if and only if $d(\mathcal{L}) = 0$.

We set $\Omega = \Omega^1\mathsf{P}/M; \Omega^{(d)} = S^d(\Omega^1\mathsf{P}/M)$ for $d \geq 0$ and $\Omega^{(d)} = S^{|d|}(\mathcal{T}\mathsf{P}/M)$ for $d < 0$. For any invertible sheaf $\mathcal{L}$ we set $\mathcal{L}^\vee = \mathcal{L}^{-1} \otimes \Omega^{(-d)}$, where $d = d(\mathcal{L})$. We obviously have $d(\mathcal{L}^\vee) = d(\mathcal{L})$.

4. Lemma. *There exists a canonical pairing $\pi_*(\mathcal{L}) \otimes \pi_*(\mathcal{L}^\vee) \longrightarrow \mathcal{O}_M$ which is functorial relative to base change and isomorphisms of the pair $\mathsf{P} \to M$ and induces a duality between $\pi_*(\mathcal{L})$ and $\pi_*(\mathcal{L}^\vee)$.*

Proof. We set $\mathcal{K} = \pi_*(\mathcal{L}(-d))$. Then canonically $\mathcal{L} = \pi^*\mathcal{K}(d)$ and $\mathcal{L}^\vee = \pi^*\mathcal{K}^{-1}(d) \otimes (\wedge^2\mathcal{T})^d$, since $\Omega^{-d} = \pi^*(\wedge^2\mathcal{T})^d(2d)$ by virtue of (3). Using part (a) of Theorem 2, we find (we are assuming that $d \geq 0$ — otherwise, both sheaves are zero):

$$\pi_*(\mathcal{L}) = \mathcal{K} \otimes S^d(\mathcal{T}^*),$$
$$\pi_*(\mathcal{L}^\vee) = \mathcal{K}^{-1} \otimes S^d(\mathcal{T}^*) \otimes (\wedge^2\mathcal{T})^d.$$

The desired pairing is obtained by composing the multiplication $\mathcal{K} \otimes \mathcal{K}^{-1} \longrightarrow \mathcal{O}_M$, the d-th symmetric power of exterior multiplication $S^d\mathcal{T}^* \otimes S^d\mathcal{T}^* \longrightarrow (\wedge^2\mathcal{T}^*)^d$, and convolution with $(\wedge^2\mathcal{T})^d$. □

5. Proposition. *In the diagram $\mathsf{P}\xrightarrow{\pi_1}M\xrightarrow{\pi_2}N$, suppose that π_1 is a relative projective line, $\pi = \pi_2 \circ \pi_1$, $\mathcal{L}$ is an invertible sheaf over P, and $d = d(\mathcal{L}) \geq -1$. Then for any $n \geq 0$ one has a canonical isomorphism over N*

$$R^n\pi_*\mathcal{L}\xrightarrow{\sim}R^{n+1}\pi_*(\mathcal{L} \otimes \Omega^{(d+1)}), \qquad \Omega = \Omega^1\mathsf{P}/M.$$

These isomorphisms are compatible with base change over N and with isomorphisms of triples (P, M, $\mathcal{L}$) over N.

Proof. We first note that $d(\mathcal{L} \otimes \Omega^{(d+1)}) = -d(\mathcal{L}) - 2$. Hence $R^j\pi_{1*}\mathcal{L} = 0$ for $j \neq 0$, and $R^j\pi_{1*}(\mathcal{L} \otimes \Omega^{(d+1)}) = 0$ for $j \neq 1$. The spectral sequence for composition of morphisms then gives

$$\begin{aligned} R^n\pi_*\mathcal{L} &= R^n\pi_{2*}(\pi_{1*}\mathcal{L}), \\ R^{n+1}\pi_*(\mathcal{L} \otimes \Omega^{(d+1)}) &= R^n\pi_{2*}(R^1\pi_{1*}(\mathcal{L} \otimes \Omega^{(d+1)})). \end{aligned} \tag{4}$$

We now observe that the Serre pairing for π_1 has the form

$$R^1\pi_{1*}(\mathcal{L} \otimes \Omega^{(d+1)}) \otimes \pi_{1*}(\mathcal{L}^\vee) \longrightarrow \mathcal{O}_M.$$

Comparing this with the Demazure pairing, we obtain a canonical isomorphism

$$R^1\pi_{1*}(\mathcal{L}\otimes\Omega^{(d+1)}) = \pi_{1*}\mathcal{L},$$

which together with (4) completes the proof. □

6. Flag spaces. We return to the general case $\operatorname{rk}\mathcal{T} = m+1$. We set $F = F_M(1,2,\dots,m;\ \mathcal{T})\xrightarrow{\pi}M$. Let $\epsilon_i = (0,\dots,0,1,0,\dots,0)\in\mathbf{Z}^{m+1}$ (with 1 in the i-th place). For any $a=(a_i)\in\mathbf{Z}^{m+1}$, we define the following invertible sheaf on F:

$$\mathcal{L}(a) = (S_F^1)^{-a_1}\otimes(S_F^2/S_F^1)^{-a_2}\otimes\cdots\otimes(S_F^{m+1}/S_F^m)^{-a_{m+1}},\qquad S_F^{m+1}=\pi^*(\mathcal{T}).$$

We set $F_i = F_M(1,\dots,i-1,\ i+1,\dots,m;\ \mathcal{T})$, and we let $\pi_i\colon F\longrightarrow F_i$ be the standard projection. This projection makes F into a relative projective line: $F=\mathbf{P}_{F_i}(S_{F_i}^{i+1}/S_{F_i}^{i-1})$. We set $\Omega_i = \Omega^1F/F_i, d_i(\mathcal{L})$ = the degree of the invertible sheaf $\mathcal{L}$ relative to $\pi_i, \Omega=\Omega^{\dim F/M}F/M$.

7. Proposition. (a) *The classes $l(a)$ of invertible sheaves of the form $\mathcal{L}(a)$ generate the relative Picard group* $\operatorname{Pic}F/M = \operatorname{Pic}F/\pi^*(\operatorname{Pic}M)$, *which is isomorphic to* $\mathbf{Z}^m=\mathbf{Z}^{m+1}/\mathbf{Z}(1,\dots,1)$.

(b) $\Omega\xrightarrow{\sim}\mathcal{L}(\beta)$, *where* $\beta=(-m,\ -m+2,\dots,\ m)$.

(c) $\Omega_i\xrightarrow{\sim}\mathcal{L}(-\alpha_i)$, *where* $\alpha_i=\epsilon_i-\epsilon_{i+1};\ d_i(\mathcal{L}(a))=a_i-a_{i+1}$.

Proof. We proceed by induction on m, making use of the fact that the relative Picard group relative to projective space is generated by the sheaf $\mathcal{O}(1)$. This gives the first step in the induction ($m=1$). In order to make the step from $m-1$ to m, it is sufficient to consider the diagram $F\xrightarrow{\lambda}F_M(1;\ \mathcal{T})=\mathbf{P}$. Then $F=F_{\mathbf{P}}(1,\dots,m-1;\ \mathcal{T}_{\mathbf{P}}/\mathcal{O}_{\mathbf{P}}(-1))$, and by the induction assumption $\operatorname{Pic}F/\mathbf{P}$ is generated by the sheaves $\mathcal{L}(\epsilon_i), i\ge 2$; in addition, $\operatorname{Pic}\mathbf{P}/M$ is generated by the sheaf $\mathcal{O}(1)$, and $\lambda^*\mathcal{O}(1)=\mathcal{L}(\epsilon_1)$. In order for the sheaf $\mathcal{L}$ to be the result of a lifting from M it is necessary that $d_i(\mathcal{L})=0$ for all i. The sufficiency of this condition is proved by a similar inductive argument. Next, we have $d_i(\mathcal{L}(\epsilon_i))=1, d_i(\mathcal{L}(\epsilon_{i+1}))=-1$; and $d_i(\mathcal{L}(\epsilon_k))=0$ for $k\neq i,\ i+1$, because $\mathcal{L}(\epsilon_k)$ is lifted from F_i. According to Theorem 1.6,

$$\Omega_i = S_F^i/S_F^{i-1}\otimes(S_F^{i+1}/S_F^i)^* = \mathcal{L}(\epsilon_{i+1}-\epsilon_i),$$

since we have $S=\mathcal{L}(\epsilon_{i+1})$ and $\tilde{S}^*=\mathcal{L}(\epsilon_i)$ on the projective line π_i.

This implies that $d_i(\mathcal{L}(a))=0$ for all i only if a is proportional to $(1,\dots,1)$. Since the $\mathcal{L}(\epsilon_i)$ are successive quotients of the sheaf $\pi^*(\mathcal{T}^*)$, we have $\mathcal{L}(1,\dots,1)\simeq\wedge^{m+1}\pi^*(\mathcal{T}^*)$.

Finally, Ω can be computed by induction using the same diagram $F \xrightarrow{\lambda} \mathbf{P}_M(1;\mathcal{T}) = \mathbf{P}$. In fact, $\Omega^m\mathbf{P}/M \simeq \wedge^{m+1}\mathcal{T}^*_{\mathbf{P}}(-m-1)$ by (3); and $\Omega^{\dim(F/\mathbf{P})}F/\mathbf{P} \simeq \mathcal{L}(0, -(m-1), \ldots, m-1)$ by the induction assumption. Thus,

$$\begin{aligned}\Omega &\simeq \Omega^{\dim(F/\mathbf{P})}F/\mathbf{P} \otimes \lambda^*\Omega^m\mathbf{P}/M = \\ &= \mathcal{L}(-(m+1), -(m-1), \ldots, m-1) \otimes \wedge^{m+1}\mathcal{T}^*_{\mathbf{P}}.\end{aligned}$$

Recalling that $\wedge^{m+1}\mathcal{T}^*_{\mathbf{P}} = \mathcal{L}(1,\ldots,1)$, we obtain the required conclusion. □

8. Positive sheaves and the $S^{(a)}$ functors. A sheaf $\mathcal{L}(a)$ is said to be "positive" if $a_1 \geq a_2 \geq \cdots \geq a_{m+1}$, or, equivalently, if $d_i(\mathcal{L}(a)) \geq 0$ for all i. We set

$$S^{(a)}(\mathcal{T}^*) = \pi_*(\mathcal{L}(a)).$$

The mappings $S^{(a)}$ extend to a functor on the category of locally free sheaves on M with isomorphisms as the morphisms of the category. The following properties are easily verified:

(a) If $(b) = (a) + k(1,\ldots,1)$, then $S^{(b)}(\mathcal{T}^*) = S^{(a)}(\mathcal{T}^*) \otimes (\wedge^{m+1}\mathcal{T}^*)^k$.

(b) $S^{(a,0,\ldots,0)}(\mathcal{T}^*) = S^a(\mathcal{T}^*)$. (Here use the projection $F \xrightarrow{\lambda} \mathbf{P}(\mathcal{T})$ and Theorem 2.)

(c) $S^{(\epsilon_1+\cdots+\epsilon_r)}(\mathcal{T}^*) = \wedge^r(\mathcal{T}^*)$. (Here use induction on r.)

More generally, as a representation space over $GL(\mathcal{T}^*)$, $S^{(a)}(\mathcal{T}^*)$ corresponds to the Young diagram (a).

9. Acyclic sheaves. The sheaf $\mathcal{L}(a)$ is said to be "acyclic" if $d_i(\mathcal{L}(a)) = -1$ for some i. We recall that on a projective line it is precisely the sheaves of degree -1 which have all cohomology trivial. Below we prove the same for acyclic sheaves: $R^n\pi_*\mathcal{L}(a) = 0$ for all n.

As is customary in representation theory, we introduce the weight $\rho = (m, m-1, \ldots, 1, 0)$; it is uniquely determined (up to a multiple of $\sum \epsilon_i$) by the condition that $d_i(\mathcal{L}(\rho)) = 1$ for all i. Thus, $\mathcal{L}(a)$ is acyclic if and only if $d_i(\mathcal{L}(a+\rho)) = 0$ for a suitable i.

10. Direct images and reflections. Proposition 5 gives us a relationship between the direct images of different sheaves $\mathcal{L}(a)$. We shall want to apply this proposition to the diagrams $F \xrightarrow{\pi_i} F_i \longrightarrow M$. For this it is important to compute the effect of the operation $\mathcal{L} \longrightarrow \mathcal{L} \otimes \Omega_i^{(d_i(\mathcal{L})+1)}$. Using Proposition 7, we can write this operation as

$$\mathcal{L}(a_1,\ldots,a_n) \longrightarrow \mathcal{L}(a_1,\ldots,a_{i-1}, a_{i+1}-1, a_i+1, a_{i+2},\ldots).$$

An abbreviated way of writing this action is as follows: if $s_i\colon \mathbb{Z}^{m+1} \longrightarrow \mathbb{Z}^{m+1}$ is the transposition of the i-th and $(i+1)$-th coordinates, then

$$\mathcal{L}(a) \longrightarrow \mathcal{L}(s_i(a+\rho)-\rho). \tag{5}$$

11. Theorem. *For any invertible sheaf $\mathcal{L}$ on F there is at most one value of n for which $R^n\pi_*\mathcal{L} \neq 0$. These direct images can be computed as follows (where, using Proposition 7, we limit ourselves to sheaves of the form $\mathcal{L} = \mathcal{L}(a)$):*

(a) If $\mathcal{L}(a)$ is acyclic, then $R^n\pi_*\mathcal{L}(a) = 0$ for all n.

(b) If $\mathcal{L}(a)$ is positive, then $R^n\pi_*\mathcal{L}(a) = 0$ for all $n \geq 1$; |see subsection |8 *concerning $\pi_*\mathcal{L}(a)$.*

(c) If $\mathcal{L}(a)$ is neither acyclic nor positive, then there exists a unique nontrivial coordinate permutation $w\colon \mathbb{Z}^{m+1} \to \mathbb{Z}^{m+1}$ and a uniquely determined positive sheaf $\mathcal{L}(b)$ such that $b+\rho = w(a+\rho)$. Let $w = s_{i_1}\cdots s_{i_n}$, where $n = l(w)$ is the length of w written in reduced form as a product of transpositions. Then

$$R^{n(w)}\pi_*\mathcal{L}(a) \simeq \pi_*\mathcal{L}(b) = S^{(b)}(\mathcal{T}^*). \tag{6}$$

Proof. (a) If $\mathcal{L}(a)$ is acyclic, with $d_i(\mathcal{L}(a)) = -1$, then $\mathcal{L}(s_i(a+\rho)-\rho)$ is isomorphic to $\mathcal{L}(a)$. Proposition 5 applied to $F \to F_i \to M$ shows that the index of the higher direct image can be made arbitrarily large without changing the image itself. Thus, $R^n\pi_*\mathcal{L} = 0$ for all n.

(b) We consider the permutation $w\colon (1,\ldots,m+1) \mapsto (m+1,\ldots,1)$. It reduces to the following product of transpositions:

$$w = s_1(s_2s_1)\cdots(s_{m-1}\cdots s_1)(s_m\cdots s_2s_1). \tag{7}$$

In fact, reading from right to left, we find that this permutation first puts 1 in the $(m+1)$-th place and moves each other number over one place, then puts 2 in the m-th place, and so on. Consequently, we have $l(w) = \frac{m(m+1)}{2} = N$. But this number is equal to the dimension of F over M (this dimension is easy to compute using induction on m, as in the proof of Proposition 7). Iterating the isomorphisms in Proposition 5 and taking (5) into account, we now obtain:

$$R^n\pi_*\mathcal{L}(a) \xrightarrow{\sim} R^{n+N}\pi_*\mathcal{L}(w(a+\rho)-\rho) = 0 \qquad \text{for} \qquad n \geq 1.$$

We note that Proposition 5 can be applied only if $d(\mathcal{L}) \geq -1$. The positivity of $\mathcal{L}(a)$ ensures that this condition holds at every step in the decomposition (7), because every time one of the transpositions s_i is applied the number in the i-th place in the sheaf's set of indices is greater than or equal to the number in the $(i+1)$-th place.

(c) Clearly, $\mathcal{L}(b)$ is positive if and only if the coordinates of $b+\rho$ are strictly decreasing. Hence, w is a permutation of $a+\rho$ which arranges it in decreasing order. There is only one such permutation, unless the coordinates of $b+\rho = w(a+\rho)$ have some identical neighbors; but then $\mathcal{L}(b)$ is acyclic, and hence $\mathcal{L}(a)$ is acyclic. If this is not the case, then we can expand w^{-1} as a product of transpositions of neighboring coordinates which arrange them in increasing order, and then, as above, we obtain the isomorphism

$$\pi_*\mathcal{L}(b) \xrightarrow{\sim} R^n\pi_*\mathcal{L}(w^{-1}(b+\rho)-\rho) = R^n\pi_*\mathcal{L}(a), \qquad n = l(w). \qquad \square$$

§ 3. The Klein Quadric and Minkowski Space

1. Twistors. In this section T will denote a four-dimensional complex vector space, which Penrose calls the "twistor" space. We set $M = G(2; T)$. As in § 1.1, we choose a basis (t_i) in T and take the corresponding basis $\{v_{ij} \mid 1 \le i < j \le 4\}$ of decomposable vectors in $\wedge^2 T$.

2. Lemma. *The Plücker mapping $p: G(2; T) \longrightarrow \mathsf{P}(\wedge^2 T) = \mathsf{P}^5$ gives an identification of M with a nonsingular quadric given by the following equation in the coordinates y^{ij} which are dual to the basis (v_{ij}):*

$$y^{12}y^{34} - y^{13}y^{24} + y^{14}y^{23} = 0.$$

Proof. The mapping p associates to a plane $S(x) \subset T$ the line $\mathbf{C}\, s_1 \wedge s_2$, where $S(x) = \mathbf{C}s_1 + \mathbf{C}s_2$. But a bivector $v \in \wedge^2 T$ is decomposable if and only if $v \wedge v = 0$. And it is $(\sum y^{ij} v_{ij})^{\wedge 2} = 0$ which is the Plücker–Klein equation. This argument shows that p gives a set-theoretic bijection of M with the quadric; by passing to the standard covering, we can easily verify that it is actually an isomorphism of analytic manifolds. $\square$

Remarks. (a) On $\wedge^2 T$ there is a canonical non-degenerate symmetric scalar product with values in the one-dimensional complex vector space $\wedge^4 T$, namely the exterior product. In any scalar product space one can define the isotropy subspaces, i.e., the subspaces on which the scalar product restricts identically to zero, and one can also define grassmannians GI of isotropy subspaces and isotropy flag spaces FI. Lemma 2 shows that $G(2; T) = GI(1; \wedge^2 T)$. We shall return to isotropy grassmannians later.

(b) Since $SL(T)$ acts as the identity on $\wedge^4 T$, this scalar product is $SL(T)$-invariant. This gives a homomorphism

$$SL(T) = SL(4;\ \mathbb{C}) \longrightarrow SO(\wedge^2 T) = SO(6;\ \mathbb{C}).$$

Its kernel is $\{\pm 1\}$. Since $SO(6;\ \mathbb{C})$ has the same dimension as $SL(4;\ \mathbb{C})$, it follows that this homomorphism is surjective and is a universal covering, i.e.,

$$SL(4;\ \mathbb{C}) = \mathrm{Spin}(6;\ \mathbb{C}).$$

The twistors are a space for the fundamental representation of the universal covering "complex conformal group" (see below) in the same way as the spinors are a space for the fundamental representation of the group $SL(2;\ \mathbb{C})$, which is the universal covering of the Lorentz group. This is the origin of the name "twistor" (note the semantic similarity of spin/twist). In both cases a completely symmetric role is played by a second fundamental representation, which is a representation in the dual space. The sum of the two fundamental representations is called the two-component spinors or two-component twistors, respectively.

We shall later study the real forms of these constructions.

3. Conformal metric on M. By a conformal metric on a complex manifold N we mean an invertible (i.e., locally free rank 1) local direct summand $\mathcal{L} \subset S^2(\Omega^1 N)$. Such a metric is said to be non-degenerate if the corresponding map $\Omega^1 N \longrightarrow \mathcal{L} \otimes \mathcal{T} N$ is an isomorphism.

In a local coordinate system (x^a) on N, a nonvanishing local section of $\mathcal{L}$ can be written in the form $g_{ab}dx^a dx^b$. It is determined up to multiplication by a nonvanishing local function. Thus, our definition is compatible with the classical one (more precisely, with the analytic version of the classical definition). The metric on $\wedge^2 T$ that was defined in § 3.2 is also a conformal metric.

There is a canonical conformal metric on $M = G(2;\ T)$, namely the subsheaf $\wedge^2 S \otimes \wedge^2 \tilde{S}$; as in § 1.9, we have a canonical decomposition

$$S^2(\Omega^1 M) = S^2(S \otimes \tilde{S}) = \wedge^2(S) \otimes \wedge^2(\tilde{S}) \oplus S^2(S) \otimes S^2(\tilde{S}).$$

We note that $\mathrm{rank}(\wedge^2 S \otimes \wedge^2 \tilde{S}) = 1$ only in the case when $\mathrm{rank}\, S = \mathrm{rank}\, \tilde{S} = 2$. This distinguishes $G(2;\ T)$ among all the grassmannians.

More generally, by a (symmetric or skew-symmetric) conformal scalar product on a locally free sheaf ϵ we mean an invertible subsheaf of $S^2\epsilon^*$ or $\wedge^2\epsilon^*$ which is a local direct summand. Since the sheaves S and $\tilde{S}$ are of rank two, they have the trivial skew-symmetric scalar products $\wedge^2 S^*$ and $\wedge^2 \tilde{S}^*$, respectively. We shall define the conformal metric g on M to be the tensor product of these two pairings:

$$(s \otimes \tilde{s},\ s' \otimes \tilde{s}') = 4(s \wedge s') \otimes (\tilde{s} \wedge \tilde{s}'),$$

where $s, s', \tilde{s}, \tilde{s}'$ are local sections of S and $\tilde{S}$, respectively. (Here we are regarding a metric as a morphism $S^2(\mathcal{T}N) \longrightarrow \mathcal{L}^*$.) From this it is clear that the tangent vectors which are null in the sense of § 1.7 are also null for our metric; the converse is also true.

Suppose that $\mathcal{L} \subset S^2(\Omega^1 N)$ is a conformal metric. A nonvanishing local section $g \in \Gamma(U, \mathcal{L})$ is called a metric on U in the given conformal class.

Since $\wedge^2 \mathcal{L} = p^*(\mathcal{O}(-1))$ (see § 1.4) and similarly for $\wedge^2 \tilde{\mathcal{L}}$, the sheaf of values of the metric $\wedge^2 S \otimes \wedge^2 \tilde{S}$ on M does not, in general, have any global sections on M. But it has sections over any big cell in M, since S and $\tilde{S}$ are trivialized over a big cell.

4. Light rays. Any projective line in $\mathsf{P}(\wedge^2 T)$ which lies entirely on the Klein quadric is called a "light ray" on M. In $\wedge^2 T$ such a projective line corresponds to a plane consisting of decomposable bivectors. Let $p, q \in \wedge^2 T$ be a basis for this plane, and let $x, y \in M$ be the corresponding points. The conditions $p \wedge p = p \wedge q = q \wedge q = 0$ are equivalent to requiring that $\dim S(x) \cap S(y) = 1$, or to requiring that $\dim(S(x) + S(y)) = 3$. In fact, if $p \wedge q = 0$ with p and q decomposable, then we can write $p = t_0 \wedge t_1$ and $q = t_0 \wedge t_2, t_i \in T$; then $S(x) \cap S(y)$ is generated by t_0, and $S(x) + S(y)$ is generated by t_0, t_1 and t_2. Moreover, the subspaces $S(x) \cap S(y)$ and $S(x) + S(y)$ do not depend on the choice of p and q in a fixed plane of decomposable vectors, i.e., they do not depend on the choice of x and y on a fixed line in M. This is easy to see if one replaces p and q by linearly independent linear combinations of p, q. We have thus obtained a map

$$\{\text{light rays}\} \longrightarrow \{(1,3)\text{-flags in } T\}.$$

5. Proposition. (a) *The above map is a bijection between the space L of light rays and the space of $(1,3)$-flags.*

(b) *The natural mapping $F(1,3;\, T) \longrightarrow G(1;\, T) \times G(3;\, T)$ gives an isomorphism of L with the "incidence quadric," which is defined by the equation $\sum_{i=1}^4 z^i w_i = 0$ in dual homogeneous coordinates z^i and w_i for the spaces $\mathsf{P}(T)$ and $\mathsf{P}(T^*)$.*

Proof. (a) If t_0 generates the 1-component of a flag and t_0, t_1, t_2 generate its 3-component, then the plane $\mathbf{C}t_0 \wedge t_1 + \mathbf{C}t_0 \wedge t_2$ in $\wedge^2 T$ determines a light ray. This map is the inverse of the one constructed above.

(b) is obvious. □

Remark. Each point x on a ray l determines a one-dimensional subspace in $S^*(x)$ and a one-dimensional subspace in $\tilde{S}^*(x) = T/S(x)$. Namely, we choose another point $y \in l$ and construct the annihilators of $S(x) \cap S(y)$ and the quotient space $(S(x) + S(y))/S(x)$. As mentioned above, these spaces do not depend upon the choice of y. It is not hard to verify that the tensor product of these subspaces in $\tilde{S}^*(x) \otimes S^*(x) = TM(x)$ is the tangent direction to l. Hence, a light ray, in agreement with its name, has null direction at every point.

6. Light cones. By the "light cone" $C(x)$ of a point $x \in M$ we mean the union of all of the light rays passing through x. We let $L(x)$ denote this set of light rays, i.e., the base of the cone $C(x)$. According to the description above, we have $L(x) = \mathsf{P}(S^*(x)) \times \mathsf{P}(\tilde{S}^*(x))$.

Thus, the "complex heavens" for a point $x \in M$ is the direct product of two Riemann spheres. We shall later see that its real part at a real point x of Minkowski space, i.e., the visible heavens, is a single Riemann sphere, because we can identify $S^*(x)$ with a space that is the complex conjugate of $\tilde{S}^*(x)$. The two complex structures for the heavens that arise here correspond to the difference between the incoming and outgoing halves of the light cone, i.e., in the last analysis to a choice of direction of time. We see one of the heavens, and we send signals to the other one.

Returning to complex geometry, we observe that $C(x)$ is the intersection of the Klein quadric with the hyperplane P^4 which is tangent to M at x. This intersection is a three-dimensional quadric with one singularity at x, i.e., it is a cone with vertex at x. The base of this cone is then a nonsingular two-dimensional quadric having two systems of generatrices, i.e., it is isomorphic to $\mathsf{P}^1 \times \mathsf{P}^1$, in agreement with what was said above.

7. Null planes. The light rays through x in all directions corresponding to one of the systems of generatrices of the base of $C(x)$, sweep out a complex projective plane $\mathsf{P}^2 \subset M$. In this plane any line — not only lines through x — is a light ray. That is, any tangent direction to this plane is a null direction. It is easy to see that all planes having this property can be obtained by this constuction.

The null planes form two disjoint connected families. Just as in the case of light rays, the planes in each family can be represented as π_2-projections of the π_1-fibres of a double fibration on M (compare with § 1.19):

$$\alpha\colon\ \mathsf{P}(T) = G(1;\, T) \xleftarrow{\pi_1} F(1,2;\, T) \xrightarrow{\pi_2} G(2;\, T),$$

$$\beta\colon\ \mathsf{P}(T^*) = G(3;\, T) \xleftarrow{\pi_1} F(2,3;\, T) \xrightarrow{\pi_2} G(2;\, T).$$

These double fibrations are sometimes called the self-dual and anti-self-dual Penrose diagrams. The same terminology is used for the corresponding families of planes; sometimes these planes are called α- and β-planes. Replacing T by T^* reverses these labels; the specification of which planes have which labels is equivalent to the choice of one of the two spinor bundles S or $\tilde{S}$ as the fundamental one.

8. Lemma. *A light plane in M is a β-plane (respectively, an α-plane) if and only if any local section of the sheaf*

$$\Omega^2_+(M) = S^2(S) \otimes \wedge^2(\tilde{S}) \qquad (\text{resp. } \Omega^2_-(M) = \wedge^2(S) \otimes S^2(\tilde{S}))$$

vanishes when restricted to the tangent directions along the given plane.

Proof. In the notation of § 1.18 we have $\Omega^1 F/\mathrm{P} = S^2_F/S^1_F \otimes \tilde{S}^2_F$, where $F = F(1,2;T), \mathrm{P} = \mathrm{P}(T)$. In fact, F/P is the relative grassmannian $G_{\mathrm{P}}(1;(T \otimes \mathcal{O}_F)/S^1_F)$. Here the tautological sheaf in this representation is S^2_F/S^1_F, so that the orthogonal sheaf is $\tilde{S}^2_F$, and it remains to apply Theorem 1.6. Thus, $\Omega^2 F/\mathrm{P} = S^2(S^2_F/S^1_F) \otimes \wedge^2(\tilde{S}^2_F)$. But $\tilde{S}^2_F = \pi_2^*(\tilde{S}^2_M)$. From this it is clear that the sections of $\Omega^2_- M$ vanish on the π_2-fibres of an α-bundle, and conversely: for any bivector in the α-plane there exists a section of $\Omega^2_+ M$ which is not equal to zero on the given bivector. □

9. Complex Minkowski space. By a complex Minkowski space we mean a big cell U in $G(2;T)$ with a metric $g = \epsilon \otimes \tilde{\epsilon}$ which is the tensor product of two spinor metrics $\epsilon \in H^0(U, \wedge^2 S), \tilde{\epsilon} \in H^0(U, \wedge^2 \tilde{S})$; the sections ϵ and $\tilde{\epsilon}$ are constant in the standard trivialization of S and $\tilde{S}$.

We now give some further details.

(a) Suppose that the big cell U is determined by the subspace $S = S^2 \subset T$ (see § 1.3). By definition, the points "at infinity" are the points

$$M \setminus U = \{x \in M \mid \dim S(x) \cap S \geq 1\}.$$

Consequently, for all $x \in M \setminus U$ except for the point ∞ (which corresponds to S), we have $\dim(S(x) \cap S) = 1$. This means that $M \setminus U = C(\infty)$, by § 2.4: every point of $M \setminus U$ lies on a light ray through ∞. In Penrose's model, complex Minkowski space is compactified by means of a light cone.

(b) According to § 1.3, the Minkowski space U is the affine space which is associated to the vector space $S(\infty) \otimes (T/S(\infty))^* = S(\infty) \otimes \tilde{S}(\infty)$. Choosing an origin $0 \in U$ is equivalent to choosing a second plane $S(0) \subset T$ and a direct sum decomposition $T = S(0) \oplus S(\infty)$, or else (by symmetry) choosing $\tilde{S}(0)$ and a decomposition $T^* = \tilde{S}(0) \oplus \tilde{S}(\infty)$. Such a choice fixes trivializations of S and $\tilde{S}$ over U using the projection of any plane onto $S(0)$ and $\tilde{S}(0)$ parallel to $S(\infty)$ and $\tilde{S}(\infty)$:

$$S|_U = S(0) \otimes \mathcal{O}_U, \qquad \tilde{S}|_U = \tilde{S}(0) \otimes \mathcal{O}_U.$$

(c) We choose nonzero elements $\epsilon \in \wedge^2(S(0))$ and $\tilde{\epsilon} \in \wedge^2(\tilde{S}(0))$, which we regard as sections of $\wedge^2 S$ and $\wedge^2 \tilde{S}$, and we use them to construct the complex metric $g = \epsilon \otimes \tilde{\epsilon} \in H^0(U, \Omega^2 U)$.

The subgroup of $SL(T)$ which preserves the structure $(U, \epsilon, \tilde{\epsilon})$ is clearly $SL(S(0)) \times SL(S(\infty)) = SL(2; \mathbf{C}) \times SL(2; \mathbf{C})$; this is the complex Lorentz group.

We can define ϵ and $\tilde{\epsilon}$ by giving coordinate systems in $S(0)$ and $\tilde{S}(0)$ which are unimodular relative to ϵ and $\tilde{\epsilon}$. Since $S(\infty)^* = \tilde{S}(0)$, we obtain a dual coordinate system in $S(\infty)$ and then in T and T^*. The notation for our basic objects in this coordinate system is called the spinor index formalism. We now describe it in more detail.

10. Coordinates in twistor spaces. We introduce the following notation:

$$\begin{array}{lll} & \text{space} & \text{coordinates} \\ T & \left\{ \begin{array}{l} S(0) \\ S(\infty) \end{array} \right. & \begin{array}{l} (z_0, z_1) \\ (z^{\dot{0}}, z^{\dot{1}}) \end{array} \\ T^* & \left\{ \begin{array}{l} \tilde{S}(0) \\ \tilde{S}(\infty) \end{array} \right. & \begin{array}{l} (w_{\dot{0}}, w_{\dot{1}}) \\ (w^0, w^1) \end{array} \end{array}$$

Small Greek letters are used for the spinor indices: $\alpha, \beta, \ldots \in \{0, 1\}$ and $\dot{\alpha}, \dot{\beta}, \ldots \in \{\dot{0}, \dot{1}\}$. As usual with index notation, a symbol like z_α can have two meanings: if it is an abbreviation for a pair of numbers (z_1, z_2), then z_α denotes an element in $S(0)$; on the other hand, if we have in mind one of the two coordinate functions on $S(0)$, then z_α denotes an element in $S(0)^*$. Moreover, one must remember that we are regarding an element of $S(0)$ simultaneously as a section of the bundle S over U which is constant in a given trivialization. When it becomes necessary to refer to non-constant sections, for example the spinors in the Dirac equations, we shall use the letter ψ instead of z or w, as is customary in the physics journals.

One has the following mnemonic rules for remembering the index game. We start with the principle that the coordinate functions x^a on Minkowski space are written with upper indices, while the coordinates of 1-forms $(f_a) = f_a dx^a$ are written with lower indices; then, since $\Omega^1 = S \otimes \tilde{S}$, the spinor indices of coordinates in $S(0)$ and $\tilde{S}(0)$ are written as subscripts, with a dot above the index replacing the tilde. It is natural to raise the indices when passing to the dual spaces $\tilde{S}(\infty) = S(0)^*$ and $\tilde{S}(0) = S(\infty)^*$. In other words, we write the invariant scalar products as follows:

$$\text{between } S(0) \text{ and } \tilde{S}(\infty): \; z_0 w^0 + z_1 w^1 = z_\alpha w^\alpha,$$

$$\text{between } S(\infty) \text{ and } \tilde{S}(0): \; z^{\dot{0}} w_{\dot{0}} + z^{\dot{1}} w_{\dot{1}} = z^{\dot{\alpha}} w_{\dot{\alpha}},$$

$$\text{between } T \text{ and } T^*: \; z_\alpha w^\alpha - z^{\dot{\alpha}} w_{\dot{\alpha}}.$$

We now write out the sections ϵ and $\tilde{\epsilon}$. We choose $\epsilon = 2w^0 \wedge w^1 = w^0 \otimes w^1 - w^1 \otimes w^0 \in \wedge^2 S(0), \tilde{\epsilon} = 2z^{\dot{0}} \wedge z^{\dot{1}}$. Thus, in coordinates we have:

$$\begin{aligned} \epsilon &= \epsilon_{\alpha\beta} w^\alpha \otimes w^\beta; & \epsilon_{01} &= -\epsilon_{10} = 1, & \epsilon_{00} &= \epsilon_{11} = 0; \\ \tilde{\epsilon} &= \epsilon_{\dot{\alpha}\dot{\beta}} z^{\dot{\alpha}} \otimes z^{\dot{\beta}}; & \epsilon_{\dot{0}\dot{1}} &= -\epsilon_{\dot{1}\dot{0}} = 1, & \epsilon_{\dot{0}\dot{0}} &= \epsilon_{\dot{1}\dot{1}} = 0. \end{aligned}$$

The exterior product $S \times S \longrightarrow \wedge^2 S$ and the analogous product for $\tilde{S}$ may be regarded as (skew-symmetric) spinor metrics, where one can identify S with S^* by means of multiplication by ϵ and $\tilde{S}$ with $\tilde{S}^*$ by means of multiplication by $\tilde{\epsilon}$. According to our conventions, this also leads to a change of notation for the coordinates (the use of z and w is reversed). We shall normalize the sign of this identification by stipulating that

$$z_\alpha = \epsilon_{\alpha\beta} w^\beta, \qquad w_{\dot{\alpha}} = \epsilon_{\dot{\alpha}\dot{\beta}} z^{\dot{\beta}}.$$

Recall that when we raise the indices in $\epsilon_{\alpha\beta}$ and $\epsilon_{\dot{\alpha}\dot{\beta}}$, i.e., when we invert these matrices, the signs again change: $\epsilon^{01} = -\epsilon^{10} = -1, \epsilon^{00} = \epsilon^{11} = 0$:

$$w^\alpha = \epsilon^{\alpha\beta} z_\beta, \qquad z^{\dot{\alpha}} = \epsilon^{\dot{\alpha}\dot{\beta}} w_{\dot{\beta}}.$$

Finally, we note that in much of the literature on the twistor technique A and B' are used instead of α and $\dot{\beta}$. We have chosen the notation which can be more conveniently extended to supersymmetry and supergravity.

11. Coordinates in complex Minkowski space. In accordance with the construction in § 1.3, the standard coordinates in the big cell U occupy a 2×2–block in a matrix of the following form:

	$S(\infty)$		$S(0)$	
basis of T:	$w_{\dot{0}}$	$w_{\dot{1}}$	w^0	w^1
coordinates in U:	$x^{0\dot{0}}$	$x^{0\dot{1}}$	1	0
	$x^{1\dot{0}}$	$x^{1\dot{1}}$	0	1

We recall the meaning of these coordinates: the point $x = (x^{\alpha\dot{\beta}})$ corresponds to the subspace of T spanned by the rows of the matrix, i.e., spanned by the vectors $x^{\alpha\dot{\beta}} w_{\dot{\beta}} + w^\alpha \in T$, $\alpha = 0, 1$. Then the constant section (in our chosen trivialization of S) with coordinates z_α as a vector in T will depend linearly on $x^{\alpha\dot{\beta}}$, since it has the form $x^{\alpha\dot{\beta}} z_\alpha w_{\dot{\beta}} + z_\alpha w^\alpha$.

We also trivialize $\mathcal{O}_U \times T/S$ and $\tilde{S}$ over U using the isomorphism $S(\infty) \longrightarrow T/S(x)$ and the dual isomorphism. After that we identify $\Omega^1 U$ with $S \otimes \tilde{S} = \mathcal{O}_U \otimes (S(0) \otimes S(\infty)^*)$ using the construction in § 1.5. We now describe the basic structures on U in the spinor coordinates.

12. Proposition. (a) *$dx^{\alpha\dot{\beta}} = w^\alpha \otimes z^{\dot{\beta}}$ under the above identification; in particular, the $dx^{\alpha\dot{\beta}}$ are decomposable and are constant in the trivialization.*

(b) *The scalar product giving the conformal Minkowski metric has the form*

$$(dx^{\alpha\dot{\beta}}, dx^{\gamma\dot{\delta}}) = \epsilon^{\alpha\gamma}\epsilon^{\dot{\beta}\dot{\delta}}(\epsilon \otimes \tilde{\epsilon}),$$

i.e., the coefficients of the metric are $g_{\alpha\dot{\beta}\gamma\dot{\delta}} = \epsilon_{\alpha\gamma}\epsilon_{\dot{\beta}\dot{\delta}}$.

(c) *The exterior 2-form $F_{\alpha\dot{\beta}\gamma\dot{\delta}} dx^{\alpha\dot{\beta}} \wedge dx^{\gamma\dot{\delta}}$ can be uniquely written as*

$$F_{\alpha\dot{\beta}\gamma\dot{\delta}} = \epsilon_{\dot{\beta}\dot{\delta}} F_{\alpha\gamma} + \epsilon_{\alpha\gamma} F_{\dot{\beta}\dot{\delta}},$$

and this decomposition agrees with the decomposition into $\pm$-components in §§ 1.9 and 2.8.

Proof. (a) The construction in § 1.5 identifies $\Omega^1 U$ with $S \otimes \tilde{S}$ in the following way: one constructs the differential $d\colon S \longrightarrow (\mathcal{O}_U \otimes T) \otimes \Omega^1 U$, which kills the constant sections (the elements of T), and then one composes it with $\mathcal{O}_U \otimes T \longrightarrow \mathcal{O}_U \otimes T/S$. In our notation $d(x^{\alpha\dot{\beta}} z_\alpha w_{\dot{\beta}} + z_\alpha w^\alpha) \bmod S \otimes \Omega^1 U = dx^{\alpha\dot{\beta}} z_\alpha w_{\dot{\beta}}$, from which we have $dx^{\alpha\dot{\beta}} = w^\alpha \otimes z^{\dot{\beta}}$.

(b) $(dx^{\alpha\dot{\beta}}, dx^{\gamma\dot{\delta}}) = 4(w^\alpha \wedge w^\gamma) \otimes (z^{\dot{\beta}} \wedge z^{\dot{\delta}}) = \epsilon^{\alpha\gamma}\epsilon^{\dot{\beta}\dot{\delta}}(\epsilon \otimes \tilde{\epsilon})$ (here the four appears because of the conventions $g = \epsilon \otimes \tilde{\epsilon}$ and $\epsilon = 2w^0 \wedge w^1, \tilde{\epsilon} = 2z^{\dot{1}} \wedge z^{\dot{1}}$).

(c) Set $F_{\alpha\gamma} = F_{\alpha\dot{0}\gamma\dot{1}}, F_{\dot{\beta}\dot{\delta}} = F_{0\dot{\beta}1\dot{\delta}}$. □

13. The diagram $L \xleftarrow{\pi_1} F \xrightarrow{\pi_2} M$ in coordinates. Here we set $L = F(1,3;T)$ and $F = F(1, 2, 3; T)$. The following equation describes $L \subset \mathrm{P}(T) \times \mathrm{P}(T^*)$ in terms of homogeneous coordinates:

$$z_\alpha w^\alpha - z^{\dot{\alpha}} w_{\dot{\alpha}} = 0. \tag{1}$$

We now consider the system of equations

$$x^{\alpha\dot{\beta}} w_{\dot{\beta}} - w^\alpha = 0, \qquad \alpha = 0, 1; \tag{2}$$

$$x^{\alpha\dot{\beta}} z_\alpha - z^{\dot{\beta}} = 0, \qquad \dot{\beta} = \dot{0}, \dot{1}. \tag{3}$$

The equations in (2) mean that $S(x^{\alpha\dot\beta})$ lies in the hyperplane T defined by equation (1) (the w are fixed); the equations in (3) mean that $S(x^{\alpha\dot\beta})$ contains the line in T which is spanned by the vector $(z_\alpha, z^{\dot\beta})$. It is then clear from geometrical considerations that (2) and (3) cannot both be solved for x unless the condition in (1) is fulfilled. In that case we have a one-dimensional system of equations which corresponds to the points of the light ray whose coordinates are given by (1).

Finally, we note that, just as the usual local coordinates on a manifold are sections of the structure sheaf, the "homogeneous coordinates" in the sense of § 1.3 are sections of the tautological sheaves and sheaves associated to them. We shall illustrate how this works in the case of our double fibration diagram, where all of the essential sheaves are invertible.

On L we have $S_L^1 \subset T \otimes \mathcal{O}_L$, and hence a surjective map $T^* \otimes \mathcal{O}_L \longrightarrow (S_L^1)^*$, which induces an isomorphism $T^* \xrightarrow{\sim} H^0(L, (S_L^1)^*)$. We similarly obtain $T \xrightarrow{\sim} H^0(L, (\tilde{S}_L^1)^*)$. It is convenient to let $\mathcal{O}_L(-a, -b)$ denote $(S_L^1)^{\otimes a} \otimes (\tilde{S}_L^1)^{\otimes b}$. Thus z_α and $z^{\dot\beta}$ can be interpreted as sections of the sheaf $\mathcal{O}_L(0, 1)$, and $w_{\dot\alpha}$ and w^β as sections of the sheaf $\mathcal{O}_L(1, 0)$.

The canonical isomorphisms $\pi_1^*(S_L^1) = S_F^1$ and $\pi_1^*(\tilde{S}_L^1) = \tilde{S}_F^1$ make it possible to regard z and w respectively as sections of $(\tilde{S}_F^1)^*$ and $(S_F^1)^*$, or $\mathcal{O}_F(0, 1)$ and $\mathcal{O}_F(1, 0)$.

14. Real Minkowski space. We shall interpret the traditional realization of Minkowski space by hermitian 2×2–matrices in the following way. On the big cell U we have an anti-holomorphic involution $\rho: U \longrightarrow U$ which is hermitian conjugation in the $x^{\alpha\dot\beta}$-coordinates:

$$\rho^*(x^{\alpha\dot\beta}) = (x^{\alpha\dot\beta})^+ = (\bar{x}^{\alpha\dot\beta})^t$$

(the bar denotes complex conjugation, and t denotes the transpose). It is the set M_0 of fixed points under this involution which is the classical real Minkowski space.

However, our own point of view is rather to regard the entire complex space together with the additional structure ρ, and not just the ρ-invariant points, as the real Minkowski space. Thus, we shall extend the action of ρ to all of the twistor objects.

15. Proposition. (a) *The hermitian conjugation on U is induced by an anti-holomorphic involution on the space L of rays which in homogeneous coordinates can be written as follows:*

$$\rho_T: T \times T^* \longrightarrow T \times T^*, \qquad \rho_T(z_\alpha, z^{\dot\beta}; w^\alpha, w_{\dot\beta}) = (\bar{w}_{\dot\alpha}, \bar{w}^\beta; \bar{z}^{\dot\alpha}, \bar{z}_\beta).$$

(b) *The involution ρ_T also determines an anti-holomorphic automorphism of the structure $(M,\ S,\ \tilde{S})$ which interchanges the tautological bundles, i.e., it gives identifications*

$$\rho^*(S) = \overline{\tilde{S}}, \qquad \rho^*(\tilde{S}) = \overline{S},$$

or, considered on points, $S(\rho(x)) = \overline{\tilde{S}}(\rho(x)), \tilde{S}(\rho(x)) = \overline{S}(x)$. *In particular, the tautological bundles are the complex conjugates of one another on real Minkowski space.*

Proof. (a) Let $(x^{\alpha\dot{\beta}})$ be the coordinates of a point $x \in M$. Then $\rho(x)^{\alpha\dot{\beta}} = \bar{x}^{\beta\dot{\alpha}}$. If we apply complex conjugation to equations (2) and (3) above, then rename the indices $\alpha \longleftrightarrow \beta$ and reverse the order of the equalities, we obtain:

$$\rho(x)^{\alpha\dot{\beta}}\, \bar{z}_\beta \; - \; \bar{z}^{\dot{\alpha}} = 0,$$

$$\rho(x)^{\alpha\dot{\beta}}\, \bar{w}_{\dot{\alpha}} \; - \; \bar{w}^{\beta} = 0.$$

Comparing with (2) and (3), we find that, if the ray $(z,\ w)$ passes through x, then the ray $\rho_T(z,\ w)$ passes through the point $\rho(x)$. It remains to observe that a point in M is uniquely determined by the set of rays which pass through it.

(b) We identify $S(x)$ with the subspace $S(x) \times \{0\} \subset T \times T^*$, and we identify $\tilde{S}(x)$ with $\{0\} \times \tilde{S}(x) \subset T \times T^*$. The previous computation then shows that ρ_T induces an antilinear isomorphism between $S(x)$ and $\tilde{S}(\rho(x))$. □

16. Pauli matrices. Since the coordinates $x^{\alpha\dot{\beta}}$ are not generally real on the real part of U, one replaces them with "single index" real coordinates by expanding $(x^{\alpha\dot{\beta}})$ with respect to the Pauli basis:

$$x^{\alpha\dot{\beta}} = \sigma_a^{\alpha\dot{\beta}} x^a, \text{ where } \sigma_0 = \begin{pmatrix} 1 & 0 \\ 0 & 1 \end{pmatrix}, \sigma_1 = \begin{pmatrix} 0 & 1 \\ 1 & 0 \end{pmatrix},$$

$$\sigma_2 = \begin{pmatrix} 0 & -i \\ i & 0 \end{pmatrix}, \sigma_3 = \begin{pmatrix} 1 & 0 \\ 0 & -1 \end{pmatrix}.$$

A more invariant treatment, which can be generalized to the curved case, consists in looking at the corresponding 1-forms:

$$\sigma_a^{\alpha\dot{\beta}} dx^a = dx^{\alpha\dot{\beta}}.$$

According to Proposition 12, our general interpretation of the right side of this equality is that these are decomposable 1-forms (in the spinor representation), so that the $\sigma_a^{\alpha\dot\beta}$ are transition matrices from the coordinate basis of 1-forms to a decomposable basis.

Returning to the flat case, from part (b) of Proposition 12 we find the standard form of the metric: $g_{\alpha\dot\beta\gamma\dot\delta}dx^{\alpha\dot\beta}dx^{\gamma\dot\delta} = \det(dx^{\alpha\dot\beta}) = (dx^0)^2 - dx^a dx^a, a = 1, 2, 3.$

17. The Euclidean form. Besides the real Minkowski space $M_0 \subset U$ and its compactification (its closure in U) by means of the real part of the light cone $C(\infty)$, another object which plays an essential role in field theory is the "real section" of M, on which the induced conformal metric has fixed sign. We shall introduce this concept using antilinear mappings on T and T^* which are not involutions, but rather have a quaternionic structure, i.e., they have square -1.

18. Proposition. *Define a quaternionic structure on T and T^* under which $S(0), S(\infty), \tilde S(0)$, and $\tilde S(\infty)$ are invariant by means of the formulas $j(z_0, z_1) = (-\bar z_1, \bar z_0)$ and similarly for $z^{\dot\alpha}$, $w_{\dot\alpha}$, w^α. Then:*

(a) *A real structure is induced on $\mathsf{P}(T)$ and $\mathsf{P}(T^*)$ which does not have real points but does have a manifold of real (j-invariant) pairwise disjoint lines passing through each point of $\mathsf{P}(T)$ and $\mathsf{P}(T^*)$. They form the fibres of a smooth bundle with base S^4.*

(b) *A real structure is induced on $M = G(2; T) = G(2; T^*)$ which in the $x^{\alpha\dot\beta}$-coordinates corresponds to the involution $j(x)^{\alpha\dot\beta} = \epsilon_{\alpha\gamma}\epsilon_{\dot\beta\dot\delta}\bar x^{\gamma\dot\delta}$. Its real points form a real sphere S^4 in the Klein quadric. The families of α- and β-planes through each real point are invariant, and they correspond to real lines on $\mathsf{P}(T)$ and $\mathsf{P}(T^*)$.*

(c) *Each of the bundles S and $\tilde S$ acquire a quaternionic structure over the real points of M, and the induced conformal metric has Euclidean signature.*

Proof. If $(az_0, az_1) = (-\bar z_1, \bar z_0)$, then $z_1 = -\overline{a}\bar z_0 = -|a|^2 z_1$, and hence $z_0 = z_1 = 0$; so there are no j-fixed points in $\mathsf{P}(T)$ or $\mathsf{P}(T^*)$. The line which passes through any point and its conjugate is a real line; conversely, a real line which passes through a point must pass through its conjugate, and so is uniquely determined.

It is easy to see that the following relations are equivalent:

$$\begin{pmatrix} x^{0\dot 0} & x^{0\dot 1} \\ x^{1\dot 0} & x^{1\dot 1}\end{pmatrix}\begin{pmatrix} w_{\dot 0} \\ w_{\dot 1}\end{pmatrix} = \begin{pmatrix} w^0 \\ w^1\end{pmatrix} \quad \text{and} \quad \begin{pmatrix} \bar x^{1\dot 1} & -\bar x^{1\dot 0} \\ -\bar x^{0\dot 1} & \bar x^{0\dot 0}\end{pmatrix}\begin{pmatrix} -\bar w_{\dot 1} \\ \bar w_{\dot 0}\end{pmatrix} = \begin{pmatrix} -\bar w^1 \\ \bar w^0\end{pmatrix},$$

and similarly for z. But this means that $j(x)^{\alpha\dot\beta} = \epsilon_{\alpha\gamma}\epsilon_{\dot\beta\dot\delta}\bar x^{\gamma\dot\delta}$. Now it is clear that j induces a quaternionic structure on the fibres of S and $\tilde S$ over the real points, since here (unlike in the case of ρ_T) S and $\tilde S$ do not change places under conjugation.

The involution j acts as follows in the decomposable basis of $\wedge^2 T$:

$$j(y^{12}, y^{13}, y^{14}, y^{23}, y^{24}, y^{34}) = (\bar y^{12}, \bar y^{24}, -\bar y^{23}, -\bar y^{14}, \bar y^{13}, \bar y^{34}).$$

In these coordinates the real points of $G(2;\,T)$ are determined by the relations $y^{12}, y^{34} \in \mathbb{R}, y^{13} = \bar{y}^{24}, y^{14} = -\bar{y}^{23}$. The restriction of the quadric's equation to these points (see Lemma 2) takes the form

$$y_{12}y_{34} - |y_{13}|^2 - |y_{14}|^2 = 0.$$

Thus, the real part of M is S^4. The intersection $S^4 \cap U$ has a natural $\mathbb{R}^4$ structure, and $S^4 \cap C(\infty)$ consists of the single point ∞.

In the $x^{\alpha\dot{\beta}}$-coordinates the reality condition is equivalent to requiring that the coefficients y^a be real in the decomposition

$$x^{\alpha\dot{\beta}} = y^0\sigma_0^{\alpha\dot{\beta}} + i\sigma_a^{\alpha\dot{\beta}}y^a.$$

The metric $\det(dx^{\alpha\dot{\beta}})$ in these coordinates is equal to $(dy^0)^2 + dy^a dy^a$. □

We note that with our choice of involutions the relative location in U of the Minkowski space and its Euclidean form is such that they have a common "time axis" (y^0 and x^0), while the space sections are rotated from one another by multiplication by i. Physicists usually use a pair of real structures in which the time axes are related by multiplication by i (the "Wick rotation").

19. Pseudo-hermitian twistor metric. The involution ρ_T in § 3.15 identifies T^* and $\overline{T}$, and so enables one to define a scalar product on T: $\frac{1}{2i}(t_1,\, \rho_T(t_2))$. In the standard coordinates with $t = (z_\alpha,\, z^{\dot{\alpha}})$ and $|t|^2 = \frac{1}{2i}(t,\, \rho_T(t))$ we find that

$$|t|^2 = \operatorname{Im} z_\alpha \bar{z}^{\dot{\alpha}}.$$

We thereby see that this "Penrose form" has the following properties:

(a) It is pseudo-hermitian with signature (2, 2).

(b) A point $(t,\, t^*) \in T \times T^*$ corresponds to a light cone in the completed Minkowski space if and only if $|t|^2 = 0$ and $t^* = \rho(t)$.

Conversely, one can define a pseudo-hermitian form with signature (2, 2) on T and determine a map $\rho_T\colon T \longrightarrow T^*$ in such a way that the form is given by $\frac{1}{2i}(t_1,\, \rho_T(t_2))$. This uniquely determines ρ_T. The choice of two subspaces in general position in T which are isotropic for this form enables one then to recover all of the remaining structures.

§ 4. Distributions and Connections

1. Definition. A *(holomorphic) distribution on a manifold F* is a subsheaf $\mathcal{T} \subset \mathcal{T}F$ which is a local direct summand. It is said to be *integrable* if $\mathcal{T}$ is a subsheaf of Lie algebras. □

We call the rank of $\mathcal{T}$ the dimension of the distribution. It is constant on the connected components of F. A local section is called a vector field tangent to the distribution. A distribution can also be defined by giving either the subsheaf $\mathcal{T}^\perp \subset \Omega^1 F$ of 1-forms which vanish on $\mathcal{T}$, or the "differential along the distribution" $\partial: \mathcal{O}_F \longrightarrow \mathcal{T}^*$, where $\partial f = df (\mathrm{mod}\, \mathcal{T}^\perp)$. We can recover $\mathcal{T}^\perp$ from the differential as the kernel of the induced map $\Omega^1 F \longrightarrow \mathcal{T}^*$.

As a rule, we shall consider distributions on a manifold which is the total space of a fibration, perhaps with some additional structure. Such distributions, assumed to be compatible in some way with the fibration structure and whatever additional structure is present, will be referred to as connections of various kinds. We begin with the simplest type of connection.

2. Definition. (a) By a *fibration* $\pi: F \longrightarrow M$ we mean a morphism which is a submersion of complex manifolds.

(b) By a *connection on a fibration* (F, π) we mean a distribution $\mathcal{T} \subset \mathcal{T}F$ for which the morphism $d\pi$ in the exact sequence

$$0 \longrightarrow \mathcal{T}F/M \longrightarrow \mathcal{T}F \xrightarrow{d\pi} \pi^*(\mathcal{T}M) \longrightarrow 0 \tag{1}$$

induces an isomorphism $\mathcal{T} \xrightarrow{\sim} \pi^*(\mathcal{T}M)$. □

The differential along a connection on a fibration maps $\mathcal{O}_F$ to $\pi^*(\Omega^1 M)$. More precisely, it follows from (1) that a connection on a fibration amounts to giving direct sum decompositions $\mathcal{T}F = \mathcal{T}F/M \oplus \pi^*(\mathcal{T}M)$ and $\Omega^1 F = \Omega^1 F/M \oplus \pi^*(\Omega^1 M)$. Such a decomposition corresponds to splitting $d: \mathcal{O}_F \longrightarrow \Omega^1 F$ into differentials in the horizontal and vertical directions: $d_h = \partial: \mathcal{O}_F \longrightarrow \pi^*(\Omega^1 M), d_v: \mathcal{O}_F \longrightarrow \Omega^1 F/M$. A morphism $h: \pi^*(\mathcal{T}M) \longrightarrow \mathcal{T}F$ which splits (1) is called a lifting of vector fields.

Speaking geometrically, at each point $x \in F$ a connection singles out a d-dimensional tangent subspace of horizontal directions, which $d\pi$ projects isomorphically onto the tangent space at $\pi(x) \in M$, where $d = \dim M$. In order to understand the structure of the set of connections, we now turn to a more general situation.

3. Sheaf extensions and splittings. We recall the elementary formalism which enables one to determine the existence of connections and describe the set of connections on a fibration.

In general, let $\mathcal{E}\colon 0 \longrightarrow \mathcal{S} \xrightarrow{i} \mathcal{T} \xrightarrow{j} \mathcal{S}' \longrightarrow 0$ be an exact sequence of coherent sheaves on F. By a splitting of the sequence we mean a morphism $h\colon \mathcal{S}' \longrightarrow \mathcal{T}$ such that $j \circ h = id_{\mathcal{S}'}$; then $\mathcal{T} = \mathcal{S} \oplus h(\mathcal{S}')$, and j is an isomorphism on $h(\mathcal{S}')$. The difference $h_1 - h_2\colon \mathcal{S}' \longrightarrow \mathcal{S}$ of two splittings maps $\mathcal{S}'$ to the kernel of j. If we identify this kernel with $\mathcal{S}$, we may regard $h_1 - h_2 \in \mathrm{Hom}(\mathcal{S}', \mathcal{S})$. Conversely, if h_1 is a splitting and $f \in \mathrm{Hom}(\mathcal{S}', \mathcal{S})$, then $h_1 + i \circ f$ is another splitting. Thus, the set of splittings either is empty or else is a principal homogeneous space for the group $\mathrm{Hom}(\mathcal{S}', \mathcal{S})$.

These notions can clearly be localized. If the morphism i is a direct sum imbedding, then there is a sheaf of splittings which is a principal homogeneous space for the sheaf $\mathcal{H}om(\mathcal{S}', \mathcal{S})$. It is well known that this sheaf can be used to construct the characteristic class $c(\mathcal{E}) \in H^1(F, \mathcal{H}om(\mathcal{S}', \mathcal{S}))$ which is the obstruction to a global splitting of $\mathcal{E}$. To determine this class explicitly, we consider the sequence $\mathcal{H}om(\mathcal{S}', \mathcal{E})\colon 0 \longrightarrow \mathcal{H}om(\mathcal{S}', \mathcal{S}) \longrightarrow \mathcal{H}om(\mathcal{S}', \mathcal{T}) \longrightarrow \mathcal{H}om(\mathcal{S}', \mathcal{S}') \longrightarrow 0$. This sequence is exact when $\mathcal{E}$ splits locally. We set $c(\mathcal{E}) = \delta(\mathrm{id}_{\mathcal{S}'})$, where $\delta\colon H^0(F, \mathcal{H}om(\mathcal{S}', \mathcal{S}')) \longrightarrow H^1(F, \mathcal{H}om(\mathcal{S}', \mathcal{S}))$ is the boundary homomorphism. If we have splittings $h_i\colon \mathcal{S}'|_{U_i} \longrightarrow \mathcal{T}|_{U_i}$ on the pieces of an open covering $F = \bigcup U_i$, then the Čech cocycle $(h_i|_{U_i \cap U_j} - h_j|_{U_i \cap U_j})$ represents the class $c(\mathcal{E})$. If $c(\mathcal{E}) = 0$, then $H^0(M, \mathcal{H}om(\mathcal{S}', \mathcal{S}))$ acts transitively and effectively on the set of splittings.

4. Examples. (a) Let $\Delta \subset F \times F$ be the diagonal, let $\pi_{1,2}\colon F \times F \longrightarrow F$ be the two projections, and let $I_\Delta \subset \mathcal{O}_{F\times F}$ be the sheaf of equations of the diagonal. We set $\mathcal{O}_F^{(n)} = \mathcal{O}_{F\times F}/I_\Delta^{n+1}$. The space $(\Delta, \mathcal{O}_F^{(n)})$ is the n-th infinitesimal neighborhood of the diagonal. We can identify Δ with F using either of the two projections π_1, π_2, with the same result from a set-theoretic point of view. However, the $\mathcal{O}_F$-algebra structures on $\mathcal{O}_F^{(n)}$ turn out to be different. We consider the map $\partial\colon \mathcal{O}_F \longrightarrow \mathcal{O}_F^{(1)}$ given by $\partial f = (\pi_1^* f - \pi_2^* f) \bmod I_\Delta^2$. One can check that this map is a derivation, and that it determines an imbedding of $\Omega^1 F$ into $\mathcal{O}_F^{(1)}$ which identifies $\Omega^1 F$ with the kernel of the map reduction mod I_Δ from $\mathcal{O}_F^{(1)}$ to $\mathcal{O}_F$. We thus have an exact sequence of rings $0 \longrightarrow \Omega^1 F \longrightarrow \mathcal{O}_F^{(1)} \longrightarrow \mathcal{O}_F \longrightarrow 0$, in which $\Omega^1 F$ is an ideal with square zero. This sequence comes with two splittings; the difference between them is d.

For any coherent sheaf $\mathcal{T}$ on F we set $\mathrm{Jet}^n \mathcal{T} = \pi_{2*}(\pi_1^* \mathcal{T} / I_\Delta^{n+1} \pi_1^* \mathcal{T})$. This is called the sheaf of n-th order jets for $\mathcal{T}$. The argument in the last paragraph implies that we have the following exact sequence of $\mathcal{O}_F$-modules which splits locally whenever $\mathcal{T}$ is locally free:

$$0 \longrightarrow \mathcal{T} \otimes \Omega^1 F \longrightarrow \mathrm{Jet}^1 \mathcal{T} \longrightarrow \mathcal{T} \longrightarrow 0. \tag{2}$$

The obstruction $c(\mathcal{T})$ to global splitting lies in $H^1(F, \mathcal{H}om(\mathcal{T}, \mathcal{T} \otimes \Omega^1 F))$. The method of Chern–Weil can be used to construct from $c(\mathcal{T})$ certain characteristic classes $c_i(\mathcal{T}) \in H^i(R, \Omega^i F)$ with values in sheaf cohomology groups which no longer depend on $\mathcal{T}$.

(b) In particular, let $F = \mathsf{P}(T)$. The standard exact sequence (see § 2.1)

$$0 \longrightarrow \Omega^1 F(1) \longrightarrow \mathcal{O}_F \otimes T^* \longrightarrow \mathcal{O}(1) \longrightarrow 0$$

is isomorphic to the sequence (2) when $\mathcal{T} = \mathcal{O}(1)$. As shown in § 2, the class $c_1 \in H^1(F, \Omega^1 F)$ is a generator of this cohomology group (and its powers $c_1^i \in H^i(F, \Omega^i F)$ are generators of those one-dimensional higher cohomology groups).

(c) More generally, on any grassmannian $F = G(d; T)$ the exact sequence

$$0 \longrightarrow S \longrightarrow \mathcal{O}_F \otimes T \longrightarrow \tilde{S}^* \longrightarrow 0$$

determines a class $c(F) \in H^1(F, S \otimes \tilde{S}) = H^1(F, \Omega^1 F)$, which can easily be shown to be nontrivial. This class also generates H^1.

Returning to the exact sequence (1), we have the following result.

5. Proposition. (a) *The obstruction to the existence of a connection on the fibration* (F, π) *is the class* $c(F, \pi) \in H^1(F, \pi^*\Omega^1 M \otimes \mathcal{T}F/M)$.

(b) *If* $c(F, \pi) = 0$, *then the group* $H^0(F, \pi^*\Omega^1 M \otimes \mathcal{T}F/M) = H^0(M, \Omega^1 M \otimes \pi_*(\mathcal{T}F/M))$ *acts transitively and effectively on the set of all connections.* □

6. The sheaf of connection coefficients on a fibration. By the sheaf of connection coefficients we mean the sheaf $\Omega^1 M \otimes \pi_*(\mathcal{T}F/M)$. The computations below will show that for important classes of fibrations F the local sections of this sheaf can be expressed in coordinates using generalized Christoffel symbols.

We now look at two classes of connections on a fibration which have the property of compatibility with some additional structure.

7. Connections on a *G*-fibration. Suppose that a complex Lie group G acts on $\pi: F \longrightarrow M$ over M, i.e., we are given a map $F \times G \longrightarrow F, (f, g) \mapsto fg$, with the usual properties and with $\pi(fg) = \pi(f)$ for all f and g. We say that a connection $\mathcal{T} \subset \mathcal{T}F$ is compatible with this structure if the subsheaf $\mathcal{T}$ is G-invariant.

8. Connections on a vector bundle. Suppose that $\pi: F \longrightarrow M$ is a vector bundle. This structure can be given as follows. Let $\mathcal{F}$ be the locally free sheaf of holomorphic sections of π. Then the sections of $\mathcal{F}^*$ are functions on F. At every point of F we have a local coordinate system, part of which is lifted from M

and the other part of which consists of a basis of sections of $\mathcal{F}^*$ which are linearly independent at the point. On F we consider the sheaf $S(\mathcal{F}^*)$ of functions which are polynomial along the fibres of π. Any connection on F is uniquely determined by its action on $S(\mathcal{F}^*)$. We say that the connection is compatible with the vector bundle structure if any local vector field X of the connection takes $S^i(\mathcal{F})^*$ to $S^i(\mathcal{F}^*)$ for all $i \geq 0$.

Usually, instead of the vector bundle F we shall study the corresponding sheaf of sections $\mathcal{F}$.

9. Connections on a fibration along a distribution on the base. Let $\pi: F \longrightarrow M$ be a fibration, let $\mathcal{T} \subset \mathcal{T}M$ be a distribution on the base, and set $\mathcal{T}_0 F = (d\pi)^{-1}(\pi^*\mathcal{T})$. By a connection on F along $\mathcal{T}$ we mean a splitting of the exact sequence

$$0 \longrightarrow \mathcal{T}F/M \longrightarrow \mathcal{T}_0 F \xrightarrow{d\pi} \mathcal{T} \longrightarrow 0.$$

As above, one can introduce the sheaf of coefficients of such a connection, i.e., $\pi_*(\mathcal{T}^* \otimes \mathcal{T}F/M)$, and also define compatibility with various additional structures, such as a vector bundle structure on F.

We now illustrate all of these concepts using some simple computations. Suppose that $F \longrightarrow M$ is a vector bundle, $\mathcal{F}$ is the dual sheaf of the sheaf of holomorphic sections of F, and $\mathrm{P}_M(\mathcal{F})$ is the corresponding relative projective space. Further suppose that we have local coordinates (x^a) on M, and that in the domain of definition of these coordinates the sheaf $\mathcal{F}$ is trivialized by a basis of sections (w^α). A trivialization of any vector bundle, i.e., a choice of isomorphism $F \xrightarrow{\sim} F_0 \times M$ compatible with π, automatically determines a connection $\mathcal{T} = \mathcal{T}F/F_0$ on the bundle (these are the vector fields which are vertical relative to projection onto the fibre). Using this connection as our "origin," we can describe all of the other connections by giving a section of the sheaf of coefficients. In our situation the sheaf of coefficients is also trivialized by the choice of (x^a) and (w^α), and it is the resulting expansion which leads to the generalized Christoffel symbols.

10. Proposition. *The following structures are equivalent:*

(a) *A connection on a vector bundle $F \longrightarrow M$ which is compatible with the vector bundle structure.*

(b) *A covariant differential $\nabla: \mathcal{F} \longrightarrow \mathcal{F} \otimes \Omega^1 M$, i.e., a* C*-linear morphism of sheaves satisfying Leibniz' formula $\nabla(af) = a\nabla f + f \otimes da$, where a is a local function and f is a local section of $\mathcal{F}$.*

(c) *A splitting of the exact sequence $0 \longrightarrow \mathcal{F} \otimes \Omega^1 M \longrightarrow \mathrm{Jet}^1 \mathcal{F} \longrightarrow \mathcal{F} \longrightarrow 0$ (see § 4.4(a)).*

(d) *A pair consisting of a connection $\mathcal{T}$ on the bundle $\mathrm{P}_M(\mathcal{F}^*) \longrightarrow M$ and a connection on the vector bundle $F \longrightarrow \mathrm{P}_M(\mathcal{F}^*)$ along the distribution $\mathcal{T}$.*

Locally, a connection on $\pi: F \longrightarrow M$ can be described by a Christoffel symbol $(\Gamma^{\alpha}_{\beta a} \in H^0(M, \mathcal{F}^* \otimes \mathcal{F} \otimes \Omega^1)$, and the induced connection on $\mathsf{P}_M(\mathcal{F}^*)$ depends only on the traceless part of the symbol: ${}^0\Gamma^{\alpha}_{\beta a} = \Gamma^{\alpha}_{\beta a} - \frac{1}{d}\delta^{\alpha}_{\beta}\Gamma^{\gamma}_{\gamma a}, d = \operatorname{rank}\mathcal{F}$.

Proof. Let $\partial_a = \partial/\partial x^a$, and let $X = h(\partial_a)$ be a lifting of ∂_a to F by means of some fixed connection h on $F \longrightarrow M$. According to § 4.8, X_a takes $S^{\cdot}(\mathcal{F}) \subset \mathcal{O}_F$ to itself; in particular, $X_a w^{\alpha} = \Gamma^{\alpha}_{\beta a} w^{\beta}$ for suitable functions $\Gamma^{\alpha}_{\beta a}(x)$ on M. Conversely, the assignment of values to $X_a w^{\alpha}$ by means of such formulas with the condition $X_a \pi^*(x^b) = \delta^b_a$ uniquely determines lifted vector fields X_a which generate distributions on F that are compatible with the vector bundle structure.

We further set $\nabla w^{\alpha} = X_a w^{\alpha} \otimes dx^a = \Gamma^{\alpha}_{\beta a} w^{\beta} \otimes dx^a$, and we extend ∇ to the local sections of $\mathcal{F}$ using the Leibniz formula. The differential ∇ is characterized uniquely by the lifting h together with the following property: $X \lrcorner \nabla w = h(X)w$ for any local field X on M and any local section w of the sheaf $\mathcal{F}$. This along with the local formulas in coordinates proves that the structures (a) and (b) are equivalent.

Since ∇ is a first order differential operator, the well-known universal property of $\mathrm{Jet}^1\mathcal{F}$ enables one to factor ∇ as the composition $\mathcal{F} \xrightarrow{j} \mathrm{Jet}^1\mathcal{F} \xrightarrow{k} \mathcal{F} \otimes \Omega^1 M$, where j is a universal operator — obtained by descending the morphism $\pi_1^*\mathcal{F} \longrightarrow \pi_1^*\mathcal{F}/I^2_{\nabla}\pi_1^*\mathcal{F}$ by means of π_2, in the notation of § 4.4 — and k depends on ∇. It is k that gives the splitting of the exact sequence in (c). This argument can be reversed. The set of splittings $\mathrm{Jet}^1\mathcal{F}$, if it is non-empty, is a principal homogeneous space over the sections of $\mathcal{F}^* \otimes \mathcal{F} \otimes \Omega^1 M$. Locally, these splittings can now be described by the coefficients $\Gamma^{\alpha}_{\beta a}$. The characteristic class of the sheaf $\mathcal{F}$ in the sense of § 4.4(a) is the obstruction to the existence of ∇.

We now proceed to consider connections on $\mathsf{P} = \mathsf{P}_M(\mathcal{F}^*) \xrightarrow{\lambda} M$. We first compute the sheaf of coefficients. A connection splits the sequence

$$0 \longrightarrow \mathcal{T}\mathsf{P}/M \longrightarrow \mathcal{T}\mathsf{P} \longrightarrow \lambda^*(\mathcal{T}M) \longrightarrow 0.$$

From the exact sequence $0 \longrightarrow \mathcal{O}_{\mathsf{P}} \longrightarrow \lambda^*(\mathcal{F}^*)(1) \longrightarrow \mathcal{T}\mathsf{P}/M \longrightarrow 0$ it follows that $\lambda_*(\mathcal{T}\mathsf{P}/M) \simeq (F * \otimes F)/\mathcal{O}_M = (\mathcal{F}^* \otimes \mathcal{F})_0$ (see § 2). This implies that the sheaf of connection coefficients on $(\mathsf{P}_M(\mathcal{F}^*), \lambda)$ is $(\mathcal{F}^* \otimes \mathcal{F})_0 \otimes \Omega^1 M$, i.e., when written out in coordinates it is the sheaf of traceless Christoffel symbols.

In order to compare connections on P with connections on F, we observe the following. The functions w^{α}/w^{γ} along with x^a (more precisely, $\lambda^*(x^a)$) make up a coordinate system on P wherever $w^{\gamma} \neq 0$. Given a connection $X_a w^{\alpha} = \Gamma^{\alpha}_{\beta a} w^{\beta}$ on F, we construct a connection on P by defining the action of the lifted vector fields on these coordinates by the natural formulas

$$X_a(w^\alpha/w^\gamma) = (X_a w^\alpha)/w^\gamma - (w^\alpha X_a w^\gamma)/(w^\gamma)^2 = \\ = \Gamma^\alpha_{\beta a} w^\beta/w^\gamma - w^\alpha \Gamma^\gamma_{\beta a} w^\beta/(w^\gamma)^2. \tag{3}$$

The correctness of this definition is not hard to check. The right side of (3) vanishes for the purely diagonal tensor $\Gamma^\alpha_{\beta a} = \delta^\alpha_\beta \Gamma_a$. One sees that this computation is equivalent to the preceding description of the vertical vector fields $\lambda_*(\mathcal{T}\mathbf{P}/M) = (\mathcal{F}^* \otimes \mathcal{F})_0$.

(We note in passing that in the smooth category the vector fields on a projective space are by no means necessarily linear, but in our situation we are seeing the effect of the rigidity of the holomorphic category.)

The information that is lost in going from F to $\mathbf{P}$ can be recovered as follows. The sheaf $\mathcal{O}(1) \subset \pi^*(\mathcal{F})$ is generated by its global sections, which can be identified with w^α. Then the formulas $X_a w^\alpha = \Gamma^\alpha_{\beta a} w^\beta$ can be interpreted, if one takes (3) into account, as the connection on $\mathcal{O}(1)$ along the distribution on $\mathbf{P}$ which corresponds to (3). □

11. Connections and the tensor algebra. The locally free sheaves with a covariant differential (or connection) $(\mathcal{F}, \nabla)$ form a category with tensor products and an internal $\mathcal{H}om$ functor. Using the following two conditions, one can uniquely determine formulas for the action of ∇ on sheaves which are obtained from a given set of sheaves using the tensor algebra operations: (1) Leibniz' formula must hold for tensor multiplication; (2) the standard tensor algebra morphisms (of which the most important is the convolution: $\mathcal{F}^* \otimes \mathcal{F} \longrightarrow \mathcal{O}_M$) are compatible with ∇. From this we obtain

$$(\mathcal{E}_1, \nabla_1) \otimes (\mathcal{E}_2, \nabla_2) = (\mathcal{E}_1 \otimes \mathcal{E}_2, \nabla_1 \otimes 1 + 1 \otimes \nabla_2),$$

$$\mathcal{H}om((\mathcal{E}_1, \nabla_1), (\mathcal{E}_2, \nabla_2)) = (\mathcal{H}om(\mathcal{E}_1, \mathcal{E}_2), \nabla_1^* \otimes 1 + 1 \otimes \nabla_2),$$

or, in more detail,

$$(\nabla f)(e_1) = \nabla_2(f(e_1)) - (f \otimes 1)(\nabla_1 e_1).$$

Since $\mathrm{Hom}((\mathcal{E}_1, \nabla_1), (\mathcal{E}_2, \nabla_2))$ is the set of morphisms $f\colon \mathcal{E}_1 \longrightarrow \mathcal{E}_2$ which commute with ∇_1 and ∇_2, it can be identified with the set of sections of $\mathcal{H}om(\mathcal{E}_1, \mathcal{E}_2)$ which are horizontal (i.e., annihilate ∇).

§ 5. Integrability and Curvature

1. The Frobenius form. By the Frobenius form of a distribution $\mathcal{T} \subset \mathcal{T}F$ we mean the map $\Phi: \mathcal{T} \times \mathcal{T} \longrightarrow \mathcal{T}F/\mathcal{T}$ which is given by $\Phi(X, Y) = [X, Y] \bmod \mathcal{T}$. We obviously have $\Phi(X, Y) = -\Phi(Y, X)$. In addition, Leibniz' formula gives us bilinearity of Φ:

$$[aX, Y] = aXY - Y(aX) = aXY - aYX - (Ya)X \equiv a[X, Y] \bmod \mathcal{T}.$$

Hence, Φ may be regarded as a mapping from $\wedge^2\mathcal{T}$ to $\mathcal{T}F/\mathcal{T}$, or as a section of the corresponding sheaf: $\Phi \in H^0(F, \wedge^2\mathcal{T}^* \otimes \mathcal{T}F/\mathcal{T})$. If $\mathcal{T}$ is a connection on $\pi: F \longrightarrow M$, then we call $\pi_*(\wedge^2\mathcal{T}^* \otimes \mathcal{T}F/\mathcal{T})$ the curvature sheaf, and we call $\pi_*(\Phi)$ the curvature of $\mathcal{T}$.

Integrability of $\mathcal{T}$ is equivalent to the vanishing of Φ. We recall some classical results about the integrable case. The first fact is the holomorphic Frobenius theorem.

2. Theorem. *The following conditions are equivalent:*

(a) *The distribution $\mathcal{T} \subset \mathcal{T}F$ is integrable.*

(b) *Each point $x \in F$ has a neighborhood with local coordinate system (x^a), $a = 1, \ldots, m$, such that $\mathcal{T}$ is freely generated in this neighborhood by a subset of the coordinate vector fields (i.e., by $\partial/\partial x^a$, $a = 1, \ldots, d = \operatorname{rank} \mathcal{T}$).* □

A polycylinder $\{(x^a) \mid |x^i| < c_i\}$ satisfying property (b) is called a Frobenius neighborhood (for $\mathcal{T}$).

3. Fiberings. Suppose that $\mathcal{T}$ is an integrable distribution, and $U \subset F$ is a Frobenius neighborhood. Then the trivial fibration $\pi: U \longrightarrow V$, where $V = \{(x^{d+1}, \ldots, x^m) \mid |x^i| < c_i\}$, has the property that $\mathcal{T}|_U = \mathcal{T}U/V$. The fibres of π are the polycylinders $x^i = \text{const}^i, d+1 \leq i \leq m$. Globally, $\mathcal{T}$ determines a fibering on F, i.e., a partition of F into immersed d-dimensional manifolds (fibres) which are tangent to $\mathcal{T}$ at each point.

More formally speaking, we equip F with the fine topology, in which the fibres $\pi^{-1}(v), v \in V$, in the Frobenius neighborhoods U form a fundamental system of open neighborhoods. Clearly, in this topology the neighborhood U_{fine} itself is the direct product of the fibre by the base V considered in the discrete topology: U "breaks up into fibres." Similarly, F_{fine} has a canonical complex manifold structure (we are given neighborhoods and coordinate functions), and the identity map $F_{\text{fine}} \longrightarrow F$ is an immersion. It is the connected components of F_{fine} which we call the fibres.

Let L denote the space of fibres with the topology induced by the quotient map $\pi: F \longrightarrow L$. It is not hard to check that π is an open mapping. We shall say that $\mathcal{T}$ is integrable up to a fibration if L can be given a complex manifold structure for which π is a holomorphic mapping. The next lemma gives a criterion for this which is often useful.

4. Lemma. *T is integrable up to a fibration if and only if the following conditions hold:*

(a) *L is a topological manifold.*

(b) *Any closed $(m-d)$-dimensional submanifold in an open subset $U \subset F$ which is transversal to the fibres of π projects locally homeomorphically to L.*

Proof. The necessity of these conditions is obvious. To prove sufficiency, we cover L with the images of Frobenius polycylinders (see § 5.3) and declare the coordinates $x_U^{d+1}, \dots, x_U^m$ on these neighborhoods to be locally analytic functions. We must verify that the transition from one coordinate system to another in a neighborhood of $l \in L$ is given by an analytically invertible mapping. In fact, suppose that x, $y \in \pi^{-1}(l)$, and $U \ni x, U' \ni y$ are small polycylinders around x and y. Let $x^{d+1}, \dots, x^m$ and $y^{d+1}, \dots, y^m$ be the corresponding coordinates. In order to express the y^a in terms of the x^a, we join U and U' by a chain of neighborhoods of points $z_1, \dots, z_k \in \pi^{-1}(l)$, where $z_i \in U_i, U \cap U_1 \neq \emptyset, U_1 \cap U_2 \neq \emptyset, \dots, U_k \cap U' \neq \emptyset$, in which manifolds transversal to $\mathcal{T}$ project homeomorphically to L. This gives a sequence of bihomeomorphic maps connecting y^a to x^a. □

Now suppose that $\mathcal{T}$ is an integrable distribution on F, and $\pi: F \longrightarrow M$ is a fibration with rank $\mathcal{T} = \dim M$ such that the fibres of π are transversal to $\mathcal{T}$, i.e., $\mathcal{T}F = \mathcal{T}F/M \oplus \mathcal{T}$. The following theorem of Ehresmann can be obtained easily from the Frobenius theorem.

5. Theorem. *In the above situation, suppose that all of the fibres are compact. Then the π-projection onto M of any fibre of $\mathcal{T}$ is an unramified covering of M.* □

This theorem can be applied to integrable connections on a fibration $\pi: F \longrightarrow M$. The fibres of the resulting fibering are "multi-valued horizontal sections of F."

6. The de Rham complex of a fibering. Let $\mathcal{T}$ be an integrable distribution on the manifold F (possibly $\mathcal{T} = \mathcal{T}F$). If it is integrable up to a fibration with base L, then $\mathcal{T} = \mathcal{T}F/L$. We shall sometimes make use of this notation in the general case, but we should remember that the space L of sheets is not necessarily a complex manifold. However, as we saw in § 5.3, in cases when F and $\mathcal{T}$ can be treated locally this notation is suggestive and appropriate.

We set $\Omega^1 F/L = \mathcal{T}^*$; $\Omega^i F/L = \wedge^i \Omega^1 F/L$, $\Omega^\cdot F/L = \bigoplus_{i \geq 0} \Omega^i F/L$. The next proposition can be proved in exactly the same way as in the classical case $\mathcal{T} = \mathcal{T}F$.

7. Proposition. *There exists a unique first order differential operator $d = d_{F/L}: \Omega^\cdot F/L \longrightarrow \Omega^\cdot F/L$ which increases the degree of a form by 1 and which satisfies the following properties:*

(a) $d_{F/L}\colon \mathcal{O}_F \longrightarrow \Omega^1 F/L$ *is the composition* $\mathcal{O}_F \xrightarrow{d} \Omega^1 F \longrightarrow \mathcal{T}^*$;

(b) $d(\omega^p \wedge \omega^q) = d\omega^p \wedge \omega^q + (-1)^p \omega^p \wedge d\omega^q, \quad \omega^p \in \Omega^p F/L$;

(c) $d^2 = 0$.

Let $\rho^{-1}(\mathcal{O}_L) \subset \mathcal{O}_F$ *denote the sheaf of holomorphic functions which are constant along the fibres of the fibering. Then the de Rham complex*

$$0 \longrightarrow \rho^{-1}(\mathcal{O}_L) \longrightarrow \mathcal{O}_F \xrightarrow{d_{F/L}} \Omega^1 F/L \longrightarrow \cdots$$

is exact, i.e., $(\Omega^{\cdot} F/L, d)$ *is a resolution of* $\rho^{-1}(\mathcal{O}_L)$. □

We recall that uniqueness of $d_{F/L}$ follows immediately from properties (a) and (b), and existence is proved using the explicit formula

$$\begin{aligned} d_{F/L}\omega^p(X_1,\ldots,X_{p+1}) &= \sum_{i=1}^{p+1}(-1)^{i+1}X_i[\omega^p(X_1,\ldots,\hat{X}_i,\ldots,X_{p+1})] + \\ &+ \sum_{i<j}(-1)^{i+j}\omega^p([X_i, X_j], X_1,\ldots,\hat{X}_i,\ldots,\hat{X}_j,\ldots,X_{p+1}). \end{aligned}$$

It is here where one uses the integrability of the connection, since the second sum involves commutators.

The exactness is established by an explicit construction of the homotopy (Poincaré's lemma).

8. Covariant differentials along a fibering. Now suppose that we are given a locally free sheaf $\mathcal{E}$ on a manifold F with fibering $\mathcal{T}F/L$. By a covariant differential (or connection) on $\mathcal{E}$ along the fibering we mean a $\mathbb{C}$-linear map $\nabla = \nabla_{F/L}\colon \mathcal{E} \longrightarrow \mathcal{E} \otimes \Omega^1 F/L$ satisfying Leibniz' formula $\nabla(ae) = a\nabla e + e \otimes d_{F/L}a$, where e is a local section of $\mathcal{E}$ and a is a local function. We leave it to the reader to state and prove a relative analog (over L) of Proposition 4.10. In particular, one finds that giving $\nabla_{F/L}$ is equivalent to giving a connection along the distribution $\mathcal{T}F/L$ on the vector bundle E associated to $\mathcal{E}$.

We now introduce the de Rham sequence of the connection $\nabla_{F/L}$ on $\mathcal{E}$.

9. Theorem. (a) *There exists a unique first order differential operator* $\nabla = \nabla_{F/L}\colon \mathcal{E} \otimes \Omega^{\cdot} F/L \longrightarrow \mathcal{E} \otimes \Omega^{\cdot} F/L$ *with the two properties*

(1) $\nabla_{F/L} = \nabla\colon \mathcal{E} \longrightarrow \mathcal{E} \otimes \Omega^1 F/L$ *is a covariant differential; and*

(2) $\nabla(e \otimes \omega^p) = \nabla_{F/L}\, e \wedge \omega^p + e \otimes d_{F/L}\omega^p$.

(b) *There exists a unique element* $\Phi = \Phi(\nabla_{F/L}) \in H^0(F, \mathcal{E}nd\, \mathcal{E} \otimes \Omega^2 F/L)$ *such that* $\nabla^2\colon \mathcal{E} \otimes \Omega^{\cdot} F/L \longrightarrow \mathcal{E} \otimes \Omega^{\cdot} F/L$ *is an* $\mathcal{O}_F$*-linear multiplication operator on* Φ, *i.e., the composition of the map* $\mathcal{E}nd\, \mathcal{E} \times \mathcal{E} \longrightarrow \mathcal{E}$ *and exterior multiplication on* $\Omega^{\cdot} F/L$.

(c) *The connection on E is integrable if and only if $\Phi = 0$. In that case the de Rham complex*

$$\mathcal{E} \xrightarrow{\nabla} \mathcal{E} \otimes \Omega^1 F/L \xrightarrow{\nabla} \mathcal{E} \otimes \Omega^2 F/L \longrightarrow \cdots$$

is a resolution of the sheaf of ($\nabla_{F/L}$)-horizontal sections of $\mathcal{E}$ (which will be denoted $\rho^{-1}(\mathcal{E}_L)$ by analogy with $\rho^{-1}(\mathcal{O}_L)$).

(d) *Let $\tilde{\nabla}\colon \mathcal{E}nd\, \mathcal{E} \otimes \Omega^\cdot F/L \longrightarrow \mathcal{E}nd\, \mathcal{E} \otimes \Omega^\cdot F/L$ be the covariant differential constructed from the same connection on $\mathcal{E}nd\, \mathcal{E}$ that is induced by $\nabla_{F/L}$ in accordance with § 4.10. Then $\tilde{\nabla}(\Phi(\nabla)) = 0$ (the Bianchi identity).*

Proof. Uniqueness follows from the identities in (a) by induction on the degree of p. To show existence, one verifies that the following operator has the properties in (a) (here $\omega^p_{\mathcal{E}}$ is a section of $\mathcal{E} \otimes \Omega^p F/L$):

$$(\nabla\omega^p_{\mathcal{E}})(X_1,\ldots,X_{p+1}) = \sum_{i=1}^{p+1}(-1)^{i+1}\, X_i \lrcorner \nabla(\omega^p_{\mathcal{E}}(X_1,\ldots,\hat{X}_i,\ldots,X_{p+1})) +$$
$$+\sum_{i<j}(-1)^{i+j}\omega^p_{\mathcal{E}}([X_i,\, X_j],\, X_1,\ldots,\hat{X}_i,\ldots,X_{p+1}).$$

In fact, ∇ is a derivation of $\mathcal{E} \otimes \Omega^\cdot F/L$ as a right ($\Omega^\cdot F/L$)-module, i.e., we have $\nabla(\omega^p_{\mathcal{E}} \wedge \omega^q) = \nabla\omega^p_{\mathcal{E}} \wedge \omega^q + (-1)^p \omega^p_{\mathcal{E}} \wedge d_{F/L}\,\omega^q$; we shall omit the verification, which is straightforward.

We apply this formula once more:

$$\nabla^2(\omega^p_{\mathcal{E}} \wedge \omega^q) = \nabla^2\omega^p_{\mathcal{E}} \wedge \omega^q + (-1)^{p+1}\nabla\omega^p_{\mathcal{E}} \wedge d_{F/L}\,\omega^q + (-1)^p\nabla\omega^p_{\mathcal{E}} \wedge d_{F/L}\,\omega^q$$
$$= \nabla^2\omega^p_{\mathcal{E}} \wedge \omega^q.$$

Thus, ∇^2 is actually ($\Omega^\cdot F/L$)-linear. In particular, its component $\mathcal{E} \longrightarrow \mathcal{E}\otimes\Omega^2 F/L$ is $\mathcal{O}_F$-linear. But this is the curvature $\Phi(\nabla_{F/L})$, as we shall now verify.

Namely, we compute the value of this map on a pair of vector fields in $\mathcal{T}F/L$ and a section e of the sheaf $\mathcal{E}$. Writing ∇_X in place of $X \lrcorner \nabla$, we have

$$\nabla e(X) = \nabla_X e,$$
$$(\nabla^2 e)(X_1,\, X_2) = \nabla_{X_1}\nabla_{X_2} e - \nabla_{X_2}\nabla_{X_1} e - \nabla_{[X_1,\,X_2]}\, e =$$
$$= ([\nabla_{X_1},\, \nabla_{X_2}] - \nabla_{[X_1,\,X_2]})e.$$

On the other hand, according to the proof of Lemma 8, the right side of the last equality gives the action of the operator $[h(X_1),\, h(X_2)] - h[X_1,\, X_2]$ on the coordinate functions e of the dual fibration $E^* \xrightarrow{\pi} F$, where $h\colon \pi^*(\mathcal{T}F/L) \longrightarrow \mathcal{T}E^*$ is a lifting of the vector fields corresponding to $\nabla_{F/L}$. Since this operator depends $\mathcal{O}_F$-bilinearly on (X_1, X_2), in essence it represents a Frobenius form and vanishes simultaneously with that form. (If one insists upon working with the fibration E rather than E^*; one can begin with a covariant differential ∇^* on $\mathcal{E}^*$. The curvatures of ∇ and ∇^* are the same when one makes the standard identification of $\mathcal{E}nd\ \mathcal{E}$ with $\mathcal{E}nd\ \mathcal{E}^*$.)

If $\Phi = 0$, then it follows from the Frobenius theorem that locally on F the sheaf $\mathcal{E}$ has a basis of sections which are horizontal along the sheets of the fibering $\mathcal{T}F/L$. Hence, the de Rham complex $(\mathcal{E}, \nabla_{F/L})$ is locally isomorphic to the direct sum of the usual de Rham complexes of the fibering, and this implies that the complex is exact at all terms except for the first. The kernel of $\nabla_{F/L}$ on $\mathcal{E}$ is $\rho^{-1}(\mathcal{E}_L)$, by definition.

It remains to prove the Bianchi identity. For any section $\omega = \omega^p_{\mathcal{E}}$ we have $\nabla(\nabla^2\omega) = \nabla^2(\nabla\omega)$, i.e., $\nabla(\omega \wedge \Phi) = (\nabla\omega) \wedge \Phi$ (where we are using the convolution $\mathcal{E} \times \mathcal{E}nd\ \mathcal{E} \longrightarrow \mathcal{E}$ and exterior multiplication). On the other hand, by the generalized Leibniz formula in the tensor algebra $\mathcal{E}$, we have $\nabla(\omega \wedge \Phi) = (\nabla\omega) \wedge \Phi + (-1)^p\omega \wedge \tilde{\nabla}\Phi$, from which it follows that $\tilde{\nabla}\Phi = 0$. □

§ 6. Conic Structures and Conic Connections

1. Definition. Let M be a complex manifold with tangent sheaf $\mathcal{T}M$, and let $d \geq 0$ be an integer. By a *d-conic structure* on M we mean a smooth closed submanifold $F \subset G_M(d;\ \mathcal{T}M)$ such that the projection $\pi\colon F \longrightarrow M$ is a submersion. □

In other words, at every point $x \in M$ such an F determines a set of d-dimensional (complex) tangent directions in $\mathcal{T}M(x)$ corresponding to the points $\pi^{-1}(x) \subset G(d;\ \mathcal{T}M(x))$. For $d = 1$ these tangent directions sweep out a cone. In general, the system of directions forms a closed smooth analytic (and hence algebraic) subspace in the grassmannian. Moreover, the set $\{\pi^{-1}(x)\}$ of these manifolds depends analytically on x. The conic structure $F = G(d;\ \mathcal{T}M(x))$ is called the "full" conic structure, since it includes all of the d-dimensional directions.

2. Examples. (a) Suppose that $M = G(d;\ T)$, S and $\tilde{S}$ are the tautological sheaves on M, and $F = \mathbf{P}_M(\tilde{S}^*) \underset{M}{\times} \mathbf{P}_M(S^*)$. We recall that $\mathbf{P}_M(S^*) = G_M(1;\ S^*)$ (here it might be useful to refresh one's memory by reviewing the notation in § 1.18). A single point of F is a pair of one-dimensional subspaces in $\tilde{S}^*(x)$ and $S^*(x), x \in M$. The tensor product of these lines is a one-dimensional subspace in $\tilde{S}^*(x) \otimes S^*(x) = \mathcal{T}M(x)$, by Theorem 1.6. The corresponding map $F \longrightarrow \mathsf{P}_M(\mathcal{T}M)$, which is called the Veronese morphism, is a closed imbedding. This gives the 1-conic structure of null directions on the grassmannian M.

(b) Again suppose that $M = G(d;\, T)$, $d = \operatorname{rank} S$, $c = \operatorname{rank} \tilde{S}$ and that $F = \mathsf{P}_M(S^*) \xrightarrow{\pi} M$, $\tilde{F} = \mathsf{P}_M(\tilde{S}^*) \xrightarrow{\tilde{\pi}} M$. To every point in F, i.e., to a ray in $S^*(x)$, we associate its tensor product with $\tilde{S}^*(x)$, which is a c-dimensional subspace of $\mathcal{T}M(x)$. This determines a c-structure $i\colon F \longrightarrow G_M(c;\, \mathcal{T}M) = G$ and a d-structure $\tilde{\imath}\colon \tilde{F} \longrightarrow G_M(d;\, \mathcal{T}M) = \tilde{G}$.

To define the morphisms i and $\tilde{\imath}$ formally is equivalent to giving the sheaves $i^*(S_G^c)$ and $\tilde{\imath}^*(S_{\tilde{G}}^d)$. Recalling the definitions, we easily see that

$$i^*(S_G^c) = \pi^*(\tilde{S}^*) \otimes \mathcal{O}_F(-1),$$

$$\tilde{\imath}^*(S_{\tilde{G}}^d) = \tilde{\pi}^*(S^*) \otimes \mathcal{O}_{\tilde{F}}(-1).$$

When $c = d = 2$, these 2-conic structures give systems of tangent directions to the α- and β-planes on the Klein quadric (see § 3.7).

(c) Suppose we are given a conformal metric $\mathcal{L} \subset S^2(\Omega^1 M)$ on the manifold M (see § 3.3). If the metric is non-degenerate, then at every point $x \in M$ it determines a non-degenerate quadratic cone of null directions, i.e., it gives a 1-conic structure F which is a relative quadric over M. The converse is also true: given a 1-conic structure $F \subset \mathsf{P}(\mathcal{T}M)$ which is a non-degenerate quadric of codimension one, one can uniquely recover a corresponding conformal metric. Namely, suppose that $\mathsf{P} = \mathsf{P}_M(\mathcal{T}M) \xrightarrow{\pi} M$, and $J_F(2) \subset \mathcal{O}_F(2)$ is the ideal of equations of F in P. We set $\mathcal{L} = \pi_*(J_F(2)) \longrightarrow \pi_*(\mathcal{O}_F(2)) = S^2(\mathcal{T}^*M)$. The sheaf $\mathcal{L}$ is invertible, since all of the quadratic equations of the quadric are proportional to one another. Next, we find that $\mathcal{O}_{\mathsf{P}}(2)/J_F(2) = \mathcal{O}_F(2)$, and $\pi_*\mathcal{O}_F(2)$ is locally free, because the dimension of its fibres does not change. Finally, $R^1\pi_{2*}J_F(2) = 0$. Consequently, $\mathcal{L} \subset S^2(\mathcal{T}^*M)$ is a local direct summand. This is then the conformal metric corresponding to the given system of null directions.

3. Definition. Let F be a d-conic structure on the manifold M.

(a) We say that a distribution of c-dimensional tangent planes in F *is tangent to the conic structure* if at each $x \in F$ the projection onto $\mathcal{T}M(x)$ of the tangent plane at this point is the d-dimensional subspace corresponding to x.

(b) By a *conic connection* on F we mean a distribution of d-dimensional tangent planes which is tangent to the conic structure.

(c) A conic connection on F is said to be *integrable* if it is integrable as a distribution.

(d) The sheets of a fibering which correspond to an integrable conic connection (and also their images in M) will be called the *geodesic manifolds* of the connection. □

4. Examples. (a) Any connection on a 1-conic structure is integrable.

By a "projective connection" on M we mean a connection on the full 1-conic structure $\mathbf{P}_M(\mathcal{T}M)$. Its geodesic curves on M pass through every point in every direction. In the general case of a 1-conic structure, the geodesics pass through every point but only in the directions which are contained in the structure. On projective space there is a canonical projective connection, whose geodesics consist of all lines. A smooth quadric has a canonical 1-conic structure, corresponding to the structure of asymptotic directions at each point. Its geodesics are the lines of the ambient projective space which lie on the quadric.

(b) In examples 2(a) and (b), along with the conic structures one also has geodesic manifolds of an integrable connection. Namely, if $M = G(d;\, T)$, then in example 2(a) we have $F = F(d-1,\, d,\, d+1;\, T)$ and the fibres of the projection $\pi\colon F(d-1,\, d,\, d+1;\, T) \longrightarrow F(d-1,\, d+1;\, T)$ are the geodesic manifolds. In example 2(b) we have $F = F(d-1,\, d;\, T)$ or $F(d,\, d+1;\, T)$, and the geodesics are the fibres of the projections onto $G(d \pm 1;\, T)$.

(c) We next look at the class of examples of "flat" integrable connections which correspond to double fibrations $L \xleftarrow{\pi_1} F \xrightarrow{\pi_2} M$ which are compact homogeneous manifolds. In general, we consider such diagrams of manifolds which have the following properties: the fibres of π_2 are compact; π_1 and π_2 are submersions of smooth manifolds; and the tangent directions to the fibres of π_1 and π_2 at any $x \in F$ have zero intersection. Then F becomes a conic structure on M by means of the following distribution (here $d = \dim F - \dim L$):

$$F \ni y \mapsto \operatorname{Ker}(d\pi_1\colon \mathcal{T}F(y) \to \mathcal{T}L(\pi_1(y))) \in G(d;\, \mathcal{T}M(\pi_2(x))).$$

The distribution $\operatorname{Ker} d\pi_1$ is an integrable conic connection, and the fibres of π_1 are its geodesic manifolds.

We shall later study how to define conic connections by means of Christoffel type coefficients. The integrability conditions then become nonlinear differential equations for these coefficients. Examples of such differential equations include the Einstein self-dual equations and various "constraints" that arise in supergravity.

5. Conic connections as splittings. The definition in § 6.3 can be reformulated as follows. Suppose that S is the tautological sheaf on $G_M(d;\, \mathcal{T}M)$ and S_F is its restriction to F, $S_F \subset \pi^*(\mathcal{T}M)$, $\pi\colon F \longrightarrow M$. We set $\mathcal{T}_cF = (d\pi)^{-1} S_F$, and we consider the exact sequence

$$0 \longrightarrow \mathcal{T}F/M \longrightarrow \mathcal{T}_cF \xrightarrow{d\pi} S_F \longrightarrow 0. \qquad (1)_F$$

A conic connection on F is a splitting of this sequence. It is therefore natural to call $\pi_*(S_F^* \otimes \mathcal{T}F/M)$ the sheaf of conic connection coefficients on F.

Let us compare three sets of connections and their coefficient sheaves:

(a) Connections on the fibration $\pi: G_M(d;\ \mathcal{T}M) \longrightarrow M$.

(b) Conic connections on the full conic structure $G = G_M(d;\ \mathcal{T}M)$.

(c) Conic connections on any conic structure $F \subset G_M(d;\ \mathcal{T}M)$.

We first note that there is a natural map from (a) to (b), since a splitting of $\mathcal{T}F$ induces a splitting of $\mathcal{T}_cF$. This map gives the obvious map of coefficient sheaves.

A choice of local coordinates (x^a) in M trivializes $\mathcal{T}M, F$, and the coefficient sheaves. As in § 4, we regard the corresponding connections as our "origin," and identify the other connections with their coefficients. We suppose that $1 \leq d <$ dim M.

6. Proposition. (a) *The sheaf of connection coefficients on the fibration $G_M(d;\ \mathcal{T}M) \longrightarrow M$ is $\Omega^1 M \otimes sl(\mathcal{T}M) = \Omega^1 M \otimes (\Omega^1 M \otimes \mathcal{T}M)_0$. Its local sections are characterized by the Christoffel symbols "with no second trace":*

$$(\Gamma^c_{ab}) = \Gamma^c_{ab} dx^a \otimes dx^b \otimes \frac{\partial}{\partial x^c}, \qquad \Gamma^b_{ab} = 0.$$

(b) *The map*

$$\{\text{connections on the fibration}\} \longrightarrow \{\text{full conic connections}\}$$

is surjective and has the following appearance on the coefficients:

$d = 1$:

$$\Omega^1 M \otimes (\Omega^1 M \otimes \mathcal{T}M)_0 \longrightarrow (S^2(\Omega^1 M) \otimes \mathcal{T}M)_0$$

(*projection onto the traceless symmetric part*);

$d > 1$:

$$\Omega^1 M \otimes (\Omega^1 M \otimes \mathcal{T}M)_0 \longrightarrow (S^2(\Omega^1 M) \otimes \mathcal{T}M)_0 \oplus (\wedge^2(\Omega^1 M) \otimes \mathcal{T}M)_0.$$

Proof. First of all, we have $\mathcal{T}G/M = S^* \otimes \tilde{S}^*$. We would like to show that $\pi_*(\mathcal{T}G/M) = sl(\mathcal{T}M)$; more precisely, that there is a natural morphism $\pi^*(\mathcal{T}^*M) \otimes \pi^*(\mathcal{T}M) \longrightarrow S^* \otimes \tilde{S}^*$ (compare with § 2), which induces an exact sequence

$$0 \longrightarrow \mathcal{O}_M \longrightarrow gl(\mathcal{T}M) \longrightarrow \pi_*(S^* \otimes \tilde{S}^*) \longrightarrow 0$$

(the imbedding of $\mathcal{O}_M$ takes 1 to id). This can be shown as follows, in the style of § 2.

We first introduce the full flag manifolds $F_1 = F_G(1,\ldots;\ S)\xrightarrow{\pi_1}G$ and $F_2 = F_G(1,\ldots;\ \tilde{S})\xrightarrow{\pi_2}G$. Then $S^* = \pi_{1*}\mathcal{L}_1(1,\,0,\ldots)$ and $\tilde{S}^* = \pi_{2*}\mathcal{L}_2(1,\,0,\ldots)$ in the notation of § 2, where the indices $i = 1,\,2$ refer to the F_i. We now consider the morphism $\rho\colon F_1 \underset{G}{\times} F_2 \longrightarrow G \longrightarrow M$. Then $\pi_*(S^*\otimes\tilde{S}^*) = \rho_*(\mathcal{L}_1(\epsilon_1)\boxtimes\mathcal{L}_2(\epsilon_2))$. On the other hand, ρ is also the composition $F_1 \underset{G}{\times} F_2 \longrightarrow \mathsf{P}(\mathcal{T}M) \underset{M}{\times} \mathsf{P}(\mathcal{T}^*M) \longrightarrow M$. Descending along the first arrow leads to the sheaf $\mathcal{T}M^*\otimes\mathcal{T}M$ lifted to $\mathsf{P}_M(\mathcal{T}M) \underset{M}{\times} \mathsf{P}_M(\mathcal{T}^*M)$ and then restricted to the incidence space $F_M(1,\,m-1;\ \mathcal{T}M) \subset \mathsf{P}(\mathcal{T}M) \underset{M}{\times} \mathsf{P}(\mathcal{T}^*M)$. The sections proportional to $\mathrm{id}_{\mathcal{T}(M)}$ vanish under this restriction. The final descent leads to the sheaf $sl(\mathcal{T}M)$ on M. This implies part (a).

Next, the sheaf of coefficients of the full conic connection is

$$\pi_*(S^*\otimes S^*\otimes\tilde{S}^*) = \pi_*(S^2(S^*)\otimes\tilde{S}^*)\oplus\pi_*(\wedge^2(S^*)\otimes\tilde{S}^*).$$

Proceeding as above, we reduce the computation to determining the direct image of the sheaf $\mathcal{L}(2,\,0,\ldots)\boxtimes\mathcal{L}(1,\,0,\ldots)$ for S^2 and of $\mathcal{L}(1,\,1,\,0,\ldots)\boxtimes\mathcal{L}(1,\,0,\ldots)$ for $\wedge^2$. One obtains $S^2(\mathcal{T}^*M)\otimes\mathcal{T}M$ and $\wedge^2(\mathcal{T}^*M)\otimes\mathcal{T}M$, respectively, on the fibre product; and when one restricts to the incidence flag the traceless parts remain. □

7. The coefficient sheaf of a general conic connection. Besides writing the exact sequence (1) for F, we can also write it for the full conic structure G and then restrict the result to F. Clearly, the sequence $(1)_F$ is imbedded in this restriction:

$$\begin{array}{ccccccccc}
0 & \longrightarrow & \varphi^*(\mathcal{T}G/M) & \longrightarrow & \varphi^*(\mathcal{T}_cG) & \xrightarrow{d\pi} & \varphi^*(S) & \longrightarrow & 0\\
 & & \uparrow & & \uparrow & & \| & & \\
0 & \longrightarrow & \mathcal{T}F/M & \longrightarrow & \mathcal{T}_cF & \longrightarrow & S_F & \longrightarrow & 0
\end{array}$$

Not every splitting of $(1)_G$ induces a splitting of $(1)_F$, because the horizontal d-dimensional patch that is chosen at a point of F might not be tangent to F. The coefficient sheaves for the full conic connection and the connection on F can be compared if we use an intermediate sheaf $\pi_*(S_F^*\otimes\varphi^*(\mathcal{T}G/M))$ along with the diagram

$$\pi_*(S_F^*\otimes\mathcal{T}F/M)\xrightarrow{\alpha}\pi_*(S_F^*\otimes\varphi^*(\mathcal{T}G/M))\xleftarrow{\beta}\pi_*(S^*\otimes\mathcal{T}G/M).$$

In order to analyze the structure of β and α, we let J_F denote the sheaf of ideals of F in $G = G_M(d;\ \mathcal{T}M)$, and we use the fact that for any coherent sheaf $\mathcal{E}$ on

G one has the exact sequence $0 \longrightarrow J_F\mathcal{E} \longrightarrow \mathcal{E} \longrightarrow \varphi^*(\mathcal{E}) \longrightarrow 0$ (for brevity we have used $\varphi^*(\mathcal{E})$ to denote the sheaf extended by zero to G). In particular, on G we have:

$$0 \longrightarrow J_F(S^* \otimes \mathcal{T}G/M) \longrightarrow S^* \otimes \mathcal{T}G/M \xrightarrow{\beta'} \varphi^*(S_F^* \otimes \mathcal{T}G/M) \longrightarrow 0, \quad (2)$$

and $\beta = \pi_*(\beta')$.

Now let $\mathcal{N}_F = (J_F/J_F^2)^*$ be the normal sheaf of F in G. Then we have the following exact sequence on F:

$$0 \longrightarrow S_F^* \otimes \mathcal{T}F/M \xrightarrow{\alpha'} S_F^* \otimes \varphi^*(\mathcal{T}G/M) \longrightarrow S_F^* \otimes \mathcal{N} \longrightarrow 0, \quad (3)$$

and $\alpha = \pi_*(\alpha')$.

Using (2) and (3), we obtain the following results.

8. Proposition. (a) *The map α is injective, and its cokernel is isomorphic to*

$$\operatorname{Ker}(\pi_*(S_F^* \otimes \mathcal{N}) \longrightarrow R^1\pi_*(S_F^* \otimes \mathcal{T}F/M)).$$

(b) *The kernel and cokernel of β are the following sheaves, respectively:*

$$\pi_*(J_F(S^* \otimes \mathcal{T}G/M)),$$

$$\operatorname{Ker}(R^1\pi_*(J_F(S^* \otimes \mathcal{T}G/M)) \longrightarrow R^1\pi_*(S^* \otimes \mathcal{T}/M)). \quad \square$$

In particular, if $R^1\pi_*(J_F(S^* \otimes \mathcal{T}G/M)) = 0$, then β is surjective, and hence the *d*-conic connections can, as before, be described locally by Christoffel symbols, perhaps subject to some additional conditions (characterizing the image of α) and considered modulo some equivalence relation (corresponding to the kernel of β).

9. Curvature and curvature sheaves. In accordance with § 5.1, conic connections, since they are distributions on a conic structure space $F \longrightarrow M$, have Frobenius forms. When descended to M, these forms become the curvature, i.e., sections of curvature sheaves. We now briefly describe these sheaves.

(a) *Connections on a fibration.* Let $\mathcal{T} \subset \mathcal{T}F, F \xrightarrow{\pi} M$ be a connection on a fibration, i.e., we have $d\pi\colon \mathcal{T} \xrightarrow{\sim} \pi^*\mathcal{T}M$. Then $\Phi \in \pi_*(\wedge^2\mathcal{T}^*M \otimes \mathcal{T}F/\mathcal{T}) \simeq \Omega^2 M \otimes \pi_*\mathcal{T}F/\mathcal{T}$. In other words, the curvature is a 2-form on the base with values in the vertical vector fields of the fibration, just as in differential geometry.

(b) *Full conic connections.* Let $G = G_M(d;\, \mathcal{T}M) \longrightarrow M$. The sheaf $\wedge^2 S^* \otimes \mathcal{T}G/\mathcal{T}$ which contains the curvature of the full conic connection $\mathcal{T} \subset \mathcal{T}G$ seems explicitly to depend on $\mathcal{T}$. However, it can be represented as an extension of two sheaves which no longer depend on $\mathcal{T}$: after factoring $\mathcal{T}G \xrightarrow{d\pi} \pi^*(\mathcal{T}M)$ by $\mathcal{T} \xrightarrow{d\pi} S$, we obtain the following exact sequence (the zero sequence if $d = 1$):

$$0 \longrightarrow \wedge^2 S^* \otimes \mathcal{T}G/M \longrightarrow \wedge^2 S^* \otimes \mathcal{T}G/\mathcal{T} \xrightarrow{\beta} \wedge^2 S^* \otimes \tilde{S}^* \longrightarrow 0.$$

For $d > 1$ the integrability condition splits into two parts. Let $\Phi_0 \in H^0(M, \wedge^2 S^* \otimes \tilde{S}^*) = (\Omega^2 M \otimes \mathcal{T}M)_0$ be the $\pi_*(\beta)$-image of the Frobenius form. The first condition for integrability is that $\Phi_0 = 0$.

If this condition holds, then $\Phi \in H^0(M, \pi_*(\wedge^2 S^* \otimes \mathcal{T}G/M)) \subset \Omega^2 M \otimes sl(\mathcal{T}M)$. This element, which is the true curvature, must also vanish.

(c) *General conic connections.* As in § 6.7, if we replace G by F and S, $\tilde{S}$ by S_F, $\tilde{S}_F$, we can write an exact sequence for the sheaf of Frobenius forms and introduce the section $\Phi_0 \in H^0(M, \pi_*(\wedge^2 S_F^* \otimes \tilde{S}_F^*))$. If $\Phi_0 = 0$, then the true curvature Φ lies in $H^0(M, \pi_*(\wedge^2 S_F^* \otimes \mathcal{T}F/M))$.

One can compare these sheaves with the corresponding sheaves in the case $F = G$, following the argument in § 6.7. In the next section we shall carry out the detailed calculations in an important special case.

§ 7. Grassmannian Spinors and Generalized Self-Duality Equations

1. Definition. A *manifold with grassmannian spinor structure* (or, more briefly, a *GS-manifold*) is a four-tuple $(M, S, \tilde{S}, \sigma)$, in which M is a complex manifold, S and $\tilde{S}$ are locally free sheaves ("spinor sheaves") on M of rank $d > 1$ and $c > 1$, respectively, and $\sigma: S \otimes \tilde{S} \longrightarrow \Omega^1 M$ is an isomorphism ("spinor decomposition of 1-forms"). □

GS-manifolds form a category in which a morphism $(M, S, \tilde{S}, \sigma) \longrightarrow (M', S', \tilde{S}', \sigma')$ consists of an open imbedding $\varphi: M \longrightarrow M'$ and a pair of isomorphisms $S \xrightarrow{\alpha} \varphi^*(S')$, $\tilde{S} \xrightarrow{\beta} \varphi^*(\tilde{S}')$ which are compatible with σ and σ' in the sense that the diagram

$$\begin{array}{ccc} S \otimes \tilde{S} & \xrightarrow{\alpha} & \Omega^1 M \\ \alpha \otimes \beta \Big\downarrow & & \Big\downarrow d\varphi \\ \varphi^*(S' \otimes \tilde{S}') & \xrightarrow{\varphi^*(\sigma')} & \varphi^*(\Omega^1 M') \end{array}$$

is commutative.

Let $\sigma^t: \tilde{S} \otimes S \longrightarrow \Omega^1 M$ be the composition of σ, reversal of the factors, and change of sign. We shall say that the GS-manifold $(M, \tilde{S}, S, \sigma^t)$ is obtained from $(M, S, \tilde{S}, \sigma)$ by reversal of orientation.

According to Theorem 1.6, the grassmannian $G(d;\, T^{d+c})$ has a canonical GS-structure. We shall regard any other GS-structure as a "curved" version of this canonical "flat" GS-structure. All of the properties of a grassmannian which use only the spinor decomposition of 1-forms carry over to a general GS-structure. Such properties include decomposition of 2-forms (and, more generally, the de Rham complex), as in § 1.9, and the possibility of defining three canonical decomposable tangent direction structures (of dimension 1, d, and c, see § 6.2).

Let h be a connection on the standard c-conic structure $F \xrightarrow{\pi} M$ of a given GS-manifold.

2. Definition. A differential equation for the coefficients of a connection h which expresses the integrability condition is called a *self-duality equation* for the GS-manifold $(M,\, S,\, \tilde{S},\, \sigma)$.

A GS-manifold together with a c-conic integrable connection on it will be called a *self-dual* or *left-flat* GS-manifold. □

An anti-self-dual or right-flat GS-manifold is a GS-manifold together with an integrable conic connection on its d-conic structure. Reversing the orientation changes a left-flat manifold to a right-flat manifold, and *vice-versa*. We can also call such manifolds "semi-flat," without specifying the orientation.

The second type of self-duality equation which we shall consider here relates to connections on locally free sheaves (or vector bundles) over a GS-manifold. Let $\nabla\colon \mathcal{E} \longrightarrow \mathcal{E} \otimes \Omega^1 M$ be a covariant connection differential, and let $F(\nabla) = F_+(\nabla) + F_-(\nabla) \in \mathcal{E}nd\, \mathcal{E} \otimes \Omega^2 M$ be its curvature, where $F_\pm(\nabla) \in \mathcal{E}nd\, \mathcal{E} \otimes \Omega^2_\pm M$, and the $\Omega^2_\pm M$ are defined as in § 1.9.

3. Definition. A differential equation for the coefficients of a connection ∇ which expresses the vanishing of $F_+(\nabla)$ is called a *Yang–Mills self-duality equation* on the GS-manifold. □

We shall later describe the structure of self-duality equations, and in particular, give their form in coordinates. In addition, we shall show that on a self-dual GS-manifold the Yang–Mills self-duality equations themselves can be represented in the form of integrability conditions, namely, the condition that the connection $\pi^*(\nabla)$ be integrable along the fibres of a fibering tangent to h.

4. Coordinates and coefficients. As in § 2, we choose local coordinates (x^a) on M and a local trivialization of the sheaves S and $\tilde{S}$ by sections w^α and $z^{\dot\beta}$, respectively, where $\alpha = 1, \ldots, d; \beta = 1, \ldots, c$. A GS-structure is determined by the $(cd)^2$ functions e on M which describe the spinor decomposition:

$$\sigma^{-1}(dx^a) = e^a_{\alpha\dot\beta} w^\alpha \otimes z^{\dot\beta},$$

or, in the dual bases,

$$e^a_{\alpha\dot\beta} \frac{\partial}{\partial x^a} = (\sigma^*)^{-1}(w_\alpha \otimes z_{\dot\beta}).$$

The choice of coordinates trivializes several fibrations and, as before, gives us a reference point, or 'origin,' for describing all possible connections by means of their coefficients.

In particular, a covariant differential $\nabla\colon \mathcal{S} \longrightarrow \mathcal{S} \otimes \Omega^1 M$ on $\mathcal{S}$ can be defined using the coefficients $\omega^\alpha_{\beta a}$ or $\omega^\alpha_{\beta\gamma\dot\delta} = e^a_{\gamma\dot\delta}\omega^\alpha_{\beta a}$, as follows:

$$\nabla w^\alpha = \omega^\alpha_{\beta a} w^\beta \otimes dx^a,$$

$$(\mathrm{id}_{\mathcal{S}} \otimes \sigma^{-1})(\nabla w^\alpha) = \omega^\alpha_{\beta\gamma\dot\delta} \otimes w^\beta \otimes w^\gamma \otimes z^{\dot\delta}.$$

According to Proposition 4.10, the differential ∇ induces a connection on the fibration $F = \mathrm{P}_M(\mathcal{S}^*) \xrightarrow{\pi} M$. This connection depends only on the traceless part of $\omega^\alpha_{\beta a}$ (the convolution with respect to α and β); and all connections on this fibration are obtained in this way.

A connection on F in turn induces a c-conic connection on M. In fact, according to § 4.2, a c-conic structure is determined by a closed imbedding $F \longrightarrow G_M(c;\, \mathcal{T}M)$ under which the grassmannian's tautological flag induces a flag on F:

$$\pi^*(\tilde{\mathcal{S}}^*)(-1) \subset \pi^*(\tilde{\mathcal{S}}^*) \otimes \pi^*(\mathcal{S}^*) = \pi^*(\mathcal{T}M).$$

The conic connection is the lifting to $\mathcal{T}F$ of the subsheaf $\pi^*(\tilde{\mathcal{S}}^*)(-1)$:

$$\begin{array}{ccccccccc} 0 & \longrightarrow & \mathcal{T}F/M & \longrightarrow & \mathcal{T}F & \xrightarrow{d\pi} & \pi^*(\mathcal{T}M) & \longrightarrow & 0. \\ & & & & & & \cup & & \\ & & & & & & \pi^*(\tilde{\mathcal{S}}^*)(-1) & & \end{array}$$

Thus, the projective connection on F, which gives a lifting of all of $\pi^*(\mathcal{T}M)$, in particular gives a lifting of the subsheaf. The corresponding mapping on connection coefficient sheaves is surjective; it can be described as a symmetrization with respect to $\beta\gamma$ in the coordinates. In fact, this mapping is actually $\pi_*(\varphi)$, where φ is the morphism in the exact sequence

$$0 \longrightarrow \mathcal{T}F/M \otimes \Omega^1 F/M(1) \otimes \pi^*(\tilde{\mathcal{S}}) \longrightarrow$$
$$\longrightarrow \mathcal{T}F/M \otimes \pi^*(\mathcal{S} \otimes \tilde{\mathcal{S}}) \xrightarrow{\varphi} \mathcal{T}F/M \otimes \pi^*(\tilde{\mathcal{S}})(1) \longrightarrow 0.$$

The surjectivity of $\pi_*(\varphi)$ follows because $R^1\pi_*(\mathcal{T}F/M \otimes \Omega^1 F/M(1)) = 0$, which, in turn, is a consequence of Theorem 2.2 applied to the exact sequence $0 \longrightarrow \mathcal{O}(-1) \longrightarrow \pi^*(\mathcal{S}^*) \longrightarrow \mathcal{T}F/M(-1) \longrightarrow 0$ tensor-multiplied by $\Omega^1 F/M(2)$:

$$0 \longrightarrow \Omega^1 F/M(1) \longrightarrow \pi^*(\mathcal{S}^*) \otimes \Omega^1 F/M(2) \longrightarrow \mathcal{T}F/M \otimes \Omega^1 F/M(1) \longrightarrow 0.$$

It is this sequence which enables one to compute $\pi_*(\mathcal{T}F/M \otimes \Omega^1 F/M(1))$ and then the kernel of $\pi_*(\varphi)$. Namely, applying Theorem 2.2 again, we have

$$0 \longrightarrow S^* \otimes \pi_*(\Omega^1 F/M(2)) \xrightarrow{\sim} \pi_*(\mathcal{T}F/M \otimes \Omega^1 F/M(1)) \longrightarrow 0,$$

and then, using the sequence $0 \longrightarrow \Omega^1 F/M(2) \longrightarrow \pi^*(S)(1) \longrightarrow \mathcal{O}(2) \longrightarrow 0$, we obtain

$$\pi_*(\Omega^1 F/M(2)) = \operatorname{Ker}(S \otimes S \longrightarrow S^2(S)) = \wedge^2 S.$$

Finally, the kernel of $\pi_*(\varphi)$ consists of the traceless symbols $\omega^\alpha_{\beta\gamma\dot{\delta}}$ which are anti-symmetric in $\beta\gamma$, and this means that $\pi_*(\varphi)$ can be identified with symmetrization.

We now summarize the description we have derived. If we stipulate that the obstructions to the existence of connections vanish, we obtain the following surjective maps, all of which are defined invariantly:

$$\begin{pmatrix}\text{connections}\\ \text{on } S\end{pmatrix} \longrightarrow \begin{pmatrix}\text{connections on the}\\ \text{fibration } \mathrm{P}_M(S^*)\end{pmatrix} \longrightarrow \begin{pmatrix}c\text{-conic connections}\\ \text{on } M\end{pmatrix},$$

$$\left(\omega^\alpha_{\beta\gamma\dot{\delta}}\right) \longrightarrow \left(\omega^\alpha_{\beta\gamma\dot{\delta}} \bmod \delta^\alpha_\beta \Gamma_{\gamma\dot{\delta}}\right) \longrightarrow \left(\omega^\alpha_{\beta\gamma\dot{\delta}} \bmod \delta^\alpha_\beta \Gamma_{\gamma\dot{\delta}} + \omega^\alpha_{[\beta\gamma]\dot{\delta}}\right).$$

However, the arrows in the reverse direction (for example, if one wants to describe c-conic connections by choosing traceless symmetric symbols which determine a connection on S) depend on the coordinate system.

5. Torsion in a GS-structure. Let $h\colon \pi^*(\tilde{S}^*)(-1) \longrightarrow \mathcal{T}F$ be a c-conic connection. According to § 6.9, its Frobenius form Φ has a canonically defined quotient $\Phi_0(h) = d\pi(\Phi(h))$. This quotient $\Phi_0(h)$ maps $\pi^*(\wedge^2\tilde{S}^*(-2))$ to $\pi^*(\mathcal{T}M)/\pi^*(\tilde{S}^*)(-1)$:

$$\Phi_0(h)(X, Y) = d\pi[(h(X), h(Y)] \bmod \pi^*(\tilde{S}^*)(-1),$$

where X and Y are local sections of $\tilde{S}^*(-1)$.

It is not hard to compute the sheaf on M whose sections can be identified with Φ_0. We multiply the standard exact sequence $0 \longrightarrow \mathcal{O}(-1) \longrightarrow \pi^* S^* \longrightarrow \mathcal{T}F/M(-1) \longrightarrow 0$ by $\pi^*\tilde{S}^*$. The result can be written in the form

$$0 \longrightarrow \pi^*\tilde{S}^*(-1) \longrightarrow \pi^*(\mathcal{T}M) \longrightarrow \mathcal{T}F/M(-1) \otimes \pi^*\tilde{S}^* \longrightarrow 0. \tag{1}$$

Hence, $\Phi_0(h)$ lies in $H^0(F, \mathcal{T}F/M(1) \otimes \pi^*(\tilde{S}^* \otimes \wedge^2\tilde{S})) = H^0(M, \tilde{S}^* \otimes \wedge^2\tilde{S} \otimes (S^2 S \otimes S^*)_0)$. We split $\Phi_0(h)$ into two irreducible components:

$$\Phi_0(h) = \Phi_0^{(1)}(h) + \Phi_0^{(2)}(h),$$

$$\Phi_0^{(1)}(h) \in (\tilde{S}^* \otimes \wedge^2\tilde{S})_0 \otimes (S^2 S \otimes S^*)_0,$$

$$\Phi_0^{(2)}(h) \in i(\tilde{S}) \otimes (S^2 S \otimes S^*)_0,$$

where $i\colon \tilde{S} \longrightarrow \tilde{S}^* \otimes \wedge^2 \tilde{S}$ is determined by the formula $i(z^{\dot\alpha}) = z_{\dot\beta} \otimes z^{\dot\beta} \wedge z^{\dot\alpha}$. (Here we write $\Phi \in \mathcal{T}$ instead of $\Phi \in H^0(\mathcal{T})$, for brevity.)

6. Theorem. (a) *$\Phi_0^{(1)}(h)$ does not depend on the choice of h. The element*

$$t_+(M) = \Phi_0^{(1)}(h) \in \Omega_+^2 M \otimes \mathcal{T}M$$

is called the "left torsion" of the GS-structure σ on M. One similarly defines the right torsion $t_-(M) \in \Omega_-^2 M \otimes \mathcal{T}M$ by choosing a connection on the d-conic structure.

(b) *There exists a unique c-conic connection h for which $\Phi_0^{(2)}(h) = 0$. It is the only c-conic connection which can be integrable; in addition to the condition $t_+(M) = 0$, a necessary and sufficient condition for it to be integrable is that its lifting to a connection on $\mathbf{P}_M(S^*)$ have curvature contained entirely in $\Omega_-^2 M \otimes sl(S)$.*

Proof. We begin by determining how $\Phi_0(h)$ changes when h is replaced by $h + \omega$, where $\omega\colon \pi^*(\tilde{S}^*)(-1) \longrightarrow \mathcal{T}F/M$ is any section of the coefficient sheaf of the c-conic connection. In the first place, by the definition of Φ_0 we have

$$[\Phi_0(h+\omega) - \Phi_0(h)](X, Y) = d\pi([\omega(X), h(Y)] - [\omega(Y), h(X)]) \bmod \pi^*(\tilde{S}^*)(-1).$$

The term $[\omega(X), \omega(Y)]$ is omitted on the right, since $\mathcal{T}F/M = \operatorname{Ker} d\pi$ is an integrable distribution. For the same reason the right side of the formula does not change if we replace h by $h' = h + \omega'$. We next observe that the right side depends $\mathcal{O}_F$-linearly on ω. In fact,

$$[f\omega(X), h(Y)] - f[\omega(X), h(Y)] = -(h(Y)f)\omega(X) \in \operatorname{Ker} d\pi.$$

Thus, on F we have the morphism of sheaves

$$\omega \mapsto \Phi_0(h+\omega) - \Phi_0(h)\colon \pi^*(\tilde{S}) \otimes \mathcal{T}F/M(1) \longrightarrow$$
$$\longrightarrow \mathcal{H}om(\wedge^2 \pi^* \tilde{S}^*(-2), \mathcal{T}F/M(-1) \otimes \pi^* \tilde{S}^*).$$

We do not lose anything if we descend to M:

$$\lambda\colon \tilde{S} \otimes \pi_*(\mathcal{T}F/M(1)) \longrightarrow \tilde{S}^* \otimes \wedge^2 \tilde{S} \otimes \pi_*(\mathcal{T}F/M(1)).$$

Using considerations of functoriality in S and $\tilde{S}$ and elementary facts from representation theory, we see that λ must be proportional to $i \otimes \mathrm{id}$, where i was defined at the end of the last subsection. It is essential to check that $\lambda \neq 0$, after which we can start with any connection h and uniquely choose a "correction" ω such that $\Phi_0^{(2)}(h+\omega)=0$. It is now clear that $\Phi_0^{(1)}(h)$ does not depend on h.

For the coordinate computations we first determine h. In order to avoid denominators, it is more convenient to give coordinate expressions for $h(1) = h \otimes \mathrm{id}_{\mathcal{O}(1)}$ by the formulas

$$h(1)(z_{\dot\beta}) = e^a_{\alpha\dot\beta} w^\alpha \frac{\partial}{\partial x^a}.$$

Then the other maps and sections also turn out to be twisted:

$$(h+\omega)(1)(z_{\dot\delta}) = e^a_{\alpha\dot\delta} w^\alpha \frac{\partial}{\partial x^a} + \omega^a_{\beta\gamma\dot\delta} w^\beta w^\gamma \otimes \frac{\partial}{\partial w^\alpha}.$$

In order to compute $t_+(M)$, we must first commute the liftings of $z_{\dot\delta}$ and $z_{\dot\epsilon}$:

$$\begin{aligned}\Phi_0(h)(2)(z_{\dot\delta}, z_{\dot\epsilon}) &= [e^a_{\alpha\dot\delta} w^\alpha \frac{\partial}{\partial x^a}, e^b_{\beta\dot\epsilon} w^\beta \frac{\partial}{\partial x^b}] \bmod \pi^*(\tilde S^*)(1)\\ &= 2e^a_{\alpha[\dot\delta}\partial_a e^b_{\beta\dot\epsilon]} w^\alpha w^\beta \partial_b \bmod \pi^*(\tilde S^*)(1),\end{aligned}$$

where we have written ∂_a instead of $\partial/\partial x^a$. Now, in order to descend the right side to M in accordance with (1), we must replace ∂_b by $e_b^{\rho\dot\sigma}\frac{\partial}{\partial w^\rho}\otimes z_{\dot\sigma}$ and then replace $\frac{\partial}{\partial w^\rho}$ here by w_ρ. Thus,

$$\begin{aligned}\Phi_0(h)(2) &= e^a_{\alpha[\dot\delta}\partial_a\sigma^b_{\beta\dot\epsilon]} e_b^{\rho\dot\sigma} w^\alpha w^\beta \otimes w_\rho \otimes z_{\dot\sigma} \otimes z^{\dot\delta} \otimes z^{\dot\epsilon}\\ &= -e^a_{\alpha[\dot\delta} e^b_{\beta\dot\epsilon]}\partial_a \sigma_b^{\rho\dot\sigma} w^\alpha w^\beta \otimes w_\rho \otimes z_{\dot\sigma} \otimes z^{\dot\delta} \otimes z^{\dot\epsilon}.\end{aligned}$$

Similarly, we have

$$\lambda(\omega)(2)(z_{\dot\delta}, z_{\dot\epsilon}) = \omega^\alpha_{\beta\gamma[\dot\delta} e^a_{\alpha\dot\epsilon]} w^\beta \otimes w^\gamma \frac{\partial}{\partial x^a} \bmod \pi^*(\tilde S^*)(1),$$

and, after descending to M,

$$\begin{aligned}\lambda(\omega) &= \omega^\alpha_{\beta\gamma[\dot\delta} e^a_{\alpha\dot\epsilon]} e_a^{\rho\dot\sigma} w^\beta \otimes w^\gamma w_\rho \otimes z_{\dot\sigma} \otimes z^{\dot\delta} \otimes z^{\dot\epsilon}\\ &= \omega^\alpha_{\beta\gamma[\dot\delta} \delta^{\rho\dot\sigma}_{\alpha\dot\epsilon]} w^\beta \otimes w^\gamma \otimes w_\rho \otimes z_{\dot\sigma} \otimes z^{\dot\delta} \otimes z^{\dot\epsilon}.\end{aligned}$$

Hence, ω can be found from the condition $\lambda(\omega) = -\Phi_0^{(2)}(h)$, i.e.,

$$\omega^\alpha_{\beta\gamma\dot\delta} = \frac{1}{c} e^a_{\beta\dot\delta}\sigma^b_{\gamma\dot\epsilon}\partial_a e_b^{\alpha\dot\epsilon}.$$

Now let h be the unique c-conic connection for which $\Phi_0^{(2)}(h) = 0$ (before this was $h + \omega$). If $t_+(M) \neq 0$, then clearly this connection cannot be integrable. On the other hand, if $t_+(M) = 0$, it follows that the Frobenius form $\Phi(h)$ takes values in the sheaf $(d\pi)^{-1}(\tilde{S}^*(-1)) \bmod h(\tilde{S}^*(-1))$. This sheaf on F can be identified with $\mathcal{T}F/M$, with the identification chosen compatibly with the corresponding identification for the connection $\tilde{h}$ on $\mathsf{P}_M(S^*)$ which extends h.

Thus, the Frobenius form for h

$$\Phi: \wedge^2 \pi^* \tilde{S}(-2) \longrightarrow \mathcal{T}F/M$$

will be the restriction of the Frobenius form for $\tilde{h}$

$$\tilde{\Phi}: \wedge^2 \pi^* \mathcal{T}M \longrightarrow \mathcal{T}F/M.$$

Descending these forms to M, we find that this restriction coincides with the $\Omega_+^2 M = S^2(S) \otimes \wedge^2(\tilde{S})$-component of the curvature of $\tilde{h}$. Consequently, the last condition for h to be integrable and the GS-structure M to be semi-flat is that this restriction vanish. □

7. The left de Rham complex of a self-dual manifold. Suppose that $(M, S, \tilde{S}, \sigma)$ is a self-dual manifold, $F = \mathsf{P}_M(S^*)$, and $\mathcal{T}F/L \subset \mathcal{T}F$ is the integrable distribution corresponding to the unique integrable c-connection on the manifold. By definition, $d\pi$ induces an isomorphism $\mathcal{T}F/L \xrightarrow{\sim} \pi^*(\tilde{S}^*)(-1)$, and so we can identify $\Omega^i F/L$ canonically with $\pi^*(\wedge^i \tilde{S})(i)$. We consider the de Rham complex of the distribution $\mathcal{T}F/L$:

$$\mathcal{O}_F \xrightarrow{d_{F/L}} \pi^* \tilde{S}(1) \xrightarrow{d_{F/L}} \pi^* \wedge^2 \tilde{S}(2) \longrightarrow \cdots$$

and the descended complex on M:

$$\mathcal{O}_M \xrightarrow{\pi_*(d_{F/L})} S \otimes \tilde{S} = \Omega^1 M \xrightarrow{\pi_*(d_{F/L})} S^2 S \otimes \wedge^2 \tilde{S} \longrightarrow \cdots.$$

The sheaves in the descended complex can be naturally identified with direct sum quotients of the de Rham complex on M. Namely, we have the standard tensor algebra formula

$$\wedge^i(S \otimes \tilde{S}) = \bigoplus_{|a|=i} S^{(a)}(S) \otimes S^{(a^t)}(\tilde{S}),$$

where $(a) = (a_1, \ldots, a_k), a_1 \geq \cdots \geq a_k \geq 0$, $|a| = \sum_{j=1}^k a_j$, and (a^t) is the sequence corresponding to the dual Young diagram; the functors $S^{(a)}$ were defined in § 2.8. Actually, all we need is an imbedding $S^i(S)\otimes\wedge^i(\tilde S)\longrightarrow\wedge^i(S\otimes\tilde S)$, which can be constructed in a completely obvious way, and then we take the projection p_i onto the image.

8. Proposition. *The de Rham complex of an integrable c-connection, when descended to M, becomes isomorphic to the quotient de Rham complex*

$$(S^i(S)\otimes\wedge^i(\tilde S), p_i\circ d\circ p_i).$$

Proof. The quotient map $p_i\colon\Omega^iM\longrightarrow\pi_*\Omega^iF/L$ is constructed as follows. We lift a form ω^i to F and regard it as a semilinear function only on fields in $\mathcal{T}F/L$. A glance at the explicit formulas for the exterior differential in § 5.7 will convince us that restriction to an integrable horizontal distribution commutes with exterior differentiation. □

Now suppose that $(\mathcal{E},\nabla)$ is a locally free sheaf with connection on a self-dual GS-manifold M. On $\pi^*(\mathcal{E})$ one can define a covariant differential $\nabla_{F/L}$ along the fibering $\mathcal{T}F/L$. Namely, we first define a lifting $\pi^*(\nabla)\colon\pi^*(\mathcal{E})\longrightarrow\pi^*(\mathcal{E})\otimes\Omega^1F$, by setting $\pi^*(\nabla)(\sum f_ie_i)=\sum(f_i\nabla e_i+e_i\otimes df_i)$, where the e_i are local sections of $\pi^*(\mathcal{E})$ that have been lifted from $\mathcal{E}$ and the f_i are local functions on F. After this we form the composition of $\pi^*(\nabla)$ with restriction to Ω^1F/L; the result is $\nabla_{F/L}$.

9. Theorem. (a) *$(\mathcal{E},\nabla)$ satisfies the Yang–Mills self-duality equation on the self-dual GS-manifold M if and only if $(\pi^*(\mathcal{E}),\nabla_{F/L})$ is integrable along the fibres of the fibering $\mathcal{T}F/L$.*

(b) *If the condition in part (a) is satisfied, then the relative de Rham complex $(\pi^*(\mathcal{E})\otimes\Omega^iF/L,\nabla_{F/L})$, when descended to M, has the form $(\mathcal{E}\otimes S^i(S)\otimes\wedge^i(\tilde S),(\mathrm{id}_\epsilon\otimes p_i)\circ\nabla)$, where p_i is defined as in § 7.8.*

Corollary. *Let $\nabla\colon S\longrightarrow S\otimes\Omega^1M$ be a connection on S whose curvature is contained in $sl(S)\otimes\Omega^2M$. This connection induces the canonical integrable connection on the self-dual GS-manifold $(M,S,\tilde S,\sigma)$ if and only if (S,∇) satisfies the Yang–Mills self-duality equation on the manifold.*

Proof. The same argument as in the preceding subsection, except considering forms with values in $\mathcal{E}$ rather than "scalar" forms, immediately gives part (b). Therefore, if the morphism $\nabla^2_{F/L}\colon\pi^*(\mathcal{E})\longrightarrow\pi^*(\mathcal{E})\otimes\Omega^2F/L$, which represents the relative curvature, is descended to M, we obtain the composition of the curvature $\nabla^2\colon\mathcal{E}\longrightarrow\mathcal{E}\otimes\Omega^2M$ and the projection $\mathrm{id}_{\mathcal{E}}\otimes p_2$. But p_2 projects Ω^2M onto Ω^2_+M. This gives us part (a). The corollary follows from this if one uses Theorem 6(b) and the fact that the curvature of ∇ can be identified with the curvature of the induced connection on $\mathrm{P}_M(S^*)$. □

10. Dirac equations for grassmannian spinors. The left Dirac operator corresponding to a connection $\nabla: S^* \longrightarrow S^* \otimes \Omega^1 M$ on a GS-manifold $(M, S, \tilde{S}, \sigma)$ is defined to be the composition

$$D: S^* \xrightarrow{\nabla} S^* \otimes \Omega^1 M \xrightarrow{\mathrm{id}\otimes\sigma^{-1}} S^* \otimes S \otimes \tilde{S} \longrightarrow \tilde{S},$$

where the last arrow is the convolution of S and S^*. Similarly, starting with $(\tilde{S}^*, \tilde{\nabla})$, we can define the right Dirac operator:

$$\tilde{D}: \tilde{S}^* \xrightarrow{\tilde{\nabla}} \tilde{S}^* \otimes \Omega^1 M \xrightarrow{\mathrm{id}\otimes\sigma^{-1}} \tilde{S}^* \otimes S \otimes \tilde{S} \longrightarrow S.$$

Suppose that we have two "mass matrices," i.e., morphisms of sheaves $\tilde{M}: \tilde{S}^* \longrightarrow \tilde{S}$ and $M: S^* \longrightarrow S$. We can then write the Dirac equations for a pair of local sections ψ, $\tilde{\psi}$ of the sheaves S^* and $\tilde{S}^*$:

$$D\psi = \tilde{M}\tilde{\psi},$$
$$\tilde{D}\tilde{\psi} = M\psi.$$

In coordinates these equations take the following form. We set

$$\psi = \psi^\alpha w_\alpha, \tilde{\psi} = \psi^{\dot\beta} z_{\dot\beta};$$

$$\nabla w_\alpha = -\omega^\beta_{\alpha a} w_\beta \otimes dx^a, \tilde{\nabla} z_{\dot\alpha} = -\omega^{\dot\gamma}_{\dot\alpha a} z_{\dot\gamma} \otimes dx^a.$$

Further let $\partial_{\alpha\dot\beta} = e^a_{\alpha\dot\beta}\partial_a$, and let $\omega_{\alpha\dot\beta} = \omega^\gamma_{\alpha a} e^a_{\gamma\dot\beta}, \tilde{\omega}_{\alpha\dot\beta} = \tilde{\omega}^{\dot\gamma}_{\dot\beta a} e^a_{\alpha\dot\gamma}$. With this notation, the Dirac equations can be written as follows:

$$\partial_{\alpha\dot\beta}\psi^\alpha - \omega_{\alpha\dot\beta}\psi^\alpha = \tilde{M}_{\dot\gamma\dot\beta}\psi^{\dot\gamma},$$
$$\partial_{\alpha\dot\beta}\psi^{\dot\beta} - \tilde{\omega}_{\alpha\dot\beta}\psi^{\dot\beta} = M_{\alpha\beta}\psi^\beta.$$

References for Chapter 1

The textbook [111] contains almost everything that one needs to know about complex manifolds and their cohomology in order to understand the basic material in this book. In § 2 we presented part of a large theory (compare with [17]), based on Demazure's article [23]. The material in §§ 4–5 consists of standard ideas from differential geometry adapted to holomorphic geometry. The basic subject in § 3 is the twistor model of Minkowski space. For physical motivation for the twistor program, first of all see the articles [91] and [93] by its inventor, Penrose. Starting in 1976, Penrose's group at Oxford has been putting out notes under the title of "Twistor Newsletter," some of which are contained in [57] in an edited version. The twistor program has attracted widespread interest for one reason in particular — its effectiveness from a technical standpoint in classifying instantons (see Chapter 2 and the references for that chapter). Penrose himself by no means limits his approach to methods of solving dynamic equations. In fact, his program stresses the connection between space-time and spin degrees of freedom, it is very compatible with the philosophy of conformal-invariant field theory and dynamic mass formation, and, finally, with the addition of odd coordinates the twistor approach almost inevitably produces the basic models of supersymmetry and supergravity, as will be shown in Chapter 5. See also [103] and the bibliography in [57].

CHAPTER 2

THE RADON–PENROSE TRANSFORM

In this chapter we give several basic applications of cohomological techniques to solving nonlinear field theory equations. In § 1 we describe some geometrical structures: complex space-time, with its curvature encoded in the spinor decomposition of the tangent sheaf; fields as sections and connections; and finally, the Lagrangians and the dynamic equations that follow from them. The reader should supplement our brief presentation with more traditional and detailed differential geometric versions; see [30], the very informative survey [32], and also [36].

In § 2 we describe the Radon–Penrose transform of a fibration or a cohomology class in its complex-analytic variant. We also introduce double fibrations, which we call "self-duality diagrams." §§ 3–4 are devoted to the classification theorem for instantons/self-dual connections on S^4.

The rest of the chapter is concerned with non-self-duality equations, which are closely related to the theory of extensions and obstructions to extending geometrical objects from a space to a larger ambient space (in our context, this is the "infinitesimal extension"). The formalism for this is described in § 6, after we introduce in § 5 the fundamental class of double fibrations to which the formalism will be applied. § 8 contains computations of the necessary cohomology groups. In §§ 7 and 9 we give the fundamental theorem on Yang–Mills non-self-duality equations in the flat case. In §§ 10–11 we give some further results on dynamic equations with sketches of proofs.

§ 1. Complex Space-Time

1. Fundamental structures. When we say "complex space-time," we shall mean a four-dimensional complex manifold M on which some or all of the following structures are defined:

(a) A "grassmannian spinor structure," i.e., two locally free sheaves S (left spinors) and $\tilde{S}$ (right spinors) together with a spinor decomposition $\sigma\colon S \otimes \tilde{S} \longrightarrow \Omega^1 M$ of 1-forms on M. The sheaves S and $\tilde{S}$ are necessarily of rank two.

(b) "Spinor connections" $\nabla_l\colon S \longrightarrow S \otimes \Omega^1 M$ and $\nabla_r\colon \tilde{S} \longrightarrow \tilde{S} \otimes \Omega^1 M$.

(c) "Spinor metrics," i.e., nonzero sections $\epsilon \in H^0(M, \wedge^2 S)$ and $\tilde{\epsilon} \in H^0(M, \wedge^2 \tilde{S})$.

The sheaves $\wedge^2 S$ and $\wedge^2 \tilde{S}$ are invertible, and the sections ϵ and $\tilde{\epsilon}$ enable one to identify them with $\mathcal{O}_M$ wherever the sections are nonzero. We shall usually assume implicitly that this condition holds on all of M.

(d) A "real structure" $\rho: M \longrightarrow M$, i.e., an involution on the set of points of M (so that $\rho^2 = \mathrm{id}_M$) for which $\rho^*(\mathcal{O}_M)$ is the sheaf of antiholomorphic functions. In other words, if (x^a) is a holomorphic local coordinate system in the region U, then $\overline{(\rho^*(x^a))}$ is a holomorphic local coordinate system in $\rho^{-1}(U)$, where the bar denotes complex conjugation of the function's values.

We shall be interested only in real structures which have a four-dimensional manifold of real points, i.e., fixed points of ρ. Moreover, we shall assume that the involution ρ extends to the spinor decomposition $S \otimes \tilde{S}$ and is compatible with ϵ and $\tilde{\epsilon}$ in a certain specific sense. We shall now briefly explain what we mean by compatibility in general.

2. Derivative structures and compatibility conditions. The choice of a spinor decomposition σ reduces the structure group of the tangent bundle on M from $GL(4)$ to $GL(2) \times GL(2)$; and the choice of ϵ and $\tilde{\epsilon}$ further reduces this structure group to $SL(2) \times SL(2)$. We shall usually make use of this reduction in the form of an imbedding of the tensor algebra in the spinor algebra, with a corresponding decomposition of tensors. In terms of coordinates, this corresponds to the two-subscript formalism which we used earlier.

The covariant spinor differentials ∇_l and ∇_r induce a covariant differential on the entire spinor algebra and, in particular, on its tensor subalgebra. We shall usually denote the latter differential by writing simply ∇, without specifying the sheaf component on which ∇ acts.

The spinor metrics ϵ and $\tilde{\epsilon}$ induce a metric $g = \epsilon \otimes \tilde{\epsilon} \in H^0(M, \wedge^2 S \otimes \wedge^2 \tilde{S}) \subset H^0(M, S^2(\Omega^1 M))$, where $\wedge^2 S \otimes \wedge^2 \tilde{S}$ is imbedded in $S^2(\Omega^1)$ by means of σ.

In general, the connection ∇ is not riemannian relative to this metric, for any of a variety of reasons: it might have torsion, it might not annihilate the metric, and it might even fail to annihilate any metric in the conformal class of g. A stipulation that some or all of these unpleasant circumstances do not occur is what we mean by a compatibility condition on the fundamental structures.

Once we choose an extension of ρ to $S \otimes \tilde{S}$, the condition that the real structure be compatible with the other structures means that ρ leaves g invariant. In that case g becomes a real-analytic (pseudo-) riemannian metric on the manifold of real points of M; its signature depends upon the action of ρ on the spinors.

3. Coordinates. The notational principles in §§ 3 and 7 of Chapter 1 are still in force. In the first place, (x^a), $a = 0, 1, 2, 3$, are local coordinate systems which take real values at points that are fixed by ρ. Next, w^0 and w^1 are basis sections of S; $z^{\dot{0}}$ and $z^{\dot{1}}$ are basis sections of $\tilde{S}$. The spinor decomposition is given by formulas

$$\sigma(w^\alpha \otimes z^{\dot\beta}) = e_a^{\alpha\dot\beta} dx^a.$$

As in § 7 of Chapter 1, the spinor connections are defined by symbols ω, which are functions on M:

$$\nabla w^\alpha = \omega^\alpha_{\beta a} w^\beta \otimes dx^a, \qquad \nabla z^{\dot\alpha} = \omega^{\dot\alpha}_{\dot\beta a} z^{\dot\beta} \otimes dx^a.$$

The sections w^α and $z^{\dot\alpha}$ are chosen in such a way that

$$\epsilon = 2w^0 \wedge w^1 = \epsilon_{\alpha\beta} w^\alpha \otimes w^\beta,$$

$$\tilde\epsilon = 2z^{\dot 0} \wedge z^{\dot 1} = \epsilon_{\dot\alpha\dot\beta} z^{\dot\alpha} \otimes z^{\dot\beta}.$$

The symbols $\epsilon_{\alpha\beta}$ are the same as in § 3 of Chapter 1. The metric has the form

$$g_{ab} dx^a dx^b = \epsilon_{\alpha\beta}\epsilon_{\dot\gamma\dot\delta} e_a^{\alpha\dot\gamma} e_b^{\beta\dot\delta} dx^a dx^b.$$

4. The Weyl electromagnetic field. Given ∇, ϵ and $\tilde\epsilon$, one has two uniquely determined 1-forms A and $\tilde A$ on M:

$$\nabla\epsilon = \epsilon \otimes A, \qquad \nabla\tilde\epsilon = \tilde\epsilon \otimes \tilde A.$$

Their differentials are the curvature forms for ∇ on $\wedge^2 S$ and $\wedge^2 \tilde S$:

$$F = dA, \qquad \tilde F = d\tilde A.$$

The metric g is horizontal relative to ∇ if and only if $A + \tilde A = 0$, since

$$\nabla(\epsilon \otimes \tilde\epsilon) = (\nabla\epsilon) \otimes \tilde\epsilon + \epsilon \otimes (\nabla\tilde\epsilon) = (\epsilon \otimes \tilde\epsilon) \otimes (A + \tilde A).$$

In terms of coordinates, we have

$$\nabla^2(w^\alpha) = (\partial_a \omega^\alpha_{\beta b} + \omega^\gamma_{\beta a}\omega^\alpha_{\gamma b}) w^\beta \otimes dx^a \wedge dx^b,$$

and hence

$$F = (\partial_a \omega^\alpha_{\alpha b} + \omega^\gamma_{\beta a}\omega^\beta_{\gamma b}) dx^a \wedge dx^b.$$

Similarly,

$$\nabla^2(z^{\dot\alpha}) = (\partial_a \omega^{\dot\alpha}_{\dot\beta b} + \omega^{\dot\gamma}_{\dot\beta a}\omega^{\dot\alpha}_{\dot\gamma b}) z^{\dot\beta} \otimes dx^a \wedge dx^b,$$

$$\tilde{F} = (\partial_a \omega^{\dot\alpha}_{\dot\alpha b} + \omega^{\dot\gamma}_{\dot\beta a}\omega^{\dot\beta}_{\dot\gamma b})dx^a \wedge dx^b.$$

If ϵ is replaced by $f\epsilon$, then A changes to $A + f^{-1}df$, and of course F does not change at all. For given S, $\tilde{S}$, ∇_l and ∇_r, the forms F and $\tilde{F}$ are the obstruction to the existence of horizontal spinor metrics. The form $F + \tilde{F}$ is the obstruction to the existence of a horizontal metric compatible with S, $\tilde{S}$, ∇_l, ∇_r. If one uses the Levi–Civita connection to describe the gravitational field, then such a horizontal metric must exist, because in that case one always has $F + \tilde{F} = 0$. But the closed 2-form F itself is not necessarily zero. It is determined up to sign by the GS-structure σ and the spinor connection ∇ (the sign changes if the orientation is switched).

In 1918, Weyl proposed interpreting F as an electromagnetic field built into the system of fundamental space-time structures. This was one of the first modern attempts at a unified theory of fundamental interactions.

5. The Christoffel coefficients and torsion. We define the connection coefficients Γ^a_{bc} by the formulas $\nabla(dx^a) = \Gamma^a_{bc} dx^b \otimes dx^c$. They can be computed in terms of σ and ω. Using σ to identify $\Omega^1 M$ with $S \otimes \tilde{S}$, we have:

$$\nabla(e^{\alpha\dot\beta}_c dx^c) = \nabla(w^\alpha \otimes z^{\dot\beta}) = \omega^\alpha_{\gamma c} w^\gamma \otimes z^{\dot\beta} \otimes dx^c \, + \, \omega^{\dot\beta}_{\dot\gamma c} w^\alpha \otimes z^{\dot\gamma} \otimes dx^c =$$

$$= \omega^\alpha_{\gamma c} e^{\gamma\dot\beta}_b dx^b \otimes dx^c \, + \, \omega^{\dot\beta}_{\dot\gamma c} e^{\alpha\dot\gamma}_b dx^b \otimes dx^c,$$

or

$$e^{\alpha\dot\beta}_c \nabla(dx^c) = (-\partial_b e^{\alpha\dot\beta}_c + \omega^\alpha_{\gamma c} e^{\gamma\dot\beta}_b + \omega^{\dot\beta}_{\dot\gamma c} e^{\alpha\dot\gamma}_b) dx^b \otimes dx^c.$$

Multiplying both sides by $e^a_{\alpha\dot\beta} = (-e)^a_{\alpha\dot\beta}$ and summing over $\alpha\dot\beta$, we finally obtain

$$\Gamma^a_{bc} = (-\partial_b e^{\alpha\dot\beta}_c + \omega^\alpha_{\gamma c} e^{\gamma\dot\beta}_b + \omega^{\dot\beta}_{\dot\gamma c} e^{\alpha\dot\gamma}_b) e^a_{\alpha\dot\beta}.$$

Recall that the skew-symmetric part of Γ is called the "torsion tensor:" $t^a_{bc} = \Gamma^a_{[bc]}$, and that a connection is said to be symmetric if $t^a_{bc} = 0$.

We now give an invariant definition of t, which immediately shows its tensor character (this is not the case for the coefficients Γ^a_{bc}). Let p_s and p_a be the projections $(\Omega^1 M)^{\otimes 2}$ onto $S^2(\Omega^1 M)$ and $\Omega^2 M$, respectively. We set $\nabla_s = p_s \circ \nabla, \nabla_a = p_a \circ \nabla, t = \nabla_a - d$. It is not hard to see that t is $\mathcal{O}_M$-linear:

$$t(f\nu) = \nabla_a(f\nu) - d(f\nu) = p_a(df \otimes \nu + f\nabla\nu) - (df \wedge \nu + f d\nu) = ft(\nu).$$

But this t is a torsion tensor, since

$$t(dx^a) = p_a(\Gamma^a_{bc} dx^b \otimes dx^c) = \Gamma^a_{[bc]} dx^b \otimes dx^c.$$

We are now ready to explain the relationship between our structures and the classical ones.

6. Proposition. (a) *Suppose that M is a four-dimensional GS-manifold with spinor connections which satisfies the two conditions*

$$F_{ab} + \tilde{F}_{ab} = 0, \qquad \Gamma^{a}_{[bc]} = 0,$$

where F, $\tilde{F}$ and Γ are as in §§ 4–5 of Chapter 1. Then the covariant differential $\nabla: \Omega^1 \longrightarrow \Omega^1 \otimes \Omega^1$ is a Levi–Civita connection. More precisely, for every point of M there is a neighborhood of the point and a non-zero holomorphic metric g in this neighborhood which is a section of $\wedge^2 S \otimes \wedge^2 \tilde{S}$ defined up to multiplication by a constant, for which ∇ is the Levi–Civita connection.

(b) *Conversely, suppose that M is a four-dimensional manifold with a holomorphic metric g. Then in a neighborhood of any point of M one can introduce a GS-manifold structure with spinor connections in such a way that g has a decomposition $\epsilon \otimes \tilde{\epsilon}$ and the Levi–Civita connection is induced by the spinor connections. This structure then satisfies the conditions in part (a).*

Proof. Part (a) has essentially already been proved: one takes g to be a local horizontal section of $\wedge^2 S \otimes \wedge^2 \tilde{S}$, which exists in a neighborhood of any point. The symmetrical connection for which g is horizontal must be the Levi–Civita connection, because of the uniqueness of the Levi–Civita connection. We note that the only obstruction to the existence of a single global metric g is nontriviality of the holonomy group; in particular, if M is simply connected, then g can be constructed on all of M.

To prove part (b) we set $\mathbf{P} = \mathbf{P}_M(\mathcal{T}M) \xrightarrow{\pi} M$. Since $S^2(\Omega^1 M) = \pi_* \mathcal{O}(2)$, we can interpret g as a section of $\mathcal{O}(2)$ on $\mathbf{P}$. Let $F \subset \mathbf{P}$ be the zeros of this section; of course, the 1-conic structure on F comes from the null directions in the metric g. Over each point of M the base of the null cone is a two-dimensional conic $\mathbf{CP}^1 \times \mathbf{CP}^1 \subset \mathbf{CP}^3$. Hence, g determines a holomorphic double covering $M' \longrightarrow M$: a point over $x \in M$ is one of the two systems of generators of the base of the null cone at x. We suppose that this covering splits (it certainly splits locally, and the global obstruction is a "Stiefel class" in $H^1(M, \mathbf{Z}_2)$). Then $F = F_l \underset{M}{\times} F_r$, where F_l and F_r are vector bundles of projective lines over M.

We now consider the sheaf $\mathcal{T}F_l/M$. Its restrictions to the fibres have degree two; hence it has a square root locally on M (again the global obstruction is a class in $H^1(M, \mathbf{Z}_2)$). We let $\mathcal{O}_l(1)$ denote this square root, and we set $S = \pi_* \mathcal{O}_l(1)$. We then similarly construct $\tilde{S} = \pi_* \mathcal{O}_r(1)$. The universal property of a grassmannian enables us to identify F_l and F_r with $\mathbf{P}_M(S)$ and $\mathbf{P}_M(\tilde{S})$, respectively. Let $\mathcal{O}_F(a,b) = \mathcal{O}_l(a) \otimes \mathcal{O}_r(b)$ on F (the exterior tensor product). The imbedding $F = \mathbf{P}_M(\mathcal{S}) \times_M \mathbf{P}_M(\tilde{\mathcal{S}}) \subset \mathbf{P}_M(\mathcal{T}M)$ is determined by stipulating that $\mathcal{O}_\mathbf{P}(1)$ induces the sheaf $\mathcal{O}_F(1,1)$ on F. This gives a spinor decomposition $\pi_*\mathcal{O}_\mathbf{P}(1) = \Omega^1 M \longrightarrow \pi_*\mathcal{O}_F(1,1) = \mathcal{S} \times \tilde{\mathcal{S}}$.

Furthermore, $\mathcal{O}_\mathbf{P}(2)$ induces the sheaf $\mathcal{O}_F(2,2)$. Since the section $\pi^*(g)$ of the sheaf $\mathcal{O}_\mathbf{P}(2)$ vanishes on F, it lies in the subgroup of sections of $\pi^*(\wedge^2\mathcal{S} \otimes \wedge^2\tilde{\mathcal{S}}) \subset \mathcal{O}_\mathbf{P}(2)$. Hence, localizing further on M if necessary, we may assume that $g = \epsilon \otimes \tilde{\epsilon}$.

It remains to construct the spinor connections. According to § 4 of Chapter 1, the Levi–Civita connection on $\mathcal{T}M$ induces a connection on the vector bundle $\mathbf{P}_M(\mathcal{T}M)$ and a connection on the sheaf $\mathcal{O}(1)$ along it. Since the section $\pi^*(g)$ of the sheaf $\mathcal{O}(2)$ is horizontal, this distribution is tangent to the zeros of $\pi^*(g)$, i.e., to F. The connection on F induces connections on F_l and F_r because $\mathcal{T}F/M = \mathcal{T}F_l/M \oplus \mathcal{T}F_r/M$. Since F_l and F_r are relative projective lines, we already know that connections on the vector bundles can be extended to spinor connections ∇_l and ∇_r. To ensure that $\nabla_l \otimes 1 + 1 \otimes \nabla_r$ coincides with ∇, it suffices to choose the extensions in such a way that $\nabla_l\,\epsilon = 0$ and $\nabla_r\,\tilde{\epsilon} = 0$. □

7. Complexification of real-analytic manifolds, and real structures. In § 3 of Chapter 1 we described the standard real structures on a big cell of the grassmannian $G(2; T^4)$, i.e., on flat complex space-time. Before extending this description to the curved case, we shall make some definitions and discuss them in enough generality for our later use in supergeometry.

(a) Complexification. Let M_0 be an m-dimensional real-analytic manifold, covered by coordinate neighborhoods U_{0j} with coordinates $(x^a_{0j}), a = 1,\ldots,m$. By letting x^a_{0j} take any complex values with real part in U_{0j}, we complexify U_{0j}, obtaining the cylinder $U_{0j} + i\mathbf{R}^m$. We next consider the real-analytic transition functions $x^a_{0j} = x^a_{0j}(x^1_{0k},\ldots,x^m_{0k})$ for all pairs j, k with $U_{0j} \cap U_{0k} \neq \emptyset$. The power series for these functions converge in some neighborhood of $U_{0j} \cap U_{0k}$ in the complexification. Thus, if we assume that the covering (U_{0j}) is locally finite, we can choose complex regions $U_j \supset U_{0j}$ such that the transition functions exist and satisfy the usual conditions on $U_k \cap U_j$. Let $M \supset M_0$ be the complex-analytic manifold obtained by gluing together these regions. We call M the complexification of M_0. Because of the choices made in the construction, it is not uniquely defined; however, given two complexifications of M_0, there is a neighborhood of M_0 in each such that the identity isomorphism of M_0 is induced by a unique isomorphism of these neighborhoods. If one wants, one has a well-defined and functorial *germ* of the complexification.

We have a complex conjugation defined on the cylinders $U_{0j} + i\mathbf{R}^m$. It induces an antiholomorphic involution $\rho\colon M \longrightarrow M$ on the complexification which does not depend on the choices made in the construction. Given (M, ρ), one can recover M_0 as the set of fixed points of the involution. The sheaf $\rho^*(\mathcal{O}_M)$ consists of the germs of antiholomorphic functions, and $\overline{\rho^*(\mathcal{O}_M)} = \mathcal{O}_M$. The map $f \mapsto f^\rho(x) = \overline{f(\rho(x))}$ is an antiholomorphic involution on $\mathcal{O}_M$ which extends ρ. For real coordinates we have $(x^a_{0j})^\rho = x^a_{0j}$. To apply ρ to a power series in $x^a_{0j} - c^a_{0j}$, one takes the complex conjugates of the series coefficients and of the initial point c^a_{0j}. The functions $\bar{f}$ and f^ρ are the same on M_0, but in general the notation $\bar{f}$ is ambiguous, since it can be taken to mean complex conjugation of the values.

When a complex field ψ and its conjugate $\bar{\psi}$ occur in a Lagrangian in quantum field theory, the correct way to extend the Lagrangian holomorphically is to replace $\bar{\psi}$ by ψ^ρ.

(b) The real structure. By a real structure on a complex manifold M we mean an antiholomorphic involution $\rho: M \longrightarrow M$ and its extension $f \mapsto f^\rho$ to $\mathcal{O}_M$ with the property that $f^\rho(x) = \overline{f(\rho(x))}$. Not every pair (M, ρ) can be obtained by complexifying a real-analytic manifold; it might not have enough real points (points fixed by ρ), and in fact it might not have any at all (see below).

(c) The real and quaternion structures on a sheaf. Let $\mathcal{E}$ be a coherent sheaf on M, and let ρ be a real structure. An antilinear map $\mathcal{E} \longrightarrow \mathcal{E}: e \mapsto e^\rho$ extending ρ from M and $\mathcal{O}_M$ (in particular, $(fe)^\rho = f^\rho e^\rho$) can exist only if $\mathcal{E}$ and $\overline{\rho^*(\mathcal{E})}$ are isomorphic. Such a mapping with the property that $(e^\rho)^\rho = e$ is called a real structure on $\mathcal{E}$; if it has the property that $(e^\rho)^\rho = -e$, then it is called a quaternion structure. The latter terminology is explained by the fact that one can introduce a left action on $\mathcal{E}$ by the division algebra $\mathbf{C}[j]$ of quaternions by means of the formula $je = e^\rho$. The action of ρ carries over to the tensor algebra according to the rule $(e_1 \otimes e_2)^\rho = e_1^\rho \otimes e_2^\rho$.

(d) Complexification of a metric. Let g_0 be a real-analytic pseudo-riemannian metric on M_0. Its coefficients in the coordinates (x^a_{0j}) converge in some neighborhood of M_0. Hence, if one takes a sufficiently small complexification of M_0, it is equipped with a holomorphic metric g which extends g_0. Obviously, $g^\rho = g$. Conversely, any ρ-invariant holomorphic metric on a complexification (M, ρ) induces a real pseudo-riemannian metric on M_0.

As an example which will be useful to us later, we now classify the real structures on $\mathbf{CP}^1$ and $\mathbf{CP}^1 \times \mathbf{CP}^1$. Let $S = \mathbf{C}^2$. We consider the following two antiholomorphic maps from S to S: $\rho(z_1, z_2) = (\bar{z}_1, \bar{z}_2)$ and $j(z_1, z_2) = (-\bar{z}_2, \bar{z}_1)$ (see § 3 of Chapter 1). They both induce real structures on $\mathbf{CP}^1 = \mathbf{P}(S)$. If we use the real structure coming from ρ, then $\mathbf{CP}^1$ is a complexification of $\mathbf{RP}^1$; in that case $\mathcal{O}(1)$ is also equipped with a real structure.

If we use the second antiholomorphic map $\jmath$, then $\mathbf{CP}^1$ does not have any fixed points; and $\mathcal{O}(1)$ is equipped with a quaternion structure.

8. Lemma. (a) *Any real structure on* $\mathbf{CP}^1$ *is isomorphic to one of these two.*

(b) *Any real structure on* $\mathbf{CP}^1 \times \mathbf{CP}^1$ *either is isomorphic to the direct product of real structures on the factors, or else has the form* $\rho(x,y) = (\tau^{-1}(y), \tau^{-1}(x))$, *where* $\tau\colon \mathbf{CP}^1 \longrightarrow \mathbf{CP}^1$ *is an antiholomorphic isomorphism. In this case,* $\mathbf{CP}^1 \times \mathbf{CP}^1$ *is a complexification of the graph of* τ.

Proof. (a) Let ρ be an antiholomorphic involution on $\mathbf{CP}^1$. We choose a point O which is not invariant under ρ, and we set $\rho(O) = \infty$. We then construct a function w which is meromorphic on $\mathbf{CP}^1$, has a simple pole at O, and has a simple pole at ∞. Then $w^\rho = aw^{-1}$. Since $w^{\rho^2} = w$, it follows that $\bar{a} = a$. Replacing w by bw changes a to $|b|^2 a$. Thus, we may suppose that $a = 1$ or -1. The two cases $a = 1$ and $a = -1$ give the two real structures described above.

(b) Let ρ be an antiholomorphic involution of $\mathbf{CP}^1 \times \mathbf{CP}^1$. When ρ acts on the Picard group, it either fixes the classes of the two standard generators $\mathcal{O}(1,0)$ and $\mathcal{O}(0,1)$, or else it interchanges them. In the first case, ρ splits into a direct product. In the second case, if we take a meromorphic function w_l on the first factor with a single zero and a single pole, lift it to $\mathbf{CP}^1 \times \mathbf{CP}^1$, and apply ρ, we obtain an analytic function w_ρ which is a lifting from the second factor. The mapping on points $\tau\colon w_l^0 \mapsto \overline{w}_r^0$ (where the zero superscript indicates the value of the function) has the required property. □

9. Application to space-time. Suppose that M_0 is a four-dimensional real-analytic manifold with (pseudo-) riemannian metric g_0; (M, ρ, g) is a complexification of this structure; and $F \xrightarrow{\pi} M$ is the space of null directions, as in § 6 of Chapter 1. Clearly, ρ induces an involution on F and also on the relative complex quadric $\pi^{-1}(M)$. Thus, over each point of M_0 we have the base of the cone of complex null directions with real structure whose type depends upon the signature of the metric g_0. There are only two cases which interest us.

(a) The Lorentz signature. This signature, which corresponds to part (b) of Lemma 8, gives us the only case where the real heavens is a sphere. In all other cases it is either empty or a torus $\mathbf{RP}^1 \times \mathbf{RP}^1$.

(b) The Riemann signature. This signature gives us the direct product of two copies of $\mathbf{CP}^1$ with no real points.

We shall consider real structures on complex space-time which extend to the spinor bundles in such a way that at each real point we have one of these two pictures.

10. The involution in coordinates: the Lorentz case. The coordinates x^a are real, and ρ gives antilinear isomorphisms between S and $\tilde{S}$; from this we obtain a real structure on the sheaf $S \oplus \tilde{S}$. The invariant sections are called Mayorana spinors. We choose bases for the sections in such a way that $(w^\alpha)^\rho = z^{\dot\alpha}, (z^{\dot\alpha})^\rho = w^\alpha$. The fact that we have a real structure means that

$$e^a_{\alpha\dot\beta} w^\alpha \otimes z^{\dot\beta} = \left(e^a_{\alpha\dot\beta} w^\alpha \otimes z^{\dot\beta}\right)^\rho = (e^a_{\alpha\dot\beta})^\rho w^\beta \otimes z^{\dot\alpha},$$

i.e., $(e^a)^\rho = (e^a)^t$, where t denotes the transpose. Thus, the e^a are hermitian matrices at the real points.

11. The involution in coordinates: the Riemann case. The coordinates x^a are real; we have $(w^\alpha)^\rho = \epsilon_{\alpha\beta} w^\beta$, and similarly for z^α. The fact that e is real means that

$$(e^a_{\alpha\dot\beta})^\rho = \epsilon_{\alpha\gamma}\epsilon_{\dot\beta\dot\delta} e^a_{\gamma\dot\delta},$$

as in § 3 of Chapter 1.

12. Compatibility with the other structures. We require that $\epsilon^\rho = \tilde\epsilon$ in the Minkowski case, and that $\epsilon^\rho = \epsilon$ and $\tilde\epsilon^\rho = \tilde\epsilon$ in the Euclidean case. Furthermore, the spinor connections must be compatible with ρ. In the Minkowski case, the condition $\nabla(w^\alpha)^\rho = (\nabla w^\alpha)^\rho$ can be rewritten in the form

$$(\omega^\alpha_{\beta a})^\rho = \omega^{\dot\alpha}_{\dot\beta a}, \qquad \left(\omega^{\dot\alpha}_{\dot\beta a}\right)^\rho = \omega^\alpha_{\beta a}.$$

In particular, on the "real section" M_0 the addition or removal of an upper dot in the spinor indices of the connection coefficients is equivalent to complex conjugation. It is clear from the formulas in § 4 of Chapter 1 that $\tilde F = F^\rho$. Thus, the Weyl electromagnetic field must be purely imaginary on the real points in order for there to be a symmetric Riemann connection. This is compatible with the field's quantum role in causing the rotation of the phase factor in the U(1)-bundle of electric charge.

In the case of a Riemann metric, the reality conditions take the form

$$(\omega^\alpha_{\beta a})^\rho = \epsilon^{\gamma\beta}\epsilon_{\alpha\delta}\omega^\delta_{\gamma a}, \qquad \left(\omega^{\dot\alpha}_{\dot\beta a}\right)^\rho = \epsilon^{\dot\gamma\dot\beta}\epsilon_{\dot\alpha\dot\delta}\omega^{\dot\delta}_{\dot\gamma a}.$$

13. Self-dual and anti-self-dual 2-forms. From the decomposition

$$\Omega^2 M = S^2 S \otimes \wedge^2 \tilde S \,\oplus\, \wedge^2 S \otimes S^2 \tilde S,$$

it is clear that $(\Omega^2_\pm M)^\rho = \Omega^2_\mp M$ in the case of the Minkowski signature. Hence, a real (ρ-invariant) connection on a sheaf with real structure cannot have a self-dual (or anti-self-dual) curvature form. In particular, a GS-manifold itself cannot both be semiflat and also have a compatible real structure with Minkowski signature.

On the other hand, we have $(\Omega^2_\pm M)^\rho = \Omega^2_\pm M$ for the Riemann signature. More precisely, the quaternion structure on S and $\tilde{S}$ induces a real structure on $\Omega^2_\pm M$. Semiflat GS-modules and self-dual Yang–Mills sheaves on them with riemannian real structure exist and are of considerable interest.

14. Fields, Lagrangians, and dynamic equations. The classical fields on the space-time M are either sections of vector bundles on M or else connections on such vector bundles. Given a set of fields $\{\Psi\}$ in some fixed physical model, one constructs the Lagrangian — the volume form on M - as a function of the components of Ψ and their derivatives. The Euler–Lagrange equations are the dynamic equations for the fields. In a geometrical treatment of field theory these equations can be defined invariantly; however, to actually write them down one must choose "functional coordinates" Ψ for the geometrical picture in order that the variations $\delta\Psi$ make sense. If there are several natural choices for these coordinates, then one obtains representations of the theory in different formalisms. For example, instead of the usual way of giving a gravitational field by means of the metric g_{ab}, one can define it by a four-tuple $(e^m_a dx^a)$ of orthonormal 1-forms, or by a four-tuple and a connection, etc. The solutions of the dynamic equations, i.e., the realizations of the stationary points of the action functional, are all the same geometrically. However, the quantum fluctuations which occur against the background of these solutions can be essentially different in different formalisms, and one must use physical considerations to choose between them. We now enumerate some of the basic fields and describe their contributions to the Lagrangian.

15. The gravitational field. Here we shall describe the gravitational field by means of a metric g on M. From the metric one constructs the standard volume form $v = \sqrt{|\det\ g|}\, d^4x$ in the usual notation (its square can be constructed purely algebraically; see § 7 of Chapter 4). All of the Lagrangians discussed below are of the form $L(\Psi)v$, where L is the Lagrangian density, i.e., a function of the fields and their derivatives whose values are functions on M (the physical terminology for this is a "scalar").

According to Hilbert–Einstein, the Lagrangian density for the gravitational field itself is the scalar curvature R (more precisely: κR, where κ is a constant; we shall usually ignore constants that come from a choice of units or a measurement of some constant of interaction, charge, etc.). Recall that, if the curvature of the Levi-Civita connection $\nabla^2 : \Omega^1 M \longrightarrow \Omega^1 M \otimes \Omega^2 M$ is $\nabla^2(dx^a) = R^a_{bcd}dx^b \otimes dx^c \otimes dx^d$, then the Ricci tensor is equal to $R_{ik} = R^a_{iak}$, and the scalar curvature is $R = R^i_i$. The dynamic equations for gravity in a vacuum (with no sources) have the form

$G = 0$, where $G = R_{ij} - \frac{1}{2}Rg_{ij}$ is the Einstein tensor; or equivalently, they have the form

$$\mathrm{Ric}^0 = 0, \qquad R = 0,$$

where Ric^0 is the traceless part of the Ricci tensor, i.e., $\mathrm{Ric}^0_{ik} = R_{ik} - \frac{1}{4}Rg_{ik}$.

In the language of spinor decomposition, the tensors R_{ik} and R can be naturally interpreted as the curvature components associated with the decomposition $\nabla = \nabla_l \otimes 1 + 1 \otimes \nabla_r$. In particular, the curvature Φ_l of the connection ∇_l has left and right components $\Phi_l = \Phi_{ll} + \Phi_{lr}$; and similarly, $\Phi_r = \Phi_{rl} + \Phi_{rr}$. The equation $\mathrm{Ric}^0 = 0$ is equivalent to either of the two self-duality conditions $\Phi_{lr} = 0, \Phi_{rl} = 0$ for the spinor connections. This means that, if the connection ∇_l (or ∇_r) is flat and if also the scalar curvature R vanishes, then M satisfies Einstein's vacuum equations. (In this discussion we are supposing that ∇_l and ∇_r together induce the Levi-Civita connection.)

If our set of basic fields includes other fields Ψ besides the gravitational field, then the dynamic equations for the gravitational field take the form $G = T(\Psi)$, where $T = T_{ik}$ is called the energy-momentum tensor for the fields Ψ. The energy-momentum tensor acts like a "gravity source;" however, this tensor may be non-trivial even when there are no sources.

16. The electromagnetic field. We shall regard the electromagnetic field as a connection ∇ on a one-dimensional vector bundle $\mathcal{E}$ over the space-time M. A local choice of basis for $\mathcal{E}$ enables one to give a covariant differential $\nabla\colon \mathcal{E} \longrightarrow \mathcal{E} \otimes \Omega^1$ by means of the potential 1-form $A_m dx^m$; changing the basis changes the form to a gauge-equivalent form. Since $\mathcal{E}nd\,\mathcal{E}$ is canonically identified with $\mathcal{O}_M$, the connection's curvature in this case is simply the tension 2-form $\Phi(\nabla) = \Phi_{ab}dx^a \otimes dx^b, \Phi_{ab} = \partial A_b/\partial x^a - \partial A_a/\partial x^b$. The contribution of the electromagnetic field to the Lagrangian density is $(\Phi(\nabla), \Phi(\nabla))$ (the scalar product on $\Omega^2 M$ is the one induced by the metric). One usually rewrites the Lagrangian in terms of the Hodge "star" operation $*\colon \Omega^i M \longrightarrow \Omega^{4-i} M$, which is defined by the formula $(\Phi_1, \Phi_2)v = \Phi_1 \wedge *\Phi_2$. Both ways of writing the Lagrangian somewhat obscure the fact that $(\Phi(\nabla), \Phi(\nabla))$ depends on the metric, and hence, strictly speaking, describes the electromagnetic field not by itself but rather interacting with gravity.

Maxwell's equations "in a vacuum" have the form

$$\begin{cases} d\Phi = 0 & \text{(the Bianchi identity)}, \\ d*\Phi = 0. \end{cases}$$

As usual, if we decompose $\Phi = \Phi_l + \Phi_r$ into its self-dual and anti-self-dual parts, we obtain $\Phi_{lr} = 0$, from which the Maxwell equations follow. In the presence of charged fields Ψ, the second Maxwell equation takes the form $d*\Phi = J$, where $J = J(\Psi)$ is the axial current 3-form.

17. The Yang–Mills field. Speaking formally, the difference between a Yang–Mills field and a Maxwell field is that the former is a connection ∇ on a vector bundle $\mathcal{E}$ of rank greater than one. The curvature form $\Phi(\nabla)$ is a section of $\mathcal{E}nd\,\mathcal{E} \otimes \Omega^2 M$. The contribution to the Lagrangian density has the form $(\Phi(\nabla), \Phi(\nabla))$ as before, except that now the scalar product involves the canonical scalar product $\mathrm{tr}(ab)$ on $\mathcal{E}nd\,\mathcal{E}$. The equations have the form $\tilde{\nabla}\Phi = 0$ (the Bianchi identity) and $\tilde{\nabla}(*\Phi) = 0$, where $\tilde{\nabla}$ is the extension of ∇ to $\mathcal{E}nd\,\mathcal{E} \otimes \Omega^{\cdot} M$. In the presence of sources, the second equation takes the form $\tilde{\nabla}(*\Phi) = J(\Psi)$. Unlike the case of Maxwell fields, there are no known classical Yang–Mills fields.

18. Matter fields of spin zero. A matter field of spin zero is a section of a vector bundle $\mathcal{E}$ on which a connection ∇ is defined. The contribution of the matter field to the Lagrangian density may include two types of terms: a potential $V(\Psi)$, often a polynomial in the coordinates of Ψ, and a kinetic term $(\nabla\Psi, \nabla\Psi)$. The kinetic term describes the interaction of Ψ with both the connection field and the gravitational field.

19. Matter fields of spin $\frac{1}{2}$. A matter field of spin $\frac{1}{2}$ is a section of the vector bundle $\mathcal{E} \otimes (S \oplus \tilde{S})$. The mass term in the Lagrangian is proportional to (Ψ^{ρ}, Ψ), where the scalar product is constructed from the metric on $\mathcal{E}$ and the spinor metrics. The kinetic term is proportional to $\mathrm{Re}(\Psi^{\rho}, D\Psi)$, where D is the Dirac operator, i.e., the composition of the covariant differential $\mathcal{E} \otimes (S \oplus \tilde{S}) \longrightarrow \mathcal{E} \otimes (S \oplus \tilde{S}) \otimes \Omega^1 M$ and the convolution $(S \oplus \tilde{S}) \otimes \Omega^1 M \longrightarrow \tilde{S} \oplus S$.

Here we shall not describe the Lagrangians for matter fields with higher spin. However, we note that the Rarita–Schwinger Lagrangian for a matter field of spin 3/2 plays an essential role in models of supergravity (see § 7 of Chapter 5).

§ 2. The Self-Duality Diagram and the Radon–Penrose Transform

1. The Radon–Penrose Transform. We consider the diagram

$$Z \xleftarrow{\pi_1} F \xrightarrow{\pi_2} M,$$

where π_1 and π_2 are submersive morphisms of manifolds. The general meaning of the Radon–Penrose transform is a transfer of various geometric objects from the base Z to the base M, or *vice-versa*, by first lifting to F and then descending.

Here is the simplest example. Let ω be a form on Z of dimension equal to the dimension of the fibres of π_2, and suppose that the fibres of π_2 are compact. Then the lifting of ω to F is the usual preimage $\pi_1^*(\omega)$, and when we descend to M we obtain a function on M which comes from integrating $\pi_1^*(\omega)$ over the fibres of π_2. One version of the classical Radon transform can clearly be described in this way, with Z a projective space, M the grassmannian of k-dimensional subspaces in it, and F the graph of the incidence relation.

In general, the manifolds Z, F, and M form a double fibration, and the geometrical objects to be moved in the direction of the arrows or the opposite direction can belong to various geometrical categories. We are working with complex-analytic objects, and for us one of the most important constructions is to transfer a locally free sheaf from Z to M. We now describe a typical situation.

Suppose that the fibres of π_2 are compact and connected. We shall say that a locally free sheaf $\mathcal{E}_Z$ on Z is "M-trivial" if the restriction of $\pi_1^*(\mathcal{E}_Z)$ to any of the fibres of π_2 is free. We set $\mathcal{N} = \mathrm{Ker}(\mathrm{res}\colon \pi_2^*\Omega^1 M \longrightarrow \Omega^1 F/Z)$, where res denotes restriction to the vector fields tangent to the fibres of π_1. We shall assume that $\pi_{2*}\mathcal{N} = 0$ and $R^1\pi_{2*}\mathcal{N} = 0$.

2. Proposition. *Under these conditions one can define a functor*

$$(M\text{-trivial sheaves on } Z) \longrightarrow (\text{sheaves with connection on } M),$$

$$\mathcal{E}_Z \mapsto (\mathcal{E}, \nabla),$$

where $\mathcal{E} = \pi_{2*}\pi_1^*(\mathcal{E}_Z)$, $\nabla = \pi_{2*}(\nabla_{F/Z})$, *and* $\nabla_{F/Z}\colon \pi_1^*(\mathcal{E}) \longrightarrow \pi_1^*(\mathcal{E}) \otimes \Omega^1 F/Z$ *is the relative connection which annihilates the subsheaf* $\pi_1^{-1}(\mathcal{E}) \subset \pi_1^*(\mathcal{E})$.

Proof. Since $\mathcal{E}_Z$ is M-trivial and the fibres of π_2 are compact and connected, it follows that we have a canonical isomorphism $\pi_2^*\mathcal{E} = \pi_1^*\mathcal{E}_Z$. We let $\mathcal{E}_F$ denote this sheaf. The equalities $\pi_{2*}\mathcal{N} = 0 = R^1\pi_{2*}\mathcal{N}$ imply that the map $\pi_{2*}(\mathrm{res})\colon \Omega^1 M \longrightarrow \pi_{2*}\Omega^1 F/Z$ is an isomorphism. Thus, $\pi_{2*}(\mathcal{E}_F \otimes \Omega^1 F/Z)$ can be canonically identified with $\mathcal{E} \otimes \Omega^1 M$, and when we descend the diagram $\nabla_{F/Z}\colon \mathcal{E}_F \longrightarrow \mathcal{E}_F \otimes \Omega^1 F/Z$ we obtain a first order differential operator $\nabla\colon \mathcal{E} \longrightarrow \mathcal{E} \otimes \Omega^1 M$. From the definitions it immediately follows that this operator is a connection. All of these constructions are clearly functorial. □

If we attempt to reverse this construction, we run up against an obstacle. Let $(\mathcal{E}, \nabla)$ be a sheaf with connection on M. We set $\mathcal{E}_F = \pi_2^*(\mathcal{E})$, and we let $\nabla_{F/Z}$ denote the composition

$$\nabla_{F/Z}\colon \mathcal{E}_F \xrightarrow{\pi_2^*(\nabla)} \mathcal{E}_F \otimes \pi_2^*\Omega^1 M \xrightarrow{\mathrm{id}\otimes\mathrm{res}} \mathcal{E}_F \otimes \Omega^1 F/Z.$$

Let $\mathcal{E}_F' = \mathrm{Ker}\, \nabla_{F/Z}$. This is the sheaf of local coefficient systems along the fibres of π_1. In principle, the connection ∇ could have nontrivial curvature or monodromy along the fibres of π_1. However, if the fibres of π_1 are connected, and if the curvature and monodromy of ∇ along these fibres are trivial, then the sheaf $\mathcal{E}_Z = \pi_{2*}(\mathcal{E}'_F)$ is locally free on Z with the same rank as $\mathcal{E}$. Combining these two constructions, we obtain the following general result.

3. Theorem. *If the fibres of π_1 are connected, the fibres of π_2 are compact and connected, and $\pi_{2*}\mathcal{N} = 0 = R^1\pi_{2*}\mathcal{N}$, then the direct and inverse Radon-Penrose transforms give quasi-inverse equivalences of the following categories:*

(a) *the category of M-trivial locally free sheaves on Z;*

(b) *the category of pairs $(\mathcal{E}, \nabla)$, where $\mathcal{E}$ is a locally free sheaf on M and ∇ is a connection on $\mathcal{E}$ with trivial curvature and monodromy along the fibres of π_1.* □

The formal proof that the correspondence between $\mathcal{E}_Z$ and $(\mathcal{E}_F, \nabla_{F/Z})$ is an equivalence of categories relies upon a theorem of Deligne ([22], p. 14, Theorem 2.23), to which we refer the reader for details.

If the sheaf $\mathcal{E}_Z$ is not M-trivial, then we do not obtain a connection by descending the diagram $\nabla_{F/Z}: \pi_1^*\mathcal{E} \longrightarrow \pi_1^*\mathcal{E}_Z \otimes \Omega^1 F/Z$; nor, in general, can we even recover the sheaf $\pi_1^*\mathcal{E}$ from $\pi_{2*}\pi_1^*\mathcal{E}$. The correct general point of view on this situation is as follows. The relative de Rham complex

$$\pi_1^*\mathcal{E}_Z \longrightarrow \pi_1^*\mathcal{E}_Z \otimes \Omega^1 F/Z \longrightarrow \pi_1^*\mathcal{E}_Z \otimes \Omega^2 F/Z \longrightarrow \cdots$$

is a resolution of $\pi_1^{-1}(\mathcal{E}_Z)$, i.e., it is quasi-isomorphic to $\pi_1^{-1}(\mathcal{E}_Z)$. Hence, the natural direct image of $\pi_1^{-1}(\mathcal{E}_Z)$ lies in the derived category of sheaves of C-spaces on M, and is isomorphic to $R\pi_{2*}(\pi_1^*\mathcal{E}_Z \otimes \Omega^\cdot F/Z)$. The fundamental computable invariants of this direct image are the relative hypercohomology sheaves $R^i\pi_{2*}(\pi_1^*\mathcal{E}_Z \otimes \Omega^\cdot F/Z)$, and also the terms and differentials in the spectral sequences which converge to these sheaves. In the concrete situations we shall later encounter, these differentials turn out to be the operators which arise naturally in the dynamical equations of field theory on the space-time M: the equations of Klein–Gordon, Dirac, and so on.

We shall now prove a technical result which is useful for cohomology computations.

4. Proposition. *Let $\pi: F \longrightarrow G$ be a submersion of complex manifolds, and let $\mathcal{E}$ be a locally free sheaf on G. Suppose that the fibres of π are connected and $H^i(\pi^{-1}(x), \mathbb{C}) = 0$ for $1 \le i \le p$. Then the natural map $H^i(G, \mathcal{E}) \longrightarrow H^i(F, \pi^{-1}(\mathcal{E}))$ is an isomorphism for $0 \le i \le p$ and a monomorphism for $i = p+1$.*

5. Lemma. *Under the conditions of the proposition, let $\mathcal{O}_{\infty,G}$ be the sheaf of smooth complex functions on G, and let $\mathcal{F}_\infty$ be a locally free sheaf of $\mathcal{O}_{\infty,G}$-modules. Then $H^i(\mathcal{F}, \pi^{-1}(\mathcal{F}_\infty)) = 0$ for $1 \le i \le p$.*

Proof of the proposition from the lemma. We consider the Dolbeault resolution $0 \longrightarrow \mathcal{E} \longrightarrow \Omega^{0,\cdot}_{\infty}G \otimes \mathcal{E}_{\infty}$ on G, where $\mathcal{E}_{\infty} = \mathcal{E} \otimes \mathcal{O}_{\infty,G}$, and we lift this resolution to F. It is well known that the spectral sequence $E_2^{ij} = H^i(H^j(F, \pi^{-1}(\Omega^{0,\cdot}_{\infty} \otimes \mathcal{E}_{\infty})))$ converges to $H^{\cdot}(F, \pi^{-1}(\mathcal{E}))$. From Lemma 5 it follows that $H^j(F, \pi^{-1}(\Omega^{0,\cdot}_{\infty} \otimes \mathcal{E}_{\infty}))$ vanishes for $1 \leq j \leq p$. Consequently, $H^i(F, \pi^{-1}(\mathcal{E})) = H^i(\Gamma(F, \pi^{-1}(\Omega^{0,\cdot}_{\infty}G \otimes \mathcal{E}_{\infty}))) = H^i(\Gamma(G, \Omega^{0,\cdot}_{\infty}G \otimes \mathcal{E}_{\infty}))$ for $0 \leq i \leq p$ (the second equality follows because the fibres of π are connected). But this last cohomology group coincides with $H^i(G, \mathcal{E}_{\infty})$. It is also clear that for $i = p + 1$ one can once again see from the spectral sequence that $H^{p+1}(G, \mathcal{E}) \longrightarrow H^{p+1}(F, \pi^{-1}(\mathcal{E}))$ is a monomorphism.

Proof of the lemma. We shall show that if $H^{i-1}(\pi^{-1}(x), \mathbb{C}) = H^i(\pi^{-1}(x), \mathbb{C}) = 0$ then $H^i(F, \pi^{-1}(\mathcal{F})) = 0$ (when $i = 1$ the H^0-condition should be replaced by connectedness of the fibre).

The key step is to show existence of an open covering $F = \bigcup_{a=1}^{\infty} F_a$ such that $F_a \Subset F_{a+1}$ and the maps $r^j_{a+1,a}: H^j(F_{a+1}, \pi^{-1}(\mathcal{F})) \longrightarrow H^j(F_a, \pi^{-1}(\mathcal{F}))$ are zero for $j = i - 1$, i (when $j = 0$ we instead require that F_a be connected). Suppose that such a covering has been constructed. Let ω be a cohomology class in $H^i(F, \pi^{-1}(\mathcal{F}_{\infty}))$. This class can obviously be represented by a $(0, i)$-form in the relative Dolbeault complex $\Omega^{0,\cdot}_{\infty}F \otimes \pi^*_{\infty}(\mathcal{F}_{\infty})$. We also use ω to denote this form. We would like to solve the equation $\omega = d_{F/G}\, \nu$ in the Dolbeault complex. To do this, we construct solutions to the equations $\omega|_{F_a} = d_{F/G}\, \nu_a$, which exist because $r^i_{a+1,a} = 0$. We then change ν_a by a coboundary, using truncated functions, in order to ensure the existence of the limit $\nu = \lim \nu_a$, $\omega = d_{F/G}\, \nu$. This can be done because $r^{i-1}_{a+1,a} = 0$.

We now construct the covering (F_a). We fix a point $x \in G$ and a compact set $K \Subset F$. The vanishing of $H^i(\pi^{-1}(x), \mathbb{C})$ implies that the fibre $\pi^{-1}(x)$ contains open subsets U and V such that $\pi^{-1}(x) \cap K \Subset U \Subset V \subset \pi^{-1}(x)$ and the restriction map $H^i(V, \mathbb{C}) \longrightarrow H^i(U, \mathbb{C})$ is zero. To construct these subsets one takes a locally finite open covering of $\pi^{-1}(x)$ by relatively compact open sets of which any finite intersection is contractible. Then U and V can be taken to be finite unions of the elements of this covering.

There exist neighborhoods $W \ni x$ and $\tilde{U} \supset U$ in F such that $\tilde{U}$ is diffeomorphic to $V \times W$ (with $\pi|_{\tilde{U}}$ the projection onto W). We cover $V \times W$ with open sets of the form $S \times W$, where S is the covering constructed earlier when we found V. The cohomology $H^{\cdot}(\tilde{U}, \pi^{-1}(\mathcal{F}))$ can be computed relative to this covering, using the Leray spectral sequence. In this manner one easily finds that $H^i(V \times$

$W, \pi^{-1}(\mathcal{F})) \longrightarrow H^i(U \times W, \pi^{-1}(\mathcal{F}))$ is the zero mapping, because this is true over every point $x \in W$. As a result we conclude that there exist open sets U' and V' in F with the properties: $K \Subset U' \Subset F$, and the map $H^i(V'; \pi^{-1}(\mathcal{F})) \longrightarrow H^i(U'; \pi^{-1}(\mathcal{F}))$ is zero. Hence, a full covering (F_a) can be constructed by induction on a. □

6. The self-duality diagram. Until now we have assumed that we have already been given the double fibration $Z \xleftarrow{\pi_1} F \xrightarrow{\pi_2} M$. However, it is also possible, under certain assumptions, to construct the double fibration from Z, M, and some additional structure on these manifolds. In particular, in his construction of "nonlinear gravitons," Penrose had the idea of starting with a self-dual four-dimensional GS-manifold M and constructing its "twistor space" Z, which parametrizes the geodesic surfaces in M; and he discovered how to recover M given Z. We shall define an appropriate class of diagrams axiomatically, and shall show how to go from one base space to the other in various situations. It is natural to consider the process of going from M to Z and back as a kind of nonlinear Radon–Penrose transform (as distinguished from the above "fibre-by-fibre integration" of fibrations or differential forms, which is essentially a linear operation).

By a *self-duality diagram* we mean a double fibration $Z \xleftarrow{\pi_1} F \xrightarrow{\pi_2} M$ with the following properties:

(a) The dimensions of Z, F, and M are 3, 5, and 4, respectively.

(b) F/M is a relative projective line: $F = \mathsf{P}_M(S^*)$, where S^* is a locally free rank two sheaf on M.

(c) The fibres of π_1 are transversal to the fibres of π_2. More precisely, the map $d\pi_2 \colon \mathcal{T}F/Z \longrightarrow \pi_2^*\mathcal{T}M$ is a local direct sum imbedding, and the corresponding morphism $i \colon F \longrightarrow G_M(2;\, \mathcal{T}M)$ is a closed imbedding.

By a real structure on a self-duality diagram we mean a set of real structures on M, F, and Z relative to which π_1 and π_2 are real, M has a four-dimensional submanifold of real points, and S^* has a quaternionic structure which induces the given real structure on F.

We shall describe several geometrical situations which lead in a natural way to self-duality diagrams.

7. Self-dual GS-manifolds. The grassmannian spinors that we studied in § 1 of Chapter 1 will give us a self-duality diagram under the following assumptions: dim $M = 4$, the spinor fibrations on M have rank two, and the integrable conic connection h on the standard 2-conic structure $F \xrightarrow{\pi_2} M$ integrates to give a fibration $F \xrightarrow{\pi_1} Z$. Then the double fibration $Z \xleftarrow{\pi_1} F \xrightarrow{\pi_2} M$ is a self-duality diagram.

We verify the conditions for Theorem 3 to be applicable to this diagram. The fibres of π_2 are compact and connected. Connectedness of the fibres of π_1 is a separate condition. In the case of the flat self-duality diagram

$$\mathsf{P}(T) \xleftarrow{\pi_1} F(1,2;\, T) \xrightarrow{\pi_2} G(2;\, T), \qquad T = \mathbf{C}^4,$$

the fibres of π_1 become complex projective planes under the Plücker imbedding $G(2;\,T) \subset \mathrm{P}(\wedge^2 T)$. Hence they are connected, as are their intersections with convex neighborhoods of points $x \in G(2;T)$ in $\mathrm{P}(\wedge^2 T)$. Thus, the fibres π_1 are also connected for curved GS-manifolds which locally are a rather small deformation of flat GS-manifolds.

Finally, by the computations in Chapter 1 § 7, there is an isomorphism of the sheaf $\mathcal{N} = \mathrm{Ker}(\pi_2^*\Omega^1 M \xrightarrow{\mathrm{res}} \Omega^1 F/Z)$ with the kernel of $\pi_2^*(S) \otimes \pi_2^*(\tilde{S}) \xrightarrow{r\otimes \mathrm{id}} \mathcal{O}_F(1) \otimes \pi_2^*(\tilde{S})$, where $r\colon \pi_2^*(S) \longrightarrow \mathcal{O}_F(1)$ is the standard morphism. Therefore, $\mathcal{N} = \Omega^1 F/M(1) \otimes \pi_2^*(\tilde{S})$, and we obviously have $\pi_{2*}\mathcal{N} = 0 = R^1\pi_{2*}\mathcal{N}$. Thus, before applying Theorem 3 it is enough to check that the fibres of π_1 are connected.

8. Deformations of a standard imbedding of a projective line. If Z is the first base in a self-duality diagram, then Z contains a four-dimensional family of imbedded projective lines, namely, the images of the fibres of π_2. We set $\mathrm{P}(x) = \pi_1\pi_2^{-1}(x)$ for $x \in M$. The normal sheaf to the imbedding $\mathrm{P}^1 \subset \mathrm{P}^3$ in a flat diagram is isomorphic to $\mathcal{O}(1) \oplus \mathcal{O}(1)$. In the diagram induced by a semi-flat GS-manifold, this normal sheaf is isomorphic to $\mathcal{N}^*|_{\pi_2^{-1}(x)}$, i.e., according to the above calculation, it is again isomorphic to $\mathcal{O}(1) \oplus \mathcal{O}(1)$.

An imbedding of a projective line in a three-dimensional complex manifold will be called a "standard imbedding" if its normal sheaf $\mathcal{N}_0^*$ is $\mathcal{O}(1)\oplus\mathcal{O}(1)$. Let Z be a manifold (not necessarily compact) which has a standard projective line imbedding $\mathrm{P}_0^1 \subset Z$. Using Kodaira-Spencer deformation theory, one has the following method of Penrose for constructing a self-duality diagram from (Z, P_0^1).

(a) Include P_0^1 in the four-dimensional analytic manifold M of its standard imbedding deformations. We now explain this in more detail.

According to a general theorem of Douady, P_0^1 is contained in a universal family of deformations inside Z, whose base is a complex space, generally with singularities, nilpotents, etc. According to Kodaira-Spencer theory, this base has an open neighborhood of the point x_0 corresponding to P_0^1 where it is actually a four-dimensional manifold, since dim $H^0(\mathrm{P}^1;\,\mathcal{N}_0^*) = 4$ (this is the space of infinitesimal deformations), and dim $H^1(\mathrm{P}^1;\,\mathcal{N}_0^*) = 0$ (there are no obstructions to extending the deformations). When applied to the family of complex structures on P_0^1, this theory also tells us that an open set in the family consists of projective lines (Riemann spheres), because this structure cannot be deformed: $H^1(\mathrm{P}^1;\,\mathcal{T}\mathrm{P}^1) = 0$. Finally, if we use this theory to examine the normal sheaf of the imbedding, we find that there is an open set in the family of deformations consisting of open imbeddings.

(b) Let $F \subset Z \times M$ be the graph of the universal family of standard imbedding deformations of P_0^1. Then the fibres of the projection $F \longrightarrow M$ are projective lines. Thus, every point of M has a neighborhood over which F has the form $\mathrm{P}(S^*)$, where

S^* is a suitable locally free sheaf of rank two. If we restrict ourselves to a sufficiently small neighborhood M' of a point $x_0 \in M$ and replace Z by $Z' = \bigcup_{x \in M'} \mathbf{P}(x)$, then we obtain a self-duality diagram $Z' \xleftarrow{\pi_1} F' \xrightarrow{\pi_2} M'$.

We note that there is a canonical self-dual GS-structure on M'. In fact, let $\mathcal{O}(1)$ be the sheaf on F corresponding to the realization $F = \mathbf{P}(S^*)$, and let $\mathcal{N}$ be the sheaf defined above, whose restrictions to the fibres of π_2 are conormal sheaves to the members of the family of deformations of $\mathbf{P}_0^1$. We set $\tilde{S} = \pi_{2*}(\Omega^1 F/Z(-1))$. From the exact sequence $0 \longrightarrow \mathcal{N}(-1) \longrightarrow \pi_2^*(\Omega^1 M)(-1) \longrightarrow \Omega^1 F/Z(-1) \longrightarrow 0$ it follows that $\tilde{S} = R^1\pi_{2*}\mathcal{N}(-1)$ is locally free of rank two, since fibre-by-fibre $\mathcal{N}(-1)$ is isomorphic to $\mathcal{O}(-2) \otimes \mathcal{O}(-2)$. If the morphism $\pi_2^*\Omega^1 M \longrightarrow \Omega^1 F/Z(-1) \otimes \mathcal{O}(1)$ is descended to M, we similarly obtain an isomorphism of GS-structures $\Omega^1 M \xrightarrow{\sim} S \otimes \tilde{S}$. The fibres of π_1 give an integrable conic connection on the corresponding 2-conic structure.

If we compare the constructions in this and the preceding subsections, we see that, at least locally on M, they are inverse to one another. We leave it to the reader to formulate this result in the language of equivalence of the categories of germs of spaces $(Z, \mathbf{P}_0^1)$ and (M, x_0).

Penrose and Ward have shown how examples of spaces $(Z, \mathbf{P}_0^1)$ with the required properties can be constructed explicitly. In their constructions, $\mathbf{P}_0^1$ is realizable as a section of a fibration $Z \longrightarrow \mathbf{P}^1$ which is given by transition functions for the standard covering of Z. They also investigated the question of choosing a metric in the conformal class determined by a given GS-structure which has zero scalar curvature.

9. Self-dual riemannian manifolds. The third, and perhaps the most interesting source of self-duality diagrams is the riemannian geometry of four-dimensional real manifolds. We shall state some results from a paper by Atiyah, Hitchin, and Singer in which they introduce this construction.

Let M_0 be a four-dimensional oriented differentiable manifold having a differentiable conformal structure with positive-definite signature.

We let $T_{\mathbf{C}} M_0$ denote the total space of the complexified tangent bundle of M_0. The fibres of $T_{\mathbf{C}} M_0$ contain the complex cones of the complex tangent null directions. (We note that there are no real null directions when the metric is positive-definite.)

Let $Q \longrightarrow M_0$ be a fibration by complex quadrics over M_0 which are bases of complex light cones; the fibration is imbedded in the projectivization of $T_{\mathbf{C}} M_0$ over M_0. The choice of orientation on M determines a decomposition $Q = F_+ \times_M F_-$, where $F_\pm \longrightarrow M_0$ are relative complex projective lines (still in the differentiable category). We set $Z = F_- \xrightarrow{\pi_0} M_0$.

Locally in M_0 we may assume that $F_\pm = \mathrm{P}(S_\pm)$, and the differentiable spinor sheaves $S_\pm$ are equipped with connections $\nabla_\pm$ with zero curvature on $\wedge^2 S_\pm$; the tensor product of these connections gives the Levi–Civita connection on the complexified tangent bundle. We suppose that M_0 is self-dual in the sense of riemannian geometry; this means that the curvature $\Phi(\nabla_-)$ is trivial. We use ∇_- to obtain a local splitting of the following sequence of differentiable tangent sheaves $\mathcal{T}_\infty$:

$$0 \longrightarrow \mathcal{T}_\infty F_-/M_0 \longrightarrow \mathcal{T}_\infty F_- \xrightarrow{d\pi_0} \pi_0^* \mathcal{T}_\infty M_0 \longrightarrow 0.$$

This local splitting enables one to introduce an almost complex structure on F_-. In fact, $\mathcal{T}_\infty F_-/M_0$ is already equipped with a complex structure; in addition, for every point $x \in F_-$, i.e., for the one-dimensional subspace $l(x) \subset S_-(x)$, $x \in M_0$, one can use the spinor decomposition $T_{\mathrm{C}} M_0(x) = S_+^*(x) \otimes S_-^*(x)$ to introduce a complex structure on $TM_0(x)$ by means of the identification

$$TM_0(x) \hookrightarrow T_{\mathrm{C}} M_0(x) = S_+^*(x) \otimes S_-^*(x) \longrightarrow S_+^* \otimes l(x)^\perp.$$

10. Theorem. (a) *The above almost complex structure on $F_- = Z$ is integrable if and only if M_0 is self-dual, i.e., if and only if $\Phi(\nabla_-) = 0$. If $\mathcal{O}_Z$ is the sheaf of holomorphic functions on Z (with respect to this almost complex structure), then the smooth functions on M_0 whose π_0-preimage belongs to $\mathcal{O}_Z$ determine a real-analytic structure on M_0.*

(b) *Any line $\pi_0^{-1}(x), x \in M_0$, is given by a standard imbedding in Z, and the self-duality diagram $Z \xleftarrow{\pi_1} F \xrightarrow{\pi_2} M$ which is constructed from such a line, contains the diagram $Z \xleftarrow{\mathrm{id}} Z = \pi_2^{-1}(M_0) \xrightarrow{\pi_0} M_0$ in the preceding subsection (in the obvious meaning of the word "contain"); in particular, M is a complexification of M_0.*

(c) *The diagram $Z \xleftarrow{\pi_1} F \xrightarrow{\pi_2} M$ is equipped with a real structure for which M_0 is the set of real points in M.* □

The proof is based on a direct computation of the obstruction to integrability of an almost complex structure, following Newlander-Nirenberg. It turns out that this obstruction is essentially $\Phi(\nabla_-)$.

In complete analogy with Theorem 10, the next theorem codifies in holomorphic terms the fibrations with self-dual connection on self-dual manifolds in the differentiable category.

Let $\mathcal{E}_0 \longrightarrow M_0$ be a locally free sheaf (in the differentiable category). Suppose we are given a connection $\nabla: \mathcal{E}_0 \longrightarrow \mathcal{E}_0 \otimes \Omega^1 M_0$ on $\mathcal{E}_0$ for which $\Phi_-(\nabla) = 0$.

11. Theorem. (a) Set

$$\mathcal{E} = \mathrm{Ker}(\pi_0^* \nabla)^{0,1} : \mathrm{C} \otimes \pi_0^*(\mathcal{E}_0) \longrightarrow \mathrm{C} \otimes \pi_0^*(\mathcal{E}_0) \otimes \Omega_\infty^{0,1} Z,$$

where $\Omega_\infty^{0,1}$ denotes the Dolbeault (0,1)-component of smooth differentiable forms on Z. Then $\mathcal{E}$ is an $\mathcal{O}_Z$-locally free sheaf of the same rank as $\mathcal{E}_0$.

(b) *The restriction of $\mathcal{E}$ to any fibre of π_0 is holomorphically trivial. The real structure on Z that was described above extends to a real structure on $\mathcal{E}$ with the property that the fibre of the total space E_0 of the sheaf $\mathcal{E}_0$ over a point $x \in M_0$ can be identified with the space of real holomorphic sections of $\mathcal{E}$ over $\pi_0^{-1}(x)$.*

(c) *Conversely, an analytic locally free sheaf $\mathcal{E}$ on Z with the property in (b) determines a differentiable locally free sheaf $\mathcal{E}_0$ on M_0 with a connection whose curvature form is self-dual.* □

The first part of this theorem can also be established using the Newlander-Nirenberg theorem. The most important step in going from Z to M_0 is to construct the connection. The idea is first to define the analytic sheaf $\mathcal{E}_M = \pi_{2*}\pi_1^*(\mathcal{E})$ on the complexification $M \supset M_0$. A (holomorphic) connection on this sheaf can be obtained by descending to M a relative connection on $\pi_1^*(\mathcal{E})$ along the fibres of π_1. (A similar argument will be carried out below in more detail.) It then remains to verify that the connection is real, and to restrict it to the real sections of $\mathcal{E}_M$ over M_0. The self-duality of the curvature form follows from § 7 of Chapter 1.

If the structure group of $\mathcal{E}_0$ is reduced, for example, if it reduces to $O(n)$, then, by interpreting this reduction as giving a horizontal positive-definite quadratic form on $\mathcal{E}_0$, we can define an analogous complex quadratic form on $\mathcal{E}_M$.

12. Twistor manifolds. Three-dimensional complex manifolds Z which contain a standard imbedding of a projective line are called (curved) twistor manifolds, by analogy with $\mathbf{P}(T)$. Their role in the theory of self-duality equations should be clear from the above discussion. We now give some facts about twistor manifolds that were proved by Hitchin.

(a) Hitchin's fundamental theorem says that, if a compact differentiable manifold M_0 has a Kählerian twistor space Z, then M_0 is conformally equivalent either to S^4 with the Euclidean metric, or else to $\mathbf{P}^2(\mathbf{C})$ with the Fubini-Study metric. In the first case $Z = \mathbf{P}(\mathbf{C}^4)$, and in the second case $Z = F(1, 2; \mathbf{C}^3)$.

(b) Thus, the twistor spaces of the simplest conformally flat manifolds $M_0 = S^3 \times S^1$ or $M_0 = (S^1)^4$ (the flat torus) are non-Kählerian. In both cases Z admits the structure of a fibration over $\mathbf{P}^1$; so that all of the standard imbeddings of lines are sections of this fibration. In the first case, the fibres are the non-Kählerian Hopf manifolds $S^3 \times S^1$. In the second case, the fibres are Kählerian.

(c) Let M_0 be the total space of the cotangent bundle on S^2. On M_0 we have a self-dual metric satisfying the Einstein equations which was constructed by Eguchi and Hanson. One obtains a noncompact twistor space Z for M_0 as follows. Take two copies of an affine quadratic cone in $\mathbf{C}^4$; resolve the singularities at the vertex in two ways by replacing the singularity by a line from one of the two families at the base of the cone, and then glue together these two copies along the complement of the preimage of the vertex.

A similar construction was proposed earlier by Hironaka, when he established the existence of non-projective algebraic varieties.

§ 3. The Theory of Instantons

1. Definition of instantons. Suppose that G is a compact Lie group, ρ is a unitary representation of G, and S^4 denotes the four-dimensional sphere with the standard conformal metric. Further let $\mathcal{E}_0$ be the sheaf of smooth sections of the vector bundle over S^4 associated with some principal G-bundle and the representation ρ. We consider a smooth connection ∇ on $\mathcal{E}_0$ with self-dual curvature form which is compatible with the group structure (i.e., parallel translations take structure frames to structure frames). We shall call any such pair $(\mathcal{E}_0, \nabla)$ an "instanton."

Our first goal is to give an algebraic description of all instantons in the case when G is a classical group in one of the series O, U, or Sp, and ρ is its simplest representation. More precisely, the parameters on which the instanton depends in our description will be "linear algebra data." This means either a set of matrices modulo some equivalence relation, or else a set consisting of several vector spaces and linear operators.

The linear algebra data arise as follows. Given a pair $(\mathcal{E}_0, \nabla)$, we construct the locally free sheaf $\mathcal{E}_Z$ on $\mathbf{P}^3 = Z$ as explained in Theorem 10 of § 2. If we use the additional structure on $\mathcal{E}_Z$ connected with the pair $(\mathcal{E}_0, \nabla)$ that gave rise to $\mathcal{E}_Z$, we can prove that $\mathcal{E}_Z$ is a monad cohomology sheaf, where the term "monad" means a three-term complex of the form $F_{-1} \otimes \mathcal{O}_{\mathbf{P}}(-1) \xrightarrow{\alpha} F_0 \otimes \mathcal{O}_{\mathbf{P}} \xrightarrow{\beta} F_1 \otimes \mathcal{O}_{\mathbf{P}}(1)$, in which the F_i are finite dimensional vector spaces. Giving such a monad is completely equivalent to giving vector space mappings $\Gamma(\alpha(1)): F_{-1} \longrightarrow F_0 \otimes \Gamma(\mathcal{O}(1))$ and $\Gamma(\beta): F_0 \longrightarrow F_1 \otimes \Gamma(\mathcal{O}(1))$.

We now describe this correspondence in more detail.

2. Instantons in the fundamental representation of G. First let $G = O(r)$, $r = \operatorname{rank} \mathcal{E}_0$. In the case of the fundamental representation ρ, we shall give a reduction of the structure group using a positive-definite symmetric form q on $\mathcal{E}_0$. Next, it is useful to reduce to the holomorphic problem of the structure of the analytic sheaf $\mathcal{E}_Z$, where instead of $(\mathcal{E}_0, q)$ we work with the complexified pair $(\mathbf{C} \otimes \mathcal{E}_0, q_{\mathbf{C}})$, on which we have a real structure which enables us to recover $\mathcal{E}_0$ and q. After all this, we shall take an $O(r)$-instanton to be a represented locally free (over $\mathcal{O}_{\infty,\mathbf{C}}$) sheaf $\mathcal{E}$ of rank r on S^4 with a real structure and a positive-definite real form.

We shall treat instantons for the groups $U(r)$ and $\mathrm{Sp}(r)$ as $O(2r)$– and $O(4r)$-instantons, respectively, equipped with some additional structure: either an orthogonal endomorphism J with $J^2 = -1$, or else two orthogonal endomorphisms J_1 and J_2 with $J_1^2 = J_2^2 = -1$ and $J_1J_2 = -J_2J_1$.

The self-dual connection ∇ is real, and the form q is horizontal with respect to this connection.

3. Representation of instantons on $\mathbf{P}^3$. Let T^* be a four-dimensional complex vector space with $\sigma(z_1, z_2, z_3, z_4) = (-\bar{z}_2, \bar{z}_1, -\bar{z}_4, \bar{z}_3)$ defining a quaternionic structure. The quaternionic structure induces a real structure on all of the flag spaces of T. In particular, S^4 is imbedded in $G(2; T)$ as the space of real points (see § 3 of Chapter 1). The self-duality diagram for S^4 is then the standard flat diagram

$$Z = \mathbf{P}(T) = \mathbf{P}^3 \xleftarrow{\pi_1} F(1, 2; T) \xrightarrow{\pi_2} G(2; T).$$

Since π_1 induces an isomorphism $\pi_2^{-1}(S^4) \xrightarrow{\sim} \mathbf{P}^3$, the real part of the diagram reduces to the projection $\pi\colon \mathbf{P}^3 \longrightarrow S^4$, whose fibres are the lines in $\mathbf{P}^3$.

According to the construction in Theorem 11 of § 2, all of the information about $(\mathcal{E}, \nabla)$ is encoded in the complex-analytic locally free sheaf $\mathcal{E}_Z = \mathrm{Ker}(\pi^*\nabla)^{0,1}$. Moreover, the $O(r)$-structure determines a complex symmetric metric and a real structure on $\mathcal{E}_Z$, while reduction to the groups U or Sp determines the corresponding holomorphic endomorphisms J. The sheaf $\mathcal{E}_Z$ is trivial along the real lines.

Locally free sheaves $\mathcal{E}_Z$ with this set of data will also be called instantons.

One recovers $(\mathcal{E}, \nabla)$ from $\mathcal{E}_Z$ using the inverse Radon–Penrose transform described in § 2.

4. Linear algebra data. By the term "orthogonal data" we mean a triple (F_{-1}, F^0, Q), where F^0 is a finite dimensional real vector space with a symmetric bilinear form Q, and F_{-1} is a complex subspace of $F^0 \otimes_{\mathbf{R}} T^*$, such that the following conditions hold:

(a) Let $\sigma\colon T^* \longrightarrow T^*$ be the quaternionic structure; then $(\mathrm{id}_{F_0} \otimes \sigma)F_{-1} = F_{-1}$.

(b) For any complex subspace $D \subset T^*$ we set $F_D = \sum_l (\mathrm{id}_{F^0} \otimes l)F_{-1} \subset F^0 \otimes_{\mathbf{R}} \mathbf{C} = F_0$, where l runs through the linear functions on T^* which vanish on D. Then for any hyperplane D the space F_D is isotropic relative to Q, and for any σ-invariant plane D we have: $\dim F_D = 2 \dim F_{-1}$.

(c) The form Q is positive-definite on all of the subspaces $F_D^{\perp} \cap F^0$, where D runs through the σ-invariant planes in T^*. Later we shall show that (c) implies what is formally a stronger condition that is easier to use, namely:

(c′) The form Q is positive-definite.

Unitary data (respectively, symplectic data) are orthogonal data along with an operator $J'\colon F_0 \longrightarrow F_0$ (respectively, two orthogonal operators J_1' and J_2') satisfying the conditions:

(d) $J'^2 = -1$ (respectively, $J_1'^2 = J_2'^2 = -1, J_1'J_2' = -J_2'J_1'$).

(e) The subspace F_{-1} is invariant relative to $J' \otimes \mathrm{id}_{T^*}$ (respectively, $J_1' \otimes \mathrm{id}_{T^*}$ and $J_2' \otimes \mathrm{id}_{T^*}$).

5. Construction of instantons from linear algebra data. The data (F_{-1}, F^0, Q) will correspond to the following instanton bundle $(\mathcal{E}_0, \nabla)$ over the sphere S^4, which parametrizes the σ-invariant planes $D \subset T^*$: the fibre of $\mathcal{E}_0$ over the point corresponding to D is $F_D^{\perp} \cap F^0$; the orthogonal metric on the fibres is induced by the form Q; and in the case of U (respectively, Sp), a complex structure (respectively, quaternionic structure) on the fibres is induced by the operators J' (respectively, J_1' and J_2'). The connection ∇ is the orthogonal projection of the trivial connection on the trivial bundle over S^4 with fibre F^0, in which $\mathcal{E}_0$ is immersed.

6. Coordinates. The claim that the connection ∇ in the last subsection is self-dual can be verified by a direct computation. Here we shall introduce coordinates which are convenient for such computations.

Let $\mathbf{R}^4$ be Euclidean space with the metric $\sum dx_i^2$. It is convenient to imagine a point of $\mathbf{R}^4$ as a quaternion. More precisely, let X be the 2×2-matrix for which

$$X^t = \sum_{a=1}^{3} ix_a\sigma_a + x_4\begin{pmatrix} 1 & 0 \\ 0 & 1 \end{pmatrix} = \begin{pmatrix} x_4 + ix_3 & x_2 + ix_1 \\ -x_2 + ix_1 & x_4 - ix_3 \end{pmatrix};$$

where σ_a is the Pauli matrix. We let a point $x \in \mathbf{R}^4$ correspond to the plane

$$\mathbf{P}_x = \left\{(z_1, \ldots, z_4) \mid \begin{pmatrix} z_1 \\ z_2 \end{pmatrix} = X^+\begin{pmatrix} z_3 \\ z_4 \end{pmatrix}\right\} \subset T^*.$$

(The plus superscript denotes Hermitian conjugate.) We also set $\mathbf{P}_\infty = \{(z_1, z_2, 0, 0)\}$.

It is not hard to see that $\sigma(\mathbf{P}_x) = \mathbf{P}_x$ for all $x \in \mathbf{R}^4 \cup \{\infty\}$, and, conversely, any σ-invariant plane in T^* is of the form $\mathbf{P}_x$ or $\mathbf{P}_\infty$. Moreover, any point $z \in T^* \setminus \mathbf{P}_\infty$ lies in the plane $\mathbf{P}_x$ for which

$$x_2 + ix_1 = \frac{-z_2\bar{z}_3 + z_4\bar{z}_1}{|z_3|^2 + |z_4|^2}, \qquad x_4 + ix_3 = \frac{z_2\bar{z}_4 + z_3\bar{z}_1}{|z_2|^2 + |z_4|^2}.$$

This is precisely the coordinate description of the map

$$\pi\colon (T^* \setminus \{0\})/\mathbf{C}^* \longrightarrow \mathbf{R}^4 \cup \{\infty\} = S^4; \qquad \pi(z) = x \Longleftrightarrow z \in \mathbf{P}_x.$$

The space of self-dual 2-forms on $\mathbf{R}^4$ has basis $dx_1 \wedge dx_2 + dx_3 \wedge dx_4$, $dx_1 \wedge dx_3 - dx_2 \wedge dx_4$, $dx_1 \wedge dx_4 + dx_2 \wedge dx_3$. If we lift this basis to $\mathbf{P}^3 \setminus \mathbf{P}_\infty$ using π^*, we easily see that we obtain a basis of 2-forms of type (1, 1). Setting $\nabla_a = \pi^*(\nabla)(\partial/\partial z_a)$ and $\nabla_{\overline{a}} = \pi^*(\nabla)(\partial/\partial \overline{z}_a)$ (the lifting to $\mathbf{C}^4$), we hence find that the self-duality equations for $\mathcal{E}_0$ are equivalent to the equations $[\nabla_a, \nabla_b] = [\nabla_{\overline{a}}, \nabla_{\overline{b}}] = 0$ for $\pi^*(\mathbf{C} \otimes_{\mathbf{R}} \mathcal{E}_0)$, and this implies that $\mathcal{E}_Z$ is locally free over $\mathcal{O}_Z$ of the same rank as $\mathcal{E}_0$.

We can now give a precise formulation of the basic result of this section.

7. Theorem. *All instantons are obtained from some linear algebra data by means of the construction in § 3.6; if two sets of data are not isomorphic, then the corresponding instantons are not isomorphic.* □

The proof of the theorem consists of several parts. The basic steps are: (a) description of the structure of the cohomology spaces of the sheaf $\mathcal{E}_Z$ which corresponds to an instanton; (b) proof that a sheaf with this cohomology can be constructed using a monad; (c) expression of the other instanton structures in terms of the monad, and translation into the language of linear algebra; and (d) verification that the instanton sheaf which corresponds to the resulting linear algebra data is isomorphic to the original instanton.

The information we need concerning the cohomology of $\mathcal{E}_Z$ is contained in the following proposition.

8. Proposition. $H^i(\mathcal{E}_Z(k)) = 0$ *for* $i \leq 1, i+k \leq -1$ *and for* $i \geq 2, i+k \geq 0$.

9. Fundamental Lemma. $H^1(\mathcal{E}_Z(-2)) = 0$.

10. Derivation of Proposition 8 from the Fundamental Lemma. If $(\mathcal{E}_0, \nabla)$ is an instanton, then the dual pair $(\mathcal{E}_0^*, \nabla^*)$ is also an instanton, and on Z it corresponds to the sheaf $\mathcal{E}_Z^* = \mathcal{H}om(\mathcal{E}_Z, \mathcal{O})$. By the Serre duality theorem, we have $H^i(\mathcal{E}_Z(k))^* = H^{3-i}(\mathcal{E}_Z^*(-4-k))$. This means that it is enough for us to show that $H^i(\mathcal{E}_Z(k)) = 0$ for $i \leq 1, i+k \leq -1$. Since the restriction of $\mathcal{E}_Z$ to a real line is trivial, it follows that the restrictions to almost all lines (in the Zariski topology) are also trivial. Because $H^0(\mathbf{P}^1, \mathcal{O}(k)) = 0$ for $k < 0$, it follows that for $k < 0$ any section $s \in H^0(\mathcal{E}_Z(k))$ vanishes on almost all lines, and thus equals zero.

Next, suppose that $D \subset \mathbf{P}^3$ is a plane containing one of the fibres $\pi^{-1}(x), x \in S^4$. Then the restriction of $\mathcal{E}_Z$ to almost any line in D is trivial. By the earlier argument, we then have $H^0(D, \mathcal{E}_Z(k)|_D) = 0$ for $k < 0$. Using the standard exact sequence $H^0(\mathcal{E}_Z|_D) \longrightarrow H^1(\mathcal{E}_Z(k-1)) \longrightarrow H^1(\mathcal{E}_Z(k))$ and descending induction on k, beginning with $k = -2$ (the Fundamental Lemma), we obtain $H^1(\mathcal{E}_Z(k)) = 0$ for $k \leq -2$. □

11. Plan of proof of the Fundamental Lemma. The general mechanism for proving the Fundamental Lemma is to identify the space $H^1(\mathcal{E}_Z(-2))$ with the kernel of the Laplace operator of an instanton connection on S^4. This operator is conformally invariant, and standard arguments show that its kernel is zero. More generally, all of the cohomology groups of the twisted sheaves $\mathcal{E}_Z$ have a similar interpretation in terms of kernels and cokernels of invariant differential operators associated with $(\mathcal{E}_0, \nabla)$; except in the case that interests us, these operators have order 1. The cohomological reasons for this phenomenon and the exact statements of results will be given later in this chapter in the context of the non-self-dual Radon–Penrose transform. Here we shall limit ourselves to the particular computations needed in our concrete situation.

The plan of proof is as follows. $H^1(\mathcal{E}_Z(-2))$ can be computed as the cohomology group of the initial portion of the Dolbeault complex

$$\Gamma(\mathcal{E}_{Z,\infty}(-2)) \xrightarrow{\overline{\partial}} \Gamma(\mathcal{E}_{Z,\infty}(-2) \otimes \Omega^{0,1}) \xrightarrow{\overline{\partial}} \Gamma(\mathcal{E}_{Z,\infty}(-2) \otimes \Omega^{0,2}),$$

where ∞ indicates that we have passed to smooth sections of the sheaf. By virtue of our fundamental construction, $\overline{\partial}$ coincides with $\pi^*(\nabla)^{0,1}$ on $\mathcal{E}_{Z,\infty}$. We must show that, if $\omega \in \Gamma(\mathcal{E}_{Z,\infty}(-2) \otimes \Omega^{0,1})$ and $\overline{\partial}\omega = 0$, then we have $\omega = \overline{\partial}\nu$ for some $\nu \in \Gamma(\mathcal{E}_{Z,\infty}(-2))$. We let ω_v denote the image of ω in $\mathcal{T}^*_\infty \mathbf{P}^3/S^4$, i.e., the restriction of ω to the vertical vector fields. From the local calculations which we shall do below, it will follow that, if $\omega_v = 0$, then $\omega = 0$. Thus, we shall look for a section $\nu \in \Gamma(\mathcal{E}_Z(-2))$ for which $(\omega - \overline{\partial}\nu)_v = 0$. Roughly speaking, this last property means that ω becomes $\overline{\partial}$-closed after restriction to the fibres of π. Since $H^1(\mathbf{P}^1, \mathcal{O}(-2)) \neq 0$, we cannot expect this property to hold automatically. However, since $H^1(\mathbf{P}^1, \mathcal{O}(-1)) = 0$, we can fix an imbedding $\mathcal{O}(-2) \subset \mathcal{O}(-1)$ and then in each fibre $\pi^{-1}(x)$ find a section ν_x of the sheaf $\mathcal{E}_Z(-1)|_{\pi^{-1}(x)}$ such that $\omega|_{\pi^{-1}(x)} = \overline{\partial}\nu_x$. In addition, we have $H^0(\mathbf{P}^1, \mathcal{O}(-1)) = 0$, so that, once the imbedding is fixed, the forms ν_x are uniquely determined. We shall verify that they glue together to give a global section ν of the sheaf $\mathcal{E}_Z(-1)$. It then remains to show that ν comes from $\mathcal{O}(-2)$. To do this one must use local arguments. We choose a plane $D \subset \mathbf{P}^3$ and realize $\mathcal{O}(-2)$ as the sheaf of local holomorphic equations of the union $D \cup \sigma D$; we use the analogous realization of $\mathcal{E}_Z(-2) \subset \mathcal{E}_Z$. By our earlier argument, $\omega_v = \overline{\partial}\nu_{D,v} = \overline{\partial}\nu_{\sigma D,v}$, where ν_D is smoothly divisible by the local equations of D and $\nu_{\sigma D}$ is smoothly divisible by the local equations of σD. We shall show that $\nu_D = \nu_{\sigma D}$, in which case $\nu = \nu_D = \nu_{\sigma D}$ lies in $\Gamma(\mathcal{E}_Z(-2))$. The equality $\nu_D - \nu_{\sigma D} = 0$ will follow from the fact that this difference turns out to be the lifting of a smooth section of $\mathcal{E}_0 \otimes \mathbf{C}$ over S^4 which lies in the kernel of the Laplace operator of the connection ∇ and decreases rapidly at "infinity" (which here is the point $\pi(D \cap \sigma D)$). The first part of this claim reflects the fact that ω is $\bar{\partial}$-closed. (Since ν_D and $\nu_{\sigma D}$ are obtained from ω by integrating, the difference $\nu_D - \nu_{\sigma D}$ is killed by a second order differential operator.) Finally, in order to bound $\nu_D - \nu_{\sigma D}$ (in the hermitian metric $\langle \nu, \nu \rangle = (\bar{\nu}, \nu)$) near $\pi(D \cap \sigma D)$, we have to choose one more pair of planes D' and $\sigma D'$, carry out the analogous constructions for them, and compare the results.

We now proceed to implement this plan. We choose the planes $z_3 = 0$ and $z_1 = 0$ for D and D', respectively.

12. Computations in coordinates. Using the coordinates in § 3.6, we set $\xi = x_2 + ix_1$, $\eta = x_4 + ix_3$. The map π is given by the formulas

$$z_1 = \bar{\eta} z_3 + \bar{\xi} z_4, \qquad z_2 = -\xi z_3 + \eta z_4. \tag{1}$$

Let $U_i \subset \mathbf{P}^3$ be the set of points where $z_i \neq 0$. We set $\lambda = z_3/z_4$. Since $U_3 \cup U_4 = \pi^{-1}(\mathbf{R}^4)$, the functions $(\xi, \bar{\xi}, \eta, \bar{\eta}, \lambda, \bar{\lambda})$ form a smooth (complex) system of coordinates in U_4 (here we write ξ instead of $\pi^*(\xi)$, etc.). On the other hand, the functions $(z_1/z_4, z_2/z_4, \lambda)$ form a holomorphic system of coordinates in U_4, and it is clear from (1) that

$$z_1/z_4 = \bar{\eta}\lambda + \bar{\xi}, \qquad z_2/z_4 = -\xi\lambda + \eta. \tag{2}$$

Hence, $(d\bar{\lambda}, d''\bar{\eta}, d''\xi)$ form a basis for the space of (0, 1)-forms at each point of U_4.

We now fix a $\bar{\partial}$-closed form $\omega \in \Gamma(\mathcal{E}_Z(-2) \otimes \Omega^{0,1})$. Once we identify $\mathcal{O}(-2)$ with the sheaf of holomorphic equations of the pair of planes $z_3 = 0$ and $z_4 = 0$, this form will be represented by a form $\omega_{34} \in \Gamma(\mathcal{E}_{Z,\infty} \otimes \Omega^{0,1})$ which is locally smoothly divisible by these equations. We write ω_{34} on U_4 in the form

$$\omega_{34} = f\, d\bar{\lambda} + g\, d''\xi + h\, d''\bar{\eta} \tag{3}$$

and we regard f, g and h as smooth sections of $\mathcal{E}_0$ over $\mathbf{R}^4$ which depend smoothly on the parameter λ. Here covariant differentiation with respect to $\bar{\lambda}$ coincides with taking the usual partial derivative.

To say that ω_{34} vanishes holomorphically on $z_3 = 0$ means that the sections $f\lambda^{-1}$, $g\lambda^{-1}$, and $h\lambda^{-1}$ remain smooth when $\lambda = 0$.

It follows from (2) that the forms $(d\bar{\lambda}^{-1}, d''\bar{\xi}, d''\eta) = (-\bar{\lambda}^{-2} d\bar{\lambda}, -\lambda\, d''\bar{\eta}, -\lambda\, d''\xi)$ give a basis of the space of $(0,1)$-forms at every point of U_3. If we use (3) to write ω_{34} in this basis, we find that divisibility of ω_{34} by z_4 implies that the sections $f\bar{\lambda}\lambda^2, g$, and h remain smooth at $\lambda = \infty$.

We set $\nabla_\xi = \nabla(\partial/\partial\xi)$, etc. Since $\overline{\partial} = (\pi^*\nabla)^{0,1}$, we have

$$\begin{aligned}\overline{\partial} f &= \frac{\partial f}{\partial \overline{\lambda}} d\overline{\lambda} + \nabla_\xi(f) d''\xi + \nabla_{\overline{\xi}}(f) d''\overline{\xi} + \nabla_\eta(f) d''\eta + \nabla_{\overline{\eta}}(f) d''\overline{\eta} \\ &= \frac{\partial f}{\partial \overline{\lambda}} d\overline{\lambda} + (\nabla_\xi + \lambda\nabla_\eta) f\, d''\xi + (\nabla_{\overline{\eta}} - \lambda\nabla_{\overline{\xi}}) f\, d''\overline{\eta}. \end{aligned} \tag{4}$$

Similar formulas hold for g and h. Thus, the condition $\overline{\partial}\omega_{34} = 0$ means that

$$(\nabla_\xi + \lambda\nabla_\eta) f = \partial g/\partial\overline{\lambda}, \qquad (\nabla_{\overline{\eta}} - \lambda\nabla_{\overline{\xi}}) f = \partial h/\overline{\partial}\lambda, \tag{5}$$

$$(\nabla_{\overline{\eta}} - \lambda\nabla_{\overline{\xi}}) g = (\nabla_\xi + \lambda\nabla_\eta) h. \tag{6}$$

From the formulas in § 3.6 it follows that $[\nabla_\xi + \lambda\nabla_\eta,\, \nabla_{\overline{\eta}} - \lambda\nabla_{\overline{\xi}}] = 0$. We can now verify that when $\omega_v = 0$ we must have $\omega = 0$. In fact, (3) shows that when $\omega_v = 0$ we have $f = 0$. According to (5), the sections g and h then depend holomorphically on λ. Thus, they are constant along the fibres of π, since $\mathcal{E}_Z$ is holomorphically trivial along the fibres. But they vanish when $\lambda = 0$, and hence must be identically equal to zero.

13. The sections ν_3 and ν_4. As we explained earlier, for every point $x \in \mathbf{R}^4$ there exists a unique smooth section ν_3 of the sheaf $\mathcal{E}_{Z,\infty}$ along $\pi^{-1}(x)$ which vanishes at $\lambda = 0$ (i.e., at $\pi^{-1}(x) \cap (z_3 = 0)$) and has the property that $f = \partial\nu_3/\partial\overline{\lambda}$. We let ν_4 denote the analogous section with a zero at $z_4 = 0$. To see that ν_3 and ν_4 depend smoothly on x, and so can be regarded as sections of $\mathcal{E}_{Z,\infty}$ over $U_3 \cup U_4$, it suffices to consider the integral formula

$$\nu_4(\lambda, x) = \frac{1}{2\pi i} \int \frac{f(\varsigma, x)}{\varsigma - \lambda}\, d\varsigma \wedge d\overline{\varsigma},$$

and use the estimate $|f(\lambda, x)| = O((1 + |\lambda|)^{-3})$, which follows from the smoothness of $f\overline{\lambda}\lambda^2$ at $\lambda = \infty$; and similarly for ν_3. Obviously, ν_3 and ν_4 are uniquely determined by their zeros together with the relation $(\omega_{34} - \overline{\partial}\nu_3)_v = (\omega_{34} - \overline{\partial}\nu_4)_v = 0$.

We set $\gamma = \nu_3 - \nu_4$ and $\nabla_a = \nabla(\partial/\partial x_a)$.

14. Lemma. *The section γ is constant along the fibres of π, and $\nabla_a\nabla_a\gamma = 0$.*

Proof. As before, the first assertion can be derived from the fact that $\partial\gamma/\partial\overline{\lambda} = 0$. To prove the second claim, we first note that

$$g = \nabla_\xi\nu_3 + \lambda\nabla_\eta\nu_4, \qquad h = \nabla_{\overline{\eta}}\nu_3 - \lambda\nabla_{\overline{\xi}}\nu_4. \tag{7}$$

In fact, we have $\partial/\partial\overline{\lambda}(g - \nabla_\xi\nu_3 - \lambda\nabla_\eta\nu_4) = 0$ by (5). In addition, g, $\nabla_\xi\nu_3$, and $\nabla_\eta\nu_4$ are locally divisible by the holomorphic equations of the plane $z_3 = 0$, so that the section $g - \nabla_\xi\nu_3 - \lambda\nabla_\eta\nu_4$ does not depend on λ and must be zero. We similarly prove the formula for h. We now substitute the right sides of the equations in (7) into (6). If we rewrite (7) in the form $g = (\nabla_\xi + \lambda\nabla_\eta)\nu_3 - \lambda\nabla_\eta\gamma, h = (\nabla_{\overline{\eta}} - \lambda\nabla_{\overline{\xi}})\nu_3 + \lambda\nabla_{\overline{\xi}}\gamma$ and use the vanishing of $[\nabla_\xi + \lambda\nabla_\eta, \nabla_{\overline{\eta}} - \lambda\nabla_{\overline{\xi}}]$, we find that $(\nabla_\xi\nabla_{\overline{\xi}} + \nabla_\eta\nabla_{\overline{\eta}}) = 0$. We then recompute this last operator in terms of the ∇_a, taking the self-duality into account (i.e., $[\nabla_1, \nabla_2] = [\nabla_3, \nabla_4]$, etc.). We obtain $\frac{1}{4}\nabla_a\nabla_a$. □

15. Computations on $U_1 \cup U_2$. We now identify $\mathcal{O}(-2)$ with the sheaf of equations of $(z_1 = 0) \cup (z_2 = 0)$, and let $\omega_{12} \in \Gamma(\mathcal{E}_{Z,\infty} \otimes \Omega^{0,1})$ denote the corresponding image of ω. We may suppose that $\omega_{12} = (z_1z_2/z_3z_4)\omega_{34}$ (with the standard choice of transition functions). As above, we construct sections ν_1 and ν_2 in $\Gamma(U_1 \cup U_2, \mathcal{E}_{Z,\infty})$ with zeros at $z_1 = 0$ and $z_2 = 0$, respectively, and with $(\omega_{12} - \overline{\partial}\nu_1)_v = (\omega_{12} - \overline{\partial}\nu_2)_v = 0$. In order to find formulas connecting ν_1 and ν_2 with ν_3 and ν_4, we set

$$\hat{\xi} = \xi(|\xi|^2 + |\eta|^2)^{-1}, \qquad \hat{\eta} = \eta(|\xi|^2 + |\eta|^2)^{-1}.$$

The functions $(\hat{\xi}, \hat{\eta}, \overline{\hat{\xi}}, \overline{\hat{\eta}})$ are smooth coordinates on

$$(\mathbf{R}^4 \setminus \{0\}) \cup \{\infty\} = S^4 \setminus \{0\} = \pi(U_1 \cup U_2).$$

16. Lemma. *On $(U_1 \cup U_2) \cap (U_3 \cup U_4)$ we have*

$$\nu_3 = z_3z_2^{-1}\hat{\xi}\nu_2 + z_3z_1^{-1}\overline{\hat{\eta}}\nu_1,$$

$$\nu_4 = z_4z_2^{-1}\hat{\eta}\nu_2 - z_4z_1^{-1}\overline{\hat{\xi}}\nu_1.$$

Proof. The right sides of these formulas are obviously divisible by z_3 and z_4, respectively. Hence it suffices to verify that

$$[\omega_{34} - \overline{\partial}(z_3z_2^{-1}\hat{\xi}\nu_2 + z_3z_1^{-1}\overline{\hat{\eta}}\nu_1)]_v = 0,$$

and similarly for ν_4. We have

$$\begin{aligned}[\overline{\partial}(z_3z_2^{-1}\hat{\xi}\nu_2 + z_3z_1^{-1}\overline{\hat{\eta}}\nu_1)]_v &= (z_3z_2^{-1}\hat{\xi} + z_3z_1^{-1}\overline{\hat{\eta}})\omega_{12,v} \\ &= z_3(z_1z_2)^{-1}(|\xi|^2 + |\eta|^2)^{-1}(z_1\xi + z_2\overline{\eta})\omega_{12,v}.\end{aligned}$$

But by (1) we have

$$z_1\xi + z_2\overline{\eta} = (\overline{\eta}z_3 + \overline{\xi}z_4)\xi + (-\xi z_3 + \eta z_4)\overline{\eta} = (|\xi|^2 + |\eta|^2)z_4.$$

Thus, the previous expression is equal to $(z_3 z_4)(z_1 z_2)^{-1}\omega_{12,v} = \omega_{34,v}$. The identity for ν_4 is proved analogously. □

17. Corollary. *On* $(U_1 \cup U_2) \cap (U_3 \cup U_4)$ *we have*

$$\gamma = \nu_3 - \nu_4 = (|\xi|^2 + |\eta|^2)^{-1}(\nu_1 - \nu_2) = |x|^{-2}(\nu_1 - \nu_2).$$

In particular, since $\nu_1 - \nu_2$ *is a smooth section near* ∞, *the section* γ *is also smooth near* ∞, *and* $|\gamma| = O(|x|^{-1})$ *in the hermitian metric on the fibres.*

Proof. From Lemma 16 and the formulas in (1) it follows that

$$\begin{aligned}\nu_3 - \nu_4 &= (z_3 z_2^{-1}\hat{\xi} - z_4 z_2^{-1}\hat{\eta})\nu_2 + (z_3 z_1^{-1}\hat{\overline{\eta}} + z_4 z_1^{-1}\hat{\overline{\xi}})\nu_1 \\ &= (|\xi|^2 + |\eta|^2)^{-1}[(z_3\xi - z_4\eta)z_2^{-1}\nu_2 + (z_3\overline{\eta} + z_4\overline{\xi})z_1^{-1}\nu_1] \\ &= (|\xi|^2 + |\eta|^2)^{-1}(\nu_1 - \nu_2). \qquad \square\end{aligned}$$

18. End of the proof of the Fundamental Lemma. It remains for us to verify that $\gamma = 0$. We set

$$I_R = \int_{|x| \le R} \langle \nabla_a\gamma,\, \nabla_a\gamma \rangle\, d^4x.$$

From Lemma 14 and Stokes' formula we obtain

$$I_R = \int_{|x| \le R} \partial_a \langle \gamma,\, \nabla_a\gamma \rangle\, d^4x = \int_{|x| = R} \langle \gamma,\, \nabla_a\gamma \rangle\, d_a^3x,$$

where $d_a^3x = \bigwedge_{i \ne x} dx_i (-1)^{a-1}$. By Corollary 17, we have $|\gamma| = O(R^{-2})$ and $|\nabla_a\gamma| = O(R^{-3})$ on a sphere of large radius R, and hence $I_R = O(R^3 \cdot R^{-5}) = O(R^{-2})$. Since the metric $\langle\ ,\ \rangle$ is positive-definite, this implies that $\nabla_a\gamma = 0$ for all a, and so $\partial_a|\gamma| = 0$, i.e., $|\gamma|$ is constant. But $\gamma(\infty) = 0$, and this completes the proof. □

19. Special monads. We now proceed to the second stage in the proof of Theorem 7: explaining the relation between $\mathcal{E}_Z$ and monads.

A locally free sheaf $\mathcal{E}_Z$ on $\mathbf{P}^3$ will be said to be "admissible" if the cohomology groups that appear in Proposition 8 are zero. By a "special monad" we shall mean a complex of sheaves of the form $\mathcal{F}: F_{-1} \otimes \mathcal{O}_{\mathbf{P}}(-1) \xrightarrow{\alpha} F_0 \otimes \mathcal{O}_{\mathbf{P}} \xrightarrow{\beta} F_1 \otimes \mathcal{O}_{\mathbf{P}}(1)$, where α is locally a direct sum imbedding and β is a surjection. To the monad $\mathcal{F}$ we associate the sheaf $\mathcal{E}(\mathcal{F}) = \operatorname{Ker}\beta/\operatorname{Im}\alpha$.

20. Proposition. *For every special monad $\mathcal{F}$, the sheaf $\mathcal{E} = \mathcal{E}(\mathcal{F})$ is admissible.*

Proof. We set $\mathcal{K} = \operatorname{Ker}\beta$. From the sequence $0 \longrightarrow \mathcal{K} \longrightarrow F_0 \otimes \mathcal{O} \longrightarrow F_1 \otimes \mathcal{O}(1)$ we obtain $H^{i-1}(F_1 \otimes \mathcal{O}(k+1)) \longrightarrow H^i(\mathcal{K}(k)) \longrightarrow H^i(F_0 \otimes \mathcal{O}(k))$. For (i, k) in the range where the cohomology of admissible sheaves must vanish, the two groups at the ends are zero, so that $H^i(\mathcal{K}(k)) = 0$. Next, from the sequence

$$0 \longrightarrow F_{-1} \otimes \mathcal{O}(-1) \longrightarrow \mathcal{K} \longrightarrow \mathcal{E} \longrightarrow 0$$

we obtain

$$H^i(\mathcal{K}(k)) \longrightarrow H^i(\mathcal{E}(k)) \longrightarrow H^{i+1}(F_{-1} \otimes \mathcal{O}(-k-1)),$$

and hence $H^i(\mathcal{E}(k)) = 0$, because the third group in the sequence also vanishes. □

21. Proposition. *Suppose that $\mathcal{F}_1$ and $\mathcal{F}_2$ are two special monads. Then the natural map* $\operatorname{Hom}(\mathcal{F}_1, \mathcal{F}_2) \longrightarrow \operatorname{Hom}(\mathcal{E}(\mathcal{F}_1), \mathcal{E}(\mathcal{F}_2))$ *is an isomorphism.*

Proof. By a morphism of monads we mean a morphism of the complexes. We set $\mathcal{E}_i = \mathcal{E}(\mathcal{F}_i)$ and $\mathcal{K}_i = \operatorname{Ker}\beta_i, i = 1, 2$. From the exact sequence

$$0 \longrightarrow F_{-1,2} \otimes \mathcal{O}(-1) \longrightarrow \mathcal{K}_2 \longrightarrow \mathcal{E}_2 \longrightarrow 0$$

we obtain

$$0 \longrightarrow \operatorname{Hom}(\mathcal{K}_1, F_{-1,2} \otimes \mathcal{O}(-1)) \longrightarrow \operatorname{Hom}(\mathcal{K}_1, \mathcal{K}_2) \longrightarrow$$
$$\longrightarrow \operatorname{Hom}(\mathcal{K}_1, \mathcal{E}_2) \longrightarrow \operatorname{Ext}^1(\mathcal{K}_1, F_{-1,2} \otimes \mathcal{O}(-1)).$$

We now show that the two end terms vanish. From the exact sequence

$$0 \longrightarrow \mathcal{K}_1 \longrightarrow F_{0,1} \otimes \mathcal{O} \longrightarrow F_{1,1} \otimes \mathcal{O}(-1) \longrightarrow 0$$

we find that

$$\operatorname{Hom}(F_{0,1} \otimes \mathcal{O}, F_{-1,2} \otimes \mathcal{O}(-1)) \longrightarrow \operatorname{Hom}(\mathcal{K}_1, F_{-1,2} \otimes \mathcal{O}(-1)) \longrightarrow$$
$$\longrightarrow \operatorname{Ext}^1(F_{1,1} \otimes \mathcal{O}(1), F_{-1,2} \otimes \mathcal{O}(-1)),$$

$$\operatorname{Ext}^1(F_{0,1} \otimes \mathcal{O}, F_{-1,2} \otimes \mathcal{O}(-1)) \longrightarrow \operatorname{Ext}^1(\mathcal{K}_1, F_{-1,2} \otimes \mathcal{O}(-1)) \longrightarrow$$
$$\longrightarrow \operatorname{Ext}^2(F_{1,1} \otimes \mathcal{O}(1), F_{-1,2} \otimes \mathcal{O}(-1)).$$

The end terms here vanish, because $\mathrm{Ext}^i(\mathcal{H}_1, \mathcal{H}_2) = H^i(\mathcal{H}_1^* \otimes \mathcal{H}_2^*)$ for locally free sheaves $\mathcal{H}_i$. Then the middle terms must also vanish, and so we have $\mathrm{Hom}(\mathcal{K}_1, \mathcal{K}_2) \simeq \mathrm{Hom}(\mathcal{K}_1, \mathcal{E}_2)$. In particular, every morphism of sheaves $\mathcal{E}_1 \longrightarrow \mathcal{E}_2$ extends uniquely to a morphism $\mathcal{K}_1 \longrightarrow \mathcal{E}_2$ and then to a morphism $\mathcal{K}_1 \longrightarrow \mathcal{K}_2$, which clearly takes $\mathrm{Im}\,\alpha_1$ to $\mathrm{Im}\,\alpha_2$.

From the exact sequence $0 \longrightarrow \mathcal{K}_1 \longrightarrow F_{0,1} \otimes \mathcal{O} \longrightarrow F_{1,1} \otimes \mathcal{O}(1) \longrightarrow 0$ we obtain

$$0 \longrightarrow \mathrm{Hom}(F_{1,1} \otimes \mathcal{O}(1), F_{0,2} \otimes \mathcal{O}) \longrightarrow \mathrm{Hom}(F_{0,1} \otimes \mathcal{O}, F_{0,2} \otimes \mathcal{O}) \longrightarrow$$
$$\longrightarrow \mathrm{Hom}(\mathcal{K}_1, F_{0,2} \otimes \mathcal{O}) \longrightarrow \mathrm{Ext}^1(F_{1,1} \otimes \mathcal{O}(1), F_{0,2} \otimes \mathcal{O}).$$

As before, the two end sheaves vanish, and so $\mathrm{Hom}(F_{0,1} \otimes \mathcal{O}, F_{0,2} \otimes \mathcal{O}) \simeq \mathrm{Hom}(\mathcal{K}_1, F_{0,2} \otimes \mathcal{O})$. In particular, any morphism $\mathcal{K}_1 \longrightarrow \mathcal{K}_2$ induces a morphism $\mathcal{K}_1 \longrightarrow F_{0,2} \otimes \mathcal{O}$, which in turn extends uniquely to a morphism $F_{0,1} \otimes \mathcal{O} \longrightarrow F_{0,2} \otimes \mathcal{O}$. If the morphism $\mathcal{K}_1 \longrightarrow \mathcal{K}_2$ was an extension of $\mathcal{E}_1 \longrightarrow \mathcal{E}_2$, then the corresponding morphism $F_{0,1} \otimes \mathcal{O} \longrightarrow F_{0,2} \otimes \mathcal{O}$ takes $\mathrm{Ker}\,\beta_1$ to $\mathrm{Ker}\,\beta_2$, and so gives a quotient morphism $F_{1,1} \otimes \mathcal{O}(1) \longrightarrow F_{1,2} \otimes \mathcal{O}(1)$. So we have finally shown that any morphism of sheaves $\mathcal{E}_1 \longrightarrow \mathcal{E}_2$ is induced by a unique morphism of monads $\mathcal{F}_1 \longrightarrow \mathcal{F}_2$. □

22. Proposition. *The functor which to a special monad $\mathcal{F}$ associates the sheaf $\mathcal{E}(\mathcal{F})$ gives an equivalence of categories between special monads and admissible sheaves.*

Proof. In view of Propositions 20 and 21, it remains only for us to construct the inverse functor.

We shall show how, given an admissible sheaf $\mathcal{E}$, one can construct the following objects which depend functorially on $\mathcal{E}$:

(a) a special monad $(F_{-1}, F_0, F_1, \alpha, \beta)$ with $F_{-1} = H^1(\mathcal{E} \otimes \Omega^2\mathbf{P}(1))$, $F_0 = H^1(\mathcal{E} \otimes \Omega^1\mathbf{P})$, $F_1 = H^1(\mathcal{E}(-1))$;

(b) an isomorphism $\mathcal{E} \longrightarrow \mathrm{Ker}\,\beta/\mathrm{Im}\,\alpha$.

Let $\mathcal{O}_\Delta$ be the direct image of the structure sheaf on the diagonal of $\mathbf{P} \times \mathbf{P}$, and let $p_i\colon \mathbf{P} \times \mathbf{P} \longrightarrow \mathbf{P}$ be the two projections. We let $K_. \longrightarrow \mathcal{O}_\Delta$ denote the Koszul resolution. This is the complex $K_i = p_1^*\mathcal{O}(i+1) \otimes p_2^*\Omega^{-i}(-i-1)$, $i \le 0$, $K_i = 0$ for $i \le -4$. Each differential $d_i\colon K_i \longrightarrow K_{i+1}$ is multiplication (and convolution in the second factor) by the canonical element in $H^0(p_1^*\mathcal{O}(1) \otimes p_2^*T\mathbf{P}(-1)) \simeq T \otimes T^*$, i.e., by $\sum_{i=1}^4 p_1^*(z_i) \otimes p_2^*(\frac{\partial}{\partial z_i})$. The Koszul complex $K_. \otimes p_2^*\mathcal{E} \longrightarrow \mathcal{O}_\Delta \otimes p_2^*\mathcal{E}$ is a resolution of "the sheaf $\mathcal{E}$ concentrated on the diagonal" which is functorial in $\mathcal{E}$.

The rest of the argument is standard from the point of view of the theory of derived categories. The map $K_. \otimes p_2^*\mathcal{E} \longrightarrow \mathcal{O}_\Delta \otimes p_2^*\mathcal{E}$ is a quasi-isomorphism. Hence, the sheaf $p_{1*}(\mathcal{O}_\Delta \otimes p_2^*\mathcal{E}) = \mathcal{E}$ is quasi-isomorphic to the complex $R_{p_{1*}}(K_. \otimes$

$p_2^*\mathcal{E}$). There are various ways one can compute a representative of this object in the derived category. In order to avoid using more homological algebra than is absolutely necessary, we shall simply split up $K_. \otimes p_2^*\mathcal{E}$ into exact triples, and for each triple we shall write out the part of its exact sequence of higher direct images that interests us. It is convenient to organize the computations as follows.

We set $\mathcal{K} = \operatorname{Im} d_{-2} = \operatorname{Ker} d_{-1}$, $J = \operatorname{Im} d_{-1}$. The following exact sequence is the "middle of the resolution": $0 \longrightarrow \mathcal{K} \otimes p_2^*\mathcal{E} \longrightarrow p_2^*(\mathcal{E} \otimes \Omega^1) \longrightarrow J \otimes p_2^*\mathcal{E} \longrightarrow 0$. Here is part of what we obtain when we descend to $\mathbf{P}$:

$$p_{1*}(J \otimes p_2^*\mathcal{E}) \longrightarrow R^1 p_{1*}(\mathcal{K} \otimes p_2^*\mathcal{E}) \longrightarrow R^1 p_{1*}(p_2^*(\mathcal{E} \otimes \Omega^1)) \longrightarrow$$
$$\longrightarrow R^1 p_{1*}(J \otimes p_2^*\mathcal{E}) \longrightarrow R^2 p_{1*}(\mathcal{K} \otimes p_2^*\mathcal{E}). \quad (8)$$

We now compute each of the sheaves in (8).

The first sheaf. The imbedding $0 \longrightarrow J \otimes p_2^*\mathcal{E} \longrightarrow p_1^*\mathcal{O}(1) \otimes p_2^*\mathcal{E}(-1)$ induces an imbedding

$$0 \longrightarrow p_{1*}(J \otimes p_2^*\mathcal{E}) \longrightarrow p_{1*}(p_1^*\mathcal{O}(1) \otimes p_2^*\mathcal{E}(-1)).$$

In what follows we shall often make use of the projection formula $R^i p_{1*}(p_1^*\mathcal{H}_1 \otimes p_2^*\mathcal{H}_2) = \mathcal{H}_1 \otimes H^i(\mathcal{H}_2)$, which holds for any locally free sheaves on $\mathbf{P}$. In particular,

$$p_{1*}(p_1^*\mathcal{O}(1) \otimes p_2^*\mathcal{E}(-1)) = \mathcal{O}(1) \otimes H^i(\mathcal{E}(-1)) = 0,$$

because $\mathcal{E}$ is admissible, and so

$$p_{1*}(J \otimes p_2^*\mathcal{E}) = 0. \quad (9)$$

The second sheaf. The beginning of the resolution $K_. \otimes p_2^*\mathcal{E} \longrightarrow \mathcal{O}_\Delta \otimes p_2^*\mathcal{E}$ has the form

$$0 \longrightarrow p_1^*\mathcal{O}(-2) \otimes p_2^*(\mathcal{E} \otimes \Omega^3(2)) \longrightarrow p_1^*\mathcal{O}(-1) \otimes p_2^*(\mathcal{E} \otimes \Omega^2(1)) \longrightarrow \mathcal{K} \otimes p_2^*\mathcal{E} \longrightarrow 0.$$

From this we obtain the exact sequence

$$R^1 p_{1*}(p_1^*\mathcal{O}(-2) \otimes p_2^*(\mathcal{E} \otimes \Omega^3(2)) \longrightarrow R^1 p_{1*}(p_1^*\mathcal{O}(-1) \otimes p_2^*(\mathcal{E} \otimes \Omega^2(1))) \longrightarrow$$
$$\longrightarrow R^1 p_{1*}(\mathcal{K} \otimes p_2^*\mathcal{E}) \longrightarrow R^2 p_{1*}(p_1^*\mathcal{O}(-2) \otimes p_2^*(\mathcal{E} \otimes \Omega^3(2))). \quad (10)$$

The first and last terms here are zero, because $H^i(\mathcal{E} \otimes \Omega^3(2)) \simeq H^i(\mathcal{E}(-2)) = 0$ for $i = 1, 2$ when $\mathcal{E}$ is admissible. Thus, the middle terms in (10) give an isomorphism which is functorial in $\mathcal{E}$:

$$R^1 p_{1*}(\mathcal{K} \otimes p_2^*\mathcal{E}) = H^1(\mathcal{E} \otimes \Omega^2(1)) \otimes \mathcal{O}(-1). \quad (11)$$

The third sheaf. This sheaf can be computed immediately:

$$R^1 p_{1*}(p_2^*\mathcal{E} \otimes \Omega^1) = H^1(\mathcal{E} \otimes \Omega^1) \otimes \mathcal{O}. \tag{12}$$

The fourth sheaf. The end of the resolution $K_{\cdot} \otimes p_2^*\mathcal{E} \longrightarrow \mathcal{O}_\Delta \otimes p_2^*\mathcal{E}$ has the form

$$0 \longrightarrow J \otimes p_2^*\mathcal{E} \longrightarrow p_1^*\mathcal{O}(1) \otimes p_2^*(\mathcal{E}(-1)) \longrightarrow \mathcal{O}_\Delta \otimes p_2^*\mathcal{E} \longrightarrow 0.$$

This gives us the exact sequence

$$\begin{aligned} p_{1*}(p_1^*\mathcal{O}(1) \otimes p_2^*\mathcal{E}(-1)) \longrightarrow p_{1*}(\mathcal{O}_\Delta \otimes p_2^*\mathcal{E}) \longrightarrow R^1 p_{1*}(J \otimes p_2^*\mathcal{E}) \longrightarrow \\ \longrightarrow R^1 p_{1*}(p_1^*\mathcal{O}(1) \otimes p_2^*\mathcal{E}(-1)) \longrightarrow R^1 p_{1*}(\mathcal{O}_\Delta \otimes p_2^*\mathcal{E}). \end{aligned} \tag{13}$$

Here the first term vanishes, because $H^0(\mathcal{E}(-1)) = 0$. The first and fifth terms can be computed using the fact that $p_1|_\Delta : \Delta \longrightarrow \mathbf{P}$ is an isomorphism, and so $R^i p_{1*}(\mathcal{H}) = 0$ when $i > 0$ for any $\mathcal{O}_\Delta$-sheaf $\mathcal{H}$. Consequently, the last term in (13) is zero, and the second term is functorially isomorphic to $\mathcal{E}$. Finally, the fourth term in (13) is functorially isomorphic to $H^1(\mathcal{E}(-1)) \otimes \mathcal{O}(1)$. We conclude that (13) gives the exact sequence

$$0 \longrightarrow \mathcal{E} \longrightarrow R^1 p_{1*}(J \otimes p_2^*\mathcal{E}) \longrightarrow H^1(\mathcal{E}(-1)) \otimes \mathcal{O}(1) \longrightarrow 0, \tag{14}$$

which is functorial in $\mathcal{E}$.

The fifth sheaf. This sheaf can be determined from the continuation of the exact sequence (10):

$$\begin{aligned} R^2 p_{1*}(p_1^*\mathcal{O}(-1) \otimes p_2^*(\mathcal{E} \otimes \Omega^2(1))) \longrightarrow R^2 p_{1*}(K \otimes p_2^*\mathcal{E}) \longrightarrow \\ \longrightarrow R^3 p_{1*}(p_1^*\mathcal{O}(-2) \otimes p_2^*(\mathcal{E} \otimes \Omega^3(2))). \end{aligned}$$

Here the third term is zero, because $H^3(\mathcal{E} \otimes \Omega^3(2)) \simeq H^3(\mathcal{E}(-2)) = 0$, since $\mathcal{E}$ is admissible. To see that the first term also vanishes, it suffices to obtain $H^2(\mathcal{E} \otimes \Omega^2(1)) = 0$. This cohomology group can be computed as follows: multiply the standard exact sequence $0 \longrightarrow \Omega^3(3) \longrightarrow \wedge^3 T^* \otimes \mathcal{O} \longrightarrow \Omega^2(3) \longrightarrow 0$ by $\mathcal{E}(-2)$ and consider the exact cohomology sequence $H^2(\wedge^3 T^* \otimes \mathcal{E}(-2)) \longrightarrow H^2(\mathcal{E} \otimes \Omega^2(1)) \longrightarrow H^3(\Omega^3(1))$. The terms at the beginning and end of this last sequence are zero. Thus, we finally have

$$R^2 p_{1*}(K \otimes p_2^*\mathcal{E}) = 0. \tag{15}$$

We can now substitute the results (9), (11) and (15) into (8). We obtain the following exact sequence of sheaves which is functorial in $\mathcal{E}$:

$$0 \longrightarrow H^1(\mathcal{E} \otimes \Omega^2(1)) \otimes \mathcal{O}(-1) \longrightarrow H^1(\mathcal{E} \otimes \Omega^1) \otimes \mathcal{O} \longrightarrow R^1 p_{1*}(J \otimes p_2^*\mathcal{E}) \longrightarrow 0.$$

Here we replace the third arrow by its composition with the third arrow in (14). In this way we obtain a monad and an isomorphism of its cohomology sheaf with $\mathcal{E}$. It was the existence of these objects that was claimed at the beginning of the proof. □

We can now proceed to the final stage in the proof of Theorem 7: translation of the monad description into the language of linear algebra data, taking into account the additional instanton structures.

23. Scalar product. We interpret a complex symmetric scalar product on $\mathcal{E}_Z$ as a symmetric isomorphism $\mathcal{E}_Z \longrightarrow \mathcal{E}_Z^*$. If we recall that the category of monads also has a functor $\mathcal{F} \longrightarrow \mathcal{F}^*$, and that $\mathcal{E}(\mathcal{F}^*) \simeq \mathcal{E}(\mathcal{F})^*$, we see that a scalar product on $\mathcal{E}(\mathcal{F})$ is induced by a symmetric isomorphism $\Phi: \mathcal{F} \longrightarrow \mathcal{F}^*$ of the corresponding monads.

A monad equipped with such an isomorphism can be defined using a diagram of vector spaces

$$F_{-1} \otimes T \xrightarrow{\alpha} F_0 \simeq F_0^* \xrightarrow{\alpha^*} F_{-1}^* \otimes T^*,$$

where the isomorphism in the middle corresponds to the bilinear form Q which Φ induces on F_0. In the notation of § 3.4, we can say that for any hyperplane $D \subset T^*$ the kernel of the map $\alpha_D^*: F_0^* \longrightarrow F_{-1}^* \otimes T^*/D$ coincides with the orthogonal complement (relative to Q) of the image of the map $\alpha_D: F_{-1} \otimes (T^*/D)^* \longrightarrow F_0$. Hence, all of these images must be Q-isotropic, and the fibre of $E(\mathcal{F})$ over the point $z \in \mathbf{P}$ corresponding to D is canonically isomorphic to $(\operatorname{Im} \alpha_D)^{\perp}/\operatorname{Im} \alpha_D$.

We note that the Chern numbers of an admissible sheaf with a symmetric bilinear form can easily be computed from its monad: $c_1(\mathcal{E}) = c_3(\mathcal{E}) = 0, c_2(\mathcal{E}) = \dim F_{-1}$; in addition, we have $\operatorname{rank} \mathcal{E} = \dim F_0 - 2 \dim F_{-1}$.

24. Real structure. An antilinear map $\sigma: \mathcal{E}_Z \longrightarrow \mathcal{E}_Z$ which extends the real structure to $\mathbf{P}$ can be carried over to the monad, since the construction in Proposition 21 is clearly functorial relative to such maps as well (for example, one can compute all of the cohomology spaces using σ-invariant Čech coverings). Furthermore, the real structure on the monad which induces a given real structure on $\mathcal{E}_Z$ is unique. In fact, if σ' were another such structure, then the composition $\sigma(\sigma - \sigma')$ would be a linear endomorphism of the monad which induces the zero endomorphism of $\mathcal{E}_Z$. Hence, $\sigma - \sigma' = 0$ by Proposition 21. We similarly show that $\sigma^2 = 1$, since this holds on $\mathcal{E}_Z$.

It now follows that σ induces a real structure on the space $F_0 = H^1(\mathcal{E} \otimes \Omega^1)$ of the monad. We set $F^0 = \{f \in F_0 \mid \sigma f = f\}$. Then $F_0 = F^0 \otimes_{\mathbf{R}} \mathbf{C}, F_0 \otimes_{\mathbf{C}} T^* = F^0 \otimes_{\mathbf{R}} T^*$, and the first arrow in the monad $0 \longrightarrow F_{-1} \otimes \mathcal{O}(-1) \longrightarrow F_0 \otimes \mathcal{O}$ determines an imbedding $F_{-1} \subset F^0 \otimes_{\mathbf{R}} T^*$.

25. The form Q on F^0. If an orthogonal metric on an instanton sheaf $\mathcal{E}_0$ is bilinearly extended to $\pi^*(\mathcal{E}_0 \otimes \mathbf{C})$ and is then restricted to $\mathcal{E}_Z$, we showed that it induces a non-degenerate quadratic form on F_0. When σ acts on the arguments of the form, we obtain the complex conjugate value. Consequently, the restriction of the form to F^0 is a real non-degenerate quadratic form Q.

This completes the construction of the orthogonal linear algebra data from the $\mathcal{O}(r)$-instanton $\mathcal{E}_0$. By now it should be obvious to the reader that the operators J, J_1, J_2 in unitary and symplectic structures can be carried over in exactly the same way to the monad and then to F^0.

26. Reconstitution of the instanton. It remains for us to verify that the method in § 3.5 for recovering $(\mathcal{E}_0, \nabla)$ from the linear algebra data returns us to our original instanton.

The fibre of $\mathcal{E}_0 \otimes_{\mathbf{R}} \mathbf{C}$ over a point $x \in S^4$ is canonically isomorphic to the space of holomorphic sections of the restriction of $\mathcal{E}_Z$ to $\pi^{-1}(x)$: in fact, this is the definition in § 2 of the inverse Radon–Penrose transform. We let $\mathcal{E}(x)$ denote this restriction, and we use analogous notation for the restrictions of other sheaves on $\mathbf{P}$. When restricted to $\pi^{-1}(x)$, the monad $\mathcal{E}_Z$ gives two exact sequences

$$0 \longrightarrow (F_{-1} \otimes \mathcal{O}(-1))(x) \longrightarrow \mathcal{K}(x) \longrightarrow \mathcal{E}_Z(x) \longrightarrow 0$$

and

$$0 \longrightarrow \mathcal{K}(x) \longrightarrow (F_0 \otimes \mathcal{O})(x) \xrightarrow{\alpha^*(x)} F_{-1}^* \otimes \mathcal{O}(1) \longrightarrow 0,$$

which lead to the isomorphisms

$$H^0(\mathcal{E}_Z(x)) \simeq H^0(\mathcal{K}(x)) \simeq \mathrm{Ker}(F_0 \xrightarrow{\alpha^*(x)} F_{-1}^* \otimes \Gamma(\pi^{-1}(x),\ \mathcal{O}(1))).$$

If the point $x \in S^4$ corresponds to a real plane $D \subset T^*$, then $\Gamma(\pi^{-1}(x),\ \mathcal{O}(1))$ can be canonically identified with T^*/D, and $\mathrm{Ker}\,\alpha^*(x)$ can be canonically identified with the subspace $[\mathrm{Im}\, F_{-1} \otimes (T^*/D)^*]^\perp = F_D^\perp \subset F_0$, because the monad is self-dual. Since $\mathcal{E}_Z$ has rank equal to $\dim F_0 - 2 \dim F_D^\perp$, we must have

$$\dim F_D = \dim F_0 - 2 \dim F_D^\perp = 2 \dim F_{-1}.$$

Finally, the fibre of $\mathcal{E}_0$ over x consists of the σ-invariant elements of the fibre of $\mathcal{E}_0 \otimes \mathbf{C}$, i.e., it can be identified with $F_D^\perp \cap F^0$. Thus, the restriction of Q to any such subspace must be positive-definite.

More precisely, the above argument shows that, for any σ-invariant plane $D \subset T^*$ and any hyperplane B containing it, the canonical map $F_D^\perp = (F_B +$

$F_{\sigma B})^{\perp} \longrightarrow F_B^{\perp}/F_B$ is an isometry. Hence, if we take the direct summand of the trivial orthogonal fibration $S^4 \times F_0$ having $F_D^{\perp}$ as its fibre over D and lift it to P, the result can be canonically identified with $\pi^*(\mathcal{E}_0 \otimes \mathsf{C})$ with all of the structure preserved (the metric, σ, J, J_1, J_2). Hence, this fibration is isomorphic to $\mathcal{E}_0 \otimes \mathsf{C}$. The self-dual connection ∇ which gives rise to a given holomorphic structure is obviously unique. Thus, in order to identify this connection with the projection of the trivial connection onto $S^4 \times F_0$, it is sufficient to verify that this projection is self-dual. This is a simple computation which we shall omit.

We conclude this section by proving a positivity theorem: the equivalence of conditions (c) and (c′) in § 3.4.

This is an important result, because in practice it is much easier and more natural to verify (c′) than (c).

27. Theorem. *Let F^0 be a finite dimensional real vector space with non-degenerate quadratic form Q, and let $F_{-1} \subset F^0 \otimes_{\mathsf{R}} T^*$ be a complex subspace which is invariant under $\mathrm{id}_{F^0} \otimes \sigma$. Then the conditions (b)–(c) of § 3.4 imply that $Q > 0$.*

Proof. We extend Q to a hermitian form $\langle , \rangle$ on $F_0 = F^0 \otimes_{\mathsf{R}} \mathsf{C}$ by means of the formula $\langle f, f \rangle = Q((\mathrm{id} \otimes \sigma) f, f)$. Instead of the imbedding $F_{-1} \longrightarrow F_0 \otimes T^*$ we shall consider the map $\varphi: F_{-1} \otimes T \longrightarrow F_0$ and use the same symbol $\langle , \rangle$ to denote the induced form on $F_{-1} \otimes T$.

In T we choose the basis $\epsilon_a, a = 1, 2, 3, 4$, corresponding to our familiar coordinates $z_1, \dots, z_4$. The point $z = [z_1, z_2, z_3, z_4] \in \mathsf{P}^3$ corresponds to the hyperplane $D_z: \sum_{a=1}^4 z_a \epsilon_a = 0$ in T^*.

For any vector $e \in F_{-1}$ we set $e_a = e \otimes \epsilon_a, e_z = \sum_{a=1}^4 e_a z_a, F_z = \{ez \mid e \in F_{-1}\} \subset F_{-1} \otimes T$. The space F_z is determined by the point $z \in \mathsf{P}^3$. Since we have $(\mathrm{id} \otimes l)(e) = \varphi(e \otimes l) \in F_0$ for any $l \in T$, the subspace $F_{D_z} \subset F_0$ is the φ-image of the space $F_z \subset F_{-1} \otimes T$. From the definition of a hermitian form we see that F_{D_z} is isotropic relative to Q if and only if F_z and $F_{\sigma z}$ are orthogonal in the metric $\langle , \rangle$.

Any σ-invariant plane $B \subset T^*$ can be represented in the form $D_z \cap D_{\sigma z}$ for a suitable $z \in \mathsf{P}$. According to the preceding argument, we have $F_B = \varphi(F_z + F_{\sigma z})$. Hence, the equality $\dim F_B = 2 \dim F_{-1}$ implies that φ is injective on the spaces F_z and $F_z \cap F_{\sigma z} = \{0\}$. In addition, the metric is non-degenerate on $F_z + F_{\sigma z}$.

From the condition that $Q|_{F_B^{\perp} \cap F^0} > 0$ for all real planes B it follows that $\langle , \rangle$ is positive-semidefinite on all of the subspaces $(F_z + F_{\sigma z})^{\perp}$ (hermitian orthogonal complement).

Let $\pi(z) = x \in \mathsf{R}^4 = S^4 \setminus \{\infty\}$. We let the matrix in § 3.6 correspond to x, but now we denote it X_2 (for reasons which will become clear later):

$$X_2 = \begin{pmatrix} x_4 + ix_3 & -x_2 + ix_1 \\ x_2 + ix_1 & x_4 - ix_3 \end{pmatrix}. \tag{16}$$

From the description of π in § 3.6 it is clear that $F_z + F_{\sigma z} = F_{-1} \otimes \mathbf{P}_x$ if $\pi(z) = x$.

We choose a basis $(e_1, \ldots, e_n)$ in F_{-1} and set $e_{ka} = e_k \otimes \epsilon_a$. This is a basis of $F_{-1} \otimes T$, which we write in the following order: $(e_{11}, \ldots, e_{n1}; \ldots; e_{14}, \ldots, e_{n4})$.

28. Lemma. *The condition $\langle F_z, F_{\sigma z}\rangle = 0$ is equivalent to the following symmetry property of the Gram matrix of the basis (e_{ka}):*

$$\Phi = \begin{pmatrix} A & D^+ \\ D & R \end{pmatrix},$$

where

$$A = \begin{pmatrix} A_n & 0 \\ 0 & A_n \end{pmatrix}, R = \begin{pmatrix} R_n & 0 \\ 0 & R_n \end{pmatrix}, D = \begin{pmatrix} B & C \\ -C^+ & B^+ \end{pmatrix}, A^+ = A, R^+ = R$$

(the + denotes hermitian conjugation). □

The proof of this lemma is immediate.

The block A is the Gram matrix of our basis of the subspace $F_{-1} \otimes \epsilon_1 + F_{-1} \otimes \epsilon_2 = F_{-1} \otimes \mathbf{P}_\infty$. Since φ imbeds this subspace into F_0 and the metric on $\varphi(F_{-1} \otimes \mathbf{P}_\infty)$ is non-degenerate, it follows that $\det A \neq 0$. We shall later show that $A > 0$. If we use the fact that $\langle , \rangle$ is positive-semidefinite on $(F_{-1} \otimes \mathbf{P}_\infty)^\perp$, we see that this implies $\Phi \geq 0$, i.e., $\langle , \rangle > 0$ on F_0 (since the metric is non-degenerate on F_0), and, finally, $Q > 0$ on F^0.

We set

$$X = \overline{X}_2 \otimes E_n = \begin{pmatrix} (x_4 - ix_3)E_n & -(x_2 + ix_1)E_n \\ (x_2 - ix_1)E_n & (x_4 + ix_3)E_n \end{pmatrix}.$$

The rows of the matrix $(\overline{X}, E_{2n})$ obviously form a basis for the subspace $F_{-1} \otimes \mathbf{P}_x$ in the coordinates corresponding to the basis (e_{ka}) of $F_{-1} \otimes T$. We add the rows of the matrix $(E_{2n}, 0)$ to this basis, and compute the new Gram matrix of the resulting set of vectors:

$$\Phi' = \begin{pmatrix} E_{2n} & 0 \\ X & E_{2n} \end{pmatrix} \begin{pmatrix} A & D^+ \\ D & R \end{pmatrix} \begin{pmatrix} E_{2n} & X^+ \\ 0 & E_{2n} \end{pmatrix} = \begin{pmatrix} A & D^+(x) \\ D(x) & R(x) \end{pmatrix}.$$

Here $D(x) = D + AX, R(x) = |x|^2 A + DX^+ + XD^+ + R$. Since the metric on $F_{-1} \otimes \mathbf{P}_x$ is non-degenerate, we have $\det R(x) \neq 0$ for all $x \in \mathbf{R}^4$. The formula

$$(E_{2n}, -D^+(x)R(x)^{-1})\Phi' \begin{pmatrix} 0 \\ E_{2n} \end{pmatrix} = 0$$

implies that the rows of the matrix $(E_{2n}, -D^+(x)\overline{R}(x)^{-1})$ are a basis of $(F_{-1} \otimes \mathsf{P}_x)^{\perp}$ in the coordinates corresponding to the basis consisting of the rows of the matrix $\begin{pmatrix} E_{2n} & 0 \\ \overline{X} & E_{2n} \end{pmatrix}$. The corresponding Gram matrix is $A - D^+(x)R(x)^{-1}D(x)$. This matrix is semidefinite because of our assumptions.

29. Lemma. $\sum_{a=1}^4 \partial_a^2 R(x)^{-1} \leq 0.$

Proof. Since $R(x)$ is hermitian and non-degenerate, it is sufficient to verify that

$$S(x) = -R(x)\Big(\sum_{a=1}^{4} \partial_a^2 R(x)^{-1}\Big) R(x) \geq 0.$$

We have

$$S(x) = \sum_{a=1}^{4} \partial_a^2 R(x) - 2\partial_a R(x) R(x)^{-1} \partial_a R(x).$$

We shall show that

$$S(x) = 4T(A - D^+(x)R(x)^{-1}D(x)),$$

where the operator T is defined on $(2n \times 2n)$-matrices by the formula

$$T\begin{pmatrix} Z_{11} & Z_{12} \\ Z_{21} & Z_{22} \end{pmatrix} = \begin{pmatrix} Z_{11} + Z_{22} & 0 \\ 0 & Z_{11} + Z_{22} \end{pmatrix}.$$

It is clear that T takes semidefinite matrices to semidefinite matrices, and so this will imply the lemma.

We set $\Sigma_a = \partial_a X$. From (16) and the properties of the Pauli matrix it follows that $\Sigma_a \Sigma_b^+ + \Sigma_b \Sigma_a^+ = 2\delta_{ab} E_{2n}$. In addition, we have $D = \sum_{a=1}^4 D_a \Sigma_a$, where

$$D_1 = \frac{C - C^+}{2i} \otimes E_2, \qquad D_2 = -\frac{C + C^+}{2} \otimes E_2,$$

$$D_3 = \frac{B - B^+}{2i} \otimes E_2, \qquad D_4 = \frac{B + B^+}{2} \otimes E_2.$$

This implies that

$$R(x) = |x|^2 A + 2\sum_{a=1}^{4} x_a D_a + R.$$

In particular, $R(x)$ has the form $\begin{pmatrix} Y & 0 \\ 0 & Y \end{pmatrix}$. Thus,

$$S(x) = 8A - 4\sum_{a=1}^{4} (x_a A + D_a) R(x)^{-1} (x_a A + D_a),$$

$$T(A - D^+(x)R(x)^{-1}D(x)) = 2A - \sum_{a,b=1}^{4} (x_a A + D_a)R(x)^{-1}(x_b A + D_b)\Sigma_a^+\Sigma_b,$$

and the lemma follows from this, since $T(\Sigma_a^+\Sigma_b) = 2\delta_{ab}E_{2n}$. □

We can now complete the proof of the positivity theorem. Asymptotically on a sphere of large radius r in $\mathbf{R}^4$ we have $\partial_a R(x)^{-1} = -2x_a r^{-4}A^{-1} + O(r^{-4})$. If we use Stokes' formula to compute the integral $\int_{|x|\le r} d^4x \sum_{a=1}^4 \partial_a^2 R(x)^{-1}$, we find that $A^{-1} > 0$. Hence $A > 0$, as required. □

§ 4. Instantons and Modules over a Grassmannian Algebra

1. Special modules and monads. Let $\mathsf{P} = \mathsf{P}(T)$, and, as above, let

$$\mathcal{F}\colon\quad 0 \longrightarrow F_{-1} \otimes \mathcal{O}_{\mathsf{P}}(-1) \xrightarrow{\alpha} F_0 \otimes \mathcal{O}_{\mathsf{P}} \xrightarrow{\beta} F_1 \otimes \mathcal{O}_{\mathsf{P}}(1) \longrightarrow 0 \qquad (1)$$

be a special monad on P. We consider the graded vector space $F = F_{-1} \oplus F_0 \oplus F_1$, and we define an action of T on it as follows. The action is a homogeneous bilinear map of degree 1:

$$T \otimes F_i \longrightarrow F_{i+1}, \qquad \xi \otimes f \mapsto \xi f \qquad (\text{here } F_i = \{0\} \text{ for } i \ge 2),$$

where $T \otimes F_{-1} \longrightarrow F_0$ is the T-dual of the map $\Gamma(\alpha(1))\colon F_{-1} \longrightarrow F_0 \otimes T^*$, and $T \otimes F_0 \longrightarrow F_1$ is the T-dual of the map $\Gamma(\beta)\colon F_0 \longrightarrow F_1 \otimes T^*$. Since $\beta\alpha = 0$ it follows that $\xi^2 f = \xi(\xi f) = 0$ for all $\xi \in T, f \in F$. Hence, our action of T extends uniquely to a graded $\wedge(T)$-module structure on F. The modules which correspond to special monads (1) will also be called "special." They are characterized by two conditions: (a) $F_i = \{0\}$ for $|i| > 1$; (b) if $\xi \in T, \xi \ne 0$, then multiplication by ξ gives an injective map $\xi\colon F_{-1} \longrightarrow F_0$ and a surjective map $\xi\colon F_0 \longrightarrow F_1$.

The category of special $\wedge(T)$-modules (with zero degree maps as the morphisms) is obviously equivalent to the category of special monads, which, by Proposition 22 in § 3, is equivalent to the category of admissible sheaves (which contains the instantons).

Our first goal is to use our new description of admissible sheaves in order to characterize the instantons $(\mathcal{E}_0, \nabla)$ by means of the singularities of the analytic continuation of the connection ∇ to its maximum domain of definition in $G(2; T)$. Of course, the results of § 3 imply that an instanton is essentially an algebraic object, and so the analytic continuation and the singularities can be characterized purely algebraically. Moreover, the real structure and the metric are not essential here; as a result, our results will relate to a certain subclass of admissible sheaves, which we shall now describe.

2. Analytic instantons and menads. By an "analytic instanton" we mean any admissible sheaf $\mathcal{E}_Z$ on $\mathbf{P}(T)$ with the following property: through any point of $\mathbf{P}(T)$ there is a line on which the restriction of $\mathcal{E}_Z$ is trivial. The usual instantons have this property, since there is a line passing through any point of $\mathbf{P}(T)$.

Let $\mathcal{E}_Z$ be an analytic instanton, and let $U \subset G(2;\, T)$ be the Zariski open set of points which correspond to all lines in $\mathbf{P}(T)$ on which the restriction of $\mathcal{E}_Z$ is trivial. We call the complement of U the "singular set" of $\mathcal{E}_Z$. By an "α-plane" in $G(2;\, T)$ we mean the set of points which parametrize the lines through a fixed point of $\mathbf{P}(T)$; then obviously the singular set of an analytic instanton does not contain any α-planes.

We shall equip the singular set with a canonical sheaf, from which it is even possible to recover $\mathcal{E}_Z$ uniquely. To do this, we make the following definition.

A *menad of rank d* is an injective morphism of sheaves on $M = G(2;\, T)$ of the form $\mathcal{O}_M^d \xrightarrow{\gamma} \mathcal{O}_M^d(1)$ (where $\mathcal{O}(1) = \wedge^2 \mathcal{S}^*$ is the sheaf which determines the Plücker imbedding). We call the sheaf $\mathcal{D} = \operatorname{Coker} \gamma$ the "singularity sheaf" of the menad.

3. Theorem. *The following categories are equivalent:*

(a) *the category of analytic instantons having no nonzero global sections;*

(b) *the category of menads whose singularity sheaf has support which does not contain any α-planes;*

(c) *the category of singularity sheaves of such menads.*

Under this correspondence the singular set of an instanton is the support of the singularity sheaf of the corresponding menad, and the rank of the menad is equal to the second Chern number of the instanton.

Proof. We divide the proof into steps, in each of which we construct one of the functors giving the equivalences of categories.

From instantons to menads. Given an analytic instanton $\mathcal{E}_Z$, we construct the corresponding monad (1) as in § 3; then we construct the $\wedge(T)$-module corresponding to this monad, as in § 4.1; and finally, from the $\wedge(T)$-module we construct the morphism of sheaves

$$F_{-1} \otimes \mathcal{O}_M \xrightarrow{\gamma} F_1 \otimes \mathcal{O}_M(1),$$

where γ is determined by requiring the map $\Gamma(\gamma)\colon F_{-1} \longrightarrow F_1 \otimes \wedge^2 T^*$ to be dual to the multiplication $\wedge^2 T \otimes F_{-1} \longrightarrow F_1$.

We must verify that γ is injective, and that its cokernel is concentrated within the singular set of $\mathcal{E}_Z$.

We first explain how, starting with a monad, one computes the space of sections of the corresponding fibration. We consider a complex of the form (1) on a projective space of any dimension, and we rewrite it as two exact sequences, by introducing $\mathcal{K} = \operatorname{Ker}\beta$. From the first sequence we find that $H^0(\mathcal{E}_Z) = H^0(\mathcal{K})$. The second sequence gives an identification of both of these groups with the kernel of the map $\Gamma(\beta): F_0 \longrightarrow F_1 \otimes T^*$. In terms of the corresponding $\wedge(T)$-module we obtain:

$$H^0(\mathcal{E}_Z) = \{f_0 \in F_0 \mid \xi f_0 = 0 \text{ for all } \xi \in T\}. \tag{2}$$

In particular, this kernel is trivial for an analytic instanton with no nonzero sections. If we apply (2) to the restriction of (1) to the line $\mathsf{P}(S(x))$, where $S(x) \subset T$ is any plane, we obtain

$$H^0(\mathcal{E}_Z|_{\mathsf{P}(S(x))}) = \{f_0 \in F_0 \mid \xi f_0 = 0 \text{ for all } \xi \in S(x)\}. \tag{3}$$

We set $d_i = \dim F_i$. According to (3), the space of all sections of $\mathcal{E}_Z$ over $\mathsf{P}(S(x))$ is the intersection in F_0 of the two subspaces $\operatorname{Ker}\xi_1$ and $\operatorname{Ker}\xi_2$, where ξ_1 and ξ_2 form a basis of $S(x)$. Since the maps $\xi_i: F_0 \longrightarrow F_1$ are surjective, the dimension of these spaces is $d_0 - d_1$. Hence, the dimension of their intersection is at least $d_0 - 2d_1$. In addition, the rank of $\mathcal{E}_Z$ is equal to $d_0 - d_1 - d_{-1}$.

In order for the sheaf $\mathcal{E}_Z$ to be an analytic instanton, it is necessary and sufficient that the following condition be satisfied: each α-plane has a point x such that $\dim H^0(\mathcal{E}_Z|_{\mathsf{P}(S(x))}) = \operatorname{rank}\mathcal{E}_Z$ and none of the sections of $\mathcal{E}_Z$ over $\mathsf{P}(S(x))$ vanishes. In fact, this is equivalent to triviality of $\mathcal{E}_Z|_{\mathsf{P}(S(x))}$. According to our analysis above, these conditions mean that $d_1 = d_{-1}$ and the subspaces $\operatorname{Ker}\xi_1$ and $\operatorname{Ker}\xi_2$ in F_0 are in general position. But if $d_1 = d_{-1}$, then the rank of $\mathcal{E}_Z|_{\mathsf{P}(S(x))}$ at the singular points $x \notin U$ can only become less than the dimension of the space of sections, and this is possible only if there are sections which vanish. Thus, in place of the condition $H^0(\mathcal{E}_Z|_{\mathsf{P}(S(x))}) = \operatorname{rank}\mathcal{E}_Z$ we can stipulate that $d_1 = d_{-1}$.

Furthermore, by the identification in (3), the section of $\mathcal{E}_Z|_{\mathsf{P}(S(x))}$ corresponding to a vector $f_0 \in F_0$ vanishes at the point corresponding to $\eta \in S(x)$ if and only if $f_0 = \eta f_{-1}$ for some $f_{-1} \in F_{-1}$. In other words, suppose that $\xi \wedge \eta \in \wedge^2 T$ is the bivector corresponding to $S(x)$. Then $\mathsf{P}(S(x))$ is a singular line for $\mathcal{E}_Z$ if and only if the multiplication $\xi \wedge \eta: F_{-1} \longrightarrow F_1$ has nontrivial kernel. Since $d_1 = d_{-1}$, we can speak of the cokernel rather than the kernel, and the cokernel is nontrivial precisely at the points in the support of the sheaf $\mathcal{D} = \operatorname{Coker}(F_{-1} \otimes \mathcal{O}_M \xrightarrow{\gamma} F_1 \otimes \mathcal{O}_M(1))$ on M.

It remains for us to verify that the support of $\mathcal{D}$ does not contain any α-planes in M. Such a plane in M corresponds to a three-dimensional linear subspace of $\wedge^2 T$ consisting entirely of decomposable bivectors, and the subspaces $S(x) \subset T$ corresponding to its points all pass through a single line in T.

Let ξ be a twistor lying on this line, and suppose that the support of $\mathcal{D}$ contains the corresponding plane. Then the kernel of multiplication by $\eta \wedge \xi$ from F_{-1} to F_1 is nontrivial for all $\eta \in T$. But this means that all of the lines through the point of $\mathsf{P}(T)$ corresponding to ξ are singular for $\mathcal{E}_Z$, in contradiction to the definition of an analytic instanton.

The map we have constructed clearly extends to morphisms, and so is a functor. We note that the definition of this function did not make use of the condition that $\mathcal{E}_Z$ not have any nonzero global sections.

From menads to instantons. First, given a menad $F_{-1} \otimes \mathcal{O}_M \xrightarrow{\gamma} F_1 \otimes \mathcal{O}_M(1)$, we construct a graded $\wedge(T)$-module $F' = F_{-1} \oplus (T \otimes F_{-1}) \oplus F_1$ with the following action of T. On F_{-1} the action is the identity morphism $T \otimes F_{-1} \longrightarrow T \otimes F_{-1} = F_0'$; on F_0' it is the composition

$$T \otimes T \otimes F_{-1} \xrightarrow{\lambda \otimes \mathrm{id}} \wedge^2 T \otimes F_{-1} \xrightarrow{\gamma'} F_1,$$

where λ is the skew-symmetrization map and γ' is the $\wedge^2 T^*$-dual of $\Gamma(\gamma)$. In order to verify that F' is a special $\wedge(T)$-module, it suffices to show that the multiplication $\xi: F_0' \longrightarrow F_1$ is surjective for all $\xi \neq 0$. If this were not the case for some ξ, then for all $\eta \in T$ the map $\xi \wedge \eta: F_{-1} \longrightarrow F_1$ would fail to be surjective. But this would mean that the support of the singularity sheaf of the menad contains an α-plane, which does not happen, by assumption.

However, the instanton corresponding to the $\wedge(T)$-module F' could have global sections, as described in (2). To get rid of them, one must replace F' by $F = F'/K$, where

$$K = \{f_0 \in F_0' = T \otimes F_{-1} \mid \xi f_0 = 0 \text{ for all } \xi \in T\}.$$

It is easy to see that F is a special $\wedge(T)$-module, and corresponds to the same monad. In particular, the multiplication $\xi: F_{-1} \longrightarrow F_0'/K = F_0$ has zero kernel, since otherwise the α-plane corresponding to ξ would be contained in the support of $\mathcal{D}$.

We have thus constructed a functor in the opposite direction.

In order to see that these functors are inverse to one another, it is enough to show that any morphism of instantons without global sections can be uniquely recovered from the corresponding map on the degree -1 and 1 components of the $\wedge(T)$-modules, without using the 0-degree components. For this it is sufficient to check that a map which is zero in degrees -1 and 1 must also be zero in degree 0. But such a map takes the entire 0-component of the first module to the kernel of multiplication by T in the 0-component of the second module. This kernel is zero, because of the absence of global sections.

From singularity sheaves to menads and back. We have the obvious functor from the category of menads to the category of their singularity sheaves. Conversely, in order to reconstitute a menad from its singularity sheaf $\mathcal{D}$, we set

$$F_1 = H^0(M, \mathcal{D}(-1)), \qquad F_{-1} = \operatorname{Ker}(F_1 \otimes \mathcal{O}_M(1) \xrightarrow{\tilde{\gamma}} \mathcal{D}), \qquad \gamma = \tilde{\gamma}(-1),$$

where $\tilde{\gamma}$ is the tautological map. Clearly, if we start with a menad and construct the corresponding $\mathcal{D}$, then this returns us to a menad which is canonically isomorphic to the original one. In order to verify that these functors give an equivalence of categories, it is enough to see that they give a bijection between the morphisms of monads and the morphisms of their singularity sheaves. But this is almost obvious: any morphism $\mathcal{D}' \longrightarrow \mathcal{D}''$ can be uniquely lifted to a morphism $F_1' = H^0(\mathcal{D}'(-1)) \longrightarrow H^0(\mathcal{D}''(-1)) = F_1''$, which then induces a uniquely determined morphism $F_{-1}' \longrightarrow F_1''$ on the kernels of γ' and γ''. □

4. Remarks. The main drawback of Theorem 3 is that it does not give us an independent characterization of the singularity sheaves of menads. The following construction is a step in the direction of such a characterization. First, given a menad of rank d, we construct a divisor D of degree d on the d-closed subsystem of codimension 1 in M whose sheaf of ideals is the image of the determinant morphism:

$$\det(\gamma(-1)) \colon \wedge^d F_{-1} \otimes \mathcal{O}_M(-d) \longrightarrow \wedge^d F_1 \otimes \mathcal{O}_M.$$

It is easy to see that D is a support scheme of $\mathcal{D}$, i.e., $\mathcal{D}$ is a sheaf of $\mathcal{O}_D$-modules.

Now suppose that $\mathcal{E}_Z$ is the analytic instanton corresponding to a monad. Further suppose that we have a non-degenerate scalar product on $\mathcal{E}_Z$ which is given by an isomorphism $\mathcal{E}_Z \xrightarrow{\sim} \mathcal{E}_Z^* = \mathcal{H}om(\mathcal{E}_Z, \mathcal{O}_{\mathbf{P}})$. It is not hard to check that $\mathcal{E}_Z^*$ corresponds to the monad $\mathcal{F}^* = \mathcal{H}om(\mathcal{F}, \mathcal{O}_{\mathbf{P}})$, which in turn corresponds to the menad $\tilde{\mathcal{F}}^* = \mathcal{H}om(\tilde{\mathcal{F}}, \mathcal{O}_M(1)) = \mathcal{H}om(\tilde{\mathcal{F}}, \Omega^4 M(5))$ (here $\mathcal{H}om$ is a complex of sheaves of morphisms). Finally, on the singularity sheaves we obtain

$$\mathcal{D}^* = \mathcal{H}om(\mathcal{D}, \omega_D(5)),$$

where ω_D is the dualizing sheaf of the singularity divisor D. Thus, the non-degenerate scalar product $\mathcal{E}_Z \times \mathcal{E}_Z \longrightarrow \mathcal{O}_{\mathbf{P}}$ is transformed to a non-degenerate scalar product $\mathcal{D} \times \mathcal{D} \longrightarrow \omega_D(5)$ on D. If D is reduced and irreducible, then $\omega_D(5) = \mathcal{O}_D(d+1)$, where $d = d_\pm = c_2(\mathcal{E}_Z)$. However, the general problem of characterizing the D which arise from instantons and the sheaves $\mathcal{D}$ on them, is unsolved.

5. Example. Let $d = 1$. Then the instanton menads are in one-to-one correspondence with the smooth hyperplane sections D of the Klein quadric

$M \subset \mathsf{P}(\wedge^2 T)$: namely, the corresponding menad is isomorphic to the complex $\mathcal{O}_M \xrightarrow{\gamma} \mathcal{O}_M(1)$, where γ is multiplication by the equation of D. If D were not smooth, then the corresponding hyperplane would be tangent to M at some point x, and it would cut out a light cone at this point — but any light cone contains α-planes.

If we let D' denote the hyperplane in $\wedge^2 T$ whose image in $\mathsf{P}(\wedge^2 T)$ cuts out D, we can write the monad of the corresponding instanton in the form $0 \longrightarrow \mathcal{O}_{\mathsf{P}}(-1) \longrightarrow T \otimes \mathcal{O}_{\mathsf{P}} \longrightarrow (\wedge^2 T/D') \otimes \mathcal{O}_{\mathsf{P}}(1) \longrightarrow 0$. In the language of $\wedge(T)$-modules, this is $\wedge(T)/(D' \oplus \wedge^3 T \oplus \wedge^4 T)$.

The parameter space of the analytic instantons with $d = 1$ is thus the affine manifold $\mathsf{P}(\wedge^2 T^*) \setminus M^*$, where M^* is the dual quadric of M.

Any such instanton has rank two. If we add a realness condition and the stipulation that there be no singularities on S^4, we obtain the following restrictions on the divisors D:

(a) D is real;

(b) $D \cap S^4 = \emptyset$.

When rank $\mathcal{E}_Z = 2$, we note that $\mathcal{E}_Z$ has a skew-symmetric real scalar product $\mathcal{E}_Z \otimes \mathcal{E}_Z \longrightarrow \wedge^2 \mathcal{E}_Z \simeq \mathcal{O}_{\mathsf{P}}$, which gives a reduction of the structure group to $\mathrm{Sp}(1)$. We have thus obtained a geometric description of the moduli space of $\mathrm{Sp}(1)$-instantons with $c_2 = 1$. With some effort, this can be extended to the case $c_2 = 2$.

6. Scalar products and real structures on $\wedge(T)$-modules. Let $F = F_{-1} \oplus F_0 \oplus F_1$ be the special $\wedge(T)$-module corresponding to an $O(r)$-instanton. If we translate the linear algebra data in § 3.4 into the language of this structure, we obtain the following collection of maps.

(a) A non-degenerate bilinear scalar product (f, g) on F. In listing its properties, we shall write $\tilde{f} = i$ if $f \in F_i$. Then for homogeneous $f, g \in F$ we have

$$(f, g) = \begin{cases} 0, & \text{if } \tilde{f} + \tilde{g} \neq 0, \\ (-1)^{\tilde{f}\tilde{g}}(g, f), & \text{if } \tilde{f} + \tilde{g} = 0, \end{cases}$$

and

$$(\xi f, g) = (-1)^{\tilde{f}}(f, \xi g) \quad \text{for} \quad \xi \in T.$$

(b) An antilinear map $\sigma: F \longrightarrow F$ with the properties

$$\widetilde{\sigma f} = \tilde{f}, \qquad \sigma^2(f) = (-1)^{\tilde{f}} f, \qquad \sigma(\xi f) = \sigma(\xi)\sigma(f), \qquad (\sigma f, \sigma g) = \overline{(f, g)}.$$

(In §§ 5–6 of Chapter 3 we shall see that these axioms are a special case of two natural supercommutative algebra structures: a symmetric scalar product and one of the superreal structures, respectively.)

(c) If $\xi \in T \setminus \{0\}$ and $i \neq 0$, then $\mathrm{Im}(\xi: F_{i-1} \longrightarrow F_i) = \mathrm{Ker}(\xi: F_i \longrightarrow F_{i+1})$.

(d) The scalar product is positive-definite on the real (i.e., σ-invariant) elements of F_0.

In general, a finite dimensional graded $\wedge(T)$-module $F = \bigoplus_{i\in\mathbf{Z}} F_i$ with the data and conditions in (a), (b) and (c) is called a "grassmannian module." It is called an instanton grassmannian module if it also satisfies $F_i = \{0\}$ for $|i| > 1$ and condition (d) is fulfilled. The notion of an instanton module is equivalent to the notion of orthogonal linear algebra data. The notion of a grassmannian module, which is somewhat more general, is essential in order for us to learn how to compute the $\wedge(T)$-modules of instantons which correspond to certain non-fundamental representations ρ of the structure group. More precisely, given two instanton $\wedge(T)$-modules, below we shall show how to compute the instanton $\wedge(T)$-module corresponding to the tensor product of our original instantons.

7. The tensor product of grassmannian modules. Let F' and F'' be two grassmannian modules. We set ($\xi \in T, f' \in F', f'' \in F''$):

$$(F' \otimes F'')_k = \bigoplus_{i+j=k} \left(F'_i \otimes_{\mathbf{C}} F''_j\right),$$

$$\xi(f' \otimes f'') = \xi f' \otimes f'' + (-1)^{\widetilde{f'}} f' \otimes \xi f'',$$

$$(f' \otimes f'', g' \otimes g'') = (-1)^{\widetilde{f''}\widetilde{g'}}(f', g')(f'', g''),$$

$$\sigma(f' \otimes f'') = \sigma f' \otimes \sigma f''.$$

An immediate verification shows that all of the axioms of a grassmannian module are fulfilled for $F' \otimes F''$. (This is an example of the "rule of signs" in a supercommutative algebra; see § 1 of Chapter 3.)

If the modules F' and F'' are instanton modules, then $F' \otimes F''$ is not, because, for example, $(F' \otimes F'')_{-2} = F'_{-1} \otimes F''_{-1} \neq \{0\}$. It turns out that this difficulty is essentially the only one.

8. Theorem. *Let F' and F'' be two instanton modules, and set $F = (\wedge(T) \cdot (F'_{-1} \otimes F''_{-1}))^{\perp}$, where the orthogonal complement is taken with respect to the scalar product in $F' \otimes F''$. Then F is an instanton module, and the corresponding sheaf $\mathcal{E}_Z$ is isomorphic to the tensor product of the sheaves $\mathcal{E}'_Z$ and $\mathcal{E}''_Z$ corresponding to F' and F''.*

Proof. Suppose that F' is a grassmannian module. It determines a vector bundle $E(F')$ over $\mathbf{P}(T)$ with fibre $\mathrm{Ker}(\xi: F_0 \longrightarrow F_{-1})/\xi F_{-1}$ over the point corresponding to ξ. Here it is not hard to see that we have $E(F' \otimes F'') = E(F') \otimes E(F'')$ (tensor product of vector bundles).

If F' and F'' are instanton modules, then $E(F' \otimes F'')$ is the appropriate tensor product of the two corresponding vector bundles. We now make use of a general theorem of I. N. Bernshtein, I. M. Gel'fand and S. I. Gel'fand (see [14]), from which it follows that $E(F' \otimes F'') = E(F)$, where $F \subset F' \otimes F''$ is an instanton submodule, and $F' \otimes F'' = F \oplus P$ with $P \subset F' \otimes F''$ a free graded $\wedge(T)$-submodule.

We now show that $P = N$, where $N = \wedge(T) \cdot (F'_{-1} \otimes F''_{-1})$. Since F is an instanton module, we have $F_{-2} = 0$, and so P must contain N. On the other hand, we claim that none of the free generators of the module P can lie outside $(F' \otimes F'')_{-2}$. In fact, the homogeneous elements of degree > -2 in $F' \otimes F''$ are killed by multiplication by $\wedge^4 T$, since $(F' \otimes F'')_k = 0$ for $k > 2$. Thus, $P = N$; in particular, this means that N is free over $\wedge(T)$.

From the last remark it follows that the dimension of N_2 is the same as the dimension of $N_{-2} = (F' \otimes F'')_{-2}$. The scalar product gives an identification of $(F' \otimes F'')_{-2}$ with $(F' \otimes F'')_2^*$. Hence, $N_2 = (F' \otimes F'')_2$. As a result, the orthogonal complement $F = N^\perp$ does not contain any elements of degree $i \neq 0,\ 1,\ -1$. The relation $(\xi f,\ g) = (-1)^{\tilde{f}}(f,\ \xi g)$ implies that $N^\perp$ is a $\wedge(T)$-submodule. Since σ takes N to itself, and $(\sigma f,\ \sigma g) = \overline{(f,\ g)}$, it follows that σ also takes F to itself. The scalar product on $F' \otimes F''$ induces a scalar product on F. We conclude that F is the desired instanton module. □

§ 5. The Diagram of Null-Geodesics

1. Structure of the diagram. The self-duality diagram introduced in § 2.6 exists only for space-time which has a three-dimensional family of null directions. General space-time, as in § 1, does not have such diagrams. In this section as a substitute we introduce a double fibration one of whose bases is the space of null-geodesics (complex light rays). The Radon–Penrose transform that is based on the diagram of null-geodesics opens up many possibilities which make it an interesting object of study both in the self-dual case and even in the flat case. This is the theme of the remainder of the chapter.

Let M be a complex space-time with the structure that was described in § 1.1. Here we shall not be interested in the real structure — one will encounter no special difficulty if one wants to incorporate the real structure into the theorems. The choice of a metric in the conformal class given by the spinor decomposition $\Omega^1 M = S_+ \otimes S_-$ determines a 1-conic connection on $F = \mathsf{P}(S_+^*) \underset{M}{\times} \mathsf{P}(S_-^*)$, where F is the 1-conic structure of null-directions. Namely, one constructs the Levi-Civita connection from the metric, and on F one considers the fibering $\mathcal{T}F/L$ of lifted null-geodesics relative to that connection. Exactly the same computations as in ordinary differential geometry show that a different choice of metric in the same conformal class does not change $\mathcal{T}F/L$. We shall always assume that $\mathcal{T}F/L$ integrates to a fibration; in that case the double fibration

$$L \xleftarrow{\pi_1} F \xrightarrow{\pi_2} M$$

will be called a diagram of null-geodesics. The fibres of π_2 are two-dimensional quadrics; the fibres of π_1 are sheets of the fibering of lifted null-geodesics. We say that M is "small" if M is a Stein manifold, if it is convex-geodesic for a suitable metric in the conformal class, and if the fibres of π_1 are connected and simply connected. Any space-time has a basis consisting of small open sets.

The flat diagram of null-geodesics is the double fibration

$$L = F(1,\, 3;\, T) \xleftarrow{\pi_1} F = F(1,\, 2,\, 3;\, T) \xrightarrow{\pi_2} M = G(2;\, T).$$

We recall that one cannot define nonzero sections of $\wedge^2 S_\pm$ on all of M.

In the flat case, there is a very important additional structure on L: a closed imbedding $L \subset \mathbf{P}(T) \times \mathbf{P}(T^*)$, which, in the language of functors of points, associates to a (1, 3)-flag in T its two components. The existence of this imbedding is connected with the fact that M is both self-dual and anti-self-dual: the projections $L \longrightarrow \mathbf{P}(T)$ and $L \longrightarrow \mathbf{P}(T^*)$ set up a correspondence between a null-geodesic and one of the two null-directions containing it. Thus, in the general case there is no such imbedding. But one can say that the mechanism of the Radon–Penrose transform allows us essentially to use infinitesimal neighborhoods $L^{(i)}$ for such an imbedding. The problem of constructing these neighborhoods in the general curved case is not completely solved. This problem splits in two: constructing a (co)normal sheaf for the imbedding $L \subset L^{(1)}$; i.e., constructing the kernel I of the restriction $\mathcal{O}_{L^{(1)}} \longrightarrow \mathcal{O}_L$ as an $\mathcal{O}_L$-module; and constructing an extension $\mathcal{O}_{L^{(i)}} \longrightarrow \mathcal{O}_L$, filtered by powers of the kernel, in such a way that the associated graded sheaf of rings is the symmetric algebra of the conormal sheaf mod I^{i+1}. The first part of the problem is solved by the following construction of Le Brun.

2. The sheaf I. We let $I \subset \Omega^1 L$ denote the sheaf of holomorphic forms ω with the following property: ω vanishes on every tangent vector in each quadric $L(x) = \pi_1\pi_2^{-1}(x), x \in M$. Below we shall prove Le Brun's theorem, which says that I is a rank 1 local direct summand. The reader might verify as an exercise that $I \simeq \mathcal{O}(-1, -1)$ in the flat case. On the other hand, L is given in $\mathbf{P}(T) \times \mathbf{P}(T^*)$ by the equation $s = \sum_{i=1}^4 t_i \otimes t^i = 0$, where (t_i) and (t^i) are dual bases in T^* and T, and hence the conormal sheaf of the imbedding $L \subset \mathbf{P}(T) \times \mathbf{P}(T^*)$ is also isomorphic to $\mathcal{O}(-1, -1)$. This is the first indication that I is the right candidate for the role of conormal subsheaf $L \subset L^{(i)}$ in the general situation.

The geometrical meaning of I is as follows. Let $l \subset M$ be a null-geodesic. Its lift is contained in the three-dimensional manifold $\pi_2^{-1}(l) \subset F$. For simplicity, suppose that M is convex-geodesic. Then for any other null-geodesic l' either $l' \cap l$ is empty or else it consists of a single point. Thus, on this manifold the map $\pi_1: \pi_2^{-1}(l) \longrightarrow L$ contracts the lifting of l to a point, and does not pinch together anything else. Consequently, the point y corresponding to l is singular in the image of this manifold. For this reason the Zariski tangent space to $\bigcup_{L(x)\ni y} L(x)$ at the point y cannot be three-dimensional: it has dimension 4 or 5. But I is precisely the sheaf of equations of these tangent spaces; the fact that it is a rank 1 local direct summand means that all of the tangent spaces are four-dimensional.

To prepare for the proof of this theorem, we introduce the following notation. Let $\mathcal{O}_F(a, b)$ be the sheaves on F corresponding to the realization of F as a relative quadric $\mathrm{P}(S_+^*) \underset{M}{\times} \mathrm{P}(S_-^*)$. As in § 2.1, we set $\mathcal{N} = \mathrm{Ker}(\mathrm{res}: \pi_2^*\Omega^1 M \longrightarrow \Omega^1 F/L)$. We have the commutative diagram

$$\begin{array}{ccc} \pi_2^*(\Omega^1 M) & \xrightarrow{\mathrm{res}} & \Omega^1 F/L \\ \simeq \Big\uparrow \pi_2^*(\sigma) & & \simeq \Big\uparrow \\ \pi_2^*(S_+) \otimes \pi_2^*(S_-) & \xrightarrow{j_+ \otimes j_-} & \mathcal{O}_F(1, 1), \end{array}$$

where σ is the spinor decomposition of $\Omega^1 M$, and $j_\pm$ are the morphisms in the sequences

$$\begin{array}{ccccccccc} 0 & \longrightarrow & \pi_2^* \wedge^2 S_+(-1, 0) & \xrightarrow{i_+} & \pi_2^* S_+ & \xrightarrow{j_+} & \mathcal{O}_F(1, 0) & \longrightarrow & 0, \\ 0 & \longrightarrow & \pi_2^* \wedge^2 S_-(0, -1) & \xrightarrow{i_-} & \pi_2^* S_- & \xrightarrow{j_-} & \mathcal{O}_F(0, 1) & \longrightarrow & 0. \end{array} \tag{1}$$

From the identification of res and $j_+ \otimes j_-$ it follows that $\mathcal{N}$ contains the rank one local direct summand

$$\mathcal{N}_0 = \mathrm{Im}\big(i_+ \otimes i_-: \pi_2^*(\wedge^2 S_+ \otimes \wedge^2 S_-(-1, -1)) \longrightarrow \pi_2^*(\Omega^1 M)\big).$$

We now choose sections $\epsilon_\pm \in \wedge^2 S_\pm$, the metric $g = \epsilon_+ \otimes \epsilon_-$, and the corresponding Levi–Civita connection ∇^g on $\Omega^1 M$. The lifting of this connection to F induces a map $\nabla^g_{F/L}: \pi_2^*\Omega^1 M \longrightarrow \pi_2^*\Omega^1 M \otimes \Omega^1 F/L$.

3. Theorem. *The subsheaf $\mathcal{N}_0 \subset \pi_2^*(\Omega^1 M)$ is invariant relative to $\nabla^g_{F/L}$, i.e., $\nabla^g_{F/L}(\mathcal{N}_0) \subset \mathcal{N}_0 \otimes \Omega^1 F/L$, and there exists a natural isomorphism*

$$\pi_1^{-1}(I) \xrightarrow{\sim} \mathrm{Ker}(\nabla^g_{F/L}: \mathcal{N}_0 \longrightarrow \mathcal{N}_0 \otimes \Omega^1 F/L).$$

Corollary. *$I \subset \Omega^1 L$ is a rank one local direct summand; its restriction to any quadric $L(x)$ is isomorphic to $\mathcal{O}(-1, -1)$.*

Proof. Any null tangent vector X on M is a nonzero product of two spinors at the same point: $X = s_+ \otimes s_-$, where the spinors are determined up to multiplication by a constant. We shall say that these spinors are "tangent" to the given null-direction. Parallel translation along a null-geodesic of a tangent vector to the geodesic preserves the tangency of the vector. Hence, the same holds if one takes a tangent spinor instead of a tangent vector and translates it by means of the connection $\nabla_\pm$. Suppose that $y \in F$, $x = \pi_2(y)$, and $X = s_+(x) \otimes s_-(x) \in \mathcal{T}F/L(y)$ is a vector in $\mathcal{T}F/L$ at the point y. Then, by definition, $\mathcal{N}_0(y) = s_+(x)^\perp \otimes s_-(x)^\perp$. Consequently, parallel translation of a covector in $\mathcal{N}_0(y)$ by means of ∇^g does not take it outside of $\mathcal{N}_0$.

Now suppose that ω is a local section of $\mathcal{N}_0$ with the following type of cylindrical domain of definition W: W is isomorphic to $V \times U$, where $V = \pi_1(W) \subset L$, U is the unit disc in $\mathbf{C}$, and π_1 is the projection of $V \times U$ onto the first factor. The basic step in the proof is to construct a 1-form ν on V such that $\pi_1^*(\nu)|_W = \omega$. We try to construct such a form by defining its values on a vector field Y on V by the formula

$$i_Y(\nu) = i_J(\omega),$$

where J is any lifting of Y to W (a lifting exists because W is cylindrical). The right side of this formula does not change if J is replaced by J', since $J - J'$ is a section of $\mathcal{T}F/L$ and we have $\mathcal{N}_0 \subset (\mathcal{T}F/L)^\perp$. The only problem is that, in general, $i_J(\omega)$ is a section of $\mathcal{O}_F$, and we must show that it is a section of $\pi_1^{-1}(\mathcal{O}_L)$, since $i_Y(\nu)$ is a function on L. Thus, we must show that $d_{F/L}(i_J(\omega)) = 0$. For this purpose we need a small amount of computation, in which for the first time we use the fact that the connection ∇^g is torsion-free.

We construct a vector field X on W such that $i_J(\omega) = \pi_2^* g(X, J)$, i.e., we "lift the indices" of ω by means of g. The field X belongs to $\mathcal{T}F/L$. Hence, once we choose X, we can still change J by multiples of X. We make use of this fact in order to replace J by a vector field $J' = J + hX$ such that $[X, J'] = 0$. If ω does not vanish on W, then neither does X; then $[X, J] = fX$ for some f, and it is sufficient to solve the equation $Xh = -f$. So suppose that we have $[X, J] = 0$. The condition $d_{F/L}(i_J\omega) = 0$ which we would like to verify is equivalent to the condition $X\pi_2^* g(X, J) = 0$. But

$$X\pi_2^* g(X, J) = \pi_2^* g(i_X \nabla^g_{F/L} X, J) + \pi_2^* g(X, i_X \nabla_{F/L} J).$$

Here $i_X \nabla^g_{F/L} X = 0$, since $\nabla^g_{F/L}\omega = 0$. Furthermore, using general formulas, we have

$$i_X(\nabla_{F/L} J) = i_J \nabla_{F/L} X + [X, J] + t(X, J),$$

where t is the torsion tensor of ∇ lifted to F. This lifted torsion tensor is zero if $\nabla = \nabla^g$ is the Levi–Civita connection, and we have $[X, J] = 0$ by our choice of J. We conclude that

$$X\pi_2^* g(X, J) = \pi_2^* g(X, i_J \nabla_{F/L} X) = \frac{1}{2} J\pi_2^* g(X, X) = 0,$$

since X is a null field.

We have thus constructed an injective map of sheaves of $\pi_1^{-1}(\mathcal{O}_L)$-modules

$$\operatorname{Ker} \nabla^g_{F/L} \cap \mathcal{N}_0 \longrightarrow \pi_1^{-1}(I) \colon \omega \mapsto \pi_1^{-1}(\nu)|_W$$

(ν vanishes on vectors tangent to the quadrics $L(x)$, because ω was lifted from M, and so vanishes on π_2-vertical vector fields). Since $\mathcal{N}_0$ has $\mathcal{O}_F$-rank one and the $\mathcal{O}_L$-rank of I is no greater than 1, it is easy to see that this map is an isomorphism. □

4. The Radon–Penrose transform. From the results in § 5.2 it is clear that the sheaf $\mathcal{N} = \operatorname{Ker}(j_+ \otimes j_-)$ can be included in the following exact sequence:

$$0 \longrightarrow \pi_2^*(\wedge^2 S_+ \otimes \wedge^2 S_-)(-1, -1) \longrightarrow \pi_2^*(\wedge^2 S_+ \otimes S_-)(-1, 0) \oplus$$
$$\oplus\, \pi_2^*(S_+ \otimes \wedge^2 S_-)(0, -1) \longrightarrow \mathcal{N} \longrightarrow 0.$$

Here all of the sheaves besides $\mathcal{N}$ are relatively acyclic over M. Hence,

$$R^i \pi_{2*} \mathcal{N} = 0 \qquad \text{for all} \qquad i \geq 0.$$

Thus, if we assume that the fibres of π_1 are connected, we can apply Theorem 2.3 in this situation, thereby concluding that the following categories are equivalent:

(a) the category of M-trivial locally free sheaves $\mathcal{E}_L$ on L;

(b) the category of pairs $(\mathcal{E}, \nabla)$, where $\mathcal{E}$ is a locally free sheaf on M, and ∇ is a connection on $\mathcal{E}$ with trivial monodromy along the null-geodesics.

We note that, unlike in the case of self-duality diagrams, here we do not have to require triviality of the curvature along the geodesic: this condition automatically holds. Thus, while in the self-dual case we need ∇ to be self-dual in order to have any chance of carrying over $(\mathcal{E}, \nabla)$ from M to Z, here we do not impose any differential equation type restrictions on ∇. On the other hand, one is faced with the question: what conditions must be satisfied by a sheaf $\mathcal{E}_L$ on L in order for $(\mathcal{E}, \nabla)$ to be, say, a solution to Yang–Mills equations? We shall take up this matter later. For now we will say a few words about recovering the self-duality diagram from L, by analogy with § 2.8.

5. Deformations of a standard quadric imbedding. If L is the space of null-geodesics, then there is a four-dimensional family of imbedded quadrics $L(x)$ in L. The conormal sheaf of this imbedding $\mathcal{N}^*|_{L(x)} = \mathcal{N}^*(x)$ is the same as in the flat case, as follows from our computations above; in particular, up to isomorphism it does not depend on x. A quadric on a five-dimensional manifold will be called a "standard imbedding" if it has this property. From a standard quadric imbedding $L(x_0) \subset L$ we construct the manifold M of all of its standard imbedding deformations. We have the following facts from Kodaira theory:

(a) M is a four-dimensional open manifold in the space of all deformations of $L(x_0)$. As in § 2.8, this follows from the rigidity of the quadric, the rigidity of the sheaf $\mathcal{N}^*(x)$, and the fact that $H^0(\mathbf{P}^1 \times \mathbf{P}^1, \mathcal{N}_0^*)$ and $H^1(\mathbf{P}^1 \times \mathbf{P}^1, \mathcal{N}_0^*)$ have dimensions 4 and 0, respectively.

(b) Let $F \subset L \times M$ be the graph of the universal family of standard quadric imbeddings. Then the fibres of the projection $F \longrightarrow M$ are quadrics. We suppose that there exists a rank one local direct summand $I \subset \Omega^1 L$ whose restriction to $L(x_0)$ is isomorphic to $\mathcal{O}(-1, -1)$. Then its restrictions to almost all standard imbedding deformations of $L(x_0)$ are isomorphic to $\mathcal{O}(-1, -1)$, since the sheaf $\mathcal{O}(-1, -1)$ is also rigid. By making M smaller if necessary, we may assume that this holds for all $x \in M$.

A simple but important observation is that the sheaf I now automatically has the property in § 5.2. In fact, restriction of the sections of I to vectors tangent to $L(x)$ determines a scalar product $I|_{L(x)} \otimes \mathcal{T}L(x) \longrightarrow \mathbf{C}$, i.e., it determines a section of the sheaf $I^*|_{L(x)} \otimes \Omega^1 L(x) \simeq \mathcal{O}(1, 1) \otimes (\mathcal{O}(-2, 0) \oplus \mathcal{O}(0, -2))$. The sheaf I has no nonzero sections.

Consequently, the subsheaf $\pi_1^* I \subset \pi_1^* \Omega^1 L \subset \Omega^1 F$ also lies in $\pi_2^* \Omega^1 M$. If we dualize this imbedding, we obtain a surjection $\pi_2^* \mathcal{T} M \longrightarrow \pi_1^* I^*$, and if we descend this morphism to M, then we obtain the spinor decomposition of $\mathcal{T} M$. From this one can now easily conclude that the diagram of standard imbedding deformations of $L(x_0)$ is essentially the diagram of null-geodesics.

We shall refine this argument somewhat in the next theorem, which presents a picture of the structures on L which correspond to a choice of metric in the conformal class.

6. Theorem. *Let M be a small space-time, and let L be its space of null-geodesics. Then the following structures on M and L are equivalent:*

(a) a spinor decomposition $\Omega^1 M = S_+ \otimes S_-$ and a pair of nowhere vanishing spinor metrics $\epsilon_\pm \in \Gamma(\wedge^2 S_\pm)$;

(b) a decomposition $I = I_+ \otimes I_-$, where $I_\pm$ are invertible sheaves whose restrictions to any quadric $L(x)$ are isomorphic to $\mathcal{O}(-1, 0)$ and $\mathcal{O}(0, -1)$, respectively, and two cohomology classes $(\epsilon_\pm)_L \in H^1(L, I_\pm^2)$ which do not vanish when restricted to any of the $L(x)$.

The correspondence between (a) and (b) satisfies the following condition. Let $(S_\pm)_L$ be the Radon–Penrose transform for the pair $(S_\pm,\ \nabla_\pm)$, where $\nabla_\pm\epsilon_\pm = 0$. Then one has exact sequences on L

$$0 \longrightarrow I_\pm \longrightarrow (S_\pm)_L \longrightarrow I_\mp^{-1} \longrightarrow 0, \tag{2}$$

whose classes coincide with $(\epsilon_\pm)_L$.

Proof. We have already done most of the work needed for the proof; it suffices to bring everything together.

In order to go from (a) to (b), we consider the exact sequences (1) on F. In the proof of Theorem 3 we noted that these sequences are $(\nabla_\pm)_{F/L}$-invariant, i.e., parallel translation along null-geodesics preserves sections of the sheaves $\pi_2^* \wedge^2 S_+(-1,\ 0)$ and $\pi_2^* \wedge^2 S_-(0,\ -1)$. We set $I_\pm = \operatorname{Ker}\left((\nabla_\pm)_{F/L} \,\middle|\, \wedge^2 S_\pm \begin{pmatrix} -1 & 0 \\ 0 & -1 \end{pmatrix}\right)$. According to Theorem 3, we have $I = I_+ \otimes_{O_L} I_-$. (As usual, we are writing equality when we really mean a canonical isomorphism, which is this case is the composition of the following identifications: a section of $I_+ \otimes I_-$ is a section of $\pi_2^*(S_+ \otimes S_-) = \pi_2^*\Omega^1 M$ which is zero along the quadrics $L(x)$ and along the fibres of π_1, and so belongs to $\pi_1^*(I)$; in addition, the section is $\nabla^g_{F/L}$-horizontal, and so descends to L.) Since $(\nabla_\pm)_{F/L}(\pi_2^*(\epsilon_\pm)) = 0$, we can identify $\pi_2^* \wedge^2 S_+(-1,\ 0)$ with $O_F(-1,\ 0)$ and $\pi_2^* \wedge^2 S_-(0,\ -1)$ with $O_F(0,\ -1)$. For the same reason, the connection induced on the quotient sheaves in (1) is dual to the connection induced on the subsheaves. Hence, when descended to L, (1) takes the form (2). We let $(\epsilon_\pm)_L$ denote the classes of the descended exact sequences. This completes the construction of the data in (b) starting from the data in (a).

To go in the other direction, we first use $(\epsilon_\pm)_L$ to construct extensions with the structure (2), except that we let $\Sigma_\pm$ denote the middle sheaves, which have not yet been identified with $(S_\pm)_L$. The exact sequences

$$0 \longrightarrow \pi_1^* I_\pm \longrightarrow \pi_1^*\Sigma_\pm \longrightarrow \pi_1^* I_\mp^{-1} \longrightarrow 0 \tag{3}$$

are equipped with relative connections $(d_\pm)_{F/L}$. Since the $(\epsilon_\pm)_L$ are not zero on any of the quadrics, we have $\pi_1^*\Sigma_\pm|_{L(x)} \simeq O^2_{L(x)}$, as follows from the fact that on $\mathbf{P}^1$ any nontrivial extension of $O(1)$ by means of $O(-1)$ is isomorphic to O^2. Hence, we can set $S_\pm^* = \pi_{2*}\pi_1^*\Sigma_\pm$; these are locally free sheaves of rank 2, and $\pi_2^* S_\pm^* = \pi_1^*\Sigma_\pm$. Since canonically $\wedge^2\pi_1^*\Sigma_\pm = O_F$, we obtain two sections $\epsilon_\pm^* \in \wedge^2 S_\pm^*$ corresponding to the unit element. Descending $(d_\pm)_{F/L}$ gives connections on $S_\pm^*$. At the end of § 5.5 we essentially described the most important point here — recovering the spinor structure on $\Omega^1 M$. Because we have an imbedding $\pi_1^*(I_+ \otimes I_-) \subset \pi_2^*(\Omega^1 M)$, we can dualize to obtain a surjection $\pi_2^* \mathcal{T}M \longrightarrow \pi_1^*(I_+^{-1} \otimes I_-^{-1})$, which, when descended, gives an isomorphism $\mathcal{T}M \longrightarrow \pi_{2*}\pi_1^* I_+^{-1} \otimes \pi_{2*}\pi_1^* I_-^{-1}$. Finally, if we descend the sequences (3) to M, we obtain isomorphisms

$$S_\pm^* = \pi_{2*}\pi_1^* \Sigma_\pm \xrightarrow{\sim} \pi_{2*}\pi_1^* I_\pm^{-1},$$

since $\pi_{2*}\pi_1^* I_\pm = R^1\pi_{2*}\pi_1^* I_\pm = 0$. We finally have $\mathcal{T}M = S_+^* \otimes S_-^*$.

It will be left to the reader to check that the above constructions are inverse to one another. □

In § 7 we shall show how the curvature of $\nabla_\pm$ is encoded in the structure of the sheaves $\Sigma_\pm$ which are determined by the classes $(\epsilon_\pm)_L$, and hence how the curvature of the Levi–Civita connection is encoded in the structure of the sheaf $\Sigma_+ \otimes \Sigma_-$.

§ 6. Extensions and Obstructions

1. Extensions. Suppose that $L \longleftarrow F \longrightarrow M$ is a diagram of null-geodesics, $\mathcal{E}_L$ is an M-trivial sheaf, or a Yang–Mills sheaf, on L, and $(\mathcal{E}, \nabla)$ is the corresponding Yang–Mills field. Locally in M this field is completely generic. How can one find conditions on $\mathcal{E}_L$ which imply, for example, the Yang–Mills equation $\tilde{\nabla}\Phi_+(\nabla) = 0$? We shall answer this question for a flat diagram in § 9: namely, in that case the condition is that $\mathcal{E}_L$ extend to a locally free sheaf on the third infinitesimal neighborhood of L in $\mathsf{P} \times \hat{\mathsf{P}}$. Some other extension problems leading to interesting field theory equations will be discussed in § 10.

Here we shall give a survey of the cohomological theory of extensions and obstructions. We begin with the basic definitions.

(a) Let $Y \subset X$ be a pair consisting of an analytic space and a closed analytic subspace, and let $J \subset \mathcal{O}_X$ be the sheaf of ideals defining Y. Then by the "n-th infinitesimal neighborhood of Y in X," denoted $Y^{(n)}$, we mean the ringed space $(Y, \mathcal{O}_X/J^{n+1})$. We call X an extension of Y, and we say that X is an infinitesimal extension if it coincides with some infinitesimal neighborhood of Y.

Most of the extensions we shall consider (but not all) will be given in advance as infinitesimal neighborhoods.

Suppose that Y and X are analytic manifolds. Then at every point $y \in Y \subset X$ there is a local coordinate system $(y_1, \ldots, y_m;\ x_1, \ldots, x_n)$ in X such that the equations of Y in X are $x_1 = \cdots = x_n = 0$. From this one easily sees that the sheaf $\operatorname{Gr} \mathcal{O}_X = \bigoplus_{n=0}^\infty \mathcal{O}_X/J^n$ is isomorphic to the symmetric algebra $S_{\mathcal{O}_Y}(J/J^2)$, where

$\mathcal{N} = J/J^2$ is a locally free sheaf over $\mathcal{O}_Y$ which is called the "conormal sheaf" of Y in X.

(b) Let $Y \subset X$ be an extension of Y. We will be considering various objects on Y which can be induced by corresponding objects on X; when that happens we shall speak of an "extension" of the object to X. Here is a list of the basic examples.

Suppose that $Y' \subset X'$ is another extension, and $f: Y \longrightarrow Y'$ is a morphism of analytic spaces. By an extension of f we mean a morphism $g: X \longrightarrow X'$ which coincides with f on Y.

Let $\mathcal{E}$ be a locally free sheaf on Y. By an extension of $\mathcal{E}$ we mean a locally free sheaf $\mathcal{F}$ on X together with an isomorphism $\mathcal{F}|_Y \xrightarrow{\sim} \mathcal{E}$. (We shall always implicitly require $\mathcal{F}$ to be locally free; hence, the "extension by zero" is not an extension if $Y \neq X$.) Two extensions $\mathcal{F}$ and $\mathcal{F}'$ are said to be isomorphic if there is an isomorphism between them which induces the identity on $\mathcal{E}$.

Let $(\mathcal{E}, \mathcal{F})$ be a locally free sheaf and an extension of it. We define an extension of a cohomology class $h \in H^k(Y, \mathcal{E})$ to be a cohomology class $h' \in H^k(X, \mathcal{F})$ for which $i^*(h') = h$, where i^* is induced by the imbedding $Y \subset X$.

(c) The problem of extending cohomology classes can be immediately restated in terms of exact sequences. Because $\mathcal{E}$ and $\mathcal{F}$ are locally free, we have an exact sequence of sheaves on X : $0 \longrightarrow J\mathcal{F} \longrightarrow \mathcal{F} \longrightarrow \mathcal{F}/J\mathcal{F} = i_*\mathcal{E} \longrightarrow 0$; in addition, $H^k(Y, \mathcal{E}) = H^k(X, \mathcal{F}/J\mathcal{F})$. We consider the following piece of the exact cohomology sequence:

$$\cdots \longrightarrow H^{k-1}(Y, \mathcal{E}) \longrightarrow H^k(X, J\mathcal{F}) \longrightarrow H^k(X, \mathcal{F}) \longrightarrow$$
$$\longrightarrow H^k(Y, \mathcal{E}) \xrightarrow{\delta} H^{k+1}(X, J\mathcal{F}) \to \cdots.$$

The following facts follow immediately from the definitions.

A class $h \in H^k(Y, \mathcal{E})$ has an extension from Y to X if and only if a certain cohomology class

$$\delta(h) \in H^{k+1}(X, J\mathcal{F}),$$

called the *obstruction to extending* h, vanishes.

If $\delta(h) = 0$, then the group $H^k(X, J\mathcal{F})$ acts transitively on the set of extensions of h (in fact, the set of extensions is actually a coset of this group in $H^k(X, \mathcal{F})$). The action of the group is effective if $H^{k-1}(Y, \mathcal{E}) = 0$.

It turns out that a similar cohomological picture applies to extension problems for other geometric objects, at least in a "linear approximation." We shall be especially interested in locally free sheaves, for which the linear approximation coincides with the full extension problem itself when one restricts oneself to the class of *simple extensions* $Y \subset X$, which we shall define in § 6.3.

2. Differential forms on analytic spaces. Let $Y \subset X$ be an extension with X a manifold and J the sheaf of ideals defining Y in $\mathcal{O}_X$. We set

$$\Omega^\cdot Y = \Omega^\cdot X/(J\Omega^\cdot X + (\Omega^\cdot X)dJ).$$

One can expand upon this definition using the construction of a canonical isomorphism between the two sheaves $\Omega^\cdot Y$ which come from two different extensions. One sees that the isomorphisms are compatible with restriction to open subsets, and so one can extend the definition of $\Omega^\cdot Y$ to the case where Y cannot be extended to a manifold globally. The point is that locally such an extension is always possible.

The exterior differential d_X on $\Omega^\cdot X$ induces d_Y on $\Omega^\cdot Y$.

Unlike in the case of manifolds, $\Omega^\cdot Y$ might not be locally free, and the de Rham complex on Y might not be exact.

A morphism $\varphi: Y \longrightarrow Z$ determines a morphism $\varphi^*(\Omega^\cdot Z) \longrightarrow \Omega^\cdot Y$. We set $\Omega^1 Y/Z = \Omega^1 Y/\operatorname{Im}\varphi^*$ and $\Omega^k Y/Z = \wedge^k(\Omega^1 Y/Z)$. This definition has drawbacks in the general case, but it is sufficient for our purposes.

Furthermore, the definition of a connection on a locally free sheaf, the de Rham sequence of a sheaf with connection, the curvature, etc., and also the relative versions of these definitions, all carry over to the case of analytic spaces.

3. Simple extensions. An extension $Y \subset X$ of analytic spaces is said to be *simple* if the following two conditions hold:

(a) $J^2 = 0$, where $\mathcal{O}_Y = \mathcal{O}_X/J$;

(b) the sequence of sheaves of $\mathcal{O}_Y$-modules

$$0 \longrightarrow J \xrightarrow{d\otimes 1} \Omega^1 X \otimes_{\mathcal{O}_X} \mathcal{O}_Y \longrightarrow \Omega^1 Y \longrightarrow 0 \tag{1}$$

is exact.

The second condition merits some words of explanation. In the first place, the imbedding $i: Y \subset X$ induces a map of $\mathcal{O}_X$-modules $\Omega^1 X \longrightarrow \Omega^1 Y$. This map is surjective, because locally any function extends from Y to X, and so induces a surjective map $\Omega^1 X \otimes_{\mathcal{O}_X} \mathcal{O}_Y \longrightarrow \Omega^1 Y$. From the definition of the sheaf of differentials it follows that dJ falls in the kernel. Since $J^2 = 0$, we know that multiplication by $f \in \mathcal{O}_X$ in J depends only on the image of f in $\mathcal{O}_Y$, and this means that J is an $\mathcal{O}_Y$-module. Finally, we have $d(fj) = df \cdot j + f\,dj$ and $(df \cdot j) \otimes_{\mathcal{O}_X} 1_Y = 0$ for $j \in J$ and $f \in \mathcal{O}_X$, and so $d \otimes 1$ is a morphism of $\mathcal{O}_Y$-modules.

Thus, the sequence (1) is defined and is a complex when (a) holds. It is always exact at the middle term, and hence the simplicity condition reduces to the requirement that $\operatorname{Ker}(d \otimes 1) = 0$.

4. The characteristic class of a simple extension. The sequence (1) determines a class

$$b = b(X, Y) \in \operatorname{Ext}^1_{\mathcal{O}_Y}(\Omega^1 Y, J),$$

which we call the "characteristic class" of the extension $Y \subset X$. In the case when $\Omega^1 Y$ is locally free (this is equivalent to Y being a manifold), we may assume that

$$b \in H^1(Y,\, \mathcal{T}Y \otimes J).$$

Conversely, suppose that J is an $\mathcal{O}_Y$-module, and we are given a class b in $\mathrm{Ext}^1_{\mathcal{O}_Y}(\Omega^1 Y,\, J)$, or, equivalently, an exact sequence of $\mathcal{O}_Y$-modules

$$0 \longrightarrow J \xrightarrow{a} \mathcal{A} \xrightarrow{b} \Omega^1 Y \longrightarrow 0. \tag{2}$$

We show how, given this sequence, one can recover a simple extension $Y \subset X$ for which (2) is canonically isomorphic to (1), with the isomorphisms being the identity on $\Omega^1 Y$ and J.

Let $\mathcal{O}_Y \oplus \Omega^1 Y$ be the sheaf of rings with multiplication given by

$$(f,\, \omega)(g,\, \nu) = (fg,\, f\nu + g\omega).$$

We consider the map $c\colon \mathcal{O}_Y \longrightarrow \mathcal{O}_Y \oplus \Omega^1 Y, \quad f \mapsto f + df$. It follows from Leibniz' formula that this is an imbedding of rings.

Given the sequence (2), we construct an extension of the sheaf of rings $\mathcal{O}_Y \oplus \Omega^1 Y$ by means of the ideal J with square zero:

$$0 \longrightarrow J \xrightarrow{\binom{0}{a}} \mathcal{O}_Y \oplus \mathcal{A} \xrightarrow{(\mathrm{id},\, b)} \mathcal{O}_Y \oplus \Omega^1 Y \longrightarrow 0,$$

and then we set $\mathcal{O}_X = (\mathrm{id},\, b)^{-1} \circ c(\mathcal{O}_Y)$. It is not hard to see that we have found a simple extension of Y with characteristic class (2).

For us the most important class of simple extensions is given in the following straightforward lemma, whose proof will be left to the reader.

5. Lemma. *Let $Y \subset X$ be a closed imbedding of one manifold in another. Then $Y^{(n+1)}$ is a simple extension of $Y^{(n)}$ for all $n \geq 0$.* □

Suppose that $Y \subset X$ is a simple extension, and $\mathcal{E}$ is a locally free sheaf on Y. We now state the fundamental result of this section, which we shall often apply in the situation of Lemma 5.

6. Theorem. (a) *In order for there to exist a (locally free) extension of the sheaf $\mathcal{E}$ to X, it is necessary and sufficient that a certain cohomological obstruction*

$$\omega(\mathcal{E}) \in H^2(Y,\, \mathcal{E}nd\, \mathcal{E} \otimes J)$$

vanish. Here J is the ideal (and conormal sheaf) of the extension $Y \subset X$.

(b) *If $\omega(\mathcal{E}) = 0$, then the group $H^1(Y, \mathcal{E}nd\ \mathcal{E} \otimes J)$ acts transitively on the set of isomorphism classes of extensions. This action is effective if there exists an extension $\mathcal{F}$ of the sheaf $\mathcal{E}$ such that any section of $\mathcal{E}nd\ \mathcal{E}$ extends to a section of $\mathcal{E}nd\ \mathcal{F}$.*

Before proving this theorem, we discuss its connection with noncommutative cohomology.

7. Noncommutative cohomology and torsors. Any rank n locally free sheaf $\mathcal{E}$ on Y can be given by transition matrix functions in a suitable open covering $Y = \bigcup U_i$:

$$g_{ij} \in \Gamma(U_i \cap U_j,\ GL(n;\ \mathcal{O}_Y)), \qquad \text{if } U_i \cap U_j \neq \emptyset,$$

which satisfy the cocycle conditions

$$g_{ii} = 1, g_{ij}g_{ji} = 1, \qquad g_{jk}g_{ki}g_{ij} = 1 \quad \text{if } U_i \cap U_j \cap U_k \neq \emptyset. \tag{3}$$

The sheaf $\mathcal{E}$ corresponding to the cocycle (g_{ij}) is obtained by gluing together the sheaves $\mathcal{O}^n_{U_i}$ according to the following rule: a set of sections $s_i \in \Gamma(U \cap U_i,\ \mathcal{O}^n_{U_i})$ represents a section $s \in \Gamma(U,\ \mathcal{E})$ if we have $g_{ij}s_j = s_i$ for all pairs i, j with $U \cap U_i \cap U_j \neq \emptyset$. A change of coordinates of the form $s'_i = h_i s_i$, $h_i \in \Gamma(U_i,\ GL(n;\ \mathcal{O}_{U_i}))$, in the trivialization $\mathcal{O}^n_{U_i}$ leads to a new cocycle which is cohomologous to the old one:

$$g'_{ij} = h_i g_{ij} h_j^{-1}. \tag{4}$$

Finally, in order to compare two ways of defining the same sheaf using different open coverings, we look at a common refinement of the two coverings. The transition functions g_{ij} in this finer covering are obtained from the old transition functions by restricting and using (4) to relate two different local trivializations of the sheaf.

If one axiomatizes these constructions, one obtains the following definitions. Let G be a sheaf of groups on a topological space Y, and let (U_i) be a covering of Y. A 1-cocycle on (U_i) with coefficients in G is a set of sections $g_{ij} \in \Gamma(U_i \cap U_j,\ G)$ satisfying (3). We let $Z^1((U_i),\ G)$ denote the set of 1-cocycles. The 1-st Čech cohomology set in the covering (U_i) is defined to be the quotient of Z^1 by the equivalence relation B^1 given by (4), where $h_i \in \Gamma(U_i,\ G)$. This set is denoted $H^1((U_i),\ G) = Z^1((U_i),\ G)/B^1((U_i),\ G)$. The 1-st cohomology set of the space Y with coefficients in G is defined to be

$$H^1(Y,\ G) = \varinjlim H^1((U_i),\ G),$$

where the limit is taken over finer and finer coverings of Y. The elements of H^1 (or the cocycles representing them) are sometimes called "G-torsors."

8. Exact sequences of noncommutative cohomology. Since $H^1(Y, G)$ is not generally a group, its behavior relative to homomorphisms of G is harder to describe than in the case of sheaves of abelian groups. The minimal picture that we need is the following. Let

$$1 \longrightarrow A \longrightarrow G \longrightarrow H \longrightarrow 1 \tag{5}$$

be an exact sequence of sheaves of groups, where the normal subgroup A is abelian. Then any torsor $h \in H^1(Y, H)$ determines a twisted sheaf A^h as follows. Let (h_{ij}) be a Čech cocycle representing h. We glue together A^h from $A|_{U_i}$, using the action of H on A given by lifting an element of H to G and then conjugating. In other words, a set of local sections $\{a_i \in \Gamma(U \cap U_i, A)\}$ represents a section $a \in \Gamma(U, A^h)$ if we have $\tilde{h}_{ij} a_j \tilde{h}_{ij}^{-1} = a_i$ for all pairs i, j with $U \cap U_i \cap U_j \neq \emptyset$, where $\tilde{h}_{ij} \in \Gamma(U_i \cap U_j, G), \tilde{h}_{ij} \mapsto h_{ij}$. If we change our cocycle representing the torsor, we obtain the same sheaf A^h up to canonical isomorphism. We are now ready for the basic abstract constructions needed to prove Theorem 6.

Obstructions. Let $h = (h_{ij})$ be a torsor with coefficients in H. We shall show how to construct from h a cohomology class $\omega(h) \in H^2(Y, A^h)$ such that $\omega(h) = 0$ if and only if h lies in the image of the map $H^1(Y, G) \longrightarrow H^1(Y, H)$. We choose a covering (U_i) which is fine enough for the components h_{ij} in the previous paragraph to lift to $\tilde{h}_{ij} \in \Gamma(U_i \cap U_j, G)$. We may also suppose that $\tilde{h}_{ii} = 1$ and $\tilde{h}_{ij} = \tilde{h}_{ji}^{-1}$, but the triple products of the type in (3) will not generally be equal to one. We set

$$a_{ijk} = \tilde{h}_{jk}\tilde{h}_{ki}\tilde{h}_{ij} \in \Gamma(U_i \cap U_j \cap U_k, A),$$

and we consider the cochain

$$\hat{a} = \{\hat{a}_{ijk} \in \Gamma(U_i \cap U_j \cap U_k, A^h)\},$$

where $\hat{a}_{ijk}$ is a_{ijk} considered in the "j-th trivialization" of the sheaf A^h; i.e., with the identification $A^h|_{U_j} = A|_{U_j}$. We claim that $\hat{a}$ is a Čech cocycle: $\hat{a} \in Z^2((U_i), A^h)$. We shall verify skew-symmetry, for example, the relation $\hat{a}_{ijk} = (\hat{a}_{jik})^{-1}$. We have $\hat{a}_{jik} = \tilde{h}_{ik}\tilde{h}_{kj}\tilde{h}_{ji}$ in the i-th trivialization. In order to compare this with $\hat{a}_{ijk}$, we move it to the j-th trivialization by conjugation by $\tilde{h}_{ij}^{-1}$:

$$(\hat{a}_{jik} \text{ in the } j\text{–th triv.}) = \tilde{h}_{ji}\tilde{h}_{ik}\tilde{h}_{kj} = ((\hat{a}_{ijk})^{-1} \text{ in the } j\text{–th triv.}).$$

We now verify that $\hat{a}$ is a cocycle, i.e., that

$$\hat{a}_{jkl}(\hat{a}_{ikl})^{-1}\, \hat{a}_{ijl}(\hat{a}_{ijk})^{-1} = 1. \tag{6}$$

We move all of the components into the same trivialization, say the k-th. For brevity we shall write $\tilde{h}_{kl}\tilde{h}_{lj}\tilde{h}_{jk}$ in the form $(kl)(lj)(jk)$, and so on. Then the left side of (6) in the k-th trivialization has the form

$$[(kl)(lj)(jk)]^{-1}[(kl)(li)(lk)]\,[(kj)(jl)(li)(ij)(jk)]^{-1}[(ki)(ij)(jk)].$$

Using the relations $(kl)(lk) = (ik)(ki) = 1$, we rewrite the third term, by inserting products equal to 1 and then arranging the factors in groups of three:

$$(kj)(jl)(li)(ij)(jk) = [(kj)(jl)(lk)][(kl)(li)(ik)][(ki)(ij)(jk)].$$

It is now clear that everything cancels.

Thus, a is a cocycle. We let $\omega(h)$ denote its cohomology class. We leave it to the reader to verify that the construction of this class is compatible with a change of trivialization or a change of covering. We shall carry out the verification that, if $\hat{a}$ is a coboundary, then h can be lifted to a torsor with coefficients in G.

Thus, suppose that

$$\hat{a}_{ijk} = \hat{b}_{jk}(\hat{b}_{ik})^{-1}\hat{b}_{ij}, \qquad \hat{b}_{ik} = (\hat{b}_{ki})^{-1}.$$

We set $\tilde{h}'_{ij} = b_{ij}^{-1}\tilde{h}_{ij}$, where b_{ij} coincides with $\hat{b}_{ij}$ in the i-th trivialization. We check that $\tilde{h}'_{jk}\tilde{h}'_{ki}\tilde{h}'_{ij} = 1$. In fact, in the j-th trivialization the expression on the left becomes

$$b_{jk}^{-1}\tilde{h}_{jk}b_{ki}^{-1}\tilde{h}_{ki}b_{ij}^{-1}\tilde{h}_{ij} = b_{jk}^{-1}\tilde{h}_{jk}(\tilde{h}_{kj}b_{ki}^{-1}\tilde{h}_{jk})\tilde{h}_{ki}(\tilde{h}_{ij}b_{ij}^{-1}\tilde{h}_{ji})\tilde{h}_{ij} = b_{jk}^{-1}b_{ki}^{-1}a_{ijk}b_{ij}^{-1} = 1.$$

The set of liftings of a cocycle. If $(\tilde{h}_{ij})$ is one lifting of the cocycle (h_{ij}) from H to G, then, as shown by the above calculation, the other liftings have the form $(b_{ij}\tilde{h}_{ij})$, where $\hat{b} \in Z^1((U_i), A^h)$. If we replace $\hat{b}$ by a cohomologous cocycle, we change $(b_{ij}\tilde{h}_{ij})$ by a cohomologous cocycle; the same happens if we use a different cocycle in the same class as h. Passing to the limit, we obtain a transitive action of $H^1(Y, A^h)$ on the set of extensions.

The degree of ineffectiveness of the action. Let $K \subset H^1(Y, A^h)$ be the stationary subgroup of some lifting (and hence of any lifting) of the class h. We choose an arbitrary lifting $\tilde{h} \in H^1(Y, G)$, and we define sheaves $G^{\tilde{h}}$ and $H^{\tilde{h}} = H^h$ by the same construction that was used to define A^h. Then there exists an exact sequence of groups

$$1 \longrightarrow \Gamma(Y, A^h) \longrightarrow \Gamma(Y, G^{\tilde{h}}) \longrightarrow \Gamma(Y, H^h) \xrightarrow{\delta} K \longrightarrow 1.$$

The boundary homomorphism δ is defined by analogy with the abelian case. Suppose that $s \in \Gamma(Y, H^h)$ is represented by sections $s_i \in \Gamma(U_i, H)$ (of course, in the i-th trivialization). We lift these sections to $\tilde{s}_i \in \Gamma(U_i, G^{\tilde{h}})$, and we set $c_{ij} = \tilde{s}_j \tilde{s}_i^{-1}$. We now let $\delta(s)$ denote the cochain

$$(\hat{c}_{ij} = c_{ij} \text{ in the } i\text{–th trivialization}).$$

We leave it to the reader to verify the necessary properties.

9. Proof of Theorem 6. We apply the general formalism of the preceding subsections to the following situation:

$$G = GL(n;\, \mathcal{O}_X), \qquad H = GL(n;\, \mathcal{O}_Y), \qquad n = \operatorname{rank} \mathcal{E},$$

where $Y \subset X$ is a simple extension. We then have the following exact sequence of sheaves of the type (5):

$$1 \longrightarrow M(n,\, J) \xrightarrow{\alpha} GL(n;\, \mathcal{O}_X) \longrightarrow GL(n;\, \mathcal{O}_Y) \longrightarrow 1,$$

where $a(m) = 1 + m$. Let $h \in H^1((U_1),\, GL(n;\, \mathcal{O}_Y))$ be the cocycle which determines the fibration $\mathcal{E}$. In this case we have a natural identification

$$M(n,\, J)^h = \mathcal{E}nd\ \mathcal{E} \otimes_{\mathcal{O}_Y} J,$$

since the patching matrices act on $M(n,\, J)$ by conjugation. This immediately implies everything in Theorem 6, except for the last claim concerning effectiveness of the action. In order for the action to be effective, it is necessary that the map

$$\Gamma(GL(n,\, \mathcal{O}_X)^{\tilde{h}}) \longrightarrow \Gamma(GL(n;\, \mathcal{O}_Y)^h),$$

where $\tilde{h}$ is an extension of h, be surjective. But these twisted sheaves can be identified with the automorphism sheaves of the sheaf $\mathcal{E}$ and its extension. Hence, if all of the endomorphisms of $\mathcal{E}$ turn out to be extendible for one of the extensions, then the same holds for the automorphisms as well. □

10. Supplementary remarks. (a) The same formalism can be used to study extensions of locally free sheaves whose structure group is reduced to a complex Lie subgroup $G \subset GL(n)$. The corresponding exact sequence of sheaves has the form

$$1 \longrightarrow \mathcal{G} \otimes_{\mathbf{C}} J \longrightarrow G(\mathcal{O}_X) \longrightarrow G(\mathcal{O}_Y) \longrightarrow 1,$$

where $\mathcal{G}$ is the Lie algebra of G. Here $(\mathcal{G} \otimes \mathcal{O}_X)^h$ is the locally free sheaf of Lie algebras which is associated to the cocycle h and the adjoint representation of G.

(b) We can use the characteristic class of a simple extension to give an invariant expression for the obstruction $\omega(\mathcal{E})$. Namely, let $b(Y) \in \operatorname{Ext}^1_{\mathcal{O}_Y}(\Omega^1 Y, J)$ be this characteristic class, and let $a(\mathcal{E}) \in \operatorname{Ext}^1_{\mathcal{O}_Y}(\mathcal{E}, \mathcal{E} \otimes \Omega^1 Y) = \operatorname{Ext}^1_{\mathcal{O}_Y}(\mathcal{E}nd\ \mathcal{E}, \Omega^1 Y)$ be the class of the extension

$$0 \longrightarrow \mathcal{E} \otimes \Omega^1 Y \longrightarrow \operatorname{Jet}^1 \mathcal{E} \longrightarrow \mathcal{E} \longrightarrow 0.$$

Using multiplication in the Ext groups, we can construct the product

$$a(\mathcal{E}) b(Y) \in \operatorname{Ext}^2_{\mathcal{O}_Y}(\mathcal{E}nd\ \mathcal{E}, J) = H^2(Y, \mathcal{E}nd\ \mathcal{E} \otimes J).$$

Up to normalizations of the various isomorphisms, it is this that is $\omega(\mathcal{E})$.

A working expressing for $\omega(\mathcal{E})$ can also be obtained in the following way. We consider the following commutative diagram with exact rows:

$$\begin{array}{ccccccccc} 1 & \longrightarrow & M(n;\, J) & \longrightarrow & GL(n;\, \mathcal{O}_X) & \longrightarrow & GL(n;\, \mathcal{O}_Y) & \longrightarrow & 1 \\ & & \| & & \big\downarrow \tilde{D} & & \big\downarrow D & & \\ 1 & \longrightarrow & M(n;\, J) & \longrightarrow & M(n;\, \Omega^1 X \otimes \mathcal{O}_Y) & \longrightarrow & M(n \,\Omega^1 Y) & \longrightarrow & 1, \end{array}$$

where $Dg = g^{-1} dg$ and $\tilde{D}\tilde{g} = \tilde{g}^{-1} d\tilde{g} \bmod J$. The exactness of the lower row follows because the extension is simple.

The formula $D(gh) = h^{-1}(Dg)h + Dh$ implies that there is a map induced on the torsors

$$H^1(Y, GL(n;\, \mathcal{O}_Y)) \longrightarrow H^1(Y, \mathcal{E}nd\ \mathcal{E} \otimes \Omega^1 Y).$$

Here h goes to $a(\mathcal{E})$. Furthermore, the exact sequence

$$0 \longrightarrow \mathcal{E}nd\ \mathcal{E} \otimes J \longrightarrow \mathcal{E}nd\ \mathcal{E} \otimes (\Omega^1 X \otimes \mathcal{O}_Y) \longrightarrow \mathcal{E}nd\ \mathcal{E} \otimes \Omega^1 Y \longrightarrow 0$$

gives us the boundary homomorphism

$$\delta \colon H^1(Y, \mathcal{E}nd\ \mathcal{E} \otimes \Omega^1 Y) \longrightarrow H^2(Y, \mathcal{E}nd\ \mathcal{E} \otimes J),$$

and we have $\delta[a(\mathcal{E})] = \omega(\mathcal{E})$. This is merely a rephrasing of the relation $\omega(\mathcal{E}) = a(\mathcal{E})b(Y)$. (The reader should recall that we are unconcerned about normalization constants.)

(c) The difference cohomology class for two extensions $\mathcal{E}_1$ and $\mathcal{E}_2$ of a sheaf $\mathcal{E}$ can be constructed canonically even if the group H^1 does not act effectively on the set of extensions. Namely, we have an exact sequence

$$0 \longrightarrow \mathcal{E}nd\ \mathcal{E} \otimes J \longrightarrow \mathcal{H}om_{\mathcal{O}_X}(\mathcal{E}_1, \mathcal{E}_2) \longrightarrow \mathcal{E}nd\ \mathcal{E} \longrightarrow 0$$

and a boundary homomorphism

$$\delta\colon H^0(Y, \mathcal{E}nd\ \mathcal{E}) \longrightarrow H^1(\mathcal{E}nd\ \mathcal{E} \otimes J).$$

We let the class $\delta(\mathrm{id})$ correspond to the difference $[\mathcal{E}_2] - [\mathcal{E}_1]$. An easy computation makes it clear that translating by this class takes $\mathcal{E}_1$ to $\mathcal{E}_2$.

11. The analytic de Rham complex on infinitesimal neighborhoods. In § 9 we shall need to consider a relative version of the de Rham complex on infinitesimal neighborhoods. Here we shall describe a situation when its exactness properties are close to the exactness properties in the case of manifolds. Suppose we are given a commutative diagram of manifolds

$$\begin{array}{ccc} Y & \longrightarrow & X \\ {\scriptstyle\varphi}\downarrow & & \downarrow{\scriptstyle\Phi} \\ V & \longrightarrow & W, \end{array}$$

in which the vertical arrows are submersions and the horizontal arrows are closed imbeddings. Let $Y^{(n)}$ and $V^{(n)}$ be the n-th infinitesimal neighborhoods of Y in X and of V in W, and let $\varphi^{(n)}\colon Y^{(n)} \longrightarrow V^{(n)}$ be the morphism induced by Φ. It turns out that one can work with $\varphi^{(n)}$ just as well as with submersions of manifolds.

12. Proposition. (a) *The relative holomorphic de Rham complex*

$$0 \longrightarrow (\varphi^{(n)})^{-1}(\mathcal{O}_{V^{(n)}}) \longrightarrow \Omega^{\cdot}Y^{(n)}/V^{(n)}$$

is exact.

(b) *Suppose that the fibres of φ are connected, and $H^p(\varphi^{-1}(v), \mathbb{C}) = 0$ for $p = 1, 2, \dots, m$ and for all $v \in V$. Let $\mathcal{E}^{(n)}$ be a locally free sheaf on $V^{(n)}$. Then the canonical homomorphism*

$$H^p(V^{(n)}, \mathcal{E}^{(n)}) \longrightarrow H^p(Y^{(n)}, \varphi^{-1}(\mathcal{E}^{(n)}))$$

is an isomorphism for $p = 0, 1, \dots, m$ and a monomorphism for $p = m + 1$.

Proof. It follows from the definition that the imbedding $Y^{(n)} \subset Y^{(n+1)}$ induces a surjection of complexes

$$\Omega^{\cdot}Y^{(n+1)}/V^{(n+1)} \longrightarrow \Omega^{\cdot}Y^{(n)}/V^{(n)}.$$

Let $S_n^{\cdot}$ denote its kernel. In order to use induction on n, we must understand the structure of $S_n^{\cdot}$.

Let I and J denote the sheaves of ideals of Y and V in $\mathcal{O}_X$ and $\mathcal{O}_W$, respectively. We consider the following diagram with exact rows:

$$\begin{array}{ccccccc}
0 \to & I^{n+1}/I^{n+2} & \to & \mathcal{O}_{Y^{(n+1)}} & \to & \mathcal{O}_{Y^{(n)}} & \to 0 \\
& \downarrow & & \downarrow & & \downarrow d_{Y^{(n)}/V^{(n)}} & \\
0 \to & \dfrac{I^{n+1}/I^{n+2}}{\Phi^*(J^{n+1}/J^{n+2})} & \to & \Omega^1 X/W \otimes \mathcal{O}_X/I^{n+1} & \xrightarrow{a_n} & \Omega^1 Y^{(n)}/V^{(n)} & \to 0.
\end{array}$$

Taking the exterior power of the morphism a_n leads to the exact sequence

$$\Omega^{p-1} X/W \otimes (I^{n+1}/I^{n+2})/\Phi^*(J^{n+1}/J^{n+2}) \longrightarrow$$
$$\longrightarrow \Omega^p X/W \otimes \mathcal{O}_X/I^{n+1} \xrightarrow{\wedge^p a_n} \Omega^p Y^{(n)}/V^{(n)} \to 0.$$

We set $R_n^p = \operatorname{Ker}(\wedge^p a_n)$, and we apply the snake lemma to the following commutative diagram with exact rows:

$$\begin{array}{ccccccccc}
0 & \longrightarrow & R_{n+1}^p & \longrightarrow & \Omega^p X/W \otimes \mathcal{O}_X/I^{n+2} & \longrightarrow & \Omega^p Y^{(n+1)}/V^{(n+1)} & \longrightarrow & 0 \\
& & \downarrow & & \downarrow & & \downarrow & & \\
0 & \longrightarrow & R_n^p & \longrightarrow & \Omega^p X/W \otimes \mathcal{O}_X/I^{n+1} & \longrightarrow & \Omega^p Y^{(n)}/V^{(n)} & \longrightarrow & 0.
\end{array}$$

We obtain the exact sequence

$$0 \longrightarrow T_n^p \longrightarrow S_n^p \longrightarrow R_n^p \longrightarrow 0, \tag{7}$$

where

$$T_n^p = \operatorname{Coker}\left(R_{n+1}^p \longrightarrow \Omega^p X/W \otimes I^{n+1}/I^{n+2}\right).$$

We now consider the composition

$$b_n : T_n^p \longrightarrow S_n^p \xrightarrow{d} S_n^{p+1} \longrightarrow R_n^{p+1}.$$

A local coordinate computation shows that this is a morphism of $\mathcal{O}_Y$-modules, and that we have the exact sequence

$$0 \longrightarrow \Omega^p Y/V \otimes \varphi^*(J^{n+1}/J^{n+2}) \longrightarrow T_n^p \xrightarrow{b_n} R_n^{p+1} \longrightarrow 0. \tag{8}$$

On the other hand, because of the compatibility of exterior differentials, the diagram

$$\begin{array}{ccc}
\Omega^p Y/V \otimes \varphi^*(J^{n+1}/J^{n+2}) & \xrightarrow{d_{Y/V}} & \Omega^{p+1} Y/V \otimes \varphi^*(J^{n+1}/J^{n+2}) \\
\downarrow & & \downarrow \\
S_n^p & \xrightarrow{d} & S_n^{p+1}
\end{array}$$

is commutative, and one more local coordinate computation shows that the composition

$$R_n^{p+1} \longrightarrow S_n^p/\Omega^p Y/V \otimes \varphi^*(J^{n+1}/J^{n+2}) \xrightarrow{d}$$
$$\xrightarrow{d} S_n^{p+1}/\Omega^{p+1} Y/V \otimes \varphi^*(J^{n+1}/J^{n+2}) \longrightarrow R_n^{p+1}$$

is the identity map. Taking (7) and (8) into account, we now find that the following isomorphism holds for all open subsets $U \subset Y$:

$$H^*(H^p(U, S_n^{\cdot})) = H^*(H^p(U, \Omega^{\cdot} Y/V \otimes \varphi^*(J^{n+1}/J^{n+2}))).$$

Part (a) of Proposition 12 follows from this if one uses induction on n, along with the exactness of the sequence

$$0 \longrightarrow \varphi^{-1}(J^{n+1}/J^{n+2}) \longrightarrow \Omega^{\cdot} Y/V \otimes \varphi^*(J^{n+1}/J^{n+2}).$$

Part (b) can also be obtained by induction on n. The case $n = 0$ is Proposition 2.4. The inductive step uses the five homomorphism lemma, applied to the exact cohomology sequence coming from the short exact sequence

$$0 \longrightarrow \mathcal{E}^{(0)} \otimes J^{n+1}/J^{n+2} \longrightarrow \mathcal{E}^{(n+1)} \longrightarrow \mathcal{E}^{(n)} \longrightarrow 0,$$

where $\mathcal{E}^{(k)} = \mathcal{E}^{(n)}|_{V^{(k)}}$ for $k \leq n$. □

§ 7. Curvature on the Space of Null-Geodesics

1. Statement of the problem. We consider a null-geodesic diagram $L \xleftarrow{\pi_1} F \xrightarrow{\pi_2} M$. Suppose that the sheaf $\mathcal{E}_L$ on L and the sheaf with connection $(\mathcal{E}, \nabla)$ on M are related to one another by means of the Radon–Penrose transform via the pair $(\mathcal{E}_F, \nabla_{F/L})$ on F. That is, $\mathcal{E}_F = \pi_1^*(\mathcal{E}_L) = \pi_2^*(\mathcal{E})$, $\nabla_{F/L} = \pi_2^*(\nabla)|_{TF/L}$. In this section we shall show how to describe the curvature $\Phi(\nabla)$, i.e., the Yang–Mills field strength, in terms of data connected with L and F. We shall begin with computations in an infinitesimal neighborhood of a point, and shall then explain how to globalize them.

An initial observation is that the value of $\Phi(\nabla)$ at $x \in M$ belongs to $\mathcal{E}nd\ \mathcal{E}(x) \otimes \Omega^2 M(x)$ and depends on the second derivatives of a section of $\mathcal{E}$ at the point x. Hence, one might expect that $\Phi(\nabla)$ corresponds to some object which is determined by the structure of $\mathcal{E}_L$ on the second infinitesimal neighborhood $L(x)^{(2)} \subset L$.

The remarks that follow come from a pointwise analysis of the Radon–Penrose transform in a second order approximation.

We fix $x \in M$, and we set $\mathcal{E}(x) = E$ and $\tilde{E} = E \otimes_{\mathbf{C}} \mathcal{O}_{L(x)}$. The Radon–Penrose transform gives an identification $\mathcal{E}_L|_{L(x)} = \tilde{E}$. The sheaf $\tilde{E}$ obviously extends to $L(x)^{(1)}$ (for example, as $E \otimes \mathcal{O}_{L(x)^{(1)}}$). All of the extensions can be classified by elements of $H^1(L(x), \mathcal{E}nd\ E \otimes \mathcal{N}(x))$, where $\mathcal{N}(x)$ is the conormal sheaf of $L(x)$ in L (see Theorem 6.6), and this group is zero (§ 5.4). Moreover, $H^0(\mathcal{O}_{L(x)^{(1)}}) = \mathbf{C}$, so that the identification $E \otimes \mathcal{O}_{L(x)^{(1)}} \xrightarrow{\sim} \tilde{E}^{(1)}$ is unique.

$\tilde{E}^{(1)}$ also has an extension to $L(x)^{(2)}$. The distinguished trivial extension $E \otimes \mathcal{O}_{L(x)^{(2)}}$ satisfies the effectiveness condition in Theorem 6.6(b). Hence, the "difference class" map which takes $\tilde{E}^{(2)}$ to (class of $\tilde{E}^{(2)}$ – class of $E \otimes \mathcal{O}_{L(x)^{(2)}}) \in H^1(L(x), \mathcal{E}nd\ E \otimes S^2\mathcal{N}(x))$ is a bijection.

In order to compute the group of extensions we consider the exact sequence

$$0 \longrightarrow \mathcal{N}(x) \xrightarrow{i} \Omega^1 M(x) \otimes \mathcal{O}_{L(x)} \xrightarrow{\text{res}} \Omega^1 F/L|_{L(x)} \longrightarrow 0 \tag{1}$$

and then the sequence

$$0 \longrightarrow S^2(\mathcal{N}(x)) \xrightarrow{S^2(i)} S^2(\Omega^1 M(x)) \otimes \mathcal{O}_{L(x)} \xrightarrow{a} \Omega^1 M(x) \otimes \Omega^1 F/L|_{L(x)} \longrightarrow 0. \tag{2}$$

If we take into account that $H^0(S^2\mathcal{N}(x)) = 0$ and $H^0(\Omega^1 F/L|_{L(x)}) = \Omega^1 M(x)$, then from (2) we obtain the isomorphism

$$\delta\colon\quad \Omega^1 M(x)^{\otimes 2}/S^2(\Omega^1 M(x)) = \Omega^2 M(x) \longrightarrow H^1(L(x), S^2\mathcal{N}(x)),$$

and then

$$\text{id} \otimes \delta\colon\quad \mathcal{E}nd\ E \otimes \Omega^2 M(x) \xrightarrow{\sim} H^1(L(x), \mathcal{E}nd\ E \otimes S^2\mathcal{N}(x)). \tag{3}$$

We are now ready to state the basic result of the section.

2. Theorem. *Under the conditions of the preceding subsection, let $\varphi(x) \in H^1(L(x), \mathcal{E}nd\ E \otimes S^2\mathcal{N}(x))$ denote the difference class for the sheaf $\mathcal{E}_L|_{L(x)^{(2)}}$ and the trivial extension $E \otimes \mathcal{O}_{L(x)^{(2)}}$, and let $\Phi(\nabla)(x) \in \mathcal{E}nd\ E \otimes \Omega^2 M(x)$ be the value of the curvature form $\Phi(\nabla)$ at the point $x \in M$. Then*

$$(\text{id} \otimes \delta)[\Phi(\nabla)(x)] = \varphi(x).$$

Proof. Let $U \ni x$ be a neighborhood over which $\mathcal{E}$ is trivial, and hence $\mathcal{E}_F|_{F(U)}$ is trivial. We define $\mathcal{E}_L$ on $L(U)$ by means of patching matrices g_{ij}, after choosing a covering (V_i) and trivializations of $\mathcal{E}_L$ on the V_i. If we also trivialize $\mathcal{E}_F|_{F(U)}$ using sections lifted from U, we obtain the factorization $g_{ij} = h_i h_j^{-1}$, where the h_i are matrix functions on $\pi_1^{-1}(V_i)$. Since $d_{F/L}(\pi_1^* g_{ij}) = 0$, we have $d_{F/L}h_i \cdot h_j^{-1} - h_i h_j^{-1} d_{F/L} h_j \cdot h_j^{-1} = 0$, or:

$$h_i^{-1} d_{F/L} h_i = h_j^{-1} d_{F/L} h_j \qquad \text{on } \pi_1^{-1}(V_i \cap V_j).$$

Thus, all of the $h_i^{-1} d_{F/L} h_i$ patch together to form a section of the sheaf $\pi_2^*(\mathcal{E}nd\ \mathcal{E}) \otimes \Omega^1 F/L$ on $F(U)$. Let this section be $\pi_2^*(A)$, where $A \in \Gamma(U, \mathcal{E}nd\ \mathcal{E} \otimes \Omega^1 M)$. It is not hard to see that $\nabla = d - A$ in our trivialization of $\mathcal{E}$. In fact, the rows of h_i form a basis of sections of $\pi_1^{-1}(\mathcal{E}_L)$ over $\pi_i^{-1}(V_i)$, and so it suffices to check that $(d - A)_{F/L} h_i = 0$; but

$$(d - A)_{F/L} h_i = d_{F/L} h_i - h_i \pi_2^*(A)|_{V_i} = d_{F/L} h_i - h_i (h_i^{-1} d_{F/L} h_i) = 0.$$

We now suppose that our basis of $\mathcal{E}$ at the point $x \in M$ is first-order horizontal. This means that in a local coordinate system (x^a) on M we have

$$A = \Phi_{ab} x^a dx^b + \cdots,$$

where $\Phi_{ab} \in \operatorname{End} E, E = \mathcal{E}(x)$, and the dots stand for second and higher order terms in x^a. Then the curvature has the form

$$-dA + A \wedge A = -\Phi_{ab} dx^a \wedge dx^b + \cdots,$$

i.e.,

$$\Phi(\nabla)(x) = -\Phi_{ab} dx^a \wedge dx^b + \cdots.$$

On the other hand, let us choose the trivializations of $\mathcal{E}_L$ on the V_i in such a way that in an infinitesimal neighborhood of $L(x)$ the g_{ij} differ from 1 only by terms of second order in the coordinates normal to $L(x)$. This is possible because of the results of § 7.1: $g_{ij} = 1 + \tilde{g}_{ij} + \cdots$. By modifying the bases of $\mathcal{E}(x)$ and $\mathcal{E}_L|_{L(x)}$, if necessary, we may assume that we also have $h_i = 1 - \tilde{h}_i + \cdots$. Then $\tilde{g}_{ij} = \tilde{h}_j - \tilde{h}_i$. This is an explicit splitting of the difference cocycle $\varphi(x)$ for $\mathcal{E}|_{L(x)^{(2)}}$ and $E \otimes \mathcal{O}_{L(x)^{(2)}}$ after it is mapped to the cochains with coefficients in $S^2\Omega^1 M(x) \otimes \mathcal{O}_{L(x)}$. From (2) it is then obvious that, if we let $\tilde{a}$ denote the composition of the morphism a with taking sections and skew-symmetrizing, then we have

$$(\operatorname{id} \otimes \delta)^{-1} \varphi(x) = (\text{the section represented by } \tilde{a}(\tilde{h}_i) \text{ over } V_i \cap L(x)).$$

We now recall that on $\pi_1^{-1}(V_i)$ we have

$$(1-\tilde{h}_i+\cdots)^{-1}(-d_{F/L}\tilde{h}_i+\cdots)=\pi_2^*(A)|_{\mathcal{T}F/L}=\Phi_{ab}x^a d_{F/L}x^b.$$

This implies that, if we apply a and skew-symmetrization, $\tilde{h}_i$ goes to the section $-\Phi_{ab}dx^a\wedge dx^b=\Phi(\nabla)(x)$. □

It is not difficult to formulate a global version of this result. However, it will relate to $\mathcal{E}_F$, because the conormal sheaves $\mathcal{N}(x)$ for different $L(x)$ are restrictions to the various fibres of π_2 of a single sheaf $\mathcal{N}$ only on F, not on L. We now give some preliminaries before stating the global result.

3. Cohomological description of $\mathcal{E}nd\, E\otimes\Omega^2 M$. In (1), (2) and (3) we replace $\mathcal{N}(x)$ by the sheaf $\mathcal{N}=\operatorname{Ker}\operatorname{res}(\pi_2^*\Omega^1 M\longrightarrow\Omega^1 F/L)$. Exactly as in § 7.1 we obtain an isomorphism

$$\mathrm{id}\otimes\delta\colon\quad \mathcal{E}nd\, E\otimes\Omega^2 M\longrightarrow\mathcal{E}nd\, E\otimes R^1\pi_{2*}S^2\mathcal{N}.$$

4. The difference class. We consider a closed imbedding $F\subset L\times M$, and we let $F^{(i)}$ denote the i-th infinitesimal neighborhood of F in this imbedding. Let $\pi_1^{(i)}$ and $\pi_2^{(i)}$ be the projections of $F^{(i)}$ onto L and M. We have $\pi_1^*(\mathcal{E}_L)=\mathcal{E}_F=\pi_2^*(\mathcal{E})$. Thus, $\pi_1^{(1)*}(\mathcal{E}_L)$ and $\pi_2^{(1)*}(\mathcal{E})$ are two extensions of $\mathcal{E}$ to $F^{(1)}$.

We now compute the conormal sheaf of the imbedding of F in $L\times M$. This sheaf can be identified with the kernel of the restriction of $\Omega^1(L\times M)|_F$ to the tangent fields of F, i.e., with the kernel of the map

$$\pi_1^*\Omega^1 L\oplus\pi_2^*\Omega^1 M\xrightarrow{i_1+i_2}\Omega^1 F,$$

where i_1 and i_2 are the imbeddings of $\pi_1^*\Omega^1 L$ and $\pi_2^*\Omega^1 M$ in $\Omega^1 F$. Let $\mathcal{N}'$ be this kernel. We define a map $\mathcal{N}\longrightarrow\mathcal{N}'$ as follows. If $\omega\in\mathcal{N}$, then ω is a section of $\pi_2^*\Omega^1 M$ which is zero on $\mathcal{T}F/L$; hence, there is a uniquely determined section ν of the sheaf $\pi_1^*\Omega^1 L$ with $i_1(\nu)=i_2(\omega)$. If we take ω to the reflection $(-\nu,\omega)$, we obtain an isomorphism of $\mathcal{N}$ with the desired conormal sheaf.

Thus, the difference class $[\pi_1^{(1)*}(\mathcal{E}_L)]-[\pi_2^{(1)*}(\mathcal{E})]$ is in the group $H^1(F,\mathcal{E}nd\,\mathcal{E}_F\otimes\mathcal{N})$. But this group vanishes, since $\pi_{2*}\mathcal{N}=0$ and $R^1\pi_{2*}\mathcal{N}=0$. Hence, $\pi_1^{(1)*}(\mathcal{E}_L)$ and $\pi_2^{(1)*}(\mathcal{E})$ are isomorphic, and it is not hard to see that there is a unique isomorphism which induces the identity on $\mathcal{E}_F$. So we have a canonically defined sheaf $\mathcal{E}_F^{(1)}$ along with two extensions to $F^{(2)}$: $\pi_1^{(2)*}(\mathcal{E}_L)$ and $\pi_2^{(2)*}(\mathcal{E})$. Their difference class φ is in the group $H^1(F, \mathcal{E}nd\ \mathcal{E}_F \otimes S^2\mathcal{N})$. Since $\pi_{2*}(S^2\mathcal{N}) = 0$, this group is equal to

$$H^1(F, \mathcal{E}nd\ \mathcal{E}_F \otimes S^2\mathcal{N}) = H^0(M, \mathcal{E}nd\ \mathcal{E} \otimes R^1\pi_{2*}S^2\mathcal{N}).$$

We are now in a position to give the global version of Theorem 2.

5. Theorem. $H^0(\mathrm{id} \otimes \delta)(\Phi(\nabla)) = \varphi$. □

To prove this theorem it is enough to check that the constructions in §§ 7.3–7.4 are compatible with the pointwise computations in §§ 7.1–7.2.

6. The curvature of the Levi–Civita connection. Let us consider the situation in § 5 when we have spinor decomposition data on L. Recall that such data consist of an isomorphism $I = I_+ \otimes I_-$, where $I_+|_{L(x)} \simeq \mathcal{O}(-1, 0)$ and $I_-|_{L(x)} \simeq \mathcal{O}(0, -1)$, and two cohomology classes $(\epsilon_\pm)_L \in H^1(L, I_\pm^2)$ which do not vanish when restricted to any $L(x)$. In this case the rank 2 sheaves $\Sigma_\pm$ which correspond to $(\epsilon_\pm)_L$, considered as the classes of extensions $0 \longrightarrow I_\pm \longrightarrow \Sigma_\pm \longrightarrow I_\pm^{-1} \longrightarrow 0$, are the Penrose transform of the spinor bundles $(S_\pm, \nabla_\pm)$. Thus, information about the spinor curvatures are encoded in the structure of $\Sigma_\pm|_{L(x)^{(2)}}$, and information about the curvature of the Levi–Civita connection are encoded in the structure of $\Sigma_+ \otimes \Sigma_-|_{L(x)^{(2)}}$. Unfortunately, this description is not sufficiently usable, for example, to be able to give the Einstein equations in terms of L.

§ 8. Cohomological Computations

1. Statement of the problem. We shall need to know several sheaf cohomology groups on L for our later analysis of the Radon–Penrose transform connected with diagrams of null-geodesics.

When M is a small space-time, this problem easily reduces to certain computations on M. Let $\mathcal{F}$ be a locally free sheaf on L. We let $D(i;\ \mathcal{F})$ denote the first order differential operator

$$D(i;\ \mathcal{F}) = \Gamma(R^i\pi_{2*}(\nabla_{F/L})): \Gamma(R^i\pi_{2*}\pi_1^*\mathcal{F}) \longrightarrow \Gamma(R^i\pi_{2*}(\pi_1^*\mathcal{F} \otimes \Omega^1 F/L)), \quad (1)$$

where $\nabla_{F/L}: \pi_1^*(\mathcal{F}) \longrightarrow \pi_1^*\mathcal{F} \otimes \Omega^1 F/L$ is the standard connection along the fibres of π_1.

2. Lemma. *If M is small, then the following sequence is exact for any i:*

$$0 \longrightarrow \mathrm{Coker}\ D(i-1;\ \mathcal{F}) \longrightarrow H^i(L,\ \mathcal{F}) \longrightarrow \mathrm{Ker}\ D(i;\ \mathcal{F}) \longrightarrow 0 \quad (2)$$

(the first group is taken to be zero when $i = 0$).

Proof. The exact cohomology sequence coming from the relative de Rham resolution of $\pi_1^{-1}(\mathcal{F})$ has the form

$$\dots \xrightarrow{H^{i-1}(\nabla_{F/L})} H^{i-1}(F, \pi_1^* \mathcal{F} \otimes \Omega^1 F/L) \longrightarrow H^i(F, \pi_1^{-1}\mathcal{F}) \longrightarrow$$
$$\longrightarrow H^i(F, \pi_1^* \mathcal{F}) \xrightarrow{H^i(\nabla_{F/L})} H^i(F, \pi_1^* \mathcal{F} \otimes \Omega^1 F/L).$$

By Proposition 2.4, which applies here because M is small, we have

$$H^i(F, \pi_1^{-1}\mathcal{F}) = H^i(L, \mathcal{F}).$$

On the other hand, the Leray spectral sequence of the morphism π_2 with second term $H^i(M, R^j\pi_{2*}\pi_1^* \mathcal{F})$ converges to $H^i(F, \pi_1^*\mathcal{F})$. If the sheaf $\mathcal{K}$ is coherent, then these groups vanish for $i > 0$ because M is intensive. Consequently, we have $H^i(F, \mathcal{F}) = \Gamma(M, R^i\pi_{2*}\pi_1^* \mathcal{F})$. Finally, in our situation $H^i(\nabla_{F/L})$ obviously coincides with $D(i; \mathcal{F})$. □

We now give two prototypical computations, which will be used in the next section to interpret the formalism of extensions and obstructions on L.

We shall use the notation of § 5, and suppose that compatible spinor decompositions have been chosen on L and M; in particular, on L we have sheaves $I_\pm$, $\Sigma_\pm$, and $S_\pm^* = \pi_{2*}\pi_1^* I_\pm^{-1}$. We shall write $\mathcal{F}(a, b)$ instead of $\mathcal{F} \otimes I_+^{-a} \otimes I_-^{-b}$, and we shall identify F with $\mathrm{P}(S_+) \underset{M}{\times} \mathrm{P}(S_-)$ (and not $\mathrm{P}(S_+^*) \underset{M}{\times} \mathrm{P}(S_-^*)$, as we usually did before). Then $\pi_1^*(I_+)$ and $\pi_1^*(I_-)$ go to $\mathcal{O}_F(-1, 0)$ and $\mathcal{O}_F(0, -1)$, respectively, and so $\pi_1^*(\mathcal{F}(a, b)) = \pi_1^*(\mathcal{F})(a, b)$. A change of the identification is a variant on the Legendre transform corresponding to a (conformal) metric. As shown in § 5, we have $\Omega^1 F/L = \pi_2^*(\wedge^2 S_+ \otimes \wedge^2 S_-)(1, 1)$.

The next theorem lists some cohomology groups of the form $H^i(L, \mathcal{E}_L(a, b))$, where $\mathcal{E}_L$ is the Radon–Penrose transform of the pair $(\mathcal{E}, \nabla)$ on M. We note that the proof does more than what is in the statement of the theorem — all of the groups are actually computed.

3. Theorem. *Let M be a small space-time. Then for all values of $(i; a, b)$ in the table below we have*

$$H^i(L, \mathcal{E}_L(a, b)) = \Gamma(\mathcal{E} \otimes S(i; a, b)).$$

The sheaf $S(i; a, b)$ is given in the box corresponding to the indices.

In addition, one has:

(a) $H^i(\mathcal{E}_L(a,\, b)) = 0$ *for* $i \neq 1,\, 2$ *and for all* $(a,\, b)$ *in the table.*

(b) *Let* $\tilde{\nabla}_3 \colon \Gamma(\mathcal{E} \otimes \Omega^3 M) \longrightarrow \Gamma(\mathcal{E} \otimes \Omega^4 M)$ *be the differential in the de Rham sequence of* $\mathcal{E}$ *corresponding to* ∇. *Then*

$$H^i(L,\, \mathcal{E}_L(-3,\, -3)) = \begin{cases} \operatorname{Ker} \tilde{\nabla}_3 & \text{for } i = 2, \\ \operatorname{Coker} \tilde{\nabla}_3 & \text{for } i = 3, \\ 0 & \text{for } i \neq 2,\, 3. \end{cases}$$

(c) $S(i;\, b,\, a)$ *is obtained from* $S(i;\, a,\, b)$ *by interchanging* S_+ *and* S_-.

Table of $S(i;\, a,\, b)$

i	(a,b)			
	$(-1, 0)$	$(-1, -1)$	$(-2, 0)$	$(-2, -1)$
1	$S_- \otimes \wedge^2 S_+$	$\wedge^2 S_+ \otimes \wedge^2 S_-$	$\wedge^2 S_+$	0
2	0	0	0	0

i	(a,b)		
	$(-2, -2)$	$(-3, -1)$	$(-3, -2)$
1	0	0	0
2	$\wedge^2 S_+ \otimes \wedge^2 S_-$	$(\wedge^2 S_+)^2 \otimes \wedge^2 S_-$	$S_+ \otimes \wedge^2 S_+ \otimes \wedge^2 S_-$

Proof. We apply Lemma 2. Setting $\mathcal{F} = \mathcal{E}_L(a,\, b)$, we find that

$$R^i \pi_{2*} \pi_1^* \mathcal{F} = \mathcal{E} \otimes R^i \pi_{2*} \mathcal{O}_F(a,\, b),$$

$$R^i \pi_{2*} \pi_1^* \mathcal{F} \otimes \Omega^1 F/L = \mathcal{E} \otimes R^i \pi_{2*} \mathcal{O}_F(a+1,\, b+1) \otimes \wedge^2 S_+ \otimes \wedge^2 S_-.$$

The cohomology spaces of the sheaves $\mathcal{O}_F(a,\, b)$ on the relative quadric are as follows. We divide the $(a,\, b)$-plane into four quadrants using the vertical and horizontal lines $a = -1$ and $b = -1$. On these lines all of the $R^j \pi_{2*}$ are zero. In each quadrant we have $R^j \neq 0$ for exactly one value of j. These sheaves are listed below:

$$\begin{aligned} a \geq 0, b \geq 0\colon &\quad \pi_{2*} \mathcal{O}_F(a,\, b) = S^a(S_+^*) \otimes S^b(S_-^*), \\ a \leq -2, b \geq 0\colon &\quad R^1 \pi_{2*} \mathcal{O}_F(a,\, b) = S^{|a|-2}(S_+) \otimes \wedge^2 S_+ \otimes S^b(S_-^*), \\ a \leq -2, b \leq -2\colon &\quad R^2 \pi_{2*} \mathcal{O}_F(a,\, b) = S^{|a|-2}(S_+) \otimes S^{|b|-2}(S_-) \otimes \wedge^2 S_+ \otimes \wedge^2 S_-, \\ a \geq 0, b \leq -2\colon &\quad R^1 \pi_{2*} \mathcal{O}_F(a,\, b) = S^a(S_+^*) \otimes S^{|b|-2}(S_-) \otimes \wedge^2 S_-. \end{aligned} \tag{3}$$

We shall write $D(i;\, a,\, b)$ instead of $D(i;\, \mathcal{E}_L(a,\, b))$. Since this operator takes $(a,\, b)$ to $(a+1,\, b+1)$, it is zero if either $(a,\, b)$ or $(a+1,\, b+1)$ corresponds to acyclic sheaves. By (2), we see that this leads to the following identifications:

$$a=-1\colon\quad H^i(L,\, \mathcal{E}_L(-1,\, b)) = \Gamma(\mathcal{E}\otimes R^i\pi_{2*}\mathcal{O}_F(0,\, b+1)\otimes\wedge^2 S_+\otimes\wedge^2 S_-),$$

$$a=-2\colon\quad H^i(L,\, \mathcal{E}_L(-2,\, b)) = \Gamma(\mathcal{E}\otimes R^i\pi_{2*}\mathcal{O}_F(-2,\, b)).$$

In particular, using (3) and symmetry (i.e., $(a,\, b)\longleftrightarrow(b,\, a)$), we obtain all of the groups in the table.

In the other cases, the cohomology spaces are not, in general, simply sections of coherent sheaves on space-time, but rather must be described as the kernels or cokernels of differential operators. We first make a list of the spaces which can be nonzero, ordering them by increasing i:

$i=0,\, a\geq 0,\, b\geq 0$:

$$H^0(\mathcal{E}_L(a,\, b)) = \operatorname{Ker}\Big(D(0;\, a,\, b)\colon \Gamma(\mathcal{E}\otimes S^a(S_+^*)\otimes S^b(S_-^*)) \longrightarrow$$
$$\longrightarrow \Gamma(\mathcal{E}\otimes S^{a+1}(S_+^*)\otimes S^{b+1}(S_-^*)\otimes\wedge^2 S_+\otimes\wedge^2 S_-)\Big);$$

$i=1,\, a\leq -3,\, b\geq 0$ or else $a\geq 0,\, b\leq -3$:

$$H^1(\mathcal{E}_L(a,\, b)) = \operatorname{Ker} D(1;\, a,\, b);$$

$i=2,\, a\leq -3,\, b\geq 0$ or else $a\geq 0,\, b\leq -3$:

$$H^2(\mathcal{E}_L(a,\, b)) = \operatorname{Coker} D(1;\, a,\, b);$$

$i=2,\, a\leq -3,\, b\leq -3$:

$$H^2(\mathcal{E}_L(a,\, b)) = \operatorname{Ker} D(2;\, a,\, b);$$

$i=3,\, a\leq -3,\, b\leq -3$:

$$H^3(\mathcal{E}_L(a,\, b)) = \operatorname{Coker} D(2;\, a,\, b).$$

We now say a few words about the structure of the operators $D(i;\, a,\, b)$. It is clear from (3) that all of the sheaves $\mathcal{E}\otimes R^i\pi_{2*}\mathcal{O}_F(a,\, b)$ are contained in the tensor algebra generated by $\mathcal{E}$, S_+ and S_-. The connection D generated by ∇, ∇_+ and ∇_- acts on this tensor algebra. We claim that any of the operators $D(i;\, a,\, b)$ can be defined as the composition of the covariant differential $D\colon \mathcal{F}\longrightarrow\mathcal{F}\otimes\Omega^1$ and a projection constructed from the symmetrization operators for the required structure. For example,

$$D(0;\, 0,\, 0)\colon \Gamma(\mathcal{E})\longrightarrow\Gamma(\mathcal{E}\otimes S_+^*\otimes S_-^*\otimes\wedge^2 S_+\otimes\wedge^2 S_-) = \Gamma(\mathcal{E}\otimes\Omega^1 M)$$

is simply ∇, and the operator

$$D(2;\,-3,-3)\colon \Gamma(\mathcal{E}\otimes S_+\otimes S_-\otimes\wedge^2 S_+\otimes\wedge^2 S_-)\longrightarrow\Gamma(\mathcal{E}\otimes(\wedge^2 S_+\otimes\wedge^2 S_-)^2)$$

is proportional to ∇_3, in view of the identifications

$$\Omega^3 M = S_+\otimes S_-\otimes\wedge^2 S_+\otimes\wedge^2 S_-,\qquad \Omega^4 M = (\wedge^2 S_+\otimes\wedge^2 S_-)^2.$$

(This proves part (b) of the theorem.) The verification of this makes use of the absence of torsion: an essential point is that the composition of the covariant differential $\nabla^g\colon\Omega^1 M\longrightarrow\Omega^1 M\otimes\Omega^1 M$ and anti-symmetrization coincides with the exterior differential when there is no torsion. This effect remains in force in the higher terms of the de Rham sequence. □

We now compute the cohomology group which classifies the "first order neighborhoods" of L, i.e., the simple extensions of the structure sheaf of the form (see § 6)

$$0\longrightarrow I\longrightarrow \mathcal{O}_{L^{(1)}}\longrightarrow\mathcal{O}_L\longrightarrow 0.$$

We recall that in the flat case there is a canonical system of neighborhoods of all orders.

4. Theorem. *If M is small, then one has the isomorphism*

$$H^1(L,\,\mathcal{T}L\otimes I)=\Gamma(\mathcal{O}_M\oplus\Omega^1 M\oplus\Omega^2 M).$$

Proof. We shall not actually construct a canonical isomorphism with the direct sum on the right, but rather shall find an invariant filtration with those factors; over a Stein manifold M such a filtration splits.

In order to apply Lemma 2, we must compute the direct images on M of the sheaves $\pi_1^*(\mathcal{T}L\otimes I)$ and $\pi_1^*(\mathcal{T}L\otimes I\otimes\Omega^1 F/L)=\pi_1^*(\mathcal{T}L)\otimes\pi_2^*(\wedge^2 S_+\otimes\wedge^2 S_-)$. The argument in § 7.4 shows that we have an exact sequence $0\longrightarrow\mathcal{N}\longrightarrow\pi_1^*\Omega^1 L\xrightarrow{\text{res}}\Omega^1 F/M\longrightarrow 0$, and hence

$$0\longrightarrow\mathcal{T}F/M\longrightarrow\pi_1^*\mathcal{T}L\longrightarrow\mathcal{N}^*\longrightarrow 0,$$

$$0\longrightarrow\mathcal{T}F/M\otimes I\longrightarrow\pi_1^*(\mathcal{T}L\otimes I)\longrightarrow\mathcal{N}^*\otimes I\longrightarrow 0.$$

The sheaf $\mathcal{T}F/M$ has a unique nonzero direct image:

$$\pi_{2*}\mathcal{T}F/M=(S_+^*\otimes S_+)_0\oplus(S_-^*\otimes S_-)_0,$$

where the zero subscript denotes the traceless part. Thus,

$$\pi_{2*}\mathcal{T}F/M\otimes\wedge^2 S_+\otimes\wedge^2 S_-=S^2 S_+\otimes\wedge^2 S_-\oplus\wedge^2 S_+\otimes S^2 S_-=\Omega^2 M.$$

The sheaf $\mathcal{T}F/M \otimes I$ is acyclic on M.

Furthermore, by definition, $\mathcal{N}$ occurs in the exact sequence

$$0 \longrightarrow \mathcal{T}F/L \longrightarrow \pi_2^*\mathcal{T}M \longrightarrow \mathcal{N}^* \longrightarrow 0,$$

from which we have $\mathcal{T}M = \pi_{2*}\mathcal{N}^*$, because $\mathcal{T}F/L$ is acyclic, and

$$\delta: R^1\pi_{2*}\mathcal{N}^* \otimes I \xrightarrow{\sim} R^2\pi_{2*}\mathcal{T}F/L \otimes I = R^2\pi_{2*}\mathcal{O}_F(-2,\,-2) \otimes \wedge^2 S_+^* \otimes \wedge^2 S_-^* = \mathcal{O}_M.$$

The other direct images of $\mathcal{N}^* \otimes I$ are all zero. Setting $\mathcal{F} = \mathcal{T}L \otimes I$ and $i = 1$ in (2), we finally obtain the following exact sequences and identifications, which complete the proof of the theorem:

$$0 \to \Gamma(\pi_{2*}\pi_1^*(\mathcal{T}L \otimes \Omega^1 F/L \otimes I)) \to H^1(L,\,\mathcal{T}L \otimes I) \to \Gamma(R^1\pi_{2*}\pi_1^*(\mathcal{T}L \otimes I)) \to 0,$$

$$0 \longrightarrow \Omega^2 M \longrightarrow \pi_{2*}\pi_1^*(\mathcal{T}L \otimes \Omega^1 F/L \otimes I) \longrightarrow \mathcal{T}M \otimes \wedge^2 S_+ \otimes \wedge^2 S_- = \Omega^1 M \longrightarrow 0,$$

$$R^1\pi_{2*}\pi_1^*(\mathcal{T}L \otimes I) = R^1\pi_{2*}\mathcal{N}^* \otimes I = \mathcal{O}_M. \qquad \square$$

§ 9. The Flow of a Yang–Mills Field on the Space of Null-Geodesics

1. Statement of the problem. We continue to work with the diagram of null-geodesics of a small space-time with the structures described in the last section.

Suppose that we have constructed a tower of infinitesimal extensions $L = L^{(0)} \subset L^{(1)} \subset L^{(2)} \subset L^{(3)}$, where the ideal $\tilde{J}$ which gives L in $L^{(3)}$ has the property that $\tilde{J}/\tilde{J}^2 = I$ as an $\mathcal{O}_L$-module, and $L^{(i)}$ is given by the ideal $\tilde{J}^i$. We consider the Yang–Mills sheaf $\mathcal{E}_L$ on L which is the Radon–Penrose transform of the pair $(\mathcal{E},\,\nabla)$ on M, and we pose the problem of describing its extensions to $L^{(i)}$.

The cohomology groups we need are described in Theorem 8.3.

(a) $H^2(\mathcal{E}nd\ \mathcal{E}_L \otimes I) = 0$, and hence there exist extensions of $\mathcal{E}_L$ to $L^{(1)}$. The group $H^1(\mathcal{E}nd\ \mathcal{E}_L \otimes I) = \Gamma(\mathcal{E}nd\ \mathcal{E} \otimes \wedge^2 S_+ \otimes \wedge^2 S_-)$ acts on the set of all classes of extensions. We choose some class $e^{(1)}$ as our origin, and we let $e^{(1)} + h$ denote the result of translating it by $h \in \Gamma(\mathcal{E}nd\ \mathcal{E} \otimes \wedge^2 S_+ \otimes \wedge^2 S_-)$.

(b) $H^1(\mathcal{E}nd\ \mathcal{E}_L \otimes I^2) = 0$, and hence, if an extension of $\mathcal{E}_L$ to $L^{(1)}$ extends to $L^{(2)}$, then this extension is unique. On the other hand, the group of obstructions is not zero: this group is $H^2(\mathcal{E}nd\ \mathcal{E}_L \otimes I^2) = \Gamma(\mathcal{E}nd\ \mathcal{E} \otimes \wedge^2 S_+ \otimes \wedge^2 S_-)$. It is possible for the obstruction map $\omega: H^1 \longrightarrow H^2$:

$$h \mapsto \omega(e^{(1)} + h): \mathcal{E}nd\ \mathcal{E} \otimes \wedge^2 S_+ \otimes \wedge^2 S_- \longrightarrow \mathcal{E}nd\ \mathcal{E} \otimes \wedge^2 S_+ \otimes \wedge^2 S_-$$

to be a bijection for a suitable tower of $L^{(i)}$. If this is the case, then there is a unique class h for which $\omega(e^{(1)}+h)=0$, i.e., there is a unique extension $\mathcal{E}_L^{(1)}$ which has an $L^{(2)}$-extension $\mathcal{E}_L^{(2)}$, and the latter extension is also unique.

(c) Suppose that the conditions in (b) hold, and $e^{(2)}$ is the class of $\mathcal{E}_L^{(2)}$. The obstruction $\omega(e^{(3)})$ to a third extension lies in the group

$$H^2(\mathcal{E}nd\ \mathcal{E}_L \otimes I^3) = \operatorname{Ker}\big(\tilde{\nabla}_3 : \Gamma(\mathcal{E}nd\ \mathcal{E} \otimes \Omega^3 M) \longrightarrow \Gamma(\mathcal{E}nd\ \mathcal{E} \otimes \Omega^4 M)\big).$$

There is one element in this group which can be constructed immediately from $(\mathcal{E}, \nabla)$. Namely, let $\Phi(\nabla)=\Phi_+(\nabla)+\Phi_-(\nabla)$. We set

$$j = \tilde{\nabla}\Phi_+(\nabla) = -\tilde{\nabla}\Phi_-(\nabla).$$

Here $\tilde{\nabla}$ is the differential in $\mathcal{E}nd\ \mathcal{E} \otimes \Omega^{\cdot} M$, and we have taken into account the Bianchi identity $\tilde{\nabla}\Phi(\nabla)=0$. One has $\tilde{\nabla}j=0$ because $\tilde{\nabla}^2$ is a commuting operator with $\Phi(\nabla)$. This element j is called the "axial flow" of the field $(\mathcal{E}, \nabla)$.

2. Definition. A tower of extensions $L \subset L^{(i)}$, $i \leq 3$, is called *regular* if any Yang–Mills sheaf $\mathcal{E}_L$ on M has a unique extension to $L^{(2)}$, and if the obstruction to extending it to $L^{(3)}$ coincides (up to normalization) with the axial flow of the field $(\mathcal{E}, \nabla)$. □

The question of existence and uniqueness of a regular tower for an arbitrary space of null-geodesics is an unsolved problem. Le Brun created the first extension which, to all appearances, is regular. The basic result of this section concerns the flat case, where, as mentioned above, we have a canonical tower $L^{(i)}$ corresponding to the imbedding $L \subset \mathrm{P}(T) \times \mathrm{P}(T^*)$.

3. Theorem. *Let $U \subset G(2;\, T) = M$ be a small space-time, and let $L(U) = \pi_2\pi_1^{-1}(U)$; then the neighborhoods $L(U)^{(i)} = (L(U),\ \mathcal{O}_{L^{(i)}}|_{L(U)})$ form a regular tower.*

4. Corollary. *A Yang–Mills field $(\mathcal{E}, \nabla)$ defined in some region of a flat space-time is a solution to the system of Yang–Mills equations with no sources*

$$\tilde{\nabla}\Phi_{\pm}(\nabla)=0$$

if and only if every point in its domain of definition has a neighborhood U such that the Radon–Penrose transform $\mathcal{E}_L$ on $L(U)$ of the pair $(\mathcal{E}, \nabla)|_U$ has an extension to $L(U)^{(3)}$. □

This corollary opens up the possibility of constructing solutions to a nonlinear system of differential equations by algebro-geometric methods. The construction of a class of YM-sheaves on arbitrary neighborhoods $L(U)^{(i)}$ by the monad method will be postponed until § 4 of Chapter 5, at which point we will be able to give this construction on superspaces. The reader who would like to become familiar with this construction sooner may read §§ 4–5 of Chapter 5 now, taking $N = 0$ there, i.e., simply disregarding all of the odd coordinates.

5. Proof of Theorem 5. We start by introducing the total space of the double fibration which connects $L^{(i)}$ with M. We take as the total space, denoted $F^{[i]}$, the i-th infinitesimal neighborhood of F in $\mathsf{P} \times \hat{\mathsf{P}} \times M$, where $\mathsf{P} = \mathsf{P}(T)$, $\hat{\mathsf{P}} = \mathsf{P}(T^*)$. We let $F^{(i)}$ denote the i-th infinitesimal neighborhood of F in $L \times M$. In what follows we shall continually be comparing the Radon–Penrose transform on the following two towers of double fibrations and on the parts associated with a small space-time $U \subset M$:

$$\begin{array}{ccccc} L^{(i)} & \xleftarrow{\pi_i^{[i]}} & F^{[i]} & \xrightarrow{\pi_2^{[i]}} & M \\ \cup & & \cup & & \| \\ L & \xleftarrow{\pi_1^{(i)}} & F^{(i)} & \xrightarrow{\pi_2^{(i)}} & M. \end{array}$$

We first show that there is a natural one-to-one correspondence between the extensions $\mathcal{E}_L^{(i)}$ of the sheaf $\mathcal{E}_L$ to $L(U)^{(i)}$ and the extensions $(\mathcal{E}_F^{[i]}, \nabla_{F/L}^{[i]})$ of the pair $(\mathcal{E}_F, \nabla_{F/L})$ to $F(U)^{[i]}$. Here $\nabla_{F/L}^{[i]}$ denotes the relative connection along the fibres of $\pi_1^{[i]}$ which when restricted to $\mathcal{E}_{F^{(i)}} = \pi_1^{*(i)}(\mathcal{E}_L)$ coincides with the standard connection $\nabla_{F/L}^{(i)}$, which kills liftings to $F^{(i)}(U)$ of sections of $\mathcal{E}_L$.

If we are given $\mathcal{E}_L^{(i)}$, then we take $\mathcal{E}_F^{[i]} = \pi_1^{[i]*}\mathcal{E}_L^{(i)}$, and $\nabla_{F/L}^{[i]}$ is differentiation along the fibres of the projection $\pi_1^{[i]}$.

A little more work is involved in proving correctness of the inverse map $(\mathcal{E}_F^{[i]}, \nabla_{F/L}^{[i]}) \mapsto \pi_{1*}^{[i]}(\operatorname{Ker} \nabla_{F/L}^{[i]}) = \mathcal{E}_L^{(i)}$. We have already done this when $i = 0$. We now proceed by induction from $i - 1$ to i. First, we have an exact sequence

$$0 \longrightarrow \mathcal{O}_{F^{[i-1]}}(-1, -1) \longrightarrow \mathcal{O}_{F^{[i]}} \longrightarrow \mathcal{O}_{F^{(i)}} \longrightarrow 0.$$

Using this sequence and the results in § 6, we find that, given an extension $(\mathcal{E}_F^{[i]}, \nabla_{F/L}^{[i]})$, we can construct the following commutative diagram with exact rows:

$$\begin{array}{ccccccccc} 0 & \longrightarrow & \mathcal{E}_F^{[i-1]}(-1, -1) & \longrightarrow & \mathcal{E}_F^{[i]} & \longrightarrow & \mathcal{E}_F^{(i)} & \longrightarrow & 0 \\ & & \big\downarrow \Delta & & \big\downarrow \nabla_{F/L}^{[i]} & & \big\downarrow \nabla_{F/L}^{(i)} & & \\ 0 & \longrightarrow & \dfrac{\mathcal{E}_F^{[i-1]} \otimes \Omega^1 F^{[i-1]}}{L^{[i-1]}(-1, -1)} & \longrightarrow & \mathcal{E}_F^{[i]} \otimes \Omega^1 F^{[i]}/L^{[i]} & \longrightarrow & \mathcal{E}_F^{(i)} \otimes \Omega^1 F^{(i)}/L & \longrightarrow & 0. \end{array} \tag{1}$$

Here we have let Δ denote the induced connection. If we twist it by (1, 1), we obtain the restriction $\nabla_{F/L}^{[i-1]} = \nabla_{F/L}^{[i]}|_{F^{[i-1]}}$. By the induction assumption, the pair $(\mathcal{E}_F^{[i-1]}, \nabla_{F/L}^{[i-1]})$ determines an extension $\mathcal{E}_L^{(i-1)} = \pi_{1*}^{[i-1]}(\mathrm{Ker}\ \nabla_{F/L}^{[i-1]})$ for which $\mathcal{E}_L^{[i-1]} = \pi_1^{[i-1]*}\mathcal{E}_L^{(i-1)}$, and $\nabla_{F/L}^{[i-1]}$ annihilates precisely the lifted sections of $\mathcal{E}_L^{(i-1)}$. Using this and the commutativity of the diagram (1), we obtain the exact sequence on $F^{[i]}(U)$

$$0 \longrightarrow \pi_1^{-1}(\mathcal{E}_L^{(i-1)})(-1, -1) \longrightarrow \mathrm{Ker}\ \nabla_{F/L}^{[i]} \longrightarrow \pi_1^{-1}(\mathcal{E}_L) \longrightarrow 0,$$

which implies that the following sequence on $L(U)$ is exact:

$$0 \longrightarrow \mathcal{E}_L^{(i-1)}(-1, -1) \longrightarrow \mathcal{E}_L^{(i)} \longrightarrow \mathcal{E}_L \longrightarrow 0.$$

If we use Nakayama's lemma, we can conclude from this by induction on i that $\mathcal{E}_L^{(i)}$ is locally free over $\mathcal{O}_{L^{(i)}}$, and $\mathcal{E}_F^{[i]} = \pi_1^{[i]*}\mathcal{E}_L^{(i)}$. To show that these categories are equivalent, and, in particular, their isomorphism classes of objects are in one-to-one correspondence, as usual we pass to tensor products and take sections on $L(U)$ or horizontal sections on $F(U)$, respectively.

We now look at the problem of extension to $F(U)$. In the first place, we have $H^1(F(U), \mathcal{E}nd\ \mathcal{E}_F \otimes \tilde{\mathcal{N}}) = 0$, where $\tilde{\mathcal{N}}$ is the conormal sheaf of F in $\mathbf{P} \times \hat{\mathbf{P}} \times M$. In fact, it is not hard to compute that $\tilde{\mathcal{N}} = \pi_2^* S_+(0, -1) \oplus \pi_2^* S_-(-1, 0)$, and hence $R^1_{\pi_{2*}} \tilde{\mathcal{N}} = 0$, at which point the Leray spectral sequence for π_2 gives us our claim, as in § 8. Consequently, there is a unique extension of $\mathcal{E}_F^{[1]}$ to $F(U)^{[1]}$, namely $\pi_2^{[1]*}(\mathcal{E})$. But this is also the unique extension of $\mathcal{E}_F^{(1)} = \pi_2^{(1)*}(\mathcal{E})$. Thus, the non-uniqueness of the extension of $\mathcal{E}_L$ to $L(U)^{(1)}$ that we mentioned in § 9.1 is manifested on $F(U)$ in the non-uniqueness of the extension $\nabla_{F/L}^{[1]}$ of $\nabla_{F/L}^{(1)}$. Such an extension exists because $\mathcal{E}_L^{(1)}$ exists. The difference of two extensions lies in the group

$$H^0(F(U), \mathcal{E}nd\ \mathcal{E}_F \otimes \Omega^1 F/L(-1, -1)) = \Gamma(\mathcal{E}nd\ \mathcal{E} \otimes \wedge^2 S_+ \otimes \wedge^2 S_-),$$

in accordance with the computations in § 9.1.

Furthermore, $\mathcal{E}_F^{[1]}$ has an extension to $F(U)^{[2]}$, for example $\pi_2^{[2]*}(\mathcal{E})$, and this extension satisfies the condition for lifting endomorphisms. Thus, the group $H^1(F(U), \mathcal{E}nd\ \mathcal{E}_F \otimes S^2\tilde{\mathcal{N}})$, which equals $\Gamma(\mathcal{E}nd\ \mathcal{E} \otimes \Omega^2 M)$ by our earlier computation of $\mathcal{N}$, acts effectively on the set of all extensions to $F(U)$ Since the surjection $\tilde{\mathcal{N}} \longrightarrow \mathcal{N}$ induces an isomorphism between this group and $H^1(F(U), \mathcal{E}nd\ \mathcal{E}_F \otimes$

$S^2\mathcal{N}$), we have exactly the same number of extensions of $\mathcal{E}_F^{(1)}$ to a sheaf $\mathcal{E}_F^{(2)} = \pi_1^*(\mathcal{E}_L)|_{F^{(2)}}$. As a consequence, there is exactly one extension $\mathcal{E}_F^{[2]}$ for which $\mathcal{E}_F^{[2]}|_{F^{(2)}} = \mathcal{E}_F^{(2)}$. After this one verifies that both the group of obstructions and the group of extensions of $\nabla_F^{(2)}$ to a connection $\nabla_F^{[2]}$ are zero:

$$H^i(F(U), \mathcal{E}nd\ \mathcal{E}_F^{[1]} \otimes \Omega^1 F^{[1]}/L^{[1]}(-1, -1)) = 0 \qquad \text{for} \quad i = 0, 1.$$

Thus, the pair $(\mathcal{E}_F^{[2]}, \nabla_F^{[2]})$ is unique. It is this pair which determines the unique extension of $\mathcal{E}_L$ to a sheaf $\mathcal{E}_L^{(2)}$.

It remains to compute $\omega(\mathcal{E}_L^{(2)})$. We shall carry out the calculation in several steps. First of all, lifting this obstruction to F transforms it to an element of the group $H^2(F(U), \mathcal{E}nd\ \mathcal{E}_F(-3, -3))$ (lying in the kernel of $\tilde{\nabla}_3$, by Theorem 8.3).

On the other hand, the infinitesimal extension $F^{(1)} \subset F^{(3)}$ is defined by the ideal $\mathcal{K}^2/\mathcal{K}^4$ with square zero, where $\mathcal{K}$ is the ideal of F in $L \times M$. The extension $\pi_2^{(3)*}(\mathcal{E})$ of the sheaf $\mathcal{E}_F^{(1)}$ satisfies the condition for endomorphisms to lift. Hence, the group $H^1(F(U), \mathcal{E}nd\ \mathcal{E}_F^{(1)} \otimes \mathcal{K}^2/\mathcal{K}^4)$ exactly classifies the extensions of $\mathcal{E}_F^{(1)}$ to $F^{(3)}$ by their difference classes with $\pi_2^{(3)*}(\mathcal{E})$. We consider the exact sequence (where J is the ideal of F in $\mathsf{P} \times \hat{\mathsf{P}} \times M$)

$$0 \longrightarrow \mathcal{E}nd\ \mathcal{E}_F^{[1]} \otimes J/J^3(-1, -1) \longrightarrow \mathcal{E}nd\ \mathcal{E}_F^{[1]} \otimes J^2/J^4 \longrightarrow \mathcal{E}nd\ \mathcal{E}_F^{(1)} \otimes \mathcal{K}^2/\mathcal{K}^4 \longrightarrow 0$$

and the corresponding boundary homomorphism

$$\delta\colon H^1(F(U), \mathcal{E}nd\ \mathcal{E}_F^{(1)} \otimes \mathcal{K}^2/\mathcal{K}^4) \longrightarrow H^2(\mathcal{E}nd\ \mathcal{E}_F^{[1]} \otimes J/J^3(-1, -1)).$$

This homomorphism takes the difference class of an arbitrary extension $\mathcal{E}_F^{(3)}$ to some element $\varphi(\mathcal{E}) \in H^2(\mathcal{E}nd\ \mathcal{E}_F^{[1]} \otimes J/J^3(-1, -1))$.

Finally, the composition

$$\mathcal{O}(-3, -3) \longrightarrow J^2/J^3(-1, -1) \longrightarrow J/J^3(-1, -1)$$

(the first map comes from $0 \to \mathcal{O}(-1, -1) \to \tilde{\mathcal{N}} \to \mathcal{N} \to 0$ and then $0 \to \mathcal{O}(-2, -2) \to S^2\tilde{\mathcal{N}}$) takes $\omega(\mathcal{E}_L^{(2)})$ to the same group that contains $\varphi(\mathcal{E})$. We shall show that in this group both classes are the same. In this we shall always be using the interpretation of an obstruction as the image of a difference class under a boundary homomorphism (see § 6).

The extensions of $\mathcal{E}_F^{[2]}$ to $F^{[3]}$ are classified by the elements of $H^1(F(U), \mathcal{E}nd\ \mathcal{E}_F \otimes \mathcal{K}^3/\mathcal{K}^4)$. This group is imbedded in the group $H^1(F(U), \mathcal{E}nd\ \mathcal{E}_F^{(1)} \otimes \mathcal{K}^2/\mathcal{K}^4)$, which classifies the extensions of $\mathcal{E}_F^{(1)}$ to $F^{(3)}$. Hence, we may suppose that the difference between $\mathcal{E}_F^{[3]}|_{F^{(3)}}$ and $\mathcal{E}_F^{(3)}$ as extensions of $\mathcal{E}_F^{(2)}$ lies in the group $H^1(F(U), \mathcal{E}nd\ \mathcal{E}_F \otimes \mathcal{K}^3/\mathcal{K}^4)$. We now consider the exact sequence

$$0 \longrightarrow \mathcal{E}nd\ \mathcal{E}_F \otimes J^2/J^3(-1,-1) \longrightarrow \mathcal{E}nd\ \mathcal{E}_F \otimes J^3/J^4 \longrightarrow \mathcal{E}nd\ \mathcal{E}_F \otimes \mathcal{K}^3/\mathcal{K}^4 \longrightarrow 0.$$

The corresponding boundary homomorphism

$$\begin{aligned}\delta\colon H^1(\mathcal{E}nd\ \mathcal{E}_F \otimes \mathcal{K}^3/\mathcal{K}^4) \longrightarrow H^2(\mathcal{E}nd\ \mathcal{E}_F \otimes J^2/J^3(-1,-1)) &\simeq \\ \simeq H^2(\mathcal{E}nd\ \mathcal{E}_F^{[1]} \otimes J^2/J^3(-1,-1))&\end{aligned}$$

takes the difference class $\left[\mathcal{E}_F^{[3]}|_{F^{(3)}}\right] - \left[\mathcal{E}_F^{(3)}\right]$ to the class $\left[\mathcal{E}_F^{(3)}\right] - \left[\pi_2^{(3)*}\mathcal{E}\right]$.

We consider the following commutative diagram with exact rows and columns:

$$\begin{array}{ccccccccc}
 & & 0 & & 0 & & 0 & & \\
 & & \downarrow & & \downarrow & & \downarrow & & \\
0 & \to & \begin{matrix}\pi_1^* I^3 = \\ \mathcal{O}_F(-3,-3)\end{matrix} & \to & J^3/J^4 & \to & (J^3/J^4)/\mathcal{O}_F(-3,-3) & \to & 0 \\
 & & \downarrow & & \downarrow & & \downarrow & & \\
0 & \to & \begin{matrix}\pi_{12}^*\Omega^1(\mathbf{P}\times\hat{\mathbf{P}}) \\ \times\mathcal{O}_F/J^3\end{matrix} & \to & \begin{matrix}\Omega^1(\mathbf{P}\times\hat{\mathbf{P}}\times M) \\ \otimes\mathcal{O}_F/J^3\end{matrix} & \to & \pi_2^*\Omega^1 M\otimes\mathcal{O}_F/J^3 & \to & 0 \\
 & & \downarrow & & \downarrow & & \downarrow & & \\
0 & \to & \pi^{[2]*}\Omega^1 L^{[2]} & \to & \Omega^1 F^{[2]} & \to & \Omega^1 F^{[2]}/L^{[2]} & \to & 0 \\
 & & \downarrow & & \downarrow & & \downarrow & & \\
 & & 0 & & 0 & & 0 & &
\end{array} \tag{2}$$

The cohomology groups considered below are connected by maps which are induced by the arrows in this diagram. The class of $\mathcal{E}_F^{[3]}$ in the group $H^1(F(U), \mathcal{E}nd\ \mathcal{E}_F^{[2]} \otimes \Omega^1(\mathbf{P}\times\hat{\mathbf{P}}\times M))$ goes to the class of $\mathcal{E}_F^{[2]}$ in the group $H^1(F(U), \mathcal{E}nd\ \mathcal{E}_F^{[2]} \otimes \Omega^1 F^{[2]})$. The class of $\mathcal{E}_L^{(2)}$ also goes to that class, because $\mathcal{E}_F^{[2]} = \pi_2^{[2]*}\mathcal{E}_L^{(2)}$. The relation $\mathcal{E}_F^{(3)} = \pi_2^{(3)*}\mathcal{E}_L$ implies that the image of the class of $\mathcal{E}_F^{[3]}$ in the group $H^1(F(U), \mathcal{E}nd\ \mathcal{E}_F^{(2)} \otimes \pi_2^*\Omega^1 M)$ is equal to the difference between the classes of $\mathcal{E}_F^{[3]}|_{F^{(3)}}$ and $\mathcal{E}_F^{(3)}$.

A simple computation shows that the natural homomorphism $(J^3/J^4)/\mathcal{O}(-3,-3) \longrightarrow \mathcal{K}^3/\mathcal{K}^4$ induces isomorphisms of all of the direct images of these sheaves on M, and only $R^1\pi_{2*}$ is nonzero. Hence, the difference between the classes of $\mathcal{E}_F^{[3]}|_{F(3)}$ and $\mathcal{E}_F^{(3)}$ in $H^1(F(U), \mathcal{E}nd\ \mathcal{E}_F \otimes \mathcal{K}^3/\mathcal{K}^4)$ can be carried over to $H^1(F(U), \mathcal{E}nd\ \mathcal{E}_F \otimes (J^3/J^4)/\mathcal{O}_F(-3, -3))$, and then we can apply to it the boundary homomorphism coming from the first row in the diagram (2). The result will be the image of the obstruction $\omega(\mathcal{E}_L^{(2)})$ in the group $H^2(F(U), \mathcal{E}nd\ \mathcal{E}_F(-3, -3))$. In order to verify that this becomes equal to $[\mathcal{E}_F^{[3]}|_{F(3)}] - [\mathcal{E}_F^{(3)}]$ in the group $H^2(F(U), \mathcal{E}nd\ \mathcal{E}_F \otimes J^2/J^3(-1, -1))$, it remains for us to use the following commutative diagram with exact rows:

$$\begin{array}{ccccccccc}
0 & \longrightarrow & \mathcal{O}_F(-3, -3) & \longrightarrow & J^3/J^4 & \longrightarrow & (J^3/J^4)/\mathcal{O}_F(-3, -3) & \longrightarrow & 0 \\
 & & \downarrow & & \| & & \downarrow & & \\
0 & \longrightarrow & J^2/J^3(-1, -1) & \longrightarrow & J^3/J^4 & \longrightarrow & \mathcal{K}^3/\mathcal{K}^4 & \longrightarrow & 0.
\end{array}$$

This completes the first step of our computations: we have described $\omega(\mathcal{E}_L^{(2)})$ in terms of a certain extension $\mathcal{E}_F^{[3]}$. This can be expressed using certain matrix functions g_{ij}, which are defined modulo J^4 on $\mathbf{P} \times \hat{\mathbf{P}} \times M$ and satisfy the conditions

$$g_{ij}^{-1} d_{\mathbf{P}\times\hat{\mathbf{P}}\times M/\mathbf{P}\times\hat{\mathbf{P}}}\, g_{ij} \in J\pi_2^*\Omega^1 M.$$

Let φ be the difference between $\mathcal{E}_F^{[3]}$ and $\pi_2^{[3]*}\mathcal{E}$ in the group $H^1(F(U), \mathcal{E}nd\ \mathcal{E}_F^{[1]} \otimes J^2/J^4)$, and let $d\varphi$ be its image in the group $H^1(F(U), \mathcal{E}nd\ \mathcal{E}_F^{[2]} \otimes \pi_2^{[2]*}\Omega^1 M)$. A computation using the cocycle (g_{ij}) shows that in this group $-d\varphi$ coincides with the class of $\mathcal{E}_F^{[3]}$, which is constructed by means of the composite map

$$GL(\mathcal{O}_{F[3]}) \longrightarrow M(\Omega^1(\mathbf{P} \times \hat{\mathbf{P}} \times M) \otimes \mathcal{O}_F^{[3]}) \longrightarrow M(\pi_2^*\Omega^1 M \otimes \mathcal{O}_{F[3]}).$$

We now construct the commutative diagram with exact rows

$$\begin{array}{ccccccccc}
0 & \longrightarrow & J^3/J^4 & \longrightarrow & J^2/J^4 & \longrightarrow & J^2/J^3 & \longrightarrow & 0 \\
 & & \downarrow & & \downarrow & & \downarrow & & \\
0 & \longrightarrow & \mathcal{K}^3/\mathcal{K}^4 & \longrightarrow & \pi_2^*\Omega^1 M \otimes \mathcal{K}/\mathcal{K}^3 & \longrightarrow & \mathcal{F} & \longrightarrow & 0,
\end{array} \tag{3}$$

where $\mathcal{F}$ is defined as the cokernel of the second arrow in the bottom row, the center vertical morphism is the composition $J^2/J^4 \longrightarrow \pi_2^*\Omega^1 M \otimes J/J^3 \longrightarrow \pi_2^*\Omega^1 M \otimes \mathcal{K}/\mathcal{K}^3$ and the right vertical morphism is induced by this composition. Computations show that all of the sheaves in this diagram satisfy $R^i\pi_{2*} = 0$ for $i \neq 1$. Hence, if we apply $R^1\pi_{2*}$ to the diagram, we obtain another commutative diagram with exact rows. The right column of this new diagram contains the sheaves

$$R^1\pi_{2*}J^2/J^3 = \Omega^2 M, \qquad R^1\pi_{2*}\mathcal{F} = \Omega^3 M,$$

which are connected by the exterior differential d_2. The cokernel of the left vertical arrow for the $R^1\pi_{2*}$-sheaves can be identified with $R^2\pi_{2*}J^2/J^3(-1,-1) = \Omega^3 M$. Thus, applying the snake lemma to the $R^1\pi_{2*}$-diagram gives us a map

$$\text{Ker } d_2 \longrightarrow \Omega^3 M.$$

One can compute this map directly and see that, up to a normalization, it takes $\Phi_+ + \Phi_-$ to $d\Phi_+$. If we tensor the diagram (3) with $\mathcal{E}nd\ \mathcal{E}_F^{[1]}$ and apply $R^1\pi_{2*}$, we obtain similar results with d replaced by $\tilde{\nabla}$ in the de Rham sequence for $(\mathcal{E}nd\ \mathcal{E}, \tilde{\nabla})$. It remains for us to trace what happens to the elements of particular interest in the cohomology groups here.

The snake mapping

$$\begin{aligned} s\colon \text{Ker}\big(H^1(F(U), \mathcal{E}nd\ \mathcal{E}_F \otimes J^2/J^3) &\longrightarrow H^1(F(U), \mathcal{E}nd\ \mathcal{E}_F^{[1]} \otimes \mathcal{F})\big) \\ &\longrightarrow H^2(F(U), \mathcal{E}nd\ \mathcal{E}_F \otimes J^2/J^3(-1,-1)) \end{aligned}$$

takes the difference between $\mathcal{E}_F^{[2]}$ and $\pi_2^{[2]*}\mathcal{E}$ to the image of the obstruction $\omega(\mathcal{E}_L^{(2)})$ in the second group. On the other hand, the following isomorphisms enable us to use Theorem 7.5 to identify this difference class with the curvature $\Phi(\nabla)$:

$$H^1(F(U), \mathcal{E}nd\ \mathcal{E}_F \otimes J^2/J^3) \xrightarrow{\sim} H^1(F(U), \mathcal{E}nd\ \mathcal{E}_F \otimes \mathcal{K}^2/\mathcal{K}^3) \xrightarrow{\sim} \Gamma(U, \mathcal{E}nd\ \mathcal{E} \otimes \Omega^2 M).$$

Furthermore, just as in our computations with $\mathcal{E}_F = \mathcal{O}_F$, we have:

$$H^1(F(U), \mathcal{E}nd\ \mathcal{E}_F^{(1)} \otimes \mathcal{F}) = \Gamma(U, \mathcal{E}nd\ \mathcal{E} \otimes \Omega^3 M),$$

$$\begin{aligned} H^2(F(U), \mathcal{E}nd\ \mathcal{E}_F \otimes J^2/J^3(-1,-1)) &= H^2(F(U), \mathcal{E}nd\ \mathcal{E}_F(-3,-3)) \\ &= \Gamma(U, \mathcal{E}nd\ \mathcal{E} \otimes \Omega^3 M), \end{aligned}$$

and the snake morphism s becomes

$$s(\Phi_+ + \Phi_-) = \tilde{\nabla}_2 \Phi_+.$$

Since the morphism takes $\Phi(\nabla)$ to $\omega(\mathcal{E}_L^{(2)})$, this completes the proof of Theorem 9.3. □

§ 10. Extension Problems and Dynamical Equations

1. General remarks. In § 9 we saw that in a small flat space-time the Yang–Mills equation $\tilde{\nabla}\Phi_{+}(\nabla) = j$ is encoded by the equation $\omega(\mathcal{E}_L^{(2)}) = j_L$, where j_L is the representative of the axial flow in the cohomology group $H^2(\mathcal{E}nd\ \mathcal{E}_L(-3, -3))$. In the present section we shall demonstrate that several other extension problems in the context of our two- or three-floor regular towers $\{L^{(i)}\}$ of infinitesimal extensions lead to various dynamical equations for different types of fields on M. We will encounter this theme again in §§ 3–4 of Chapter 5 in connection with the Radon–Penrose transform for supersymmetric equations.

Before going any further, we should emphasize that this is an unfinished story. In the case of curved space-time, it is not even clear what is the correct way to formulate the Einstein equations. How can one state the (anti-) self-duality of the sheaves $\Sigma_{\pm}$ in a more geometrical manner? How is the problem of existence of a regular tower of infinitesimal extensions $L^{(i)}$ connected with the Einstein equations?

In this section we shall limit ourselves to a description of some of the known results, omitting explicit computations of the obstruction operators, for which the reader is referred to the journal literature.

2. Invertible sheaves on $L^{(i)}$. In order to state the extension problems, we shall need canonical extensions not only for the YM-fields $\mathcal{E}_L$, but also for the sheaves $\mathcal{O}_L(a, b) = I_{+}^{-a} \otimes I_{-}^{-b}$. In the flat case, the imbedding $L \subset \mathbf{P} \times \hat{\mathbf{P}}$ determines a natural choice. In the curved case, a small strengthening of the regularity condition makes it possible for us to define these canonical extensions.

Namely, suppose we are given a tower of simple extensions $L \subset L^{(1)} \subset L^{(2)}$ with corresponding ideals I and I^2. Since $H^2(I) = 0$, the exact sequence $0 \longrightarrow I \xrightarrow{\exp} \mathcal{O}_{L^{(1)}}^{*} \longrightarrow \mathcal{O}_L^{*} \longrightarrow 1$ gives us

$$0 \longrightarrow H^1(I) \longrightarrow \text{Pic } L^{(1)} \longrightarrow \text{Pic } L \longrightarrow 0. \tag{1}$$

Furthermore, starting with the exact sequence $0 \longrightarrow I^2 \xrightarrow{\exp} \mathcal{O}_{L^{(2)}}^{*} \longrightarrow \mathcal{O}_{L^{(1)}}^{*} \longrightarrow 1$, and using the equalities $H^1(I^2) = 0$ and $H^2(I^2) = \Gamma(\wedge^2 S_{+} \otimes \wedge^2 S_{-})$, we obtain the exact sequence

$$0 \longrightarrow \text{Pic } L^{(2)} \longrightarrow \text{Pic } L^{(1)} \longrightarrow H^2(I). \tag{2}$$

If we now restrict the last morphism in (2) to the subgroup $H^1(I)$ in (1), we obtain the characteristic homomorphism of the tower:

$$\delta \colon H^1(I) \longrightarrow H^2(I^2). \tag{3}$$

Suppose that this is an isomorphism (this is true for the standard tower in the flat case). Then (1) and (2) give us canonical isomorphisms Pic $L^{(1)}$ = Pic $L \oplus H^1(I)$, Pic $L^{(2)}$ = Pic L. In particular, the sheaves $\mathcal{O}_L(a, b)$ have a standard extension to $L^{(1)}$ and $L^{(2)}$. Since $H^1(L, I^k) = 0$ for $k \geq 3$, the further extensions of $\mathcal{O}_L(a, b)$ in any tower are unique if they exist.

In what follows we shall suppose that (3) is an isomorphism, and that we have taken special note of the corresponding extensions $\mathcal{O}_{L^{(i)}}(a, b), i \leq 3$.

We remark that (3) is also the map which to a first extension of the trivial YM-sheaf $\mathcal{O}_L$ associates the obstruction to a second extension. For any YM-sheaf $\mathcal{E}_L$, the analogous map $\delta_L : H^1(\mathcal{E}nd\ \mathcal{E}_L \otimes I) \longrightarrow H^2(\mathcal{E}nd\ \mathcal{E}_L \otimes I^2)$ is id $\otimes\, \delta$. We have thus also ensured the existence of a canonical second extension $\mathcal{E}_L^{(2)}$.

3. Extension of cohomology classes. With these assumptions, given a YM-sheaf $\mathcal{E}_L$ we construct the sheaves $\mathcal{E}_L^{(i)}(a, b), i \leq 3$. We consider the exact sequence

$$0 \longrightarrow \mathcal{E}_L(a - i - 1,\, b - i - 1) \longrightarrow \mathcal{E}_L^{(i+1)}(a, b) \longrightarrow \mathcal{E}_L^{(i)}(a, b) \longrightarrow 0. \tag{4}$$

It induces a coboundary map

$$\delta(\mathcal{E}_L(a, b);\, i,\, k) : H^k(\mathcal{E}_L^{(i)}(a, b)) \longrightarrow H^{k+1}(\mathcal{E}_L(a - i - 1,\, b - i - 1)),$$

and also a map

$$H^k(\mathcal{E}_L^{(i+1)}(a, b)) \longrightarrow \operatorname{Ker} \delta(\mathcal{E}_L(a, b);\, i,\, k),$$

which is an isomorphism under the condition that $H^k(\mathcal{E}_L(a-i-1,\, b-i-1)) = 0$. If we choose $a, b; i, k$ in different ways, and identify the cohomology groups on L with the sections of certain sheaves on M using the results in § 8, we can obtain for δ the standard field theory operators. In what follows we shall only identify the symbols of these operators, since in the curved case we have not at all been able to connect the structure of the tower of $L^{(i)}$ with the curvature of M.

4. The Dirac operator. We set $(a, b) = (-1, 0)$ and $i = k = 1$. From the exact sequence (4) with $i = 0$ and the computations in § 8 we obtain:

$$H^1(\mathcal{E}_L^{(1)}(-1, 0)) = H^1(\mathcal{E}_L(-1, 0)) = \Gamma(\mathcal{E} \otimes S_- \otimes \wedge^2 S_+),$$
$$H^2(\mathcal{E}_L(-3, -2)) = \Gamma(\mathcal{E} \otimes S_+ \otimes \wedge^2 S_+ \otimes \wedge^2 S_-),$$
$$H^1(\mathcal{E}_L(-3, -2)) = 0.$$

Consequently,

$$H^1(\mathcal{E}_L^{(2)}(-1,\,0)) =$$
$$= \operatorname{Ker}\Big(\delta(\mathcal{E}_L(-1,\,0);\,1,\,1)\colon \mathcal{E}\otimes S_-\otimes\wedge^2 S_+ \to \mathcal{E}\otimes S_+\otimes\wedge^2 S_+\otimes\wedge^2 S_-\Big).$$

The symbol of this operator δ can be computed without any special difficulty, as a result of which one can identify it as the two-component Dirac operator in the setting of the Yang–Mills field $\mathcal{E}$. The second operator corresponds to the choice $(a,\,b) = (0,\,-1)$. The cohomological calculation of the space of Dirac null-modes can be effectively applied to an instanton sheaf $\mathcal{E}$ which is given explicitly in terms of its monad lifted to L. However, it is more convenient to do this in another way.

5. Another description of the Dirac operator. The advantage of the description which we shall now give is that it is sufficient to know the spinor sheaves $\Sigma_\pm$ on L. We tensor the exact sequence $0 \longrightarrow \mathcal{O}_L(-1,\,0) \longrightarrow \Sigma_+ \longrightarrow \mathcal{O}_L(1,\,0) \longrightarrow 0$ with $\mathcal{E}_L(-2,\,0)$ and write out the corresponding exact cohomology sequence. Taking into account the computations in § 8, we obtain

$$\begin{array}{ccc} 0 \longrightarrow H^1(\mathcal{E}_L(-3,0)) & \longrightarrow & H^1(\mathcal{E}_L\otimes\Sigma_+(-2,0)) \longrightarrow \\ & & \| \\ \operatorname{Ker} D(1;\Sigma_+(-2,0)) & = & \Gamma(\mathcal{E}\otimes S_+\otimes\wedge^2 S_+) \xrightarrow{D_+} \end{array}$$

$$\begin{array}{ccc} \longrightarrow & H^1(\mathcal{E}_L(-1,0)) & \longrightarrow H^2(\mathcal{E}_L(-3,0)) \longrightarrow 0 \\ & \| & \\ & \xrightarrow{D_+} \Gamma(\mathcal{E}\otimes S_-\otimes\wedge^2 S_+). & \end{array}$$

In order to identify D_+ as a Dirac operator, one must lift the entire picture to F and do a diagram chase in the cohomology table associated to the diagram

$$\begin{array}{ccccccccc} & & 0 & & 0 & & 0 & & \\ & & \downarrow & & \downarrow & & \downarrow & & \\ 0 & \to & \pi_1^{-1}\mathcal{O}_L(-3,0) & \to & \mathcal{O}_F(-3,0) & \xrightarrow{\nabla^+_{F/L}} & \mathcal{O}_F(-2,1) & \to & 0 \\ & & \downarrow & & \downarrow & & \downarrow & & \\ 0 & \to & \pi_1^{-1}(\Sigma_+(-2,0)) & \to & \pi_2^* S_+(-2,0) & \xrightarrow{\nabla^+_{F/L}} & \pi_2^* S_+(-1,1) & \to & 0 \\ & & \downarrow & & \downarrow & & \downarrow & & \\ 0 & \to & \pi_1^{-1}(\mathcal{O}_L(-1,0)) & \to & \mathcal{O}_F(-1,0) & \xrightarrow{\nabla^+_{F/L}} & \mathcal{O}_F(0,1) & \to & 0. \\ & & \downarrow & & \downarrow & & \downarrow & & \\ & & 0 & & 0 & & 0 & & \end{array}$$

Using this result, we obtain the following description of the null-modes and the cokernel:

$$H^1(\mathcal{E}_L(-3,0)) = \operatorname{Ker} D_+, \qquad H^2(\mathcal{E}_L(-3,0)) = \operatorname{Coker} D_+.$$

6. The Klein–Gordon operator. We set $(a, b) = (-2, 0)$ and $i = 0$, $k = 1$ in (4). Using the table in § 8, we obtain

$$H^1(\mathcal{E}_L(-2, 0)) = \Gamma(\mathcal{E} \otimes \wedge^2 S_+),$$
$$H^2(\mathcal{E}_L(-3, -1)) = \Gamma(\mathcal{E} \otimes (\wedge^2 S_+)^2 \otimes \wedge^2 S_-),$$
$$H^1(\mathcal{E}_L(-3, -1)) = 0.$$

Consequently,

$$H^1(\mathcal{E}_L^{(1)}(-2, 0)) = \\ = \operatorname{Ker}\Big(\delta(\mathcal{E}_L(-2, 0); 0, 1){:}\, \Gamma(\mathcal{E} \otimes \wedge^2 S_+) \to \Gamma(\mathcal{E} \otimes (\wedge^2 S_+)^2 \otimes \wedge^2 S_-)\Big).$$

Here δ is a second order differential operator whose symbol coincides with multiplication by the metric. We have thus computed the null-modes of the Klein–Gordon operator.

7. Some nonlinear systems. In sections 4 and 5 we represented scalar and spinor fields on M by cohomology classes on L, and then integrated the dynamical equations as extension problems for these classes. However, cohomology classes can be used to encode many different geometrical objects on L, such as extensions of fields or of infinitesimal neighborhoods. This opens up the possibility of incorporating nonlinear systems into extension problems. Here we shall describe a technique in the flat case that is due to G. M. Henkin.

Let $(\mathcal{E}_i, \nabla_i), i = 1, 2$, be two Yang–Mills fields on M, and let $\mathcal{E}_{iL}$ be their Radon–Penrose transforms. We set $\mathcal{F}_L = \mathcal{E}_{1L}(-1, 0) \oplus \mathcal{E}_{2L}(0, -1)$, and we consider the problem of extending this sheaf to the levels of a regular tower.

(a) *The first extension.* This always exists. Moreover, there is a canonical extension, and it satisfies the condition for endomorphisms to lift. We therefore have a bijection between the classes of extensions $\mathcal{F}_L^{(1)}$ and the elements of the space

$$H^1(\mathcal{E}nd\ \mathcal{F}_L(-1, -1)) = H^1\big((\mathcal{E}nd\ \mathcal{E}_{1L} \oplus \mathcal{E}nd\ \mathcal{E}_{2L})(-1, -1) \oplus \\ \oplus \mathcal{E}_{1L}^* \otimes \mathcal{E}_{2L}(0, -2) \oplus \mathcal{E}_{1L} \otimes \mathcal{E}_{2L}^*(-2, 0)\big) \\ = \Gamma\big((\mathcal{E}nd\ \mathcal{E}_1 \oplus \mathcal{E}nd\ \mathcal{E}_2) \otimes \wedge^2 S_+ \otimes \wedge^2 S_- \oplus \\ \oplus (\mathcal{E}_1^* \otimes \mathcal{E}_2) \otimes \wedge^2 S_- \oplus \mathcal{E}_1 \otimes \mathcal{E}_2^* \otimes \wedge^2 S_+\big).$$

If we trivialize $\wedge^2 S_\pm$ using $S_\pm$, we can say that the first extensions of $\mathcal{F}_L$ are nothing other than the scalar fields in the adjoint representation of the structure group of the sheaf $\mathcal{E} \oplus \mathcal{E}'$.

(b) *The second extension.* The group of obstructions to a second extension is

$$\begin{aligned}H^2(\mathcal{E}nd\ \mathcal{F}_L(-2,-2)) &= H^2((\mathcal{E}nd\ \mathcal{E}_{1L}\oplus\mathcal{E}nd\ \mathcal{E}_{2L})(-2,-2)\oplus\\ &\quad\oplus \mathcal{E}_{1L}^*\otimes\mathcal{E}_{2L}(-1,-3)\oplus\mathcal{E}_{1L}\otimes\mathcal{E}_{2L}^*(-3,-1))\\ &= \Gamma((\mathcal{E}nd\ \mathcal{E}_1\oplus\mathcal{E}nd\ \mathcal{E}_2)\otimes\wedge^2S_+\otimes\wedge^2S_-\oplus(\mathcal{E}_1^*\otimes\mathcal{E}_2)\otimes\\ &\quad\otimes\wedge^2S_+\otimes(\wedge^2S_-)^2\oplus\mathcal{E}_1\otimes\mathcal{E}_2^*\otimes(\wedge^2S_+)^2\otimes\wedge^2S_-).\end{aligned}$$

The obstruction map $\omega: H^1 \longrightarrow H^2$ on the components with values in $\mathcal{E}nd\ \mathcal{E}_1 \oplus \mathcal{E}nd\ \mathcal{E}_2$ is simply $\mathrm{id}\otimes\delta$, where δ is the map (3). Hence, the further extension of the sheaf $\mathcal{F}_L^{(1)}$ is determined by two fields

$$\varphi_+ \in \Gamma(\mathcal{E}_1\otimes\mathcal{E}_2^*\otimes\wedge^2S_+) \qquad \text{and} \qquad \varphi_- \in \Gamma(\mathcal{E}_1^*\otimes\mathcal{E}_2\otimes\wedge^2S_-),$$

whereas the first two scalars must be zero. We let $\mathcal{F}_L^{(1)}(\varphi_+,\varphi_-)$ denote this extended sheaf. Its obstruction $\omega(\mathcal{F}_L^{(1)}(\varphi_+,\varphi_-))$ consists of a field in the third component and a field in the fourth component of the group $H^2_,$ which depend on $\varphi_\pm$:

$$\psi_+ \in \Gamma(\mathcal{E}_1\otimes\mathcal{E}_2^*\otimes(\wedge^2S_+)^2\otimes\wedge^2S_-),$$
$$\psi_- \in \Gamma(\mathcal{E}_1^*\otimes\mathcal{E}_2\otimes\wedge^2S_+\otimes(\wedge^2S_-)^2).$$

Henkin computed $\psi_\pm$ in terms of $\varphi_\pm$ in the flat case. In order to state his results, we let $D_\pm$ denote the Klein–Gordon operators in $\mathcal{E}_1\otimes\mathcal{E}_2^*$ and $\mathcal{E}_1^*\otimes\mathcal{E}_2$, respectively, in the sense of the previous subsection. Next, we note that the products $\varphi_+\otimes\varphi_-\otimes\varphi_+$ and $\varphi_-\otimes\varphi_+\otimes\varphi_-$, when convoluted with respect to the indices of the middle factor, fall in the same group which contains $\psi_\pm$ and $D_\pm\varphi_\pm$. We write these products simply as $\varphi_+\varphi_-\varphi_+$ and $\varphi_-\varphi_+\varphi_-$.

8. Proposition. *There exists a nonzero constant λ, depending on the normalization of the isomorphisms, such that for a flat space-time*

$$\omega(\mathcal{F}_L^{(1)}(\varphi_+,\varphi_-)) = (D_+\varphi_+ + \lambda\varphi_+\varphi_-\varphi_+, D_-\varphi_- + \lambda\varphi_-\varphi_+\varphi_-).$$

Thus, the condition that $\mathcal{F}_L^{(1)}(\varphi_+,\varphi_-)$ extend to $L^{(2)}$ is equivalent to the system of equations

$$D_\pm\varphi_\pm + \lambda\varphi_\pm\varphi_\mp\varphi_\pm = 0. \quad \square \tag{5}$$

In the case $\mathcal{E}_1 = \mathcal{E}_2 = \mathcal{O}_M, \nabla = d$, and $\varphi_+ = \varphi_-$, we obtain the classical nonlinear Klein–Gordon equation

$$D\varphi + \lambda\varphi^3 = 0.$$

9. The third extension. We now suppose that $\omega(\mathcal{F}_L^{(1)}(\varphi_+, \varphi_-)) = 0$. It is not hard to see that the group which classifies the second extensions is zero. Thus, there is a unique second extension $\mathcal{F}_L^{(2)}(\varphi_+, \varphi_-)$. The obstruction to a third extension lies in the group

$$\begin{aligned} H^2(\mathcal{E}nd\ \mathcal{F}_L(-3, -3)) &= H^2(\mathcal{E}nd\ \mathcal{E}_{1L} \oplus \mathcal{E}nd\ \mathcal{E}_{2L}(-3, -3)\oplus \\ &\quad \oplus \mathcal{E}_{1L}^* \oplus \mathcal{E}_{2L}(-2, 4) \oplus \mathcal{E}_{1L} \otimes \mathcal{E}_{2L}^*(-4, -2)) \\ &= \Gamma(\operatorname{Ker} \tilde{\nabla}_3^{(1)} \oplus \operatorname{Ker} \tilde{\nabla}_3^{(2)} \oplus \mathcal{E}_1^* \otimes \mathcal{E}_2 \otimes S^2(S_-) \otimes \wedge^2 S_+ \otimes \\ &\quad \otimes \wedge^2 S_- \oplus \mathcal{E}_2^* \otimes \mathcal{E}_1 \otimes S^2(S_+) \otimes \wedge^2 S_+ \otimes \wedge^2 S_-). \end{aligned}$$

The first two components of $\omega(\mathcal{F}_L^{(2)}(\varphi_+, \varphi_-))$ are flows for the fields $(\mathcal{E}_1, \nabla_1)$ and $(\mathcal{E}_2, \nabla_2)$, respectively. They are constructed using $\varphi_\pm$ as follows. In general, let ∇ denote the connection constructed from ∇_1 and ∇_2 on the tensor algebra generated by $\mathcal{E}_1$ and $\mathcal{E}_2$; thus, in particular, one can construct the fields $\nabla\varphi_\pm$ and products such as $\varphi_+\nabla\varphi_-$, where the latter is understood to mean the convolution of $\varphi_+ \otimes \nabla\varphi_-$ with respect to $\mathcal{E}_1$. The result lies in $\mathcal{E}nd\ \mathcal{E}_2 \otimes \Omega^1 M \otimes \wedge^2 S_+ \otimes \wedge^2 S_-$, i.e., in $\mathcal{E}nd\ \mathcal{E}_2 \otimes \Omega^3 M$ under the standard identification.

Under these conditions we have the following result.

10. Proposition. *The sheaf $\mathcal{F}_L^{(2)}(\varphi_+, \varphi_-)$ has a third extension if and only if the flows j_1 and j_2 of the Yang–Mills fields ∇_1 and ∇_2 satisfy the equalities*

$$j_1 = (\nabla\varphi_-)\varphi_+ - \varphi_-(\nabla\varphi_+), \qquad j_2 = (\nabla\varphi_+)\varphi_- - \varphi_+(\nabla\varphi_-),$$

and also the second two components of the obstruction vanish. □

§ 11. The Green's Function of the Laplace Operator

1. Statement of the problem. The Green's function for the classical Laplace operator $D = \sum_{i=1}^4 \partial^2/(\partial x^a)^2$ is a solution $G(x, y)$ of the equation $D_x G(x, y) = \delta(x - y)$. If one imposes the boundary condition that it vanish at infinity, then $G(x, y) = \frac{1}{4\pi^2} \frac{1}{|x-y|^2}$.

More generally, let $D: \wedge^2 S_+ \longrightarrow (\wedge^2 S_+)^2 \otimes \wedge^2 S_-$ be the Klein–Gordon operator on a small space-time. If we analytically continue the function $|x - y|^2$ with respect to x (with y fixed), we obtain the "square of the complex geodesic distance" $r^2(x, y)$ — in general, this is a multivalued analytic function. It has a distinguished branch locally near y; the equation $r^2(x, y) = 0$ determines a light conoid $C(y)$ at y which is swept out by the null-geodesics through y.

Hadamard characterized a complex Green's function for D as a holomorphic solution to the equation $D_x G(x, y) = 0$ near y with a singularity on $C(y)$ of the type $A(x, y)/r^2(x, y) + B(x, y) \log r(x, y)$. The function A/r^2 is uniquely determined by the metric up to multiplication by a constant. The coefficient B is zero if G has no ramification on $C(y)$. Thus, up to multiplication by a constant and addition of a regular solution to the Klein–Gordon equation, an unramified Green's function can be characterized by the condition that it have a first order pole in x on $C(y)$.

With some small modifications the same is true of the Klein–Gordon operator in the setting of a Yang–Mills field $(\mathcal{E}, \nabla)$. In that case an unramified function $G(x, y)$ is a meromorphic section of the sheaf $\mathcal{E}nd\ \mathcal{E} \otimes \wedge^2 S_+$ which has a specified type of principal part near the conoid $C(y)$ and satisfies the equation $D_x G(x, y) = 0$ outside $C(y)$.

The purpose of this section is to use cohomological techniques to describe the Green's function (more precisely, its Radon–Penrose transform on L) for a flat space-time.

2. Preliminary constructions. First suppose that M is a small space-time which is not necessarily flat. We construct the following diagram with exact rows:

$$\begin{array}{ccccc} H^1(L, \mathcal{E}nd\ \mathcal{E}_L^{(1)}(-2,0)) & \longrightarrow & H^1(L \setminus B(y), \mathcal{E}nd\ \mathcal{E}_L^{(1)}(-2,0)) & \longrightarrow \\ \| & & \downarrow a & \\ H^0(M, \text{Ker } D) & \longrightarrow & H^0(M \setminus C(y), \text{Ker } D) & \longrightarrow \\ & & \longrightarrow H^2_{B(y)}(L, \mathcal{E}nd\ \mathcal{E}_L^{(1)}(-2,0)) & \\ & & \downarrow & \\ & & \longrightarrow \quad H^1_{C(y)}(M, \text{Ker } D). & \end{array} \tag{1}$$

In the first row $B(y)$ denotes the set of null-geodesics which belong entirely to $C(y)$; the two groups at the right of the diagram are cohomology groups whose supports were inserted formally in order to make the rows exact; and

$$D: \mathcal{E}nd\ \mathcal{E} \otimes \wedge^2 S_+ \longrightarrow \mathcal{E}nd\ \mathcal{E} \otimes (\wedge^2 S_+)^2 \otimes \wedge^2 S_-$$

is the Klein–Gordon operator. The isomorphism between the two groups at the left of the diagram was constructed in the last section. The second vertical arrow, here denoted a, would be obtained formally from the results in § 10 and would also be an isomorphism if $M \setminus C(y)$ were a small space-time with space of null-geodesics $L \setminus B(y)$. However, this is not the case: the null-geodesics in M which have non-empty zero-dimensional intersection with $C(y)$ are no longer simply connected in $M \setminus C(y)$.

Therefore, to construct a we must introduce two intermediate objects. We set

$$F_1(y) = \pi_1^{-1}(B(y)), \qquad F_2(y) = \pi_2^{-1}(C(y)),$$

and we consider the diagram

$$\begin{array}{ccccc} F \setminus F_1(y) & & \supset & F \setminus F_2(y) & \\ & \searrow \pi_{11} & & \pi_{12} \swarrow \quad \searrow \pi_{22} & \\ & L \setminus B(y) & & & M \setminus C(y) \end{array}$$

The fibre of π_{11} over any point $l \in L \setminus B(y)$ is precisely $\pi_1^{-1}(l)$, and so it is connected and simply connected. Thus, π_{11} induces an isomorphism

$$H^1(L \setminus B(y),\ \mathcal{E}nd\ \mathcal{E}_L^{(1)}(-2,0)) = H^1\big(F \setminus F_1(y),\ \pi_{11}^{-1}(\mathcal{E}nd\ \mathcal{E}_L^{(1)}(-2,0))\big). \quad (2)$$

On the other hand, the fibres of π_{22} are compact quadrics, and the same argument as in § 8 gives the isomorphism

$$H^0(M \setminus C(y),\ \mathrm{Ker}\ D) = H^1\big(F \setminus F_2(y),\ \pi_{11}^{-1}(\mathcal{E}nd\ \mathcal{E}_L^{(1)}(-2,0))\big). \quad (3)$$

The groups on the right in (2) and (3) are connected by a boundary morphism. It is this boundary morphism which is the map a in diagram (1).

The third vertical arrow in (1) is defined to be the map induced by a.

In $H^0(M \setminus C(y), \mathrm{Ker}\ D)$, we would like to construct the class of Green's function as the image under a of some cohomology class on L. We are able to do this only in the flat case (actually, self-duality is sufficient). Before describing this class, we shall explain the geometrical reason why we have to work in the flat case.

In a flat space-time, $C(y)$ is a hyperplane section of the Plücker quadric $G(2;\ T)$, and the null-geodesics are straight lines. A line lying entirely in $C(y)$ either passes through y, or else determines a plane spanned by the line and y; that plane intersects $G(2;\ T)$ at three points, and so it is also contained in $G(2;\ T)$. Thus, if we let $L \subset \mathbf{P}_1 \times \mathbf{P}_2 = \mathbf{P} \times \hat{\mathbf{P}}$, $\rho_i \colon L \longrightarrow \mathbf{P}_i$ be the projections, and $\mathbf{P}_i^1(y) \subset \mathbf{P}_i$ be the lines corresponding to $y \in M$, then we have

$$B(y) = B_1(y) \cup B_2(y) = \rho_2^{-1}(\mathbf{P}_1^1(y)) \cup \rho_2^{-1}(\mathbf{P}_1^1(y)).$$

The irreducible components of $B_1(y)$ and $B_2(y)$ are three-dimensional; and the intersection $B_1(y) \cap B_2(y) = L(y)$.

In the case of a general curved space one would expect $C(y)$ to contain only the null-geodesics which pass through y. This means that $B(y) = L(y)$. Thus, in the flat (or semi-flat) case $B(y)$ undergoes a jump in codimension — from three to two. The construction below relies in an essential way upon the circumstance that one of the components of $B(y)$ can be given by two equations.

Let $J(y)^{(1)} \subset \mathcal{O}_{L^{(1)}}$ be the ideal of the intersection $(\mathrm{P}_1^1(y) \times \mathrm{P}_2) \cap L^{(1)}$ We let u and v denote independent sections of $\mathcal{O}_L^{(1)}(1, 0)$ which vanish on $\mathrm{P}_1^1(y)$. Then we have the exact Koszul complex

$$0 \longrightarrow \mathcal{O}_L^{(1)}(-2, 0) \longrightarrow \mathcal{O}_L^{(1)}(-1, 0) \oplus \mathcal{O}_L^{(1)}(-1, 0) \xrightarrow{(u,v)} J(y)^{(1)} \longrightarrow 0.$$

We tensor this with $\mathcal{E}_L^{(1)}$ and restrict to $L \setminus B(y)$. We take into account that $J(y)^{(1)} \mid L \setminus B(y)$ and let g' and g denote the classes of the resulting extension in the groups

$$\mathrm{Ext}^1(L,\ \mathcal{E}_L^{(1)} \otimes J(y)^{(1)}, \mathcal{E}_L^{(1)}(-2, 0)) \longrightarrow H^1(L \setminus B(y),\ \mathcal{E}nd\ \mathcal{E}_L^{(1)}(-2, 0)).$$

3. Theorem. *The class $a(g) \in H^0(M \setminus C(y),\ D)$ is a meromorphic solution of the equation $D_x G = 0$ with a first order pole on $C(y)$.*

Proof. We shall regard $a(g)$ as a section of the sheaf $\mathcal{E}nd\ \mathcal{E} \otimes \wedge^2 S_+$. To determine its value at a point $x \in M \setminus C(y)$, one restricts it to the quadric $L(x) = \pi_1 \pi_2^{-1}(x)$ and makes use of the identification $H^1(L(x),\ \mathcal{E}nd\ \mathcal{E}_L(-2, 0)) = \mathcal{E}nd\ \mathcal{E}(x) \otimes \wedge^2 S_+(x)$ (compare with the computations in § 7). Let $s = 0$ be the equation of the light cone $C(y)$. Then $\pi_2^*(s) = 0$ is the equation of $F_2(y)$ in F. By (3) and the analog of this identity for M, it clearly suffices to verify that $\pi^*(s)g$ extends to an element of the group $H^1\big(F,\ \pi_{11}^{-1}(\mathcal{E}nd\ \mathcal{E}_L^{(1)}(-2,0))\big)$.

Let $J(y)^{(1)}$ be the ideal of $F_1(y) = \pi_1^{-1}(B(y))$ in $\mathcal{O}_{F^{(1)}}$. According to the definition, g' is represented on F by the class of the extension

$$0 \longrightarrow \mathcal{E}_F^{(1)}(-2, 0) \longrightarrow \mathcal{E}_F^{(1)}(-1, 0) \oplus \mathcal{E}_F^{(1)}(-1, 0) \longrightarrow \mathcal{E}_F^{(1)} \otimes J_F(y)^{(1)} \longrightarrow 0.$$

But the ideal $\tilde{J}_F(y)^{(1)}$ which is generated by $\pi_2^*(s)$ and is the defining ideal for $F_2(y)$, is contained in $J_F(y)^{(1)}$. We can interpret restriction of g' from $F \setminus F_1(y)$ to $F \setminus F_2(y)$ as a morphism of the corresponding Ext^1-groups which comes from the imbedding $\mathcal{E}_F^{(1)} \otimes \tilde{J}_F(y)^{(1)} \subset \mathcal{E}_F^{(1)} \otimes J_F(y)^{(1)}$. The element g is the image of g' under this morphism. But $\tilde{J}_F(y)^{(1)}$ is an invertible sheaf, and is generated by $\pi_2^*(s)$. Thus, g is in the group $H^1(F \setminus F_2(y),\ \mathcal{E}nd\ \mathcal{E}_F^{(1)} \otimes \big[\tilde{J}_F(y)^{(1)}\big]^{-1}(-2, 0))$, since it is obtained by restricting a suitable element $\tilde{g}$ from $H^1\big(F,\ \mathcal{E}nd\ \mathcal{E}_F^{(1)} \otimes$

$\otimes[\tilde{J}_F(y)^{(1)}]^{-1}(-2, 0))$ (namely, the element which is induced by g' when $J_F(y)^{(1)}$ is replaced by $\tilde{J}_F(y)^{(1)}$). Multiplication by $\pi_2^*(s)$ gives an isomorphism of sheaves $[\tilde{J}_F(y)^{(1)}]^{-1} \xrightarrow{\pi_2^*(s)} \mathcal{O}_{F(1)}$. The image of $\tilde{g}$ under the induced map of H^1-groups is the desired extension of $\pi_2^*(s)g$. □

We leave it to the reader to verify that our solution to the Klein–Gordon equation actually has a singularity on $C(y)$. This means that the image of g in the local cohomology group $H^2_{B(y)}(L, \mathcal{E}nd\ \mathcal{E}_L^{(1)}(-2, 0))$ is nonzero.

References for Chapter 2

There are quite a few surveys covering the material in the first part of this chapter (the self-dual case). Among them we mention the lectures by M. F. Atiyah [1] and by R. Wells [112], [114], and the seminar notes from Paris [25] and Berlin [38]. Our purpose in this book is not to give a complete presentation of self-dual geometry, and so we did not touch upon many important themes. The reader can trace the connections with the Radon–Penrose transform by looking through the articles [39], [40] and the book [46].

The Radon–Penrose transform for vector bundles was defined for a self-duality diagram by Ward [107] in the local situation on $\acute{M}$. A similar approach was presented in a paper by Belavin and Zakharov [12]. The correspondence between instantons and locally free sheaves on $\mathbf{P}^3$ is described in the article [6] by Atiyah and Ward. There they also give a method for constructing rank 2 instanton sheaves as extensions of two rank 1 sheaves (physicists call this the Atiyah–Ward *ansatz*). This method turned out not to be very convenient for instanton sheaves, because, in the first place, there is no canonical representation of the instanton sheaf in the form of an extension, and, in the second place, given an arbitrary extension it is difficult to determine whether it is an instanton extension and when two extensions are isomorphic. The monad method, described in [3] and in §§ 3–4 of this chapter, met with much greater success. This method gives a completely transparent construction of instantons; however, in treating the problem of describing instanton modules it leads to the need to study a real manifold given by matrix equations, and there is little beyond its dimension which is known about such a manifold. Various ideas about such manifolds are collected in [26]. Their topological complexity is demonstrated in the article [5] by Atiyah and Jones. It is interesting to note that it becomes simpler to describe the modules if one passes to the stable version, i.e., if one allows the rank of the group to increase as the instanton number increases (see [27]). Taking the limit as $N \longrightarrow \infty$ (where N is the rank of the gauge group) is also very popular in quantum field theory.

The construction of Atiyah and Ward turned out to be very useful for classifying self-dual $SU(2)$-bundles over $\mathbf{R}^4$ with certain other boundary conditions — the so-called "'t Hooft-Polyakov monopoles." A very clear description of their physical and mathematical properties can be found in the book [60]. A monopole sheaf is defined on $\mathbf{P}^3 \setminus \mathbf{P}^1$, and it is canonically represented there in the form of an extension (see Hitchin's papers [54], [55], which complete a cycle of studies that includes articles by Ward [110], Corrigan and Goddard [20], and Nahm [77], [78]). Nahm is the originator of the remarkable idea of constructing monopole solutions again using monads, only monads of infinite rank. The spaces F_i in Nahm's monad are spaces of functions on an interval, and the boundary operators are first order d-differential operators. The condition that $d^2 = 0$ immediates gives Lax equations of the type of Euler's equations for the rotation of a solid body. The Atiyah–Ward–Hitchin–Nahm spectral curve in the monopole problem corresponds to the spectral curve in "finite-zone" integration of the Euler equations. This was probably the first time that the long-discussed connection between problems of field theory in four-dimensional space-time and one- and two-dimensional completely integrable systems became so sharply dilineated. The paper [104] treats this connection along somewhat different lines.

A general discussion of self-dual Riemannian geometry is given in the seminal paper by Atiyah, Hitchin and Singer [4]. The role of self-dual Einstein spaces in the future quantum theory of gravity is discussed by Penrose in [92]; see also [73]. Concrete classes of such spaces are constructed in papers by Ward [108], [109] and Hitchin [51], [53]. Self-dual Einstein spaces also arise in the classical general relativity theory of isolated systems which create an asymptotically flat space-time; see [33] and [66].

Some very strong and unexpected results on the topology of four-dimensional smooth manifolds were recently obtained by Donaldson [24], who studied the moduli space of self-dual fibrations over a (necessarily self-dual) compact Riemannian manifold. He proved that, if the intersection form for the 2-cycles on a four-dimensional smooth compact simply connected and oriented manifold is positive-definite, then it can be reduced over $\mathbf{Z}$ to a sum of squares. The fundamental existence theorem which Donaldson relies upon was proved by Taubes [102]; his paper uses subtle nonlinear functional analysis.

Returning to instantons, we mention the survey [105], which is devoted to their physical aspects. It summarizes the results of the research initiated in [11], where instantons were introduced. Our exposition in §§ 3–4 was based upon the articles [28], [29], [10]. The instanton problem brought about a flourish of interest among algebraic geometers in the problem of classifying locally free sheaves on $\mathbf{P}^n$. Barth's paper [7] was a decisive breakthrough in the case $n = 2$. In the article [8] by Barth and Hulek, an analysis is given of Horrocks' monad technique. In article

[9] by A. A. Beilinson and article [14] by I. N. Bernshtein, I. M. Gel'fand and S. I. Gel'fand, a structure theorem of fundamental importance is obtained for the derived category of coherent sheaves on $\mathbf{P}^n$; this theorem is a vast generalization of the monad technique. See also the introductory text [86].

Among the work on other aspects of the Radon–Penrose transform in the self-dual and flat cases, especially the problem of solving linear (massless) equations of field theory, we mention Hitchin's article [52], where the general cohomological mechanism is described, and also [40], [90], [31], [113]. The last paper in this list gives a cohomological interpretation of generalized solutions.

A theory of the non-self-dual Radon–Penrose transform has been begun in papers by Witten [118] and by Isenberg, Yasskin and Green [59], where a basic discovery was made: solvability of the Yang–Mills equations without sources for $(\mathcal{E}, \nabla)$ on a flat space-time is equivalent to the possibility of extending $\mathcal{E}_L$ to $L^{(3)}$. In the author's paper [70] the standard techniques of cohomology were applied to the extension problem (see [43] and the very abstract generalization in [58]), and it was discovered that accounting for the obstruction to extension is equivalent to introducing the flow which generates the Yang–Mills field. After this, a systematic examination of the various extension problems in [49], [50], [74], [47] and [48] made it possible to give a geometrical reformulation of a large number of field theory equations. §§ 6–10 of this chapter are based on these papers. The results have been somewhat generalized here, in order to allow us to consider curved space-time as well. Papers by Le Brun [63]-[65] and Buchdahl [18], [19] were of great assistance to the author in this. A proof of the results in §§ 9–10 based on Dolbeault cohomology can be found in [74]. The treatment of the Green's function in § 11 is based on a paper [56] by Hoàng Lê Minh, who partially generalizes a construction of Atiyah [2].

An interpretation of the Einstein equations in terms of L is described in [75], and is reproduced in this chapter.

Concerning general aspects of the Radon–Penrose transform, see [89], [69], [22], [87]. For discussions of analysis on analytic spaces with nilpotents, see [94], [42], [35].

CHAPTER 3

INTRODUCTION TO SUPERALGEBRA

This chapter contains preparatory information for the following ones. Here we lay out the principal facts about supermanifolds, spaces on which some of the coordinate functions commute and the others anti-commute. Recently it has become clear that the whole apparatus of commutative algebra, of the theory of Lie groups, of algebraic, analytic and differential geometry permits the systematic introduction of anti-commuting (odd, fermion) quantities. This process significantly enriches these structures and results in extremely non-trivial variants of them in "supermathematics."

After laying out the basic principles of superalgebra in § 1, we consider several details of this algebraic apparatus which have immediate application to supergeometry. This is related to the tensor formalism described in § 2 in terms of the modern ideology of tensor categories. The next section, § 3, is devoted to matrix computations; here the concepts of superdeterminant and supertrace are introduced. In § 4 several important complexes are defined, including the de Rham complex, and a homological interpretation of the superdeterminant is given. Bilinear forms, scalar products, and the details of the concept of a real structure are examined in §§ 5 and 6.

§ 1. The Rule of Signs

1. General principles. The fundamental algebraic structures which occur in geometry are rings (of functions, differential operators, differential forms and vector fields), modules (sheaves of modules) and homomorphisms between them. Therefore, all the basic laws of composition can be divided naturally into additive ones and multiplicative ones. The first class encompasses addition in rings and modules and addition of morphisms of modules. The second includes multiplication in rings, forming commutators of vector fields, composition of morphisms, tensor multiplication of elements and morphisms of modules.

The first characteristic special property of superalgebra is that all the additive groups of its basic and derived structures are $\mathbf{Z}_2$-graded, where $\mathbf{Z}_2 = \mathbf{Z}/2\mathbf{Z} = \{0, 1\}$. This will be our notation for the grading on an additive group: $A = A_0 \oplus A_1$. If

$a \in A_\epsilon$ is a homogeneous element of degree ϵ, then $\tilde{a} = \epsilon$. Elements of A_0 are called even, and those of A_1 are called odd. Any multiplicative law of composition is compatible with the grading: $\widetilde{ab} = \tilde{a} + \tilde{b}$.

The second characteristic property of superalgebra is that in the definitions and axioms for the basic structures, and in the multilinear identities which are derived from them, there are (in comparison with ordinary algebra) additional factors of ± 1. Here is the general heuristic rule for computing these factors.

If in some formula of usual algebra there are monomials with interchanged terms, then in the corresponding formula in superalgebra every interchange of neighboring terms, say a and b, is accompanied by the multiplication of the monomial by the factor $(-1)^{\widetilde{a}\widetilde{b}}$.

A rather surprising mathematical principle, which can be deduced from observing how this rule works, is its internal consistency: if the law of signs is taken into account correctly in definitions, it appears of itself in theorems. Probably it is not a good idea to turn this principle into a metatheorem; for then there would be a danger of not noticing the very interesting situations when it does not apply. An example of such a situation is the formula for the superdeterminant, where the multilinear identity $\det A = \sum \det(s) a_{1s(1)} \cdots a_{ns(n)}$ changes in superalgebra to an expression with a denominator (see § 3).

The goal of this section is to give a list of definitions and simple constructions where the law of signs holds without unexpected peculiarities.

2. Rings, commutators, and supercommutativity. Let $A = A_0 \oplus A_1$ be a $\mathbb{Z}_2$-graded ring. In the axiom of associativity $a(bc) = (ab)c$ there are no interchanges of factors, so it remains in its usual form in superalgebra. The supercommutator of a pair of elements $a, b \in A$ is defined to be the element

$$[a, b] = ab - (-1)^{\widetilde{a}\widetilde{b}} ba$$

For this type of notation it is always assumed that a and b are homogeneous and that the definition is extended to nonhomogeneous elements by additivity. The sign $(-1)^{\widetilde{a}\widetilde{b}}$ is not significant in characteristic 2, and we will always assume that in our rings the element 2 is invertible and that in modules multiplication by 2 is an isomorphism.

We will say that a supercommutes with b if $[a, b] = 0$. Two odd elements supercommute if $ab = -ba$. The supercenter of A is $Z(A) = \{ a \in A \mid \forall b \in A, [a, b] = 0 \}$. A ring homomorphism $f: A \to B$ defines the structure of an A-algebra on B if $f(A) \subset Z(B)$. All ring homomorphisms preserve the grading.

A ring A is supercommutative if $[a, b] = 0$ for all a, $b \in A$. If $a \in A_1$, then $a^2 = \frac{1}{2}[a, a] = 0$, so all odd elements are nilpotent.

3. Examples. (a) Let $A = k[\xi_1, \ldots, \xi_n]$ be the Grassmann algebra over a field k of characteristic $\neq 2$ generated by elements with relations $\xi_i^2 = 0$ and $\xi_i \xi_j = -\xi_j \xi_i$ for $i \neq j$. It has a free basis made up of the 2^n elements $\xi^\alpha = \xi_1^{\alpha_1} \cdots \xi_n^{\alpha_n}$, where $\alpha_i = 0$ or 1. We set $\widetilde{\xi_i} = 1$ and $\widetilde{\xi^\alpha} = |\alpha| \bmod 2$, where $|\alpha| = \sum \alpha_i$. This algebra is supercommutative.

More generally, we can consider the ring $A = k[x_1, \ldots, x_m; \xi_1, \ldots, \xi_n]$ generated by elements with relations $x_i x_j = x_j x_i$, $x_i \xi_j = \xi_j x_i$ and $\xi_i \xi_j = -\xi_j \xi_i$. This is the ring of algebraic (polynomial) functions on affine superspace of dimension $m|n$ over the field k. Its spectrum (in the sense of Grothendieck's theory of schemes, which is easily generalized to the supercase) is an affine superscheme. Of course $\widetilde{x_i} = 0$.

In an analogous manner, one can extend the ring of smooth functions on a domain $U \subset \mathbf{R}^m$ by adding odd variables to obtain the ring $C^\infty(U^{m|n}) = C^\infty(U)[\xi_1, \ldots, \xi_n]$ or one can extend the ring of analytic functions on the domain $V \subset \mathbf{C}^m$ to obtain the ring of superanalytic functions $\mathcal{O}(V)[\xi_1, \ldots, \xi_n]$. The functions of each type (algebraic, analytic, smooth) are polynomials in the odd variables.

(b) More generally, let B be a commutative ring and let S be a B-module. If $A = \wedge_B(S) = \bigoplus_{i \geq 0} \wedge_B^i(S)$ is the exterior algebra of S, set $A_0 = \bigoplus_{i \geq 0} \wedge_B^{2i}(S)$ and $A_1 = \bigoplus_{i \geq 0} \wedge_B^{2i+1}(S)$. This is supercommutative. The $\mathbf{Z}$-grading on A is consistent with the $\mathbf{Z}_2$-grading in the sense that the parity of an element is defined by the parity of its $\mathbf{Z}$-grading. Generally speaking, on rings, modules, Lie superalgebras, etc., there is no invariantly defined $\mathbf{Z}$-grading, although when one can be defined it turns out to be extremely useful. An example is the structure sheaf on projective superspace.

(c) In any associative ring B the supercommutator satisfies two fundamental identities:

$$[a, b] = -(-1)^{\widetilde{a}\widetilde{b}}[b, a]$$

$$[a, [b, c]] + (-1)^{\widetilde{a}(\widetilde{b}+\widetilde{c})}[b, [c, a]] + (-1)^{\widetilde{c}(\widetilde{a}+\widetilde{b})}[c, [a, b]] = 0$$

The reader should verify this directly and convince himself that they are obtained from the usual Jacobi identies and the anticommutativity of the commutator by a formal application of the rule of signs. It is worth checking also that if B is an A-algebra the commutator is A-bilinear in the following sense ($a \in A$; $b, c \in B$):

$$a[b, c] = [ab, c] = (-1)^{\widetilde{a}\widetilde{b}}[ba, c] = (-1)^{\widetilde{a}\widetilde{b}}[b, ac] = (-1)^{\widetilde{a}(\widetilde{b}+\widetilde{c})}[b, c]a.$$

All these identities become in superalgebra the axioms for a Lie superalgebra.

4. Modules. The axioms for the structure of a left A-module, a right A-module and an A-bimodule in superalgebra are the same as the $\mathbf{Z}_2$-graded versions of the usual axioms; there are no additional sign changes.

If the ring A is supercommutative, then, just as in commutative algebra, a one-sided module can be transformed canonically into a bimodule. More precisely, we say that two structures on S, that of a left and of a right module, are consistent if the left product $as = (-1)^{\widetilde{a}\widetilde{b}}sa$ for $a \in A$, $s \in S$. Then:

(a) Corresponding to any left A-module structure on S, there is a unique right A-module structure which is consistent with it, and conversely.

(b) A module with consistent structures is a bimodule (check that $a(sb) = (as)b$).

Using these recipes, we can always freely pass from one structure to the other. If $S = S_0 \oplus S_1$, then S_0 and S_1 are A_0-submodules of S.

An additive mapping of A-modules $f\colon S \to T$ is called an even (homo) morphism if it preserves the grading and is A-linear. Then from $f(as) = af(s)$ it follows that $f(sb) = f(s)b$ and conversely, so that a morphism of left (right) A-modules is automatically a morphism of bimodules. We will denote the additive group of such morphisms by $\mathrm{Hom}_0(S, Y)$.

An additive mapping of A-modules $f\colon S \to T$ is called an odd homomorphism if it reverses the grading ($\widetilde{f(s)} = \widetilde{s} + 1$) and is A-linear in the sense of the rule of signs: $f(as) = (-1)^{\widetilde{a}}af(s)$, $f(sb) = f(s)b$. Again, either of these formulas follows from the other. We will denote the additive group of such morphisms by $\mathrm{Hom}_1(S,T)$ and set

$$\mathrm{Hom}(S,T) = \mathrm{Hom}_0(S,T) \oplus \mathrm{Hom}_1(S,T).$$

One can introduce the structure of an A-module on $\mathrm{Hom}(S,T)$ by means of the usual formula $(af)(s) = a(f(s))$. It is easy to verify that for the consistent right A-module structure we have $(fa)(s) = (-1)^{\widetilde{f}\widetilde{a}}(af)(s) = f(as)$, etc.

From the point of view of the ordinary algebra of categories, $\mathrm{Hom}(S,T)$ should be considered the intrinsic Hom functor. Another point of view, more natural in superalgebra, is that every additive category in superalgebra should be endowed with a superstructure; the $\mathbf{Z}_2$-grading on Hom and the signs appear in the definition of inverse categories, functors, etc. Thus $\mathrm{Hom}(S,T)$ is the correct definition of the group of morphisms of modules, even for one-sided modules.

5. Parity change functors. Let S be a left or right module. We define the module ΠS by the conditions: (a) $(\Pi S)_0 = S_1$ and $(\Pi S)_1 = S_0$; (b) addition in ΠS is the same as in S; right multiplication by A is the same as in S; left multiplication differs by sign, $a(\Pi s) = (-1)^{\tilde{a}}\Pi(as)$, where Πs is the element of ΠS corresponding to $s \in S$. Here is a mnemonic rule: consider Π to be a formal factor of degree 1. In an analogous manner, $S\Pi$ can be defined, where left multiplication is preserved.

If A is supercommutative, then this recipe translates consistent structures on S to consistent structures on ΠS. Corresponding to the morphism $f: S \to T$, we let $f^{\Pi}: \Pi S \to \Pi T$ be the morphism which agrees with f as a mapping of sets. It is easy to see that this makes the association $S \mapsto \Pi S$ a functor, with $\Pi^2 = \mathrm{id}$.

Moreover, corresponding to a morphism $f: S \to T$, one can construct morphisms $\Pi f: S \to \Pi T$, with $(\Pi f)(s) = \Pi(f(s))$, and $f\Pi: \Pi S \to T$, with $(f\Pi)(\Pi s) = f(s)$. Clearly, $f^{\Pi} = \Pi f \Pi$. There are identifications

$$\begin{array}{ccccc} \mathrm{Hom}(\Pi S, T) & \overset{\sim}{\leftarrow} & \mathrm{Hom}(S,T) & \overset{\sim}{\rightarrow} & \mathrm{Hom}(S, \Pi T) \\ \Cup & & \Cup & & \Cup \\ f\Pi & \leftarrow\!\shortmid & f & \mapsto & \Pi f \end{array}$$

which themselves are odd isomorphisms of A-modules.

6. Free A-modules. In any of the three categories of A-modules, there is the ring A itself with left, right, or two-sided multiplication. A free A-module of rank $p|q$ is defined to be an A-module isomorphic to $A^{p|q} = A^p \oplus (\Pi A)^q$. This has a free system of generators, p of which are even and q of which are odd.

If A is supercommutative, then the rank is defined uniquely in the sense that free modules of different (finite) rank are not isomorphic. This is proved just as in the commutative case: taking the quotient of the ring with respect to a maximal ideal reduces the problem to the invariance of dimension of the even and odd components of a vector space over a field.

The reader should pay attention to the fact that the decomposition of $A^{p|q}$ into an "evenly generated" part $A^{p|0}$ and an "oddly generated" part $A^{0|q}$ is, in general, not invariantly defined and does not coincide with the decomposition into even and odd parts $\left[A_0^p \oplus (\Pi A_1)^q\right] \oplus \left[A_1^p \oplus (\Pi A_0)^q\right]$. Only when $A_1 = \{0\}$ are these decompositions the same.

7. Matrices in linear superalgebra. We will write an element of the left A-module $A^{p|q}$ as a row of coordinates and an element of the right module as a column. In the case of a bimodule with consistent structures of left and right multiplication, the transition from the row of left coordinates to the column of right coordinates does not consist of simply taking the transpose: $e = \sum_{i=1}^{p+q} a_i e_i = \sum_{i=1}^{p+q} (-1)^{\tilde{a}_i \tilde{e}_i} e_i a_i$. In order to write the signs correctly, it is necessary to know the parity of e_i, or the "parity of the position i," while the parity of $\tilde{a}_i$ is defined by the parity of e and the parity of the position, namely $\tilde{a}_i = \tilde{e} + \tilde{e}_i$.

In an analogous way, morphisms between free A-modules can be given by matrices, which act on the left on columns and on the right on rows of coordinates, but here the position of each element must be furnished with an indication of the parity of the rows and columns. We will give a formal definition.

A *matrix format* consists of two finite sets which are divided into pairs of nonintersecting subsets of even and odd elements; $I = I_0 \cup I_1$ and $J = J_0 \cup J_1$. A matrix in the given format with values in the ring A is a function $a: I \times J \to A$, where the values $a(i,j) = a_{ij}$ are the elements of the matrix. The parity of the position (ij) in the given matrix format is $\tilde{i} + \tilde{j}$. A matrix of the given format is called even if $\tilde{a}_{ij} = \tilde{i} + \tilde{j}$ for all i, j and is called odd if $\tilde{a}_{ij} = \tilde{i} + \tilde{j} + 1$ for all i, j.

Standard matrix format is the format with $I_0 = (1, \ldots, m)$, $I_1 = (m+1, \ldots, m+n)$ and $J_0 = (1, \ldots, p)$, $J_1 = (p+1, \ldots, p+q)$ (if $m = 0$, then $I_0 = \emptyset$, etc.). A matrix in standard format decomposes into four blocks

$$B = \begin{array}{|c|c|c} \hline B_1 & B_2 & m \\ \hline B_3 & B_4 & n \\ \hline \multicolumn{1}{c}{p} & \multicolumn{1}{c}{q} & \end{array}$$

it is even (resp., odd) if B_1 and B_4 are filled with even (resp., odd) elements of the ring while B_2 and B_3 are filled with odd (resp., even) elements.

The set of matrices in standard format with elements from the ring A is denoted by $M(m|n; p|q; A)$. It forms a $\mathbb{Z}_2$-graded module which, with the usual matrix multiplication, is naturally isomorphic to $\mathrm{Hom}(A^{p|q}, A^{m|n})$. (Unless otherwise stipulated, the matrix acts on the left on a column of right coordinates under this mapping.) Multiplication of morphisms corresponds to multiplication of matrices

$$M(m|n; p|q; A) \times M(p|q; r|s; A) \to M(m|n; r|s; A).$$

For further information about matrix computations, see § 3.

8. Derivations and Lie superalgebras. Let A be a ring. An additive homogeneous mapping $X: A \to A$ is called a left (super)derivation if it satisfies the Leibniz identity

$$X(ab) = (Xa)b + (-1)^{\tilde{a}\tilde{X}} a(Xb) \quad \text{for all } a, b \in A.$$

If A is a B-algebra, X is called a derivation over B if $Xb = 0$ for all $b \in B$. Setting $[X, Y] = XY - (-1)^{\tilde{X}\tilde{Y}} YX$ in the ring of additive mappings, we find the derivations of A over B form a Lie superalgabra in the sense of section 3(c).

If A is supercommutative, this Lie superalgebra is an A-module with left multiplication $(aX)(b) = a(Xb)$.

Let A be supercommutative and let M be an A-module. An additive mapping $X: A \to M$ is called a derivation with values in M if

$$X(ab) = (Xa)b + (-1)^{\tilde{a}\tilde{X}} a(Xb) \quad \text{for all } a, b \in A.$$

To each such derivation corresponds a morphism of A to the ring $A \oplus M$ (for even X) or $A \oplus \Pi M$ (for odd X) with law of multiplication is $(a, m)(a', m') = (aa', ma' + am')$ (henceforth, we will write $a + m$ instead of (a, m)):

$$a \mapsto a + Xa \qquad (\tilde{X} = 0);$$
$$a \mapsto a + \Pi Xa \qquad (\tilde{X} = 1).$$

The sense of the Leibniz formula is that this mapping is compatible with the ring multiplication. In fact, in the odd case we have

$$\begin{aligned}[a + \Pi Xa]\,[b + \Pi Xb] &= ab + (\Pi Xa)b + a(\Pi Xb)\\ &= ab + \Pi[Xa \cdot b + (-1)^{\tilde{a}} aX \cdot b] = ab + \Pi(X(ab)).\end{aligned}$$

As in the commutative case, there are universal derivations, both absolute derivations and derivations of B-algebras:

$$\begin{aligned}&\text{even}: \quad d_{\text{ev}}: A \to \Omega^1_{\text{ev}} A \qquad &&(\text{with respect to } \Omega_{\text{ev}} A/B),\\ &\text{odd}: \quad d_{\text{odd}}: A \to \Omega^1_{\text{odd}} A \qquad &&(\text{with respect to } \Omega_{\text{odd}} A/B).\end{aligned}$$

These are related to the natural isomorphism $\Omega^1_{\text{odd}} A = \Pi \Omega^1_{\text{ev}} A$ which corresponds to the identification $d_1 = \Pi d_0$.

We will return to the module of derivations in § 4.

§ 2. The Tensor Algebra over a Supercommutative Ring

1. Tensor product. Let A be a supercommutative ring, and let S and T be two A-modules. Their tensor product $S \otimes_A T$ is defined in the standard way, using the right A-module structure on S and the left structure on T (i.e., $sa \otimes t = s \otimes at$) and the grading $(S \otimes T)_k = \bigoplus_{i+j=k} S_i \otimes T_j$. this product is itself an A-module with respect to the consistent left and right multiplications:

$$a(s \otimes t) = as \otimes t = (-1)^{\tilde{a}(\tilde{s}+\tilde{t})}(s \otimes t)a = (-1)^{\tilde{a}(\tilde{s}+\tilde{t})} s \otimes ta.$$

The possibility of employing the tensor product over supercommutative rings just as freely as in the commutative case is facilitated by introducing some supplementary signs in the definition of the standard isomorphisms of tensor algebra, the most important of which is the morphism of commutativity (commutativity constraint):

$$\psi_{S,T}: S \otimes T \stackrel{\sim}{\rightarrow} T \otimes S,$$

$$\psi_{S,T}(s \otimes t) = (-1)^{\widetilde{s}\widetilde{t}}\, t \otimes s.$$

The family of all the $(\psi_{S,T})$ should be viewed as a morphism of functors.

2. The category of A-modules as a tensor category. The structure of a tensor category of modules over a supercommutative ring A is defined by the following data.

(a) The associativity morphism (associativity constraint) i.e., the family of standard functorial isomorphisms

$$\phi_{R,S,T}: R \otimes (S \otimes T) \stackrel{\sim}{\rightarrow} (R \otimes S) \otimes T,$$

$$\phi_{R,S,T}[r \otimes (s \otimes t)] = (r \otimes s) \otimes t.$$

It satisfies the pentagon axiom, which says this diagram is commutative:

$$\begin{array}{ccccc} R \otimes (S \otimes (T \otimes U)) & \xrightarrow{\phi} & (R \otimes S) \otimes (T \otimes U) & \xrightarrow{\phi} & ((R \otimes S) \otimes T) \otimes U \\ \downarrow 1 \otimes \phi & & & & \uparrow \phi \otimes 1 \\ R \otimes ((S \otimes T) \otimes U) & & \xrightarrow{\qquad\phi\qquad} & & (R \otimes (S \otimes T)) \otimes U \end{array}$$

(b) The commutativity morphism ψ, described above, which is compatible with the associativity morphism in the sense that diagrams of the form

$$\begin{array}{ccccc} R \otimes (S \otimes T) & \xrightarrow{\phi} & (R \otimes S) \otimes T & \xrightarrow{\psi} & T \otimes (R \otimes S) \\ \downarrow 1 \otimes \psi & & & & \downarrow \phi \\ R \otimes (T \otimes S) & \xrightarrow{\phi} & (R \otimes T) \otimes S & \xrightarrow{\psi \oplus 1} & (T \otimes R) \otimes S \end{array}$$

are commutative.

(c) A unit object, that is a pair, consisting of a module U and an isomorphism $u: U \rightarrow U \otimes U$ such that the mapping $S \rightarrow U \otimes S$ is an equivalence of categories. One can set $U = A$ and $u(1) = 1 \otimes 1$. All unit objects are canonically isomorphic.

3. Tensor product of a family of modules. Let I be a finite set, and let $(S_i \mid i \in I)$ be a family of modules indexed by I. Then one can define the module $\bigotimes_{i \in I} S_i$ as an object of the category. It is not defined uniquely; rather, it has several different realizations which are related by canonical isomorphisms. Each realization is defined (a) by an ordering of I and (b) an arrangement of parentheses between the elements of I which corresponds to the realization of the tensor product as an iteration of pairwise products. The isomorphism between two realizations is formed from the composition of isomorphisms ϕ and ψ. The axioms of section 2, (a) and (b), guarantee independence of the arbitrary choices made.

Here is a precise formulation of these assertions. For every finite set I, one can define a functor $\bigotimes_I : \mathit{Mod}^I \to \mathit{Mod}$ (where Mod is the category of A-modules), and for every mapping of finite sets $a: I \to J$ one can define an isomorphism of functors

$$\otimes(a): \bigotimes_{i\in I} S_i \stackrel{\sim}{\to} \bigotimes_{j\in J}\left(\bigotimes_{a(i)=j} S_i\right)$$

with the following properties.

For a set with one element, $\bigotimes_I$ is the identity functor;. for a mapping a between two sets with one element, $\otimes(a)$ is the identity isomorphism. For $I = \{1,2\} = J$, if a is the interchange, then $\otimes(a)$ is induced by the commutativity morphism (on any realization). One procedes in an analogous way for $I = \{1,2,3\}$ and the morphism of "rearrangement of parentheses."

4. Action of the symmetric group. For $I = \{1,\dots,n\}$, we will write $T^{\otimes n}$ instead of $\bigotimes_I T$. According to the preceding discussion, every permutation $a: I \to I$ defines an automorphism $\otimes(a): T^{\otimes n} \to T^{\otimes n}$. This is a representation of the symmetric group S_n on $T^{\otimes n}$ which can be used to carry out the usual constructions related to Schur symmetrization.

5. Symmetric algebra. The A-module $\bigoplus_{n=0}^{\infty} T^{\otimes n} = A\langle T\rangle$, where $T^0 = A$, is a $\mathbb{Z}$-graded object in the category of associative superalgebras, with the usual multiplication $T^{\otimes n} \times T^{\otimes m} \to T^{\otimes(m+n)}$. The mapping $T \to A\langle T\rangle$ is universal for mappings of T into associative A-algebras (an A-algebra B is a morphism of superalgebras $A \to B$ such that the image of A is in the supercenter of B, i.e., it supercommutes with all the elements of B).

We denote by J_S the ideal in $A\langle T\rangle$ which is generated by all the supercommutators. It can be shown that it is generated by supercommutators $t_1 \otimes t_2 - (-1)^{\tilde{t}_1\tilde{t}_2} t_2 \otimes t_1 \in T^{\otimes 2}$, $t_i \in T$. We define the symmetric algebra of the A-module T to be the quotient ring $S_A(T) = A[T] = A\langle T\rangle/J_S$. The mapping $T \to S_A(T)$ is universal for mappings of T to supercommutative A-algebras. If A is a $\mathbb{Q}$-algebra, then the projections of the symmetrization, $\frac{1}{n!}\sum_{a\in S_n} \otimes(a): T^{\otimes n} \to T^{\otimes n}$ make it possible to single out $S_A(T)$ as a direct summand of $A\langle T\rangle$ consisting of the collection of completely symmetric tensors.

The presence of a $\mathbb{Z}_2$-grading leads to a peculiar mixture of symmetry with antisymmetry since the commutator of odd elements is, from the naive point of view, the anticommutator. Therefore, for the oddly generated A-module $T = A^{0|q}$, its symmetric algebra is the Grassmann algebra.

6. Exterior algebra. There are two variants of the definition of the exterior algebra. Let J_Λ be the ideal in $A\langle T\rangle$ generated by the (super)commutators $t_1 \otimes t_2 + (-1)^{\tilde{t}_1\tilde{t}_2} t_2 \otimes t_1 \in T^{\otimes 2}$; and, as in ordinary algebra,

$$\Lambda_A(T) = A\langle T\rangle / J_\Lambda.$$

The other variant consists of taking the definition of Bernstein-Leites:

$$E_A(T) = S_A(\Pi T).$$

The rules for commuting two elements $t_1, t_2 \in T$ in $\Lambda_A(T)$ and in $E_A(T)$ differ in sign if and only if t_1 and t_2 have different parity.

More precisely,

$$t_1 t_2 = -(-1)^{\tilde{t}_1\tilde{t}_2} t_2 t_1 \text{ in } \Lambda_A(T),$$

$$\Pi t_1 \Pi t_2 = -(-1)^{\widetilde{\Pi t_1}\,\widetilde{\Pi t_2}} \Pi t_2 \Pi t_1 = (-1)^{\tilde{t}_1 + \tilde{t}_2 + 1} (-1)^{\tilde{t}_1\tilde{t}_2} \Pi t_2 \Pi t_1 \text{ in } S_A(\Pi T),$$

since $\widetilde{\Pi t} = \tilde{t} + 1$. However, as A-modules, $\Lambda_A^n(T)$ and $S_A^n(\Pi T)$ are isomorphic.

As A-algebras, $\Lambda_A(T)$ and $E_A(T)$ are certainly not isomorphic; for the second one is supercommutative and the first one is not.

7. Superadditive categories. It is convenient to formulate the functorial properties of tensor multiplication using the following categorical concepts.

A superadditive category C is an additive category whose group of morphisms is provided with a $\mathbb{Z}_2$-grading compatible with composition of morphisms. Functors between superadditive categories are ordinary functors which are even mappings on the morphisms.

Let C be a superadditive category. Then C^t is defined to be a category with the same objects and morphisms as C but with a modified composition of arrows, namely $g \circ f$ (composition in C^t) $= (-1)^{\tilde{g}\tilde{f}} gf$ (composition in C). As usual, C^0 denotes C with the arrows reversed. In the places where the category C^0 arises in ordinary algebra (say, in the definition of contravariant functors) the category C^{0t} usually appears in superalgebra.

In the definition of the product of superadditive categories, signs also appear:

$$\mathrm{Ob}\,(C \times D) = \mathrm{Ob}\,C \times \mathrm{Ob}\,D,$$

$$\mathrm{Hom}_{C\times D}\big(\langle S_0,T_0\rangle,\langle S_1,T_1\rangle\big)=\mathrm{Hom}_C(S_0,S_1)\times\mathrm{Hom}_D(T_0,T_1),$$

$$\langle f_0,g_0\rangle\langle f_1,g_1\rangle=(-1)^{\widetilde{g_0}\widetilde{f_1}}\langle f_0f_1,g_0g_1\rangle.$$

We leave it to the reader to verify that the associativity functor for the product $(C\times D)\times E\to C\times(D\times E)$

$$\langle\langle S_0,S_1\rangle,S_2\rangle\mapsto\langle S_0,\langle S_1,S_2\rangle\rangle,$$

$$\langle\langle f,g\rangle,h\rangle\mapsto\langle f,\langle g,h\rangle\rangle$$

actually preserves the composition of morphisms.

8. Homomorphism bifunctor. Now let C_A be the superadditive category of modules over a supercommutative ring A. We define the functor

$$\mathcal{H}om\colon C_A^{0t}\times C_A\to C_A$$

by the following rules:

objects: $\mathcal{H}om(\langle S,T\rangle)=\mathrm{Hom}(S,T)$ as an A-module;

morphisms: $$\mathcal{H}om\left(\langle S_0,T_0\rangle\xrightarrow{\langle f^0,g\rangle}\langle S_1,T_1\rangle\right)=\mathrm{Hom}(S_0,T_0)\to\mathrm{Hom}(S_1,T_1),$$
$$h\mapsto ghf$$

where the diagram $S_1\xrightarrow{f}S_0\xrightarrow{h}T_0\xrightarrow{g}T_1$ defines the triple product of morphisms. In order to persuade oneself of the correctness of these definitions, one must carry out a rather large number of verifications involving the rule of signs.

9. Multilinear mappings and tensor product. Let $S_1,\dots,S_m$ and T be modules over a supercommutative ring A. As a temporary definition, for $0\le i\le n$, we will say that a homogeneous multiadditive mapping $b\colon S_1\times\ldots\times S_m\to T$ is i-polylinear and will write its value in the form $s_1\dots s_ibs_{i+1}\dots s_m$ if the following conditions are satisfied: for $j\le i$ and $a\in A$, $s_1\dots as_j\dots bs_{i+1}\dots s_m=(-1)^{\tilde a(\tilde s_1+\dots+\tilde s_{j-1})}a(s_1\dots s_ibs_{i+1}\dots s_m)$; while for $j>i$, $s_1\dots s_ib\dots as_j\dots s_m=(-1)^{\tilde a(\tilde b+\tilde s_1+\dots+\tilde s_{j-1})}a(s_1\dots s_ibs_{i+1}\dots s_m)$. We denote by $L_i(S_1,\dots,S_m;T)$ the set of all i-polylinear mappings. We introduce the structure of an A-module on the set with the obvious sum and with the multiplication such that $\dots s_i(ab)s_{i+1}\dots=\dots(s_ia)bs_{i+1}\dots$

The following facts are true; we leave the precise formulation and proof of them to the reader:

(a) The rule of signs allows the definition of mutually compatible canonical isomorphisms $L_i(S_1, \ldots, S_m; T) \to L_j(S_1, \ldots, S_m; T)$; thus a general object can be defined, $L(S_1, \ldots, S_m; T)$, the module of polylinear mappings.

(b) There is a canonical identification $L(S_1, \ldots, S_m; T) = \mathrm{Hom}(S_1 \otimes \ldots \otimes S_m; T)$.

(c) $L(S_1, \ldots, S_m; T)$ extends to a functor $C_A^{0t} \times \ldots \times C_A^{0t} \to C$.

In § 5 we will study bilinear mappings in more detail.

10. On the formalism of indices in tensor algebra. As in ordinary tensor algebra, it is convenient to employ notation for the elements of the tensor algebra of a module which describes the elements by taking the expansion with respect to a tensor basis and using multi-index notation. For this, however, it is necessary to take account of the following circumstances.

(a) The tensor algebra of a module T is not generated only by T (the "lower indices") and T^* (the "upper indices"), but also ΠT and possibly $\overline{T}$, the "adjacent modules of T" in superalgebra (see § 6 below). For all these, supplementary types of indices are needed. Also, it is sometimes convenient to separate the indices belonging to the even and the odd generators. By tradition, lower case Latin and Greek letters, respectively, are used for this.

(b) Several important constructions cease to be polylinear; the analog of the "maximal exterior power of the module T" is only a subquotient in the tensor algebra (see § 4 below).

§ 3. The Supertrace and Superdeterminant

We will continue to assume that A is a supercommutative ring in which 2 is invertible.

1. Supertranspose. Let the matrix B in standard format describe an even or odd morphism $b: S \to T$ of free A-modules. There exists a natural morphism $b^*: T^* \to S^*$ with the property

$$(b^*(t^*), s) = (-1)^{\widetilde{bt}}(t^*, b(s)) \quad \text{for all } t^* \in T^*, s \in S,$$

where the parentheses denote the canonical pairing. By B^{st} we denote the matrix of b^* with respect to the bases of T^* and S^* which are dual on the left to the bases of S and T used for defining B. These are the formulas for B^{st}:

$$\begin{pmatrix} B_1 & B_2 \\ B_3 & B_4 \end{pmatrix}^{st} = \begin{cases} \begin{pmatrix} B_1^t & B_3^t \\ -B_2^t & B_4^t \end{pmatrix} & \text{for } \widetilde{B} = 0, \\ \begin{pmatrix} B_1^t & -B_3^t \\ B_2^t & B_4^t \end{pmatrix} & \text{for } \widetilde{B} = 1. \end{cases}$$

Here t denotes the usual transpose.

Here are several properties of the operation of supertransposition:

(a) $(B+C)^{st} = B^{st} + C^{st}$.

(b) $(BC)^{st} = (-1)^{\widetilde{B}\widetilde{C}} C^{st} B^{st}$.

(c) $(st)^4 = \mathrm{id}$, $(st)^2 \neq \mathrm{id}$.

(d) If B is a row of left coordinates (in standard format), then B^{st} is a column of right coordinates of a homogeneous vector in the same basis.

2. Π-transpose. Under the same assumptions, let $b^{\Pi}: \Pi S \to \Pi T$ be the morphism which is the same set-theoretically as B; and let B^{Π} be its matrix in standard format in terms of the same bases, where the whole block of even (formerly odd) elements is placed before the block of odd (formerly even) ones. Then

$$\begin{pmatrix} B_1 & B_2 \\ B_3 & B_4 \end{pmatrix}^{\Pi} = \begin{pmatrix} B_4 & B_3 \\ B_2 & B_1 \end{pmatrix}.$$

Here are several properties of the Π-transpose:

(a) $(B+C)^{\Pi} = B^{\Pi} + C^{\Pi}$.

(b) $(BC)^{\Pi} = B^{\Pi} C^{\Pi}$.

(c) $\Pi^2 = \mathrm{id}$; $\Pi \circ st \circ \Pi = (st)^4$.

3. Supertrace. On $S^* \otimes S$ contraction is defined, this is the linear functional $\sum s_i^* \otimes s_i \mapsto \sum (s_i^*, s_i)$. On the other hand, $S^* \otimes S$ can be identified with $\operatorname{End} S = \operatorname{Hom}(S, S)$ by $\left(\sum s_i^* \otimes s_i\right)(t) = \sum (-1)^{\widetilde{s_i}\widetilde{t}} (s_i^*, t) s_i$. Writing the elements of $\operatorname{End} S$ as matrices in standard format, we obtain an expression for the contraction in the form of the supertrace functional:

$$\operatorname{str} \begin{pmatrix} B_1 & B_2 \\ B_3 & B_4 \end{pmatrix} = \begin{cases} \operatorname{tr} B_1 - \operatorname{tr} B_4 & \text{for } \widetilde{B} = 0; \\ \operatorname{tr} B_1 + \operatorname{tr} B_4 & \text{for } \widetilde{B} = 1. \end{cases}$$

Here are several properties of the supertrace:

(a) $\operatorname{str}: \operatorname{End} S \to A$ is a morphism of A-modules.

(b) $\operatorname{str}[B, C] = 0$ for any matrices in a format for which BC, CB and $[B, C]$ are defined. In particular, $\operatorname{str}(CBC^{-1}) = \operatorname{str} B$.

(c) $\operatorname{str}(B^{st}) = \operatorname{str} B$, $\operatorname{str}(B^{\Pi}) = -(-1)^{\widetilde{B}} \operatorname{str} B$.

The fundamental difference between the supertrace and the ordinary trace is evidenced in the formula $\operatorname{str}(\mathrm{id}_{p|q}) = p - q$, where $\mathrm{id}_{p|q}$ is the identity mapping of $A^{p|q}$.

4. Berezinian, or superdeterminant. Let $B = \mathrm{GL}(p|q; A)$ (the invertible even automorphisms of $A^{p|q}$. Writing B in the standard format, we set

$$\mathrm{Ber}\begin{pmatrix} B_1 & B_2 \\ B_3 & B_4 \end{pmatrix} = \det(B_1 - B_2 B_4^{-1} B_3) \det B_4^{-1}.$$

The right-hand side is well-defined, for $B_1 \in \mathrm{GL}(p|0; A_0)$ and $B_4 \in \mathrm{GL}(q|0; A_0)$. It belongs to A_0 and is invertible, i.e., it lies in $\mathrm{GL}(1|0; A_0)$.

5. Theorem. *The map* $\mathrm{Ber}\colon \mathrm{GL}(p|q; A) \to \mathrm{GL}(1|0; A_0)$ *is a group homomorphism which agrees with the determinant when* $q = 0$.

Proof. It is sufficient to verify that $\mathrm{Ber}\, BC = \mathrm{Ber}\, B \cdot \mathrm{Ber}\, C$ for all $B, C \in \mathrm{GL}(p|q; A)$. We set

$$G = \{\, C \in \mathrm{GL}(p|q; A) \mid \forall B \in \mathrm{GL},\ \mathrm{Ber}\, BC = \mathrm{Ber}\, B \cdot \mathrm{Ber}\, C \,\}.$$

The following assertions can be shown without difficulty, and together they prove the theorem:

(a) G is a group.

(b) G contains the matrices of the following form (in standard format):

$$\begin{pmatrix} 1 & 0 \\ C_3 & 1 \end{pmatrix}, \quad \begin{pmatrix} C_1 & 0 \\ 0 & C_4 \end{pmatrix}, \quad \begin{pmatrix} 1 & C_2 \\ 0 & 1 \end{pmatrix},$$

where 1 is the identity matrix and C_2 is elementary, that is, it has only one nonzero element.

(c) The matrices described in (b) above generate $\mathrm{GL}(p|q; A)$. This follows from the identity

$$\begin{pmatrix} C_1 & C_2 \\ C_3 & C_4 \end{pmatrix} = \begin{pmatrix} 1 & C_2 C_4^{-1} \\ 0 & 1 \end{pmatrix} \begin{pmatrix} C_1 - C_2 C_4^{-1} C_3 & 0 \\ 0 & C_4 \end{pmatrix} \begin{pmatrix} 1 & 0 \\ C_4^{-1} C_3 & 1 \end{pmatrix} = C_+ C_0 C_-$$

and from the following decomposition of the first matrix on the right side into a product:

$$\begin{pmatrix} 1 & C + C' \\ 0 & 1 \end{pmatrix} = \begin{pmatrix} 1 & C \\ 0 & 1 \end{pmatrix} \begin{pmatrix} 1 & C' \\ 0 & 1 \end{pmatrix}.$$

In the verification of assertion (b), only the claim that $C = \left(\begin{smallmatrix} 1 & C_2 \\ 0 & 1 \end{smallmatrix}\right) \in G$ for an elementary matrix C_2 demands anything more than a direct computation using the formula defining the berezinian. We wish to establish that $\mathrm{Ber}\, BC = \mathrm{Ber}\, B \cdot \mathrm{Ber}\, C$ for all B. We represent B in the form $B_+ B_0 B_-$ as in assertion (c). A direct computation shows that multiplication on the left by B_+ and by B_0 in multiplicative with respect to Ber, so it sufficient to consider $B = B_-$, i.e., to verify the identity $\mathrm{Ber}(B_-C_+) = 1$, where C_2 is elementary. Omitting the block indices, we have

$$B_-C_+ = \begin{pmatrix} 1 & 0 \\ B & 1 \end{pmatrix} \begin{pmatrix} 1 & C \\ 0 & 1 \end{pmatrix} = \begin{pmatrix} 1 & C \\ B & 1+BC \end{pmatrix},$$

$$\mathrm{Ber}(B_-C_+) = \det(1 - C(1+BC)^{-1}B)\det{}^{-1}(1+BC).$$

The single nonzero element c of the matrix C lies in A_1, so $c^2 = 0$. Therefore, the product of any two elements of the matrices BC and $C(1+BC)^{-1}B$ (in the commutative ring A_0) equals zero. For a matrix D with this property we have $(1+D)^{-1} = 1 - D$ and $\det(1+D) = \mathrm{tr}\, D + 1$: in the ordinary expansion of the determinant, only the diagonal product is not equal to zero. Consequently,

$$\mathrm{Ber}(B_-C_+) = \det(1 - CB)\det{}^{-1}(1+BC) = 1 - \mathrm{tr}\, CB - \mathrm{tr}\, BC = 1.$$

This completes the proof. □

6. Properties of the berezinian. (a) If S is a free A-module of finite rank and $b \in \mathrm{GL}(S)$, we set $\mathrm{Ber}\, b = \mathrm{Ber}\, B$, where B is any matrix in standard format representing b. This definition does not depend upon the choice of basis for S.

(b) $\mathrm{Ber}(B^{st}) = \mathrm{Ber}\, B$; therefore, $\mathrm{Ber}\, b^* = \mathrm{Ber}\, b$.

(c) $\mathrm{Ber}(B^\Pi) = (\mathrm{Ber}\, B)^{-1}$ and $\mathrm{Ber}(b^\Pi) = (\mathrm{Ber}\, b)^{-1}$. In fact,

$$\text{if } B = \begin{pmatrix} B_1 & B_2 \\ B_3 & B_4 \end{pmatrix}, \text{ then } B = \begin{pmatrix} 1 & 0 \\ B_3B_1^{-1} & 1 \end{pmatrix} \begin{pmatrix} B_1 & B_2 \\ 0 & B_4 - B_3B_1^{-1}B_2 \end{pmatrix},$$

whence

$$\mathrm{Ber}\, B = \det B_1 \cdot \det(B_4 - B_3B_1^{-1}B_2)^{-1}.$$

On the other hand,

$$\mathrm{Ber}\, B^\Pi = \det(B_4 - B_3B_1^{-1}B_2)\det B_1^{-1} = (\mathrm{Ber}\, B)^{-1}.$$

(d) $\mathrm{Ber}(b \oplus c) = \mathrm{Ber}\, b \cdot \mathrm{Ber}\, c$, where b and c are even automorphisms of S and T and $b \oplus c$ is an even automorphism of $S \oplus T$.

(e) $\mathrm{Ber}\, e^{BT} = e^{(\mathrm{str}\, B)t}$ in the ring of formal power series over A in the even variable t.

7. Berezinian of a free A-module. Let S be a free A-module of rank $p|q$. We define a free A-module $\mathrm{Ber}\, S$ of rank $1|0$ (if q is even) or $0|1$ (if q is odd) which is functorial with respect to even isomorphisms of A-modules and coincides with $\Lambda^p S$ for $q = 0$.

Namely, Ber S is defined along with a distinguished class of bases: any standard basis (s_i) of the module S corresponds to a basis (a single element) $D(s_i) \in \operatorname{Ber} S$ with the relations that if $(s'_i) = (s_i)B$, then $D(s'_i) = D(s_i) \operatorname{Ber} B$.

This is the law of functoriality: for any even isomorphism $b: S \to S'$, there is a corresponding isomorphism $\operatorname{Ber} B: \operatorname{Ber} S \to \operatorname{Ber} S'$ which carries $D(s_i)$ to $D(b(s_i))$. If $S = S'$, then $\operatorname{Ber} b$ can be naturally identified with multiplication by the element of A_0 which is denoted by $\operatorname{Ber} b$ in the previous subsection.

Sometimes it is convenient to use also the elements $D(s_i)$ for a basis (s_i) with non-standard ordering; we define $D(s_i) = (-1)^{\tilde{\sigma}} D(s_{\sigma(i)})$, where σ is some permutation which makes the ordering standard and $(-1)^{\tilde{\sigma}}$ is its parity. Another choice of permutation τ leads to the same answer, for $\sigma\tau^{-1}$ permutes separately the even and the odd positions in the standard format. But in this case $D(\cdot)$ is multiplied by the sign of the permutation according to the formula for the berezinian.

An analogous rule allows us to extend the definition of the berezinian to an even matrix in rectangular format with a non-standard enumeration of rows and (or) columns.

§ 4. Some Complexes in Superalgebra

1. A complex as an $A[\xi]$-module. In ordinary commutative algebra, a complex of A-modules (S^i, d^i) can be viewed as a $\mathbb{Z}$-graded module over a supercommutative ring $A[\xi]$ with one odd generator ξ; the module $S = \bigoplus_i S^i$, and d^i is the left multiplication by ξ on S^i. In an analogous way, a bicomplex is a $\mathbb{Z} \times \mathbb{Z}$-graded module over $A[\xi_1, \xi_2] = S_A(A\xi_1 \oplus A\xi_2)$; then the $\mathbb{Z}_2$-grading should be defined as the reduction of the total degree $\pmod 2$ (or as the opposite grading). The rules of signs in homological algebra then will agree with the rule of sign in superalgebra; in particular, the tensor product of complexes corresponds to the tensor product of $A[\xi]$-modules, etc.

Now let A be a supercommutative ring ($A_1 \neq \{0\}$ in general), and let (S^i, d^i) be a complex of A-modules; in particular, its components are $\mathbb{Z}_2$-graded. We will assume that all the d^i are homogeneous and have the same parity. The interpretation of $(\bigoplus S^i, \bigoplus d^i)$ as a $\mathbb{Z}$-graded module over a supercommutative ring $A[\xi]$ is possible as before, if the d^i are odd.

In the definitions of natural complexes in superalgebra, an arbitrary choice of parity for the differential often occurs; the preceding remark motivates a preference for odd d.

We will point out a general principle for constructing complexes in (super) algebra and then describe several of its realisations.

2. Koszul differential. Let S and T be two A-modules (A supercommutative), and let $\sigma_a: S \to S$ and $\tau_a: T \to T$ be two finite families of homogeneous morphisms with the same set of indices a. We will say that σ_a and σ_b (super) commute if $[\sigma_a, \sigma_b] = 0$ and that they (super) anticommute if $\{\sigma_a, \sigma_b\} = \sigma_a\sigma_b + (-1)^{\widetilde{\sigma_a}\widetilde{\sigma_b}}\sigma_b\sigma_a = 0$. We suppose that the following conditions are fulfilled: (a) $\widetilde{\sigma}_a + \widetilde{\tau}_a$ does not depend on a; (b) for any pair (a, b), the operators σ_a and σ_b either commute or anticommute, and the same is true for τ_a and τ_b.

We define the mapping $d = \sum_a \sigma_a \otimes \tau_a: S \otimes T \to S \otimes T$; we recall that $(\sigma \otimes \tau)(s \otimes t) = (-1)^{\widetilde{\tau s}}\sigma(s) \otimes \tau(t)$.

3. Lemma. *We assume that under the stated conditions one of two assumptions is satisfied:*

(a) *The operator d is even, and for all (a, b) the rules of commutation for (σ_a, σ_b) and (τ_a, τ_b) are opposite.*

(b) *The operator d is odd, and for all (a, b) the rules of commutation for (σ_a, σ_b) and (τ_a, τ_b) are the same.*

Then $d^2 = 0$.

Proof. We check that in the expression for d^2 the following pairs of sums cancel:

$$(\sigma_a \otimes \tau_b)(\sigma_b \otimes \tau_b) + (s_b \otimes \tau_b)(\sigma_a \otimes \tau_a) =$$

$$(-1)^{\widetilde{\tau_a}\widetilde{\sigma_b}}\sigma_a\sigma_b \otimes \tau_a\tau_b + (-1)^{\widetilde{\sigma_a}\widetilde{\tau_b}}\sigma_b\sigma_a \otimes \tau_b\tau_a =$$

$$\left[(-1)^{\widetilde{\tau_a}\widetilde{\sigma_b}} + (-1)^{\widetilde{\sigma_a}\widetilde{\tau_b}}(-1)^{\widetilde{\sigma_a}\widetilde{\sigma_b}+\widetilde{\tau_a}\widetilde{\tau_b}}\epsilon\eta\right] \sigma_a\sigma_b \otimes \tau_a\tau_b,$$

where ϵ and η are the signs of commutation of σ_a, σ_b and τ_a, τ_b, respectively: $\sigma_a\sigma_b = (-1)^{\widetilde{\sigma_a}\widetilde{\sigma_b}}\sigma_b\sigma_a\epsilon$. If d is even, then $\widetilde{\tau}_a = \widetilde{\sigma}_a$, $\widetilde{\tau}_b = \widetilde{\sigma}_b$, and $\epsilon\eta = -1$. If d is odd, then $\widetilde{\tau}_a = \widetilde{\sigma}_a + 1$, $\widetilde{\tau}_b = \widetilde{\sigma}_b + 1$, and $\epsilon\eta = 1$. In both cases the coefficient in brackets is equal to zero. □

4. Abstract de Rham complex. Let A be a supercommutative B-algebra, and let (X_a) be a finite family of homogeneous superderivations of A over B; $[X_a, X_b] = 0$ for all a, b; (ω_a) is a family of variables which are free over B. The construction of Lemma 3 leads to two complexes. The first is a de Rham complex with even differentials:

$$\Omega^{\cdot}_{\text{ev}}(A/B; (X_a)) = B\{\omega_a\} \otimes_B A; \quad \widetilde{\omega}_a = \widetilde{X}_a;$$

$$\{\omega_a, \omega_b\} = 0; \quad d_{\text{ev}} = \sum_a \omega_a \otimes X_a.$$

The second is a de Rham complex with odd differentials:

$$\Omega^{\cdot}_{\text{odd}}(A/B;(X_a)) = B\,[\omega_a] \otimes_B A; \quad \widetilde{\omega}_a = \widetilde{X}_a + 1;$$

$$[\omega_a, \omega_b] = 0; \quad d_{\text{odd}} = \sum_a \omega_a \otimes X_a.$$

In the formulas for d the ring element ω_a is understood to be the operator of left multiplication by this element.

A $\mathbb{Z}$-grading can be chosen in various ways; for example, $\deg A = 0$ and $\deg \omega_a = 1$ or $\deg \omega_a = (-1)^{\widetilde{\omega}_a}$. If even one among the X_a is odd, then the complex becomes infinite.

In supergeometry this construction leads to a version of the usual de Rham complex, if we set (in the smooth case)

$$B = \mathbb{R}, \quad A = C^{\infty}(x^1, \dots, x^m)[\xi^1, \dots, \xi^n],$$

$$(X_a) = \left(\frac{\partial}{\partial x^b}, \frac{\partial}{\partial \xi^\beta}\right), \quad \omega_a = \left(dx^b, d\xi^\beta\right).$$

The action of the partial derivatives is defined by the obvious formulas:

$$\frac{\partial}{\partial x^a}\left[f(x)(\xi^1)^{\beta_1}\cdots(\xi^n)^{\beta_n}\right] = \frac{\partial f}{\partial x^a}(\xi^1)^{\beta_1}\cdots(\xi^n)^{\beta_n},$$

$$\frac{\partial}{\partial \xi^b}\left[f(x)(\xi^1)^{\beta_1}\cdots(\xi^n)^{\beta_n}\right] = (-1)^{\sum_{a<b}\beta_a}\beta_b f(x)(\xi^1)^{\beta_1}\cdots(\widehat{\xi^b})\cdots(\xi^n)^{\beta_n}.$$

The differentials dx^a and $d\xi^b$ supercommute for odd d and superanticommute for even d. In a similar way, one can construct relative de Rham complexes along the fibers of a submersion, a complex-analytic de Rham complex, etc.

5. Proposition (Poincaré lemma). *The de Rham complex of the ring $C^\infty(x^a, \xi^b)$ described in the preceding paragraph, with the $\mathbb{Z}$-grading $\deg(dx^a) = \deg(d\xi^b) = 1$, is a right resolution of $\mathbb{R}$.*

Proof. We confine ourselves to the case of $\Omega^{\cdot}_{\text{odd}}$ and show that the standard proof of the Poincaré lemma carries over to our case. First of all, a form of zero $\mathbb{Z}$-degree is a function $f = \sum_\alpha f_\alpha(x^1, \dots, x^m)\xi^\alpha$, and $df = 0$ if $\partial f_\alpha/\partial x^i = 0$ for all i and $f_\alpha = 0$ if $|\alpha| > 0$. Therefore, f is constant.

Now let ω be a form of $\mathbb{Z}$-degree ≥ 1. We introduce an even auxiliary variable t and construct the form $\omega_1(t)$ in the following way: replace all the x^i in ω with $x^i t$ and all the ξ^i with $\xi^i t$; replace all the dx^i with $d(x^i t) = dx^i t + x^i dt$ and all the $d\xi^i$

with $(d\xi^i)t - \xi^i dt$. then we write the result in the form $\omega(t) = \omega_0(t) + dt \cdot \omega_1(t)$, where dt does not appear in $\omega_0(t)$. Then we set $k(\omega) = k'(\omega(t)) = \int_0^1 dt\, \omega_1(t)$. We check that $(kd + dk)\omega = \omega$; then from $d\omega = 0$ it follows that $\omega = d(k\omega)$.

More generally, if the form $\omega(t)$ depends on a parameter t, we have $(k'd + dk')(\omega(t)) = \omega(1) - \omega(0)$. Indeed, suppose first that dt does not appear in $\omega(t)$. Then $k'(\omega(t)) = 0$, and

$$(k'd + dk')(\omega(t)) = k'd(\omega(t)) = k'\left(dt\,\frac{\partial\omega(t)}{\partial t}\right) =$$

$$\int_0^1 dt\,\frac{\partial\omega(t)}{\partial t} = \omega(1) - \omega(0).$$

Now if $\omega(t) = dt\,\omega_1(t)$, then $\omega(1) - \omega(0) = 0$, while on the other hand,

$$(k'd + dk')(dt\,\omega_1(t)) = -k'(dt\,d\omega_1(t)) + d\int_0^1 dt\,\omega_1(t) =$$

$$-\int_0^1 dt\,d\omega_1(t) + d\int_0^1 dt\,\omega_1(t) = 0.$$

Similar arguments go through for the relative and the complex-analytic de Rham complexes. □

6. Koszul complexes. Let A be a $\mathbb{Z}_2$-graded ring (not necessarily supercommutative), and let (y_a) be a finite family of homogeneous elements of A which pairwise supercommute. Let (e_a) and (e^a) be two free systems of independent supercommuting variables. We will denote monomials in these variables by $e_\alpha = e_1^{\alpha_1} \cdots e_n^{\alpha_n}$ and $e^\alpha = (e^1)^{\alpha_1} \cdots (e^n)^{\alpha_n}$, respectively. We define these two complexes.

The right Koszul complex:

$$K^\cdot(A_1(y_a)) = \mathbb{Z}[e^a] \otimes_{\mathbb{Z}} A = \bigoplus_\alpha e^\alpha A, \quad \widetilde{e}_a = \widetilde{y}_a + 1,$$

$$d^\cdot = \sum e^a \otimes y_a, \quad \deg A = 0, \quad \deg e_a = 1;$$

The left Koszul complex

$$K_\cdot(A, (y_a)) = A \otimes_{\mathbb{Z}} \mathbb{Z}[e_a] = \bigoplus_\alpha A e_\alpha, \quad \widetilde{e}_a = \widetilde{y}_a + 1,$$

$$d_{\cdot} = \sum y_a \otimes \frac{\partial}{\partial e_a}, \quad \deg A = 0, \quad \deg e_a = -1.$$

In both cases the differential is odd and has degree 1 in the $\mathbb{Z}$-grading.

7. Homological interpretation of the berezinian. Let T be a free module of finite rank over a supercommutative ring A. In the symmetric algebra $S_A(\Pi T \oplus T^*)$, we consider the following element d. We choose free bases (t_a) for T and (t^a) for T^* such that $(t_a, t^b) = \delta_a^b$ and set $d = \sum_a (\Pi t_a) t^a$. Then d is independent of these bases and $d^2 = 0$. We identify d with left multiplication by d. It is not difficult to see that $K_A^{\cdot}(T) = (S_A(\Pi T \oplus T^*), d)$ is a right Koszul complex. Let $H(K_A^{\cdot}(T))$ be its cohomology.

Let (t_a) be indexed in standard format: $\widetilde{t}_1 = \ldots = \widetilde{t}_m = 0$, $\widetilde{t}_{m+1} = \ldots = \widetilde{t}_{m+n} = 1$. We consider the element $\Pi t_1 \cdots \Pi t_m t^{m+1} \cdots t^{m+n}$. Clearly, this is a cycle. We denote its cohomology class by $h(t_a)$.

Proposition. *The mapping* $\operatorname{Ber} T \to H(K_A^{\cdot}(T))$ *which maps the element* $D(t_a)$ *to the class* $h(t_a)$ *does not depend on the choice of basis* (t_a), *and it defines an isomorphism with parity* $(-1)^m$ *between* $\operatorname{Ber} T$ *and the corresponding cohomology group.*

Proof. First we verify that $H(K_A^{\cdot}(T))$ is a free A-module generated by the class $h(t_a)$. It is convenient to change notation: let (x_i) be all the even and (ξ_i) be all the odd elements of the system $(t_a) \cup (\Pi t_a)$, indexed so that $d = \sum x_i \xi_i$ (more precisely, left multiplication by this element) in the ring $K_A^{\cdot}(T) = A[x_i, \xi_i] = B$. We introduce a $\mathbb{Z}$-grading on the complex, with $\deg \xi = 1$ and $\deg x_i = 0$; then we find that it is an ordinary commutative Koszul complex, so

$$H^i(B) = 0 \quad \text{for} \quad i \leq n-1;$$

$$H^n(B) = (B\xi_1 \cdots \xi_n)/(\sum B x_i)\xi_1 \cdots \xi_n \simeq A\xi_1 \cdots \xi_n.$$

The only thing left to check is that $D(t_a)$ and $h(t_a)$ behave the same under change of basis. This is clear for diagonal changes and for elementary upper and lower triangular changes. For example, for the change from t_{m+1} to $t'_{m+1} = f t_{m+1}$, for an invertible $f \in A_0$, $D(t_a)$ is multiplied by f^{-1} (the change of berezinian) and $h(t_a)$ is also multiplied by f^{-1}, for t^{m+1} changes to $f^{-1} t^{m+1}$ in order to preserve the duality of the bases. □

§ 5. Scalar Products

1. Bilinear forms. Let A be a supercommutative ring, and let T_1 and T_2 be A-modules. An even or odd morphism of A-modules $b: T_1 \otimes_A T_2 \to A$ is called, respectively, an even or odd bilinear form. A bilinear form can be uniquely identified with the function $b(t_1, t_2) = b(t_1 \otimes t_2)$, which satisfies the following conditions;

(a) b is biadditive and homogeneous;

(b) $b(at_1, t_2) = (-1)^{\widetilde{ab}} ab(t_1, t_2)$, and $b(t_1, t_2 a) = b(t_1, t_2)a$, for $a \in A$, $t_i \in T_i$.

We associate to an element $t_1 \in T_1$ the mapping $t_2 \mapsto b(t_1, t_2)$. The mapping is right linear in t_2 and homogeneous of degree $\widetilde{b} + \widetilde{t}_1$; it depends on t_1 such that we obtain a morphism $T_1 \to T_2^*$ of the same degree as b. Usually, we will denote this morphism by the same letter b.

The form b is said to be nondegenerate if $b: T_1 \to T_2^*$ is an isomorphism. We observe that for $\widetilde{b} = 1$ we have rk $T_1 = p|q$ and rk $T_2 = q|p$.

A form b is called direct if the kernel and image of the morphism are direct summands. (This concept is important when considering superanalogs of vector bundles with fiber metrics, where instead of a single fiber one considers modules of local sections.)

2. Two involutions. Let $b: T_1 \otimes T_2 \to A$ be a bilinear form.

(a) We set $b^\tau(t_1, t_2) = (-1)^{\widetilde{t}_1 \widetilde{t}_2} b(t_2, t_1)$. It is easy to check that b^τ is a bilinear form of the same parity as b; it defines a morphism $b^\tau: T_2 \to T_1^*$. We have $(b^\tau)^\tau = b$.

(b) We set $b^\pi(\Pi t_1, \Pi t_2) = (-1)^{\widetilde{t}_1} b(t_1, t_2)$. The mapping $b^\pi: \Pi T_1 \times \Pi T_2 \to A$ is a bilinear form of the same parity as b. Here is a verification that it is linear in the first argument:

$$\begin{aligned} b^\pi(a\Pi t_1, \Pi t_2) &= b^\pi((-1)^{\widetilde{a}}\Pi(at_1), \Pi t_2) = (-1)^{\widetilde{a} + \widetilde{at_1}} b(at_1, t_2) \\ &= (-1)^{\widetilde{t}_1 + \widetilde{ab}} ab(t_1, t_2) = (-1)^{\widetilde{ab}} ab^\pi(\Pi t_1, \Pi t_2). \end{aligned}$$

We have $(b^\pi)^\pi = b$.

3. Gram matrix. If T_1 and T_2 are free, with respective bases (e_i) and (e'_j) of homogeneous elements, then we can determine a bilinear form by its Gram matrix, which by definition has entries $b_{ij} = (-1)^{\widetilde{be_i}} b(e_i, e'_j)$. Naturally, the format of the matrix is defined by the parity of the e_i and the e'_j. Writing t_1 and t_2 is terms of left and right coordinates, respectively, we find

$$b(t_1,t_2) = b\left(\sum a_i e_i, \sum e'_j a'_j\right)$$
$$= \sum_i \sum_j (-1)^{\widetilde{ba_i}} a_i b\left(e_i, e'_j\right) a'_j = (-1)^{\widetilde{bt_1}} \sum_i \sum_j a_i b_{ij} a'_j \,.$$

Let B be the Gram matrix of the form b, written in the standard format $\left(\begin{smallmatrix} B_1 & B_2 \\ B_3 & B_4 \end{smallmatrix}\right)$. Then the matrix B^τ of the form b^τ, with respect to the same bases, has the following form:

$$B^\tau = \begin{cases} \begin{pmatrix} B_1^t & B_3^t \\ B_2^t & -B_4^t \end{pmatrix} & \text{for } \widetilde{B} = 0; \\ \begin{pmatrix} B_1^t & -B_3^t \\ -B_2^t & -B_4^t \end{pmatrix} & \text{for } \widetilde{B} = 1. \end{cases}$$

In a similar way, the matrix B^π of the form b^π, with respect to the bases πe_i and $\pi e'_j$ transformed to the standard format, has the form

$$B^\pi = \begin{cases} \begin{pmatrix} B_4 & B_3 \\ -B_2 & -B_1 \end{pmatrix} & \text{for } \widetilde{B} = 0; \\ \begin{pmatrix} -B_4 & -B_3 \\ B_2 & B_1 \end{pmatrix} & \text{for } \widetilde{B} = 1. \end{cases}$$

4. Symmetry conditions. Now let $T_1 = T_2 = T$. Then along with the form b, we also have the form b^τ on the same A-module T. As in the case of commutative algebra, we say that a form is symmetric or antisymmetric if $b^\tau = b$ or $b^\tau = -b$. Below we indicate in abbreviated form the four possible combinations of parity and symmetry for a Gram matrix:

Type O Sp (even symmetric)	$\begin{pmatrix} B_1 = B_1^t & B_2 \\ B_2^t & B_4 = -B_4^t \end{pmatrix}$,
Type Sp O (even antisymmetric)	$\begin{pmatrix} B_1 = -B_1^t & B_2 \\ -B_2^t & B_4 = B_4^t \end{pmatrix}$,
Type Π O (odd symmetric)	$\begin{pmatrix} B_1 = B_1^t & B_2 \\ -B_2^t & B_4 = -B_4^t \end{pmatrix}$,
Type Π Sp (odd antisymmetric)	$\begin{pmatrix} B_1 = -B_1^t & B_2 \\ B_2^t & B_4 = B_4^t \end{pmatrix}$.

If b satisfies one of these symmetry conditions, then b^π has the same parity but the opposite symmetry: for $b^\tau = \eta b$, with $\eta = \pm 1$, we have

$$b^{\pi\tau}(\Pi t_1, \Pi t') = (-1)^{(\widetilde{t}+1)(\widetilde{t'}+1)} b^\pi(\Pi t', \Pi t) = (-1)^{\widetilde{tt'}+\widetilde{t}+1} b(t', t)$$
$$= (-1)^{\widetilde{t}+1} b^\tau(t,t') = (-1)^{\widetilde{t}+1} \eta b(t,t') \ = (-1)^{\widetilde{t}+1} \eta b(t,t').$$

5. Isotropic submodules. Let T be a free A-module with a nondegenerate scalar product b. Let S be a direct submodule of T (distinguished as a direct summand) of rank $d = d_0|d_1$ and corank $c = c_0|c_1$. Then $S^\perp = \{ f \in T^* \mid S \subset \operatorname{Ker} f \}$ is a direct submodule of T^* of rank c and corank d, while $S_b^\perp = b^{-1}(S^\perp)$ is a direct submodule of T of the same rank and corank if b is even and Π-opposite rank and corank if b is odd. The submodule $S \subset T$ is called b-isotropic if $S \subset S_b^\perp$.

From this it follow that the maximal rank of an isotropic submodule can be $r|s$ for $\mathrm{O\,Sp}(2r|2s)$, $\mathrm{O\,Sp}(2r+1|2s)$ and $\Pi\,\mathrm{Sp}(r+s|r+s)$. This rank is attained for scalar products with the following Gram matrices:

$$\mathrm{O\,Sp}(2r|2s): \quad \left(\begin{array}{cc|cc} 0 & E_r & & \\ E_r & 0 & \multicolumn{2}{c}{0} \\ \hline & & 0 & E_s \\ \multicolumn{2}{c|}{0} & -E_s & 0 \end{array}\right),$$

$$\mathrm{O\,Sp}(2r+1|2s): \quad \left(\begin{array}{c|cc|cc} 1 & 0 & 0 & & \\ \cline{1-3} 0 & 0 & E_r & \multicolumn{2}{c}{0} \\ 0 & E_r & 0 & & \\ \hline & & & 0 & E_s \\ \multicolumn{3}{c|}{0} & -E_s & 0 \end{array}\right),$$

$$\Pi\,\mathrm{Sp}(r+s|r+s): \quad \left(\begin{array}{c|c} 0 & E_{r+s} \\ \hline E_{r+s} & 0 \end{array}\right).$$

The cases $\mathrm{Sp\,O}$ and $\Pi\,\mathrm{O}$ are obtained from these by applying the operation Π.

We will say that a scalar product admitting such a Gram matrix is split and the corresponding basis will be said to be splitting.

We observe that for the type $\Pi\,\mathrm{Sp}$ maximal isotropic submodules in T can have different ranks.

§ 6. Real Structures

1. Real structures in commutative algebra. In § 3 of Chapter 1, we constructed "real sections" of complex Minkowski space as the set of fixed points of an antiholomorphic involution. In § 1 of Chapter 2, it was observed that there are real structures without fixed points which arise naturally on complex manifolds. Nevertheless, an object such as the "real-analytic circle of radius i," $x^2 + y^2 = -1$ should be a space which is a full-fledged object in the appropriate category. In fact the absence of points here is the absence of $\mathbb{R}$-points; points "with values in other real-analytic manifolds" do exist on this circle in sufficient quantity. In supergeometry all these possibilities are retained and are typically supplemented with effects due to oddness.

In commutative algebra, these are the basic definitions.

(a) A real structure on a commutative C-algebra A is a C-antilinear isomorphism of "complex conjugation," $A \to A$, $a \mapsto a^\rho$ with the properties $a^{\rho\rho} = a$, $(ab)^\rho = a^\rho b^\rho$, and $(\alpha a)^\rho = \overline{\alpha} a^\rho$ for $\alpha \in \mathrm{C}$. We set $A^0 = \{a \mid a^\rho = a\}$. Then $A = \mathrm{C} \otimes_{\mathrm{R}} A^0 = A^0 \oplus A^0 i$, with $a = \frac{1}{2}(a + a^\rho) + \frac{1}{2}(a - a^\rho)$. Then $(a + bi)^\rho = a - bi$ for $a, b \in A^0$. Therefore, it is convenient to write $\overline{a}$ instead of a^ρ. Being given ρ is equivalent to being given the R-subalgebra A^0 of real elements.

(b) An extension of ρ to a real structure on an A-module S is a mapping $S \to S$, $s \mapsto s^\rho$, for which

$$(as)^\rho = a^\rho s^\rho, \quad s^{\rho\rho} = s.$$

As before, we set $S^0 = \{s \mid s^\rho = s\}$ and get $S = \mathrm{C} \otimes_{\mathrm{R}} S^0 = S^0 \oplus S^0 i$, where $(s + ti)^\rho = s - ti$ for all $s, t \in S^0$. Again it is natural to write $s^\rho = \overline{s}$. The real elements S^0 form an A^0-submodule of S.

(c) The extension of ρ to a quaternionic structure on an A-module S is a mapping $S \to S$, $s \mapsto s^\rho$ for which $(as)^\rho = a^\rho s^\rho$ and $s^{\rho\rho} = -s$. There are no real elements in S.

(d) If A is noncommutative, then it is natural to consider also a hermitian structure on A which changes the order of multiplication:

$$a^{\rho\rho} = a, \quad (\alpha a)^\rho = \overline{\alpha} a^\rho, \quad (ab)^\rho = b^\rho a^\rho.$$

In superalgebra we make the following definition:

2. Definition.

(a) *A (generalized) real structure of type* $(\epsilon_1, \epsilon_2, \epsilon_3)$, where $\epsilon_i = \pm 1$, on a Z_2-graded C-algebra A is an even R-linear mapping $A \to A$, $a \mapsto a^\rho$ with the properties:

$$(\alpha a)^\rho = \overline{\alpha} a^\rho \text{ for } \alpha \in \mathrm{C}; \qquad a^{\rho\rho} = (\epsilon_1)^{\tilde{a}} a;$$

$$(ab)^\rho = \epsilon_3 (\epsilon_2)^{\tilde{a}\tilde{b}} b^\rho a^\rho.$$

(b) *An extension of a real structure* ρ to an A-bimodule S is a homogeneous mapping $S \to S$, $s \mapsto s^\rho$ with the properties:

$$s^{\rho\rho} = \eta \left[(-\epsilon_2)^{\tilde{\rho}} \epsilon_1 \right]^{\tilde{s}} s; \quad (as)^\rho = \epsilon_3 (\epsilon_2)^{\tilde{a}\tilde{s}} s^\rho a^\rho;$$

$$(sa)^\rho = (-1)^{\tilde{\rho}\tilde{a}} \epsilon_3 (\epsilon_2)^{\tilde{s}\tilde{a}} a^\rho s^\rho.$$

The *type* of the extension is defined to be $(\eta, \widetilde{\rho})$. □

3. Remarks.

(a) If A is supercommutative, then for $\epsilon_3 = 1$ a real structure is induced on A_0, and the A_0-module A_1 is real or quaternionic depending on the sign of ϵ_1. Under conjugation, the order of multiplication of odd elements is either preserved or changed depending on whether $\epsilon_3\epsilon_2 = -1$ or $\epsilon_3\epsilon_2 = 1$. If A is a ring with identity, then $\epsilon_3 = 1^\rho$; therefore, most of the time $\epsilon_3 = 1$. but for Lie algebras a consideration of type $(-1,-1,-1)$ is necessary.

(b) It is natural to call an extension of ρ to S superreal or superquaternionic depending on the sign of η. If A is supercommutative, the pairs of structures induced on the A_0-modules S_0 and S_1 depend on the type of the extension for $\widetilde{\rho} = 0$; for $\widetilde{\rho} = 1$ of course ρ exchanges S_0 and S_1.

The choice of signs in remark (b) can be justified if to begin with one postulates identities of the form $s^{\rho\rho} = \eta_1(\eta_2)^{\widetilde{s}}s$ and $(as)^\rho = \eta_3(\eta_4)^{\widetilde{a}\widetilde{s}}s^\rho a^\rho$. Then one can express η_i in terms of ϵ_i using the identities $[a(bs)]^\rho = [(ab)s]^\rho$ and $(as)^{\rho\rho} = \eta_1(\eta_2)^{\widetilde{a}+\widetilde{s}}as$.

4. Derived real structures. Now let a real structure ρ be given on A of type $(\epsilon_1, \epsilon_2, \epsilon_3)$ and let it have extensions to the bimodules S and S' of types $(\eta, r = \widetilde{\rho})$ and $(\eta', r' = \widetilde{\rho}')$ respectively. The table on the following page contains information about the series of real structures which can be defined from the initial ones. Below we give several comments and computations relating to the table.

5. The structure ρ^{-1}.

On A the mapping ρ^2 is an automorphism of C-algebras which is the identity for $\epsilon_1 = 1$; $\rho^4 = id$ on A and S. The mapping $\rho^{-1} = \rho^3$ on A is always a real structure, while it is on S only when $r = 0$ or $r = 1$ and $\epsilon_2 = -1$; otherwise the identity for $(as)^\rho$ does not hold.

6. Derivations. Let $X\colon A \to S$ be a homogeneous derivation over C. We set $X^\rho a = (-\epsilon_2)^{\widetilde{X}\widetilde{a}}\left(Xa^{\rho-1}\right)^\rho$. We verify the Leibnitz formula for X^ρ:

$$X^\rho(ab) = (-\epsilon_2)^{\widetilde{X}\left(\widetilde{a}+\widetilde{b}\right)}\left(X\left(\epsilon_3\epsilon_2^{\widetilde{a}\widetilde{b}}b^{\rho^{-1}}a^{\rho^{-1}}\right)\right)^\rho$$

$$= \epsilon\left[X\left(b^{\rho^{-1}}a^{\rho^{-1}}\right)\right]^\rho, \quad \text{for } \epsilon = (-1)^{\widetilde{X}\left(\widetilde{a}+\widetilde{b}\right)}\epsilon_2^{\widetilde{X}\left(\widetilde{a}+\widetilde{b}\right)+\widetilde{a}\widetilde{b}}\epsilon_3 .$$

Further,

$$\left[X\left(b^{\rho^{-1}}a^{\rho^{-1}}\right)\right]^\rho = \left[Xb^{\rho^{-1}}\cdot a^{\rho^{-1}} + (-1)^{\widetilde{X}\widetilde{b}}b^{\rho^{-1}}Xa^{\rho^{-1}}\right]^\rho$$

$$= (-1)^{\widetilde{\rho}\widetilde{a}}\epsilon_3(\epsilon_2)^{(\widetilde{b}+\widetilde{X})\widetilde{a}}a\left(Xb^{\rho^{-1}}\right)^\rho + (-1)^{\widetilde{X}\widetilde{b}}\epsilon_3(\epsilon_2)^{\widetilde{b}(\widetilde{X}+\widetilde{a})}\left(Xa^{\rho^{-1}}\right)^\rho b.$$

Table of Derived Real Structures

On What	Definition	Type	Comments
A	ρ^{-1}	$(\epsilon_1, \epsilon_2, \epsilon_3)$	$\rho^{-1} = \rho$ for $\epsilon_1 = 1$
S	ρ^{-1}	(η, r)	Only for $(-\epsilon_2)^r = 1$
$\mathrm{Der}_{\mathbb{C}}(A, S)$	$X^\rho a = (-\epsilon_2)^{\widetilde{X}\widetilde{a}}(X a^{\rho^{-1}})^\rho$	$(1, r)$ (A-module)	$X^\rho \in \mathrm{Der}_{\mathbb{C}}(A, S)$ always; the axioms for a real structure are fulfilled if A is supercommutative and $r = 0$
$\mathrm{Der}_{\mathbb{C}}(A, A)$	"	$(\epsilon_1, \epsilon_2, 1)$ (Lie algebra)	"
$\Omega^1_{\mathrm{ev}} A$ $\Omega^1_{\mathrm{odd}} A$	$(da)^\rho = da^\rho$ $(da)^\rho = (-\epsilon_2)^{\widetilde{a}} da^\rho$	$(1, 0)$ $(\epsilon_1, 0)$	A supercommutative; the structure is defined by the condition $d^\rho = d$
$S \otimes S'$ $= S' \otimes S$	$(s \otimes t)^\rho$ $= (-1)^{r\widetilde{t}} \epsilon_3 (\epsilon_2)^{\widetilde{s}\widetilde{t}} t^\rho \otimes s^\rho$	$(\eta\eta'(-\epsilon_2) r r',$ $r + r')$	A supercommutative; otherwise $S \otimes S' \neq S' \otimes S$
$A[S]$	"	$(\epsilon_1, \epsilon_2, \epsilon_3)$	Only for $\eta = 1$, $r = 0$
$\Omega^{\cdot}_{\mathrm{odd}} A$	"	$(\epsilon_1, \epsilon_2, \epsilon_3)$	Only for $\epsilon_1 = 1$

Multiplying the right side by ϵ, we obtain

$$X^\rho(ab) = X^\rho a \cdot b + (-1)^{(\widetilde{X}+\widetilde{\rho})\widetilde{a}} a \cdot X^\rho b.$$

We verify the identity of a real structure for $(bX)^\rho$:

$$\begin{aligned}(bX)^\rho a &= (-\epsilon_2)^{(\widetilde{b}+\widetilde{X})\widetilde{a}} \left(b\left(X a^{\rho^{-1}}\right)\right)^\rho \\ &= (-\epsilon_2)^{(\widetilde{b}+\widetilde{X})\widetilde{a}} \epsilon_3 (\epsilon_2)^{\widetilde{b}(\widetilde{X}+\widetilde{a})} \left(X a^{\rho^{-1}}\right)^\rho \cdot b^\rho \\ &= \epsilon_3 (\epsilon_2)^{\widetilde{b}\widetilde{X}} (-1)^{\widetilde{b}\widetilde{a}} (X^\rho a)\, b^\rho.\end{aligned}$$

Therefore, $(bX)^\rho = \epsilon_3 (\epsilon_2)^{\widetilde{b}\widetilde{X}} X^\rho \cdot b^\rho$, if the ring A is supercommutative and on $\mathrm{Der}_{\mathbb{C}}(A, S)$ the right multiplication is compatible with the left. In an analogous manner, the identity for $(X \cdot b)^\rho$ can be verified in this case. Moreover,

$$X^{\rho^2}(a) = (-\epsilon_2)^{(\widetilde{X}+\widetilde{\rho})\widetilde{a}} \left(X^\rho a^{\rho^{-1}}\right)^\rho$$
$$= (-\epsilon_2)^{\widetilde{\rho}\,\widetilde{a}} \left(X a^{\rho^{-2}}\right)^{\rho^2} = (-\epsilon_2)^{\widetilde{\rho}\,\widetilde{a}} \epsilon_1^{\widetilde{X}} X a\,,$$

Therefore, $X^{\rho^2} = \epsilon_1^{\widetilde{X}} X$ when $\widetilde{\rho} = 0$.

Finally, we check that $[X,Y]^\rho = (\epsilon_2)^{\widetilde{X}\widetilde{Y}}[Y^\rho, X^\rho]$. We have

$$Y^\rho X^\rho a = (-\epsilon_2)^{\widetilde{X}\widetilde{a}+\widetilde{Y}\widetilde{a}+\widetilde{X}\widetilde{Y}} \left(Y X a^{\rho^{-1}}\right)^\rho;$$
$$[Y^\rho, X^\rho]\, a = (-\epsilon_2)^{\widetilde{X}\widetilde{a}+\widetilde{Y}\widetilde{a}+\widetilde{X}\widetilde{Y}} \left([Y,X]\, a^{\rho^{-1}}\right)^\rho = (-\epsilon_2)^{\widetilde{X}\widetilde{Y}}\, [Y,X]^\rho\, a\,.$$

7. Differentials. Let A be supercommutative, and let $d\colon A \to \Omega^1 A$ be the universal superderivation over C, even or odd. On $\Omega^1 A$ there is no more than one extension ρ which satisfies the conditions $\widetilde{\rho} = 0$ and $d^\rho = d$. In fact, from $db = d^\rho b = (-\epsilon_2)^{\widetilde{d}\cdot\widetilde{a}} \left(d b^{\rho^{-1}}\right)^\rho$, setting $b = a^\rho$ we find $(da)^\rho = (-\epsilon_2)^{\widetilde{d}\cdot\widetilde{a}} d(a^\rho)$, which defines the action of ρ on all the generators of $\Omega^1 A$. (The same reasoning applies to the ring of smooth functions A.) The existence of this extension can be checked mechanically.

8. Tensor products. For $s \in S$ and $t \in S'$ we set

$$(s \otimes t)^\rho = (-1)^{r\widetilde{t}} \epsilon_3 (\epsilon_2)^{\widetilde{s}\,\widetilde{t}}\, t^{\rho'} \otimes s^\rho\,.$$

We again suppose that A is supercommutative and identify $S \otimes S'$ with $S' \otimes S$. Instead of ρ' we will write ρ for simplicity. Here are several verifications:

$$(s \otimes at)^\rho = (-1)^{r(\widetilde{t}+\widetilde{a})} (\epsilon_2)^{\widetilde{s}\,\widetilde{a}+\widetilde{s}\,\widetilde{t}+\widetilde{a}\,\widetilde{t}}\, t^\rho a^\rho \otimes s^\rho = (sa \otimes t)^\rho\,,$$
$$(s \otimes t)^{\rho\rho} = (-1)^{r\widetilde{t}} \epsilon_3 (\epsilon_2)^{\widetilde{a}\,\widetilde{t}} (t^\rho \otimes s^\rho)^\rho$$
$$= (-1)^{r\widetilde{t}+r'(\widetilde{s}+r)} (\epsilon_2)^{r'\widetilde{s}+r\widetilde{t}+rr'}\, \eta \left[(-\epsilon_2)^r \epsilon_1\right]^{\widetilde{s}} \eta' \left[(-\epsilon_2)^{r'} \epsilon_1\right]^{\widetilde{t}} s \otimes t$$
$$= \eta\eta'(-\epsilon_2)^{rr'} \left[(-\epsilon_2)^{r+r'} \epsilon_1\right]^{\widetilde{s}+\widetilde{t}} s \otimes t\,.$$

9. Geometric language. Let A be the ring of holomorphic functions on a complex supermanifold M; let's say it is Stein so that its points are separated by functions. (We are getting a little ahead of ourselves; for the precise definitions, look in the next chapter.) We will say that a real structure on A of type $(\epsilon_1, \epsilon_2, 1)$ is also a real structure on M. Let B be yet another supercommutative ring with a real structure of the same type. We define a B-point of M to be a homomorphism of C- algebras $x: A \to B$. We call a homomorphism which commutes with ρ a real B-point.

Let $f \in A$; then f can be considered to be a function on the B-points with values in B by setting $f(x) = x(f)$. We will denote conjugation in B by a bar. Then the condition of reality for a point x, i.e., $x(f^\rho) = \overline{x(f)}$, can be written in the form

$$f^\rho(x) = \overline{f(x)}.$$

This establishes a relation between our definitions and the standard even variant in § 1 of Chapter 2. Applications to Minkowski superspace are described in § 1 of Chapter 5.

10. Adjacency structures. Let A be a supercommutative ring. The functors Π and $*$ are natural "discrete symmetries" in the category of A-modules. If A is provided with a real structure ρ with $\epsilon_3 = 1$, then one can treat its extension to an A-module S in an analogous manner. More precisely, from A and S we define the ring A^{t} and the A^{t} module S^{t}, setting $A_i^{\mathrm{t}} = A_i$ and $S_i^{\mathrm{t}} = S_i$; $a^{\mathrm{t}} b^{\mathrm{t}} = (\epsilon_2)^{\tilde{a}\tilde{b}} (ba)^{\mathrm{t}}$, and $a^{\mathrm{t}} s^{\mathrm{t}} = (\epsilon_2)^{\tilde{a}\tilde{s}} (sa)^{\mathrm{t}}$. Then ρ on A can be viewed as a C-antilinear isomorphism $\rho: A \to A^{\mathrm{t}}$, and the extension of ρ to S can be viewed as an isomorphism of A-modules, $S \to \rho^*(S^{\mathrm{t}})$. We call the modules ΠS, S^* and $\rho^*(S^{\mathrm{t}})$, and also the modules obtained by applying these operations several times to S, *adjacent* to S. An *adjacency structure* on S is given by defining isomorphisms of S with the adjacent modules. The most important class of adjacency structures consists of those which have a large group of automorphisms; Lie supergroups of classical type can also be defined as such groups of automorphisms. Among the structures considered earlier, the following are adjacency structures: $b: S \to S^*$ (an even bilinear form), $b: S \to \Pi s^*$ (an odd bilinear form), $\rho: S \to \rho^*(S^{\mathrm{t}})$ (a real structure). Specifically for superalgebras, "Π-symmetry" is an adjacency structure $p: S \to \Pi S$.

References for Chapter 3

The "rule of signs" was of course well known in topology and homological algebra. F. A. Berezin was one of the first to systematically study supercommutative algebra as a generalization of commutative algebra (see [13]); his invention of the superdeterminant was an important discovery. For various aspects of superalgebra, see [68] and [21]; the fundamental work of Kac [61] contains the classification of simple finite-dimensional Lie superalgebras, which are enumerated in § 10 of the next chapter; this was recounted again in [95].

CHAPTER 4

INTRODUCTION TO SUPERGEOMETRY

The basic objects in supergeometry are the superspaces, "spaces on which some of the coordinates commute and the rest anticommute." A precise definition of superspaces and their mappings is given in terms of sheaves. This chapter is devoted to the exposition of the basic facts about superspaces. In § 1 the fundamental definitions are introduced, especially the concept of a supermanifold, a space with local systems of independent coordinate functions. In § 2 it is explained how supermanifolds and superfibrations are classified up to isomorphism in the smooth category. The obstructions to the validity of the analogous theorems for complex manifolds are cohomological invariants. In § 3 flag superspaces are introduced; these are important in their own right and also provide good illustrations of the general theory. In § 4 are gathered the obvious supergeometric definitions in the theory of integrability, including the Frobenius theorem and connections. In § 5 objects specific to supergeometry are introduced—integral forms. The complex of integral forms is constructed using a canonical right connection on the sheaf of Berezin volume forms of the supermanifold; the idea of a right connection on an arbitrary vector superbundle is also defined and studied here. Integration on supermanifolds is the subject of §§ 6–9. The fundamental idea here is the Berezin integral of a volume form with compact support on a supermanifold; this is defined in § 6. Various derived constructions which permit one to define integrals over immersed supermanifolds of any dimension are described in §§ 7 and 9. The Stokes formula for integral forms is proved in § 8. Finally, in § 10 there is a list of finite-dimensional simple Lie superalgebras over the field of complex numbers.

§ 1. Superspaces and Supermanifolds

1. Definition. A *superspace* is defined to be a pair $(M, \mathcal{O}_M)$ consisting of a topological space M and a sheaf of supercommutative rings $\mathcal{O}_M$ on it such that the stalk $\mathcal{O}_x = \mathcal{O}_{M,x}$ at any point $x \in M$ is a local ring. □

In the definitions, of course we imply that all the restriction morphisms of the structure sheaf $\mathcal{O}_M$ are compatible with the grading. From now on, all morphisms of structure sheaves, sheaves of modules over them, tensor products, etc., will automatically be assumed to obey the rule of signs and the other conventions of

Chapter 3. Sometimes we will not mention $\mathcal{O}_M$ explicitly but will denote the superspace simply by M. On an open subset $U \subset M$ the structure of a superspace $(U, \mathcal{O}_M|U)$ is induced; we will call this an open sub(super)space of M.

The objects of the fundamental geometric categories—differential and analytic manifolds, analytic spaces and schemes—are all superspaces with $\mathcal{O}_{M,0} = \mathcal{O}_M$ and $\mathcal{O}_{M,1} = \{0\}$. We will call such superspaces *purely even*. The stalk of the structure sheaf $\mathcal{O}_x$ at the point x on a purely even manifold M consists of germs of functions defined near x; the maximal ideal $\mathfrak{m}_x \subset \mathcal{O}_x$ consists of the germs of functions which vanish at the point x. On a general superspace $\mathfrak{m}_x$ consists of the germs of all the nilpotent sections of the structure sheaf, in particular all of the odd sections.

2. Definition. A *morphism of superspaces* $(M, \mathcal{O}_M) \to (N, \mathcal{O}_N)$ is defined to be a pair (ϕ, ψ), where $\phi: M \to N$ is a continuous mapping of topological spaces and $\psi: \mathcal{O}_N \to \phi_*(\mathcal{O}_M)$ is a morphism of sheaves of rings, such that for any point $x \in M$ the homomorphism $\psi_x: \mathcal{O}_{N,\phi(x)} \to \mathcal{O}_{M,x}$ is local: $\psi_x(\mathfrak{m}_{\phi(x)}) \subset \mathfrak{m}_x$. □

For those superspaces for which the structure sheaf consists of germs of genuine functions which separate points, there is no necessity to specify both ϕ and ψ separately. If ϕ is given, then ψ can be reconstructed from it as the pullback of (local) functions by the formula $\psi_x(f)(x) = f(\phi(x))$. If ψ is given, then ϕ can be reconstructed from it since any point of M or N can identified by its local coordinates.

In the general case, sections of the structure sheaf cannot be implicity identified with functions; for example, nilpotents (on an infinitesimal neighborhood of an analytic space) are identically equal to zero as functions on points. This effect constantly manifests itself in the general theory of superspaces, where the odd sections of the structure sheaf are nilpotent. In algebraic geometry, to the extent that $s \in \Gamma(\mathcal{O}_M)$ is a function, it takes its values in a field which depends on the point: $s(x) \in k(x) = \mathcal{O}_x/\mathfrak{m}_x$. The axiom of locality for the mapping ψ is the minimal condition providing the analog of the formula $\psi(s)(x) = s(\phi(x))$ when both sides are equal to zero. From this for differential and analytic manifolds it is not difficult to deduce that this formula is true in the general case provided ϕ is the identity map on the constant functions, i.e., technically speaking, it is a morphism over $\mathbb{R}$ or $\mathbb{C}$.

3. Some constructions. Let $(M, \mathcal{O}_M)$ be s superspace. Then $(M, \mathcal{O}_{M,0})$ is a purely even superspace and the pair $(\mathrm{id}_M, \mathcal{O}_{M,0} \hookrightarrow \mathcal{O}_M)$ is the projection of $(M, \mathcal{O}_M)$ onto its "even quotient."

We introduce the permanent notation $J_M = \mathcal{O}_{M,1} + (\mathcal{O}_{M,1})^2$. This is a sheaf of ideals in $\mathcal{O}_M$. We set $\mathrm{Gr}_i\, \mathcal{O}_M = J_M^i / J_M^{i+1}$. Then $\mathrm{Gr}_0\, \mathcal{O}_M = \mathcal{O}_M / J_M$ is a sheaf of rings, and $(M, \mathrm{Gr}_0\, \mathcal{O}_M)$ is a purely even superspace which we denote by M_{rd}. The pair $(\mathrm{id}_M, \mathcal{O}_M \to \mathcal{O}_M / J_M)$ is a closed imbedding $M_{\mathrm{rd}} \to M$.

Sometimes it is also useful to consider $M_{\rm red} = (M, \mathcal{O}_M/\mathcal{N})$, where $\mathcal{N}$ is the sheaf of all nilpotents in the structure sheaf. For supermanifolds, with which we will mostly be concerned below, $M_{\rm rd} = M_{\rm red}$.

More information about M than in $M_{\rm rd}$ is contained in the ringed space

$$\operatorname{Gr} M = (M, \operatorname{Gr} \mathcal{O}_M) = (M, \textstyle\bigoplus_{i \geq 0} \operatorname{Gr}_i M).$$

Its structure sheaf is $\mathbb{Z}$-graded. If we consider $\operatorname{Gr} M$ to be a superspace, then the $\mathbb{Z}_2$-grading is always taken to be the one obtained by reducing the $\mathbb{Z}$-grading modulo 2.

Let $\mathcal{E}$ be a sheaf of $\mathcal{O}_M$-modules. In a similar way we set $\operatorname{Gr} \mathcal{E} = \bigoplus_{i \geq 0} \operatorname{Gr}_i \mathcal{E}$ and $\operatorname{Gr}_i \mathcal{E} = \mathcal{E} J_M^i / \mathcal{E} J_M^{i+1}$. It often turns out that it is easier to compute $\operatorname{Gr} M$ and $\operatorname{Gr} \mathcal{E}$ than M and $\mathcal{E}$; and in any case, information about the graded objects is extremely useful. In physics the corresponding procedure is called component analysis of superfields and dynamical equations.

4. Definition. (a) A superspace $(M, \mathcal{O}_M)$ is said to be *decomposable over* M^0 if it is isomorphic to $(M^0, S_{\mathcal{O}_{M^0}}(\mathcal{E}))$, where $(M^0, \mathcal{O}_{M^0})$ is a purely even superspace and $\mathcal{E}$ is a locally free sheaf of $\mathcal{O}_{M^0}$-modules of purely odd rank $0|q$.

(b) A superspace $(M, \mathcal{O}_M)$ is said to be *decomposable* if it is decomposable over $M_{\rm rd}$.

(c) A superspace $(M, \mathcal{O}_M)$ is said to be *locally decomposable* if for every point $x \in M$ there is an open neighborhood such that $(U, \mathcal{O}_M|U)$ is decomposable. □

We remark that if M is decomposable, then in addition to the structural imbedding $M_{\rm rd} \to M$ there is also a projection $M \to M_{\rm rd}$: $(\mathrm{id}_M, \mathcal{O}_M/J_M = \mathcal{O}_{M_{\rm rd}} \hookrightarrow S_{\mathcal{O}_{M_{\rm rd}}}(\mathcal{E}))$.

Now we introduce the class of superspaces which is most important for the rest of the book, the supermanifolds.

5. Definition. A *supermanifold—differentiable, analytic* or *algebraic*—is defined to be a locally decomposable superspace $(M, \mathcal{O}_M)$ such that $M_{\rm rd}$ is isomorphic to an ordinary (purely even) manifold of the appropriate class, i.e., a Hausdorff space with countable basis and a sheaf of functions locally isomorphic to a domain in $\mathbb{R}^n$ with the real differentiable functions or a domain in $\mathbb{C}^n$ with the complex analytic functions. □

Manifolds differ from general purely even spaces in that for every point of the manifold there is a neighborhood possessing a finite system of local independent coordinates $(x^1, \dots, x^m)$ (we will not pause to make the additional comments needed for the algebraic case). Similarly, supermanifolds can be represented as superspaces with local systems of independent even-odd coordinates $(x^1, \dots, x^m, \xi^1, \dots, \xi^n)$.

For the (ξ^i) one must take a local basis of the sections of the sheaf realizing the local decomposition. The definition of decomposibility means that among the x^i and the ξ^j there are no other relations besides the requirements of supercommutativity. Locally, any (super)function on M, by definition a section of the structure sheaf, can be represented in the form $f = \sum_\alpha f_\alpha(x)\xi^\alpha$, and the representation is unique. Here the $f_\alpha(x)$ are local functions on M_{rd} with $\xi^\alpha = (\xi^1)^{\alpha_1}\cdots(\xi^n)^{\alpha_n}$ as usual. The value of f at the point $x \in M$ is defined to be $f \bmod \mathfrak{m}_x$. An (irreducible) supermanifold has an invariantly defined (super)dimension $m|n$.

We will encounter the most important superspaces which are not supermanifolds as the result of the following construction: let $(M, \mathcal{O}_M)$ be a supermanifold of dimension $m|n$; then for $i < n$ the superspace $M^{(i)} = (M, \mathcal{O}_M/J_M^{i+1})$ is not a supermanifold. This superspace is the i-th infinitesimal neighborhood of the underlying space. Analogously, one can construct neighborhoods of closed subsupermanifolds in more general situations. Thus the whole manifold M is an "infinitesimal neighborhood of M_{rd} in the odd directions."

In the analytic and algebraic categories, a reasonable, general class of superspaces which are not necessarily supermanifolds is introduced by the following definition.

6. Definition. An *analytic superspace* (respectively, an *algebraic superspace*, *superscheme*) is defined to be a superspace $(M, \mathcal{O}_M)$ such that $(M, \mathcal{O}_{M,0})$ is an ordinary analytic space (respectively, algebraic space, scheme), while $\mathcal{O}_{M,1}$ is a coherent sheaf of $\mathcal{O}_{M,0}$-modules. □

The neighborhood $M^{(i)}$ belongs to this class when M does.

7. Morphisms and local coordinates. We return to the case of supermanifolds. They are pasted together from superdomains, and our first goal is to show that the morphisms of pasting, as well as general morphisms, can be written in local coordinates and that we can work with them just as in the ordinary geometric categories.

More concretely, let $U^0 \subset \mathbf{R}^m$ or $U^0 \subset \mathbf{C}^m$ be a connected domain with coordinate functions $(x^1, \ldots, x^m)$ and with $\xi^1, \ldots, \xi^n$ a basis of the sections of the trivial sheaf of rank $0|n$ on the domain. We assume that $\mathcal{O}_{U^0}$ is the sheaf of C^∞ functions in the first case and holomorphic functions in the second. We call the superspace $U^{n|m} = (U^0, \mathcal{O}_{U^0}[\xi^1, \ldots, \xi^n])$ respectively a differentiable or a complex analytic superdomain, with the coordinate system (x, ξ) on it. Let $V^{p|q}$ be another superdomain of the same type as $U^{p|q}$, with coordinate system $(y^1, \ldots, y^p; \eta^1, \ldots, \eta^q)$.

8. Proposition. *There are natural bijections between the following sets:*

(a) $\mathrm{Hom}(U, V)$ *in the category of superspaces.*

(b) $\mathrm{Hom}(\mathcal{O}(V), \mathcal{O}(U))$ *(where* $\mathcal{O}(U) = \Gamma(U, \mathcal{O}_U)$*) in the category of commutative superalgebras (over* $\mathbf{R}$ *or* $\mathbf{C}$*).*

(c) *The subset of* $\mathcal{O}_0(U)^p \oplus \mathcal{O}_1(U)^q$*, where* $p|q = \dim V$*, consisting of those tuples of functions* $(y^i(x,\xi), \eta^j(x,\xi))$ *such that for* $x \in U^0$ *we have* $y(x,0) \in V^0$*. The morphism* $(\phi, \psi): U \to V$ *defined by the preimages* $(y^i(x,\xi), \eta^j(x,\xi))$ *of the coordinate functions of* V *in the ring* $\mathcal{O}(U)$ *is a local isomorphism in a neighborhood* U' *of a point of* U *if and only if at this point or, equivalently, in a neighborhood of it the matrix of partial derivatives is invertible:*

$$J(y,x) = \begin{pmatrix} \dfrac{\partial y^i}{\partial x^k} & \dfrac{\partial \eta^j}{\partial x^k} \\ \dfrac{\partial y^i}{\partial \xi^l} & \dfrac{\partial \eta^j}{\partial \xi^l} \end{pmatrix}.$$

Outline of proof. The mapping (a) $\to$ (b) associates $(\phi, \psi): U \to V$ to the ring morphism of functions $\phi^* = \Gamma(\psi): \Gamma(V, \mathcal{O}_V) \to \Gamma(V, \phi_*(\mathcal{O}_U)) = \Gamma(U, \mathcal{O}_U)$. The mapping (b) $\to$ (c) associates the homomorphism $\phi^*: \mathcal{O}(V) \to \mathcal{O}(U)$ to the image of the coordinate system $(\phi^*(y^i), \phi^*(\eta^j))$.

To prove the bijectivity of both mappings, one must use the fact that at every point functions on superdomains have Taylor series and that a function on $U^{m|n}$ is defined by its truncated Taylor series of degree $\leq n$ at each point. More precisely, we set $\widetilde{\mathfrak{m}}_x \subset \mathcal{O}(U)$ to be the ideal of functions which vanish at the point $x \in U^0$. Then the following facts are true:

(1) $\widetilde{\mathfrak{m}}_x = (x^i - x_0^i; \xi^j)$.

(2) $\mathcal{O}(U)/\widetilde{\mathfrak{m}}_x^k \simeq K\left[x^i - x_0^i; \xi^j\right] \bmod \left((x - x_0)^\alpha \xi^\beta : |\alpha| + |\beta| \geq k\right)$, where $K = \mathbf{R}$ or $\mathbf{C}$.

(3) $\bigcap_{x \in U^0} \widetilde{\mathfrak{m}}_x^{n+1} = \{0\}$. These follow immediately from the corresponding assertions for $n = 0$.

Consequently, the equation $\phi^* = \Gamma(\psi)$ uniquely determines $\phi: U^0 \to V^0$ (on the maximal ideals of the function ring) and also a mapping ψ on all possible Taylor polynomials and so on the whole structure sheaf. This means that (a) $\to$ (b) is an injection. In a similar way, we see that (b) $\to$ (c) is an injection; for knowing $\phi^*(y^i)$ and $\phi^*(\eta^j)$ means that we know the value of ϕ^* on the Taylor polynomials.

Finally, to prove the surjectivity of these mappings, the essential point is that we can explicitly define how to extend the mapping to all differentiable or analytic functions, when the mapping is initially defined only on the coordinates. Let $f(y^i, \eta^j) = \sum f_\alpha(y)\eta^\alpha$; we wish to substitute in this $\phi^*(y) = y(x,\xi)$ and $\phi^*(\eta) = \eta(x,\xi)$ in order to compute $\phi^*(f)$. It is sufficient to define $\phi^*(f_\alpha(y))$. We write $\phi^*(y^i) = y_0^i(x) + y_1^i(x,\xi)$, where y_0^i does not contain ξ explicitly and y_1^i is nilpotent. Then we set by definition

$$f_\alpha(y_0 + y_1) = \sum_\beta \frac{1}{\beta!} \frac{\partial^\beta}{\partial y^\beta} f_\alpha(y_0) y_1^\beta .$$

The correctness of this can be verified without difficulty.

We observe that such use of the Taylor series makes it possible to work with even-odd systems of coordinates and with morphisms which intermix coordinates as freely as in the purely even case. But if we wished to define, let us say, $f(x+\xi_1\xi_2)$ for a continuous but nondifferentiable function f by the formula $f(x) + f'(x)\xi_1\xi_2$, we would have to introduce generalized superfunctions.

Finally, we turn to the last assertion. The matrix of partial derivatives given in the proposition is the right transition matrix for the differentials: $(dy, d\xi) = (dx, d\xi)J(y,x)$. If both (x,ξ) and (y,η) form local systems of coordinates in a neighborhood of a point, then their differentials form two bases of the module Ω^1; therefore, the transition matrix is invertible.

Conversely, if the matrix is invertible, then we may first define $\phi_0^{-1}: V'_{\mathrm{rd}} \xrightarrow{\sim} U'_{\mathrm{rd}}$ using the classical implicit function theorem and the invertibility of $(\partial y^i/\partial x^k)$. Then ϕ_0^{-1} can be extended to an isomorphism $\phi_1^{-1}: V'^{(1)} \xrightarrow{\sim} U'^{(1)}$ of the first neighborhoods using the invertibility of $(\partial \eta^j/\partial \xi^k)$. Finally, we can check the solvability of the problem of extending $\phi_k^{-1}: V'^{(k)} \to U'^{(k)}$ to an isomorphism of the $(k+1)$-st neighborhood. We leave the details to the reader. □

9. Locally free sheaves and vector bundles. Let $(M, \mathcal{O}_M)$ be a superspace. A locally free sheaf of rank $p|q$ on M is defined to be a sheaf of $\mathcal{O}_M$-modules S ($\mathbb{Z}_2$-graded, of course) which is locally isomorphic to $\mathcal{O}_M{}^{p|q} = \mathcal{O}_M{}^p \oplus (\Pi\mathcal{O}_M)^q$.

We will use this idea as a complete replacement for the idea of a vector superbundle over M. The analog of a vector subbundle is not an arbitrary subsheaf but rather a locally direct subsheaf $S' \subset S$, i.e., it can be locally distinguished as a direct summand.

All the formalism of tensor algebra, the functor Π, bilinear forms and so forth can be transferred to general quasi-coherent sheaves of $\mathcal{O}_M$-modules and in particular to locally free sheaves.

When it is necessary to consider a vector bundle as a superspace in one of the standard categories, we do the following. The total superspace N corresponding to the sheaf S is defined along with the projection $\pi: N \to M$ and a special class of charts. Each open set $U \subset M$ over which S is trivial defines a subset $\pi^{-1}(U) \subset N$ such that $\Gamma(\pi^{-1}(U), \mathcal{O}_n) \supset S_{\Gamma(U,\mathcal{O}_M)}(\Gamma(U, S^*))$. In other words, the linear functionals on S, the sections of S^*, become the coordinate functions on N along with the functions pulled back from M. These functions are distinguished; they define on $\pi: N \to M$ the structure of a vector bundle. In the category of superschemes it is natural to write simply $N = \mathrm{Spec}_{\mathcal{O}_M} S_{\mathcal{O}_M}(\mathcal{S}^*)$. It is clear that to produce the whole sheaf $\mathcal{O}_N$, one must add the functions analytic along the fibers to the functions "polynomial along the fibers."

10. Tangent and cotangent sheaves. Let M be a supermanifold, either differentiable or analytic. We denote by $\mathcal{T}M$ the sheaf of its local vector fields, i.e, the (super)derivations (over $\mathbf{R}$ or $\mathbf{C}$) of the ring of functions. This sheaf is locally free of rank equal to the dimension of M. In a domain with local coordinates (x^i, ξ^j), the sheaf $\mathcal{T}M$ is freely generated by its sections $\left(\partial/\partial x^i, \partial/\partial \xi^j\right)$.

The cotangent sheaf $\Omega^1 M$ can be defined in two ways: as the sheaf of morphisms $\Omega^1_{\mathrm{ev}} = (\mathcal{T}M)^* = \mathcal{H}om_{\mathcal{O}_M}(\mathcal{T}M, \mathcal{O}_M)$ or as $\Omega^1_{\mathrm{odd}} M = \mathcal{H}om_{\mathcal{O}_M}(\mathcal{T}M, \Pi\mathcal{O}_M)$. There is a differential $d: \mathcal{O}_M \to \Omega^1 M$ for which $(X, df) = Xf$ (or ΠXf). As in § 4 of Chapter 1, the de Rham complexes for supermanifolds can be constructed.

As in the classical case, a morphism of supermanifolds $\pi = (\phi, \psi): M \to N$ defines a morphism of sheaves $\mathcal{T}M \xrightarrow{d\pi} \pi^*(\mathcal{T}N)$. More precisely, a (local) vector field X on M, when acting on functions lifted from N defines a derivation on N with values in the ring of functions on M. This gives a morphism of $\pi^{-1}(\mathcal{O}_N)$-modules $\pi^{-1}(\Omega^1 N) \to \mathcal{O}_M$ or, equivalently, a morphism of $\mathcal{O}_M$-modules $\pi^*(\Omega^1 N) \to \mathcal{O}_M$ or finally, a section of $\pi^*(\mathcal{T}N)$ which is said to be the image of X with respect to $d\pi$.

We set $\mathcal{T}M/N = \ker d\pi$ and $\Omega^1_{\mathrm{ev}}(M/N) = (\mathcal{T}M/N)^*$. Thus, as in ordinary geometry we obtain complexes of $\mathcal{O}_M$-modules:

$$0 \to \mathcal{T}M/N \to \mathcal{T}M \xrightarrow{d\pi} \pi^*(\mathcal{T}N),$$
$$0 \leftarrow \Omega^1_{\mathrm{ev}} M/N \xleftarrow{\mathrm{res}} \Omega^1_{\mathrm{ev}} M \leftarrow \pi^*\left(\Omega^1_{\mathrm{ev}} N\right).$$

Here res is the restriction of the differential forms to the vector fields which are vertical with respect to π.

In the category of supermanifolds there are direct products; the construction is evident for superdomains and it carries over to the general case by pasting.

11. Definition. *A morphism of supermanifolds, $\pi: M \to N$ is called:*

(a) *an immersion if it is locally isomorphic to a morphism of superdomains of the form $i: U \to U \times V$ with $i^*(u^i) = u^i$ and $i^*(v^j) = 0$, where (u^i), (v^j) are even-odd systems of coordinates in U and V respectively;*

(b) *a submersion if it is locally isomorphic to a morphism of superdomains of the form $p: U \times V \to V$, where $p^*(v^a) = v^a$ in the notation of (a).* □

12. Proposition. *In the notation of subsection 10, the following statements are true.*

(a) *A morphism* $\pi: M \to N$ *is an immersion if and only if* $\mathcal{T}M/N = 0$ *and* $d\pi$ *is a locally direct injection of sheaves.*

(b) *A morphism* $\pi: M \to N$ *is a submersion if and only if* $d\pi$ *is a surjective morphism.* □

The proof closely follows the usual reasoning in the purely even case, since we already know how morphisms and isomorphisms are written in terms of local coordinates. We omit the details.

§ 2. The Elementary Structure Theory of Supermanifolds

1. Basic invariants. Let $(M, \mathcal{O}_M)$ be a supermanifold, either differentiable or analytic; and let $J_M = \mathcal{O}_{M,1} + \mathcal{O}_{M,1}{}^2$ and $\mathcal{F} = J_M/J_M^2$. Let $\dim M = m|n$; we view $\mathcal{F}$ as a locally free sheaf of rank $0|n$ on $M_{\mathrm{rd}} = (M, \mathcal{O}_{M_{\mathrm{rd}}})$, where $\mathcal{O}_{M_{\mathrm{rd}}} = \mathrm{Gr}_0\, \mathcal{O}_M = \mathcal{O}_M/J_M$. The successive quotients of the filtration of the structure sheaf by the ideals $\mathcal{O}_M \supset J_M \supset J_M^2 \supset \cdots \supset J_M^{n+1} = \{0\}$ are $\mathrm{Gr}_i\, \mathcal{O}_M = S^i(\mathcal{F})$ (the symmetric powers over $\mathcal{O}_{M_{\mathrm{rd}}}$); thus locally on M, by definition, the sheaves $\mathcal{O}_M$ and $\mathrm{Gr}\, \mathcal{O}_M$ are isomorphic. Therefore, the distinction between $\mathcal{O}_M$ and $\mathrm{Gr}\, \mathcal{O}_M$ is related globally to some cohomological invariant. However, for $n > 1$ the associated sheaf of coefficients is a sheaf of noncommutative groups. It has a sequence of normal subgroups with commutative quotients. The goal of this section is to introduce very simple "characteristic classes" which distinguish M from $\mathrm{Gr}\, M$. They can be nontrivial only in the analytic case; for the differentiable case they are zero.

2. Theorem. *Any differentiable supermanifold M is isomorphic (not canonically) to* $\mathrm{Gr}\, M$.

The proof uses the following fact.

3. Lemma. *Any locally free sheaf* $\mathcal{E}$ *on a differentiable supermanifold M is fine. Therefore,* $H^q(M, \mathcal{E}) = 0$ *for* $q \geq 1$. □

The lemma follows immediately from the existence of a partition of unity, just as in the case of ordinary manifolds. In order to establish Theorem 2, we will only need the last case, which is well known.

The supermanifold $\mathrm{Gr}\, M$ admits a projection $\mathrm{Gr}\, M \to M_{\mathrm{rd}}$ which is the identity on $M_{\mathrm{rd}} \subset \mathrm{Gr}\, M$. This corresponds to the pair $(\mathrm{id}_M, \mathrm{Gr}_0\, \mathcal{O}_M \hookrightarrow \mathrm{Gr}\, \mathcal{O}_M)$. The following lemma proves that the cohomological obstruction to the existence of such a projection $M \to M_{\mathrm{rd}}$ is absent in the differentiable case.

4. Lemma. *There exists a morphism* $\phi: M \to M_{\mathrm{rd}}$ *which is the identity on* M_{rd}.

Proof. As in § 1 we set $M^{(i)} = \left(M, \mathcal{O}_M/J_M^{i+1}\right)$; thus $M^{(i)} \subset M^{(i+1)}$ (a closed imbedding). We set $\phi^{(0)} = \mathrm{id}: M^{(0)} = M_{\mathrm{rd}} \to M_{\mathrm{rd}}$. Now we will construct by induction on i mappings $\phi^{(i+1)}: M^{(i+1)} \to M_{\rm rd}$ such that $\phi^{(i+1)}$ is an extension of $\phi^{(i)}$. The essential point is the construction of an obstruction

$$\omega(\phi^{(i)}) \in H^1\left(M_{\rm rd}, (\mathcal{T} M_{\rm rd} \otimes S^{i+1}(\mathcal{F}))_0\right)$$

such that if $\omega(\phi^{(i)}) = 0$ then $\phi^{(i+1)}$ exists. By Lemma 3, in the differentiable case $\omega(\phi^{(i)}) = 0$.

We construct a locally–finite cover for M such that on elements U of the cover $\phi^{(i)}$ extends to $\phi_U^{(i+1)}$. We represent the $\phi_U^{(i+1)}$ as injections of rings of functions: $\phi_U^{(i+1)}: \mathcal{O}(U_{\rm rd}) \to \mathcal{O}(U^{(i+1)})$ such that their reduction mod J_M^{i+1} agrees with the restriction of $\phi^{(i)}$ to U. We set

$$\begin{aligned} \omega_{UV} = \phi_V^{(i+1)}|_{U\cap V} - \phi_U^{(i+1)}|U \cap V\colon \\ \mathcal{O}_{M_{\rm rd}}(U \cap V) \to \Gamma\left(U \cap V, S^{i+1}(\mathcal{F})\right) . \end{aligned}$$

It is not difficult to see that ω_{UV} is an even derivation of the ring $\mathcal{O}_{M_{\rm rd}}(U \cap V)$ and therefore can be identified with a section of the sheaf $(\mathcal{T} M_{\rm rd} \otimes S^{i+1}(\mathcal{F}))_0$. The family of sections (ω_{UV}) forms a Čech cocycle. If the extensions $\phi_U^{(i+1)}$ are replaced by different choices, the cocycle is changed by a coboundary; conversely, any addition of a coboundary may be obtained in this way. Therefore, the existence of a consistent system $(\phi_U^{(i+1)})$ is equivalent to the vanishing of the cohomology class (ω_{UV}). □

5. Proof of Theorem 2. Let us fix a morphism $\phi: M \to M_{\rm rd}$ as in Lemma 4 which corresponds to the inclusion $\mathcal{O}_{M_{\rm rd}} \subset \mathcal{O}_M$. If $\dim M = m|n$, then $\mathcal{O}_M$ is locally free as an $\mathcal{O}_{M_{\rm rd}}$-module of rank $2^{n-1}|2^{n-1}$, and it admits a filtration by the $\mathcal{O}_{M_{\rm rd}}$-submodules J_M^i, which are also locally free and are locally directly included. In the differentiable category, any extension of locally free modules splits, since the obstruction to the splitting of $0 \to \mathcal{E}_1 \to \mathcal{E} \to \mathcal{E}_2 \to 0$ is a cohomology class in the group $H^1(M, \mathcal{H}om_0(\mathcal{E}_2, \mathcal{E}_1))$. Therefore, in particular the quotient module ${\mathcal{O}_M}^{(1)} = \mathcal{O}_{M_{\rm rd}} \oplus \mathcal{F}$ can be injected into $\mathcal{O}_M$ as an $\mathcal{O}_{M_{\rm rd}}$-submodule. We identify ${\mathcal{O}_M}^{(1)}$ with its image under this injection. The injection extends to a homomorphism of $\mathcal{O}_{M_{\rm rd}}$-algebras $\psi: S(\mathcal{F}) \to \mathcal{O}_M$ by the universal property of the symmetric algebra. On the associated graded objects (graded by the degree of $(\mathcal{F})$ in the first sheaf and by the powers of J_M in the second) this homomorphism is an isomorphism. Therefore, ψ is an isomorphism. □

6. Theorem. *Let M be a differentiable supermanifold, and let $\mathcal{E}$ be a locally free sheaf on it. Let $\mathcal{E}_{\rm rd}$ be its restriction to $M_{\rm rd}$ and $\phi: M \to M_{\rm rd}$ be a projection* *as in Lemma 4. Then $\mathcal{E}$ is isomorphic (not canonically) to the sheaf $\phi^*(\mathcal{E}_{\rm rd})$, where $\mathcal{E}_{\rm rd} = \mathrm{Gr}_0\,\mathcal{E}$.*

Proof. Let $\phi^{(i)}$ be the restriction of ϕ to $M^{(i)}$. By induction on i we construct a sequence of isomorphisms $\psi^{(i)}: \mathcal{E}^{(i)} = \mathcal{E} \otimes \mathcal{O}_{M^{(i)}} \to \phi^{(i)*}(\mathcal{E}_{\rm rd})$, beginning with $\psi^{(0)} = \mathrm{id}$. The obstruction to the existence of the extension of $\psi^{(i)}$ to $\psi^{(i+1)}$ is the cohomology class

$$\omega(\psi^{(i)}) \in H^1\left(M_{\rm rd}, \left(\mathcal{E}nd\, \mathcal{E}_{\rm rd} \otimes S^{i+1}(\mathcal{F})\right)_0\right).$$

This is constructed exactly as in the proof of Lemma 4. On the elements U of a cover on which $\mathcal{E}$ and M are trivial, choose extensions $\psi_U^{(i+1)}: \mathcal{E} \otimes \mathcal{O}_{M^{(i+1)}}|_U \to \mathcal{E}_{\rm rd} \otimes_{\mathcal{O}_{M_{\rm rd}}} \mathcal{O}_{M^{(i+1)}}|_U$ of the isomorphism $\psi_U^{(i)}$. Two such extensions $\psi_U^{(i+1)}$ and $\psi_V^{(i+1)}$ agree modulo $\mathcal{E} \otimes J_N^{i+1}$. so their incompatibility can be characterized using the cocycle

$$\omega_{UV}: \mathcal{E}_{\rm rd}|_{U\cap V} \to \mathcal{E}_{\rm rd} \otimes S^{(i+1)}(\mathcal{F})|_{U\cap V}.$$

We omit the rest of the details. □

7. Corollary. *The following two classification problems reduce to the problem of classifying differentiable vector bundles on $M_{\rm rd}$.*

(a) *Classification up to isomorphism of all differentiable superextensions of $M_{\rm rd}$ (i.e., supermanifolds M with a given $M_{\rm rd}$).*

(b) *The classification of locally free sheaves on these superextensions.* □

Nevertheless, supergeometry, even in the differentiable case, does not allow any trivial reduction to purely even geometry because in the supercategory there are many more morphisms $M \to N$ than there are morphisms of pairs $(M_{\rm rd}, \mathcal{F}_M) \to (N_{\rm rd}, \mathcal{F}_N)$ because of the possibility of "changes of variables which are not linear in the odd coordinates."

We now turn out attention to analytic supermanifolds, where all the cohomological obstructions can be nonzero. In the proofs, we briefly recall some fragments of the formalism of obstruction theory from Chapter 2.

8. Proposition. *Let $(M, \mathcal{O}_M)$ be an analytic supermanifold of odd dimension 1.*

(a) *$(M, \mathcal{O}_M)$ is defined up to isomorphism by the pair $(M_{\rm rd}, \mathcal{F})$.*

(b) *A locally free sheaf $\mathcal{E}$ on $(M, \mathcal{O}_M)$ is defined up to isomorphism by the pair consisting of the sheaf $\mathcal{E}_{\rm rd}$ and a cohomology class $\omega \in H^1\left(M_{\rm rd}, (\mathcal{E}nd\, \mathcal{E}_{\rm rd} \otimes \mathcal{F})_0\right)$; ω can be arbitrary.*

Proof. Since $\mathrm{rk}\,\mathcal{F} = 0|1$, we have $\mathcal{O}_{M,0} = \mathcal{O}_{M_{\rm rd}}$ and $J_M = \mathcal{O}_{M,1} = \mathcal{F}$; multiplication in $\mathcal{O}_M = \mathcal{O}_{M_{\rm rd}} \oplus \mathcal{F}$ is uniquely defined by the action of $\mathcal{O}_{M_{\rm rd}}$ on $\mathcal{F}$, since $\mathcal{F}$ is an ideal with square zero. This proves the first assertion.

In particular, the $\mathcal{O}_M$-module $\mathcal{E}$ is an $\mathcal{O}_{M_{\rm rd}}$-module and as such is represented in the form of an extension of locally free modules $0 \to \mathcal{F}\otimes\mathcal{E}_{\rm rd} \to \mathcal{E} \to \mathcal{E}_{\rm rd} \to 0$. We denote by ω the class of this extension. Knowing ω we know $\mathcal{E}$ up to isomorphism, the filtration $J_M\mathcal{E} \subset \mathcal{E}$ and the identification of $\mathcal{E}/J_M\mathcal{E}$ with $\mathcal{E}_{\rm rd}$. This is sufficient to reconstruct on $\mathcal{E}$ the structure of an $\mathcal{O}_M$-module, since it is clear how to extend the definition of multiplication on J_M. □

9. Proposition. *Let $(M, \mathcal{O}_M)$ be an analytic supermanifold of odd dimension 2.*

(a) *$(M, \mathcal{O}_M)$ is defined up to isomorphism by the pair $(M_{\rm rd}, \mathcal{F})$ and a cohomology class $\omega_M \in H^1\left(M_{\rm rd}, \mathcal{T}M_{\rm rd} \otimes S^2(\mathcal{F})\right)$; the class ω_M can be arbitrary.*

(b) *A locally free sheaf $\mathcal{E}$ on $(M, \mathcal{O}_M)$ of rank $p|0$ is defined up to isomorphism by a pair consisting of the sheaf $\mathcal{E}_{\rm rd}$ and a cohomology class $\omega_{\mathcal{E}} \in H^1\left(M_{\rm rd}, \mathcal{E}nd\,\mathcal{E}_{\rm rd} \otimes S^2\mathcal{F}\right)$; this class can be arbitrary. In order to determine $\omega_{\mathcal{E}}$ uniquely, one must fix a sheaf $\mathcal{E}'$, with $\mathcal{E}'_{\rm rd} = \mathcal{E}_{\rm rd}$.*

Proof. Since rk $\mathcal{F} = 0|2$, we have $\mathcal{O}_{M,1} = \mathcal{F}$, while $\mathcal{O}_{M,0}$ is an extension of $\mathcal{O}_{M_{\rm rd}}$ by the ideal with square zero, $S^2(\mathcal{F})$. The class ω_M is defined as the obstruction to the construction of an injection $\mathcal{O}_{M_{\rm rd}} \to \mathcal{O}_{M,0}$, i.e., to the splitting of this extension, as in the proof of Lemma 4. Conversely, for any class ω such an extension $0 \to S^2(\mathcal{F}) \to \mathcal{O} \to \mathcal{O}_{M_{\rm rd}} \to 0$ can be constructed for which ω is the obstruction to splitting. To this end, on a locally finite cover $\{U\}$ on which the restriction of ω is trivial, one must construct sheaves $\mathcal{O}_U = \mathcal{O}_{U_{\rm rd}} \oplus S^2\mathcal{F}|_U$ and then paste them on the intersections using the components ω_{UV} of a cochain representing ω. Finally, it is not difficult to show that two extensions $\mathcal{O}$ and $\mathcal{O}'$ having the same obstruction are isomorphic. This proves assertion (a).

A locally free sheaf $\mathcal{E}$ on M can be reconstructed from the $\mathcal{O}_{M_0}$-module $\mathcal{E}_0$, since $\mathcal{O}_1 = \mathcal{F} = \mathcal{F}\otimes\mathcal{O}_{M,0}$ and so $\mathcal{E}_1 = \mathcal{F}\otimes\mathcal{E}_0$. The sheaf $\mathcal{E}_0$ is locally free over $\mathcal{O}_{M,0}$ if the rank of $\mathcal{E}$ is even and only in this case. It is an extension of the sheaf $\mathcal{E}_{\rm rd}$ and is defined by the extension cohomology class of $0 \to S^2\mathcal{F}\otimes\mathcal{E}_{\rm rd} \to \mathcal{E}_0 \to \mathcal{E}_{\rm rd} \to 0$ with respect to the "base point" $\mathcal{E}'_0$. This class $\omega_{\mathcal{E}}$ is the obstruction to the extension of the identity isomorphism $\mathcal{E}_{\rm rd} \to \mathcal{E}'_{\rm rd}$ to an isomorphism $\mathcal{E} \to \mathcal{E}'$. □

10. Examples. (a) Let $M_{\rm rd} = \mathbf{P}^m = \mathbf{P}(T)$, where $T = \mathbf{C}^{m+1}$ and rk $\mathcal{F} = 0|2$. There exists a k such that $S^2\mathcal{F}$ is isomorphic to $\mathcal{O}(k)$. By the Serre duality theorem the space $H^1(\mathbf{P}^m, \mathcal{T}\mathbf{P}^m(k))$ is dual to $H^{m-1}(\mathbf{P}^m, \Omega^1\mathbf{P}^m(-k-m-1))$. From Theorem 1.2.2 it is evident that this group can be different from zero only for $m = 1$ and $k \le -4$. Computing it, we obtain the following list of values of k for which there is a nontrivial superextension of $\mathbf{P}^1$ of odd dimension 2:

$$S^2(\mathcal{F}) = \mathcal{O}(k), \qquad k \le -4;$$
$$H^1 = T \otimes S^{|k|-3}(T)/S^{|k|-2}(T), \qquad \dim H^1 = |k| - 3.$$

In order to obtain a complete system of invariants of such superextensions, we consider two cases. First, if $S^2(\mathcal{F}) = \mathcal{O}(k)$, then $\mathcal{F} = \Pi(\mathcal{O}(a) \oplus \mathcal{O}(b))$, for $a + b = k$. In the second case the group Aut $\mathcal{F}$ acts on H^1, and isomorphic superextensions correspond to invariants ω which lie in a single orbit. This action reduces to multiplication by a nonzero complex number. Finally, the following list of superspaces exhausts the examples of P for which $\mathsf{P}_{\rm rd} = \mathsf{P}^1$, rk $\mathcal{F} = 0|2$ and $\mathsf{P} \not\simeq \operatorname{Gr} \mathsf{P}$:

$$\mathsf{P}^{1/2}_{\omega}(a,b) : a + b \leq -4,$$
$$\omega \in \mathsf{P}\left(H^1(\mathsf{P}^1, \mathcal{O}(a+b+2))\right) = \mathsf{P}^{|a+b|-4},$$
$$\mathrm{Cr}\ \mathsf{P}^{1/2}_{\omega} = \left(\mathsf{P}^1, \wedge(\mathcal{O}(a) \oplus \mathcal{O}(b))\right).$$

In the case $a + b = -4$ there are no continuous parameters.

(b) On $\mathsf{P} = \left(\mathsf{P}^1, \mathcal{O}_{\mathsf{P}^1} \oplus \Pi\mathcal{O}(k)\right)$ we consider locally free sheaves $\mathcal{E}$ for which $\mathcal{E}_{\rm rd} = \bigoplus_i \Pi^{\epsilon_i}\mathcal{O}(a_i)$, $\epsilon_i = 0, 1$. By Proposition 8 they are classified by cohomology classes in

$$H^1\left(\mathsf{P}^1, \bigoplus_{\epsilon_i+\epsilon_j+1=0} \mathcal{O}(a_i - a_j + k)\right).$$

(c) On $\mathsf{P} = \left(\mathsf{P}^1, \wedge(\mathcal{O}(a) \oplus \mathcal{O}(b))\right)$ we consider locally free sheaves $\mathcal{E}$ for which $\mathcal{E}_{\rm rd} = \bigoplus_i \mathcal{O}(a_i)$. By Proposition 9 they are classified by elements of the cohomology group $H^1\left(\mathsf{P}^1, \bigoplus_{i,j} \mathcal{O}(a_i - a_j + a + b)\right)$. As a base point we can make a canonical choice of $\phi^*(\mathcal{E}_{\rm rd})$, where $\phi\colon \mathsf{P} \to \mathsf{P}_{\rm rd}$ is the standard projection.

§ 3. Supergrassmannians and Flag Superspaces

1. In this section we construct some important examples of superspaces which play as fundamental a role in supersymmetry as they do in purely even geometry. We begin with grassmannians. Most of the constructions parallel the corresponding ones in § 1 of Chapter 1, and this parallelism will be preserved in the notation.

2. Initial data. Let A be a supercommutative ring, and let T be a free A-module of rank $d + c$, where $d = d_0|d_1$ and $c = c_0|c_1$. One should imagine that A is the ring of functions on a superspace M in one of the standard geometric categories—on a superscheme Spec A or on a differential supermanifold or an analytic Stein superspace—while T is the module of sections of the trivial bundle $\mathcal{T}$ over M.

The "relative grassmannian" $G_A(d;T)$ (or $G_A(d;\mathcal{T})$) will be described as a family of polynomial A-algebras and a rule for coordinate change "on the intersection of their spectra." We will explain briefly the rather obvious construction of the corresponding superspace in one of the categories.

3. Standard atlas and coordinates. We choose a basis for T in the standard format: $T = A^{d_0+c_0} \oplus (\Pi A)^{d_1+c_1}$. The elements of T will be written as rows of left coordinates with respect to this basis. Analogously, such rows with elements in the A-algebra B will be interpreted as elements of the module $B \otimes_A T$.

Let $I = I_0 \cup I_1$ be a set of $d_0|d_1$ (indices of) elements of the given basis of T.

(a) To each set I we associate a free set of even variables x_I^{ab} and odd variables ξ_I^{ab} filling in all the places in a matrix in $d_0|d_1 \times (d_0+c_0)|(d_1+c_1)$ standard format, except for the columns with indices in I, which together form a unit matrix:

$$
\begin{array}{c|c|c|c|c|}
 & c_0 & d_0 & d_1 & c_1 \\
\hline
d_0 & x_I & \begin{matrix}1 & & \\ & \ddots & \\ & & 1\end{matrix} & 0 & \xi_I \\
\hline
d_1 & \xi_I & 0 & \begin{matrix}1 & & \\ & \ddots & \\ & & 1\end{matrix} & x_I \\
\hline
\end{array}
$$
$$\underbrace{\qquad d_0 \qquad d_1 \qquad}_{I}$$

We denote by U_I one of the following superspaces.

In the category of superschemes, $U_I = \operatorname{Spec} A[x_I, \xi_I] \xrightarrow{\pi} M = \operatorname{Spec} A$, where π is induced by the standard inclusion of rings $A \to A[x_I, \xi_I]$. Clearly, $U_I = M \times \operatorname{Spec} \mathbf{Z}[x_I, \xi_I]$.

In the category of differentiable supermanifolds, $U_I \xrightarrow{\pi} M$ is the supermanifold $M \times \mathbf{R}^{c_0d_0+c_1d_1|c_1d_0+c_0d_1}$, on which x_I, ξ_I are smooth coordinates along the fibers.

In the category of analytic superspaces, $U_I \xrightarrow{\pi} M$ is the analytic superspace $M \times \mathbf{C}^{c_0d_0+c_1d_1|c_1d_0+d_0d_1}$, on which x_I, ξ_I are complex coordinates along the fibers.

(b) Now let I and J be two sets of basis elements of T of size $d_0|d_1$. We denote by B_{IJ} the submatrix of Z_I formed by columns with indices in J. We denote by $U_{IJ} \subset U_I$ the maximal open subsuperspace on which B_{IJ} is invertible as a matrix of functions. Certainly for this it is sufficient that two even functions be invertible, the determinants of the upper left $d_0 \times d_0$ submatrix and of the right lower $d_1 \times d_1$ submatrix. The structure of the matrix Z_I shows that in all three cases $U_{IJ} = M \times V_{IJ}$, where V_{IJ} is a dense open subspace in the corresponding fiber of the projection.

(c) On the superspaces U_{IJ} and U_{JI} there are common functions, the elements of A, and two systems of coordinates "over A", (x_I, ξ_I) and $(x_J.\xi_J)$. We define the rule for changing coordinates by the following pasting equations:

$$Z_J = B_{IJ}^{-1} Z_I \quad \text{on } U_{IJ}.$$

It is clear that $(x_J.\xi_J)$ are expressed in the form of rational functions in (x_I, ξ_I) with integer coefficients and invertible (on U_{IJ}) denominator. Therefore, these transition functions belong to the structure sheaves of any of the three categories under consideration.

(d) We set $G_M(d; \mathcal{T}, (t)) = (\bigcup_I U_I)/R$, where R is the equivalence relation generated by the coordinate change which has been described and (t) is the initial basis of T.

The canonical mappings $U_I \to G_M(d; \mathcal{T})$ are isomorphisms onto open subsuperspaces of the grassmannian. These are the "big cells" of the grassmannians. From this construction it is clear that our relative grassmannians are obtained by a change of basis of Spec $\mathbf{Z}\left[\frac{1}{2}\right]$, $\mathbf{R}^{1|0}$ or $\mathbf{C}^{1|0}$ on M from the corresponding absolute grassmannian in each of the three categories.

(e) Now let $t' = H(t)$ be a new basis of T and let H be the coordinate change matrix. If an element was described by a row of left coordinates (x, ξ) in the old basis, then $(x', \xi') = (x, \xi)H^{-1}$ are its new coordinates. Therefore, we identify $G_M(d; \mathcal{T}, (t))$ and $G_M(d; \mathcal{T}, (t'))$ with the help of the coordinate change rule below.

We denote by B'_{IJ} the matrix consisting of columns of $Z_I H^{-1}$ with numbers in J and by Z'_J the matrix corresponding to the cell U'_J in the standard atlas of $G_M(d; \mathcal{T}, (t'))$. On the open subset U_I where B'_{IJ} is invertible, the transition to the primed coordinates occurs according to the formula

$$(B'_{IJ})^{-1} Z_I H^{-1} = Z'_J .$$

A series of straightforward verifications shows that this defines an isomorphism of superspaces over A, $\phi_H: G_M(d; \mathcal{T}; (t)) \to G_M(d; \mathcal{T}; (t'))$, with $\phi_{H_1 H_2} = \phi_{H_1} \phi_{H_2}$. This defines a left action of the group $\mathrm{GL}(d + c; A)$ on $G_M(d; \mathcal{T}, (t))$.

We will denote by $G_M(d; \mathcal{T})$ the superspace represented by these canonically isomorphic objects. We have constructed it along with the action of $\mathrm{GL}(\mathcal{T})$; this is the group of (even) automorphisms of the A-module $\mathcal{T}$ represented variously by the matrix groups $\mathrm{GL}(d + c; A)$ for various choices of basis for $\mathcal{T}$. This is the grassmannian.

The relative dimension (over M) of the grassmannian is equal to dc, the product of the rank d by the corank, $\operatorname{rk}\mathcal{T} - d$, of the submodules of $\mathcal{T}$ which classify the points of the grassmannian (see below, § 9).

In particular, the grassmannians of purely odd relative dimension are of special interest: $G(d|0, A^{d|c})$ and $G(0|c, A^{d|c})$.

4. Example: projective superspaces. We introduce the notation

$$\mathsf{P}_A(T) = \mathsf{P}_A^{m|n} = G_A(1|0; A^{m+1|n}).$$

Here $I = \{i\}$, $1 \leq i \leq m+1$, and

$$Z_{\{i\}} = \left(x^k_{\{i\}}, k \neq i, 1 \leq k \leq m+1 \mid \xi^l_{\{i\}}, l = 1, \ldots, n\right).$$

The transition rules are $x^k_{\{j\}} = \left(x^j_{\{i\}}\right)^{-1} x^k_{\{i\}}$ and $\xi^l_{\{j\}} = \left(x^j_{\{i\}}\right)^{-1} \xi^l_{\{i\}}$. As in the purely even case, these rules are most conveniently described by introducing homogeneous coordinates (over A) on $\mathsf{P}_A(T)$: $(x^1 : \ldots : x^{m+1} \mid \xi^1 : \ldots : \xi^n)$. We set $x^k_{\{j\}} = (x^j)^{-1}x^k$ and $\xi^l_{\{j\}} = (x^j)^{-1}\xi^l$. The structure of a projective superspace is extremely simple. Let $A = \mathsf{Z}\left[\frac{1}{2}\right]$, R or C (a "point" in the appropriate category), and let $T = T_0 \oplus T_1$ be a free A-module of rank $m+1|n$.

5. Proposition. $\mathsf{P}^{m|n} = \mathsf{P}_A(T)$ *is a smooth supermanifold of dimension $m|n$ which is canonically isomorphic to* $\operatorname{Gr}\mathsf{P}^{m|n}$:

$$\mathsf{P}^{m|n} = \operatorname{Gr}\mathsf{P}^{m|n} = \left(\mathsf{P}(T_0), S\left(T_1^* \otimes \mathcal{O}_{\mathsf{P}(T_0)}(-1)\right)\right).$$

Proof. The isomorphism $\mathsf{P}(T)_{\mathrm{rd}} \xrightarrow{(\phi,\psi)} \mathsf{P}(T_0)$ is constructed in the obvious way on the standard covering $\mathsf{P}(T_0) = \bigcup U_{\{i\}}$ with coordinates $\left(x^j_{\{i\}} \mid j \neq i\right)$. An isomorphism of sheaves $\psi\colon S\left(T_1^* \otimes \mathcal{O}(-1)\right) \to \mathcal{O}_{\mathsf{P}^{m|n}}$ is induced by the mapping $\psi\left(\xi^l \otimes (x^j)^{-1}\right) = \xi^l_{\{j\}}$ where ξ^l is viewed as an element of T_1^* and $(x^j)^{-1}$ as a section of $\mathcal{O}_{\mathsf{P}(T_0)}(-1)$ over $U_{\{j\}}$. □

6. Tautological sheaf. We return to the general supergrassmannian over A. We denote by S_I the free $\mathcal{O}(U_I)$-module of rank $d_0|d_1$ generated by the rows of the matrix Z_I. (It is free since the I columns form the identity matrix.) This is the module of sections of the free sheaf $\mathcal{S}_I$. On $U_{IJ} = U_{JI}$ the restrictions $\mathcal{S}_I|U_{IJ}$ and $\mathcal{S}_J|U_{JI}$ are identified by means of the same transition matrix B_{IJ} which was used for the construction of the transition functions. This allows us to define a locally free sheaf S on $G_A(d;T) = G$ which we call the tautological sheaf. It is defined along with a canonical locally direct injection $S \to \mathcal{O}_G \otimes_A T$; the rows of the matrix Z_I are mapped to the section of $\Gamma(U_I, \mathcal{O}_G \otimes T)$ represented by this row.

7. Functoriality. The supergrassmannian $G_A(d;T)$ has the usual functorial properties with respect to A and T.

(a) $G_B(d; B \otimes_A T)$ can be identified canonically with $B \otimes_A G_A(d;T)$ for any A-algebra B. This formulation is suited for the category of superschemes. In the category of supermanifolds the corresponding fact can be formulated without difficulty by the reader.

(b) Any isomorphism of A-modules $f: T \to T'$ induces a corresponding isomorphism $G_A(d;T) \to G_A(d;T')$. More precisely, we defined $G_A(d;T)$ along with a class of distinguished models corresponding to the bases of T. This permits us to construct in an obvious way an isomorphism $G_A(f): G_A(d;T(t)) \to G_A(d;T';(f(t)))$. After this, the verification of the compatibility of these isomorphisms for different models reduces to checking whether the change of basis isomorphisms are well-defined; we have already addressed this question briefly in subsection 3(d).

(c) With respect to the identifications described in (a) and (b), the tautological sheaves are carried to each other.

8. General relative grassmannians. Now let M be a superscheme or a differentiable manifold or an analytic superspace. For any locally free sheaf J of rank $d+c$ on M, we can construct s superspace $G = G_M(d;\mathcal{T}) \xrightarrow{\pi} M$, along with a canonical projection onto M, which is in the same category as M. This superspace is defined along with a family of standard atlases. A standard atlas on G is defined by the pair consisting of

(a) an atlas $M = \bigcup V_i$ made up of affine open subschemes, superdomains or Stein open analytic superspaces on which $\mathcal{T}$ is can be trivialized and

(b) a family of trivializations of $\mathcal{T}|V_i$ over the charts of this atlas.

If such a pair is given, we set $A_i = \mathcal{O}(V_i)$ and $T_i = \mathcal{T}(V_i) = T(V_i, \mathcal{T})$; then we paste together $G_M(d;\mathcal{T})$ from the $G_{A_i}(d;T_i)$ using the functorial properties from subsection 7 for the canonical identification of $\pi_i^{-1}(V_{ij})$ with $\pi_j^{-1}(V_{ij})$, where the $\pi_i: G_{A_i}(d;\mathcal{T}_i) \to V_i$ are the structure morphisms and $V_{ij} = V_i \cap V_j$.

Along with $G_M(d;\mathcal{T})$ is constructed the tautological locally direct subsheaf $S \subset \pi^*(\mathcal{T})$ of rank d, pasted together from the $G_A(d;\mathcal{T}_i)$ by subsection 7 (c).

9. N-points of a grassmannian. Let $\phi: N \to M$ be a morphism of superspaces in one of the three categories. To any lifting of it to a morphism $\chi: N \to G_M(d;\mathcal{T})$ (i.e., to a χ such that $\pi \circ \chi = \phi$) we associate the preimage of the tautological sheaf $\chi^*(S_M)$, which is realized as a locally direct subsheaf $\chi^*(S_M) \subset \phi^*(\mathcal{T})$ of rank d. We obtain the following mapping of sets:

$$\{N\text{-points of } G_M(d;\mathcal{T}) = \mathrm{Hom}_M\,(N, G_M(d;\mathcal{T}))\}$$
$$\to \{\text{locally direct subsheaves of rank } d \text{ in } \phi^*(\mathcal{T})\}\,.$$

Here we fix M, $\mathcal{T}$ and d. As (N,ϕ) varies among all the superspaces over M in one of the three categories, the families of sets on the left and the right extend to functors in (N,ϕ) and the mapping between them extends to a morphism of functors.

10. Theorem. *The morphism of functors just described is an isomorphism in each of the three categories.*

The proof does not differ at all from the corresponding reasoning in the purely even category; a fragment of this was given in § 1 of Chapter 1. In addition to a straightforward verification of all the definitions, this proof depends upon one simple algebraic fact: if the rows of a matrix in standard format, of size $d\times(d+c)$, generate a direct submodule in A^{d+d} of rank d, then it has an invertible $k\times k$ submatrix (the converse is clear). From this it follows that the atlas (U_I) actually covers all the N-points of the grassmannian. □

11. Theorem. *Let $\mathcal{T}G_M/M$ be the relative tangent sheaf of the grassmannian $G_M(d;\mathcal{T})$, and let $\widetilde{S} \subset \pi^*(\mathcal{T}^*)$ be the sheaf which is orthogonal to the tautological sheaf. Then there is a canonical isomorphism*

$$\mathcal{T}G_M/M = \mathcal{H}om(S, \pi^*(\mathcal{T})/S) = \widetilde{S}^* \otimes S^*\,.$$

Proof. It is sufficient to repeat again the reasoning of § 1 of Chapter 1. Let X be a locally vertical field on G_M (annihilating $\pi^{-1}(\mathcal{O}_M)$). There is a natural action of X on $\pi^*(\mathcal{T})$ which annihilates all the sections of $\pi^{-1}\mathcal{T}$. Restricting this to S and then reducing the result modulo S, we obtain by the Leibniz formula an $\mathcal{O}_M$-linear mapping

$$\mathcal{T}G_M/M \to \mathcal{H}om(S, \pi^*(\mathcal{T})/S) \colon X \mapsto \overline{X},\ \overline{X}\,s = X\,s \bmod S\,.$$

On a big open cell of the type of U_I, this is an isomorphism. In fact, let $(y_I) = (x_I, \xi_I)$. Then $\mathcal{T}U_I/A$ is freely generated by the fields $\partial/\partial y_I^{ab}$; if we consider that a and b index the rows and columns of the matrix Z_I, then the y_I^{ab} are independent variables for $b \notin I$. The basis of sections of S given by a fixed basis (t_b) of the A-module T has over U_I the form $s^a = \sum_{b\notin I} y_I^{ab} \otimes t_b + t_{I(a)}$, where $I(a)$ is the entry where the one is located in the a-th row of the matrix Z_I. Therefore,

$$\frac{\partial}{y_I^{ab}} s^c \equiv \delta_a^c t_b \bmod S, \quad b \notin I.$$

Finally, the t_b mod S form a basis for the sections of $\pi^*(\mathcal{T})$ mod S over U_I for $b \notin I$, since for $c \in I$ one can express t_c in terms of the t_b, $b \in I$, and the rows of Z_I. Consequently, the linear mappings $\overline{\partial/\partial y^{ab}}$ generate freely $\mathcal{H}om(S|U_I, \pi^*(\mathcal{T})^I/S|U_I)$. □

12. Example. On the projective superspace $\mathsf{P} = \mathsf{P}_M(\mathcal{T}) = G_M(1|0; \mathcal{T})$, we will denote the tautological sheaf as usual by $\mathcal{O}(-1)$. Using Theorem 11 we obtain the exact sequences

$$0 \to \mathcal{O}(-1) \to \pi^*(\mathcal{T}) \to \mathcal{T}\mathsf{P}/M(-1) \to 0,$$
$$0 \to \Omega^1_{\text{odd}}\mathsf{P}/M(1) \to \pi^*(\Pi\mathcal{T}^*) \to \Pi\mathcal{O}(1) \to 0,$$

which are used just as in § 2 of Chapter 1. For example, we compute $\operatorname{Ber}\mathsf{P}/M = (\operatorname{Ber}\Omega^1_{\text{odd}}\mathsf{P}/m)^*$. Let rk $\mathcal{T} = m + 1|n$.

13. Proposition.

$$\operatorname{Ber}\mathsf{P}/M = \Pi^{m+n}\operatorname{Ber}(\pi^*(\mathcal{T}^*))(-m-1+n).$$

Proof. For any exact triple of locally free sheaves on a superspace, $0 \to S_1 \to S \to S_2 \to 0$ we can define a canonical isomorphism $f\colon \operatorname{Ber} S_1 \otimes \operatorname{Ber} S_2 \to \operatorname{Ber} S$. Namely, let (s_{1i}, s_{2j}) be a local basis of S such that (s_{1i}) is a local basis of S_1 and $(s_{2j} \bmod S_1)$ is a local basis of S_2. The isomorphism f is defined such that

$$f\,(D(s_{1i}) \otimes D(s_{2j} \bmod S_1)) = (-1)^{\tilde{a}} D(a(s_{1i}, s_{2j})),$$

where (s_{1i}) and (s_{2j}) have the standard order and a is the permutation which gives the standard order to the combined basis. The correctness of this definition can be verified without difficulty. Therefore

$$\operatorname{Ber}(\Omega^1_{\text{odd}}\mathsf{P}/M)(n-m) \otimes \Pi\mathcal{O}(-1) = \operatorname{Ber}(\pi^*(\Pi\mathcal{T}^*)).$$

This proves the assertion, making use of the isomorphism $\operatorname{Ber}(\Pi S^*) = \Pi^{\operatorname{rk} S}\operatorname{Ber} S$. □

From this it follows in particular that for $m + 1 = n$ the relative berezinian $\operatorname{Ber}\mathsf{P}/M$ is isomorphic to a sheaf lifted from M; thus it is trivial when M is a point.

For general grassmannians, the analog of Proposition 5 is a result which computes the associated graded space. We begin with a simple technical assertion which permits us to lift the structure of this space if the structure of the supertangent sheaf is known.

14. Proposition. *Let M be a differentiable or analytic supermanifold, and let $\mathcal{E}$ be a locally free sheaf on M. Then there exist the following isomorphisms:*

(a) $\mathrm{Gr}_i\,\mathcal{E} = \mathrm{Gr}_0\,\mathcal{E} \otimes S^i\,(\mathrm{Gr}_1\,\mathcal{O}_M)$ *(over* $\mathrm{Gr}_0\,\mathcal{O}_M = \mathcal{O}_{M_{\mathrm{rd}}}$*).*

(b) $\mathrm{Gr}_1\,\mathcal{O}_M = \left(\mathrm{Gr}_0\,\Omega^1_{\mathrm{ev}}M\right)_1$, *where the outside index refers to the $\mathbf{Z}_2$-grading.*

Proof. Let (x^a, ξ^b) be a local system of coordinates on M, and let (e^c) be a local basis of sections of $\mathcal{E}$. Then $\mathrm{Gr}_i\,\mathcal{E} = \mathcal{E}J^i_M/\mathcal{E}J^{i+1}_M$ is freely generated locally by the classes $\left\{ e^c\xi^\alpha \bmod \mathcal{E}J^{i+1}_M \mid |\alpha| = i \right\}$ over $\mathcal{O}_{M_{\mathrm{rd}}}$, while $\mathrm{Gr}_1\,\mathcal{O}_M$ is freely generated by the classes $\left\{ \xi^b \bmod J^2_M \right\}$. The first isomorphism in the proposition carries $e^c(\xi^1)^{\alpha_1}\cdots(\xi^n)^{\alpha_n}$ to $(e^c \bmod \mathcal{E}J_M \otimes \left(\xi^1 \bmod J^2_M\right)^{\alpha_1}\cdots\left(\xi^n \bmod J^2_M\right)^{\alpha_n}$. The correctness of this definition can be easily checked (taking into account that M is a supermanifold, i.e., that $\mathcal{O}_M$ is locally isomorphic to $\mathrm{Gr}\,\mathcal{O}_M$).

The second isomorphism carries $\xi^b \bmod J^2_M$ to $d_{\mathrm{ev}}\xi^b \bmod J_M\Omega^1_{\mathrm{ev}}M$. In order to check that this is well-defined, we verify that for any system of coordinates (y,η) the sections $\eta^c \bmod J^2_M$ are carried to $d_{\mathrm{ev}}\eta^c \bmod J_M\Omega^1_{\mathrm{ev}}M$. In fact, let $\eta^c \equiv \sum_b f^c_b(x)\xi^b \bmod J^2_M$. Then

$$d_{\mathrm{ev}}\eta^c = \sum_a d_{\mathrm{ev}}x^a\frac{\partial\eta^c}{\partial x^a} + \sum_b d_{\mathrm{ev}}\xi^b\frac{\partial\eta^c}{\partial\xi^b} \equiv \sum_b d_{\mathrm{ev}}\xi^b f^c_b(x) \bmod J_M\Omega^1_{\mathrm{ev}}M\,.$$

This completes the proof. □

Now let $T = T_0 \oplus T_1$ be a linear superspace over $\mathbf{R}$ (in the differentiable case) or over $\mathbf{C}$ (in the analytic case) of dimension $d + c = d_0 + c_0|d_1 + c_1$. We set

$$G = G(d;T), \quad G_0 = G(d_0|0;T_0), \quad G_1 = G(0|d_1;T_1).$$

The superspaces G_0 and G_1 are the usual purely even grassmannians with the exception of the fact that the tautological sheaf S_{G_1} has rank $0|d_1$ and not $d_1|0$. Let $\widetilde{S}_{G_0}$ and $\widetilde{S}_{G_1}$ be the sheaves orthogonal to the tautological sheaves. The following theorem gives a description of $\mathrm{Gr}\,G$ and, therefore, of G in the differentiable category.

15. Theorem. *The following canonical isomorphisms can be defined.*

(a) $G_{\mathrm{rd}} = G_0 \times G_1$.

We denote by S_0, S_1 (respectively, $\widetilde{S}_0$, $\widetilde{S}_1$) the sheaves S_{G_0}, S_{G_1} (respectively, $\widetilde{S}_{G_0}$, $\widetilde{S}_{G_1}$) lifted to G_{rd} with respect to the projection.

(b) $\mathrm{Gr}\,\mathcal{O}_G = S(S_0 \otimes \widetilde{S}_1 \oplus \widetilde{S}_0 \otimes S_1)$ *(the symmetric algebra over $\mathcal{O}_{G_{\mathrm{rd}}}$).*

(c) $\mathrm{Gr}\,S_G = (S_0 \oplus S_1) \otimes \mathrm{Gr}\,\mathcal{O}_G$ *(over $\mathcal{O}_{G_{\mathrm{rd}}}$), where S_G is the tautological sheaf on G.*

Proof. The first assertion of the theorem is a corollary of the fact that a $d_0|d_1$-dimensional subspace of $T_0 \oplus T_1$ decomposes into the direct sum of its even and odd parts. The other two allow the structure of the odd coordinates to be understood in invariant terms.

Reasoning more formally, on $G_0 \times G_1$ we consider the sheaf $\mathcal{S}_0 \oplus \mathcal{S}_1 \subset \mathcal{O}_{G_0 \times G_1} \otimes (T_0 \oplus T_1)$. By Theorem 10 this is induced by some morphism $\rho\colon G_0 \times G_1 \to G$, i.e., $\rho^*(\mathcal{S}_G) = \mathcal{S}_0 \oplus \mathcal{S}_1$. Computing in the coordinates on a big cell U_I, we find that ρ corresponds to the vanishing of the coordinates ξ_I, i.e., it defines an isomorphism of $G_0 \times G_1$ with G_{rd} and so $\mathcal{S}_0 \oplus \mathcal{S}_1 = \mathrm{Gr}_0\, \mathcal{S}_G$. In a similar way, $\widetilde{\mathcal{S}}_0 \oplus \widetilde{\mathcal{S}}_1 = \mathrm{Gr}_0\, \widetilde{\mathcal{S}}_G$. Therefore, assertion (c) follows from Proposition 14 (a). Further, by Proposition 14 (b), $\mathrm{Gr}_1\, \mathcal{O}_G = \left(\mathrm{Gr}_0\, \Omega^1_{\mathrm{ev}} G\right)_1$. Finally, according to Theorem 11, $\Omega^1_{\mathrm{ev}} G = (\mathcal{T} G^*) = \mathcal{S} \otimes \widetilde{\mathcal{S}}$. Therefore,

$$\left(\mathrm{Gr}_0\, \Omega^1_{\mathrm{ev}} G\right)_1 = \left(\mathrm{Gr}_0\, \mathcal{S} \otimes \mathrm{Gr}_0\, \widetilde{\mathcal{S}}\right)_1 = \mathcal{S}_0 \otimes \widetilde{\mathcal{S}}_1 \oplus \mathcal{S}_1 \otimes \widetilde{\mathcal{S}}_0 .$$

This completes the proof. □

16. Example. We consider in detail the structure of $G = G_{\mathbf{C}}(1|1; T^{2|2})$ and show that in the analytic category (and in the category of algebraic supermanifolds) that G is not isomorphic to $\mathrm{Gr}\, G$.

In the notation of Theorem 15 we have $G_{\mathrm{rd}} = \mathbf{P}^1_0 \times \mathbf{P}^1_1$, $\mathcal{S}_0 = \mathcal{O}(-1,0)$, $\mathcal{S}_1 = \Pi\mathcal{O}(0,-1)$; $\widetilde{\mathcal{S}}_0 = (\pi_0^*\Omega^1\mathbf{P}^1_0)\,(1,0)$, $\widetilde{\mathcal{S}}_1 = (\pi_1^*\Omega^1\mathbf{P}^1_1)\,(0,1)$, where $\pi_i\colon G_{\mathrm{rd}} \to \mathbf{P}^1_i$. Therefore,

$$\begin{aligned} \mathcal{F} &= J_G/J_G^2 = \Pi\left[(\pi_0^*\Omega^1\mathbf{P}^1_0)\,(1,-1) \oplus (\pi_1^*\Omega^1\mathbf{P}^1_1)\,(-1,1)\right], \\ J_G^2 &= \mathrm{Gr}_2\, \mathcal{O}_G = S^2(\mathcal{F}) = \Omega^2\left(\mathbf{P}^1_0 \times \mathbf{P}^1_1\right). \end{aligned}$$

According to the results of § 2, the obstruction to the existence of an isomorphism between G and $\mathrm{Gr}\, G$ is an invariant

$$\begin{aligned} \omega &\in H^1\left(\mathcal{T}\left(\mathbf{P}^1_0 \times \mathbf{P}^1_1\right) \otimes \Omega^2\left(\mathbf{P}^1_0 \times \mathbf{P}^1_1\right)\right) \\ &= H^1\left(\mathbf{P}^1_0, \Omega^1\mathbf{P}^1_0\right) \oplus H^1\left(\mathbf{P}^1_1, \Omega^1\mathbf{P}^1_1\right) = \mathbf{C} \oplus \mathbf{C} \end{aligned}$$

(the last isomorphism is canonical). We will show that $\omega = (1,1)$.

We introduce the following notation for the coordinates on big cells $U_i \subset G$:

x_1	1	ξ_1	0
η_1	0	y_1	1

,

x_2	1	0	ξ_2
η_2	0	1	y_2

,

1	x_3	ξ_3	0
0	η_3	y_3	1

,

1	x_4	0	ξ_4
0	η_4	1	y_4

.

To compute B_{ij}, we employ the formulas

$$\begin{pmatrix} 1 & \xi \\ 0 & y \end{pmatrix}^{-1} = \begin{pmatrix} 1 & -y^{-1}\xi \\ 0 & y^{-1} \end{pmatrix}, \quad \begin{pmatrix} x & 0 \\ \eta & 1 \end{pmatrix}^{-1} = \begin{pmatrix} x^{-1} & 0 \\ -x^{-1}\eta & 1 \end{pmatrix},$$

$$\begin{pmatrix} x & \xi \\ \eta & y \end{pmatrix}^{-1} = \begin{pmatrix} x^{-1} + x^{-2}y^{-1}\xi\eta & -x^{-1}y^{-1}\xi \\ -x^{-1}y^{-1}\eta & y^{-1} - x^{-1}y^{-2}\xi\eta \end{pmatrix},$$

and obtain from the equation $Z_j = B_{1j}^{-1} Z_1$ the following coordinate change formulas:

$$\begin{pmatrix} x_1 & \xi_1 \\ \eta_1 & y_1 \end{pmatrix} = \begin{pmatrix} 1 & \xi_2 \\ 0 & y_2 \end{pmatrix}^{-1} \begin{pmatrix} x_2 & 0 \\ \eta_2 & 1 \end{pmatrix} = \begin{pmatrix} x_2 - y_2^{-1}\xi_2\eta_2 & -y_2^{-1}\xi_2 \\ y_2^{-1}\eta_2 & y_2^{-1} \end{pmatrix}$$

$$= \begin{pmatrix} x_3 & 0 \\ \eta_3 & 1 \end{pmatrix}^{-1} \begin{pmatrix} 1 & \xi_3 \\ 0 & y_3 \end{pmatrix} = \begin{pmatrix} x_3^{-1} & x_3^{-1}\xi_3 \\ -x_3^{-1}\eta_3 & y_3 + x_3^{-1}\xi_3\eta_3 \end{pmatrix}$$

$$= \begin{pmatrix} x_4 & \xi_4 \\ \eta_4 & y_4 \end{pmatrix}^{-1} = \begin{pmatrix} x_4^{-1} + x_4^{-2}y_4^{-1}\xi_4\eta_4 & -x_4^{-1}y_4^{-1}\xi_4 \\ -x_4^{-1}y_4^{-1}\eta_4 & y_4^{-1} - x_4^{-1}y_4^{-2}\xi_4\eta_4 \end{pmatrix}.$$

This is a complete system; using it one can express in terms of one another any pairs of coordinate systems.

We set $X_i = x_{i,\mathrm{rd}}$ and $Y_i = y_{i,\mathrm{rd}}$. We obtain the standard four-element covering of $\mathbf{P}_0^1 \times \mathbf{P}_1^1 = G_{\mathrm{rd}}$ with coordinates

$$\begin{pmatrix} X_1 \\ Y_1 \end{pmatrix} = \begin{pmatrix} X_2 \\ Y_2^{-1} \end{pmatrix} = \begin{pmatrix} X_3^{-1} \\ Y_3 \end{pmatrix} = \begin{pmatrix} X_4^{-1} \\ Y_4^{-1} \end{pmatrix}.$$

The element ω is represented by a cocycle which measures the incompatibility of some injections $\psi_i\colon \mathcal{O}(U_{i,\mathrm{rd}}) \to \mathcal{O}(U_i)$ which agree with the identity after reducing by odd coordinates. For example, $\psi_i\binom{X_i}{Y_i} = \binom{x_i}{y_i}$. Let

$$(\psi_1 - \psi_i)\begin{pmatrix} X_1 \\ Y_1 \end{pmatrix} = \begin{pmatrix} A_i \\ B_i \end{pmatrix} \in \Gamma\left(U_{i1}, J_G^2\right).$$

Then $(\partial/\partial X_1) \otimes A_i + (\partial/\partial Y_1) \otimes B_i$ will be a component of the cocycle ω on U_{i1}. Computing using the preceding formulas, we find half (three of six) of the components of ω:

$$\omega_{21} = \frac{\partial}{\partial X_1} \otimes Y_2^{-1}\xi_2\eta_2 = -\frac{\partial}{\partial X_1} \otimes Y_1^{-1}\xi_1\eta_1,$$

$$\omega_{31} = -\frac{\partial}{\partial Y_1} \otimes X_3^{-1}\xi_3\eta_3 = \frac{\partial}{\partial Y_1} \otimes X_1^{-1}\xi_1\eta_1,$$

$$\omega_{41} = -\frac{\partial}{\partial X_1} \otimes X_4^{-2}Y_4^{-1}\xi_4\eta_4 + \frac{\partial}{\partial Y_1} \otimes X_4^{-1}Y_4^{-2}\xi_4\eta_4.$$

This is sufficient to compute the part of ω which has been lifted from $\mathbf{P}^1_1$. In fact, the isomorphism $\pi_0^*\left(\Omega^1\mathbf{P}^1_0\right)\oplus\pi_1^*\left(\Omega^1\mathbf{P}^1_1\right)\cong\mathcal{T}\left(\mathbf{P}^1_0\times\mathbf{P}^1_1\right)\otimes J_G^2$ can be given by the mappings

$$\pi_0^*(dX_1)\mapsto\frac{\partial}{\partial Y_1}\otimes\xi_1\eta_1\quad\text{on } U_1\cup U_2\,;$$
$$\pi_1^*(dY_1)\mapsto-\frac{\partial}{\partial X_1}\otimes\xi_1\eta_1\quad\text{on } U_1\cup U_3\,;$$

After this, the canonical generating element of the group $H^1\left(\mathbf{P}^1_1,\Omega^1\mathbf{P}^1_1\right)$, which is given on the Čech covering $\{\,Y_1\neq 0\,\}\cup\{\,Y_2\neq 0\,\}$ by the cocycle $Y_1^{-1}\,dY_1=-Y_2^{-1}\,dY_2$ (this is its component on the intersection of two elements of the cover), is carried to the cocycle on $\mathbf{P}^1_0\times\mathbf{P}^1_1$ with component ω_{21} on the preimage of this cover.

It is suggested that the reader compute the remaining three components of ω, as well as the part of ω which is lifted from $\mathbf{P}^1_0$, and convince himself that the difference of ω and the sum of the lifted cocycles is cohomologous to zero.

17. Flag functors and flag spaces. Now let M be a base superspace of one of three types—a superscheme, a differentiable supermanifold or a superanalytic space. Let $\mathcal{T}$ be a locally free sheaf on it of rank d, and let $0<d_1<\cdots<d_k<d$ be a sequence of superranks. It is understood that $d_i=d_i^0|d_i^1$ and that $d_i<d_{i+1}$ means that $d_i^0\le d_{i+1}^0$, that $d_i^1\le d_{i+1}^1$ and that at least one of the two inequalities is strict.

A flag of length k and type $(d_1,\dots,d_k)$ in $\mathcal{T}$ is defined to be a sequence of locally free subsheaves $S_1\subset S_2\subset\cdots\subset S_k\subset\mathcal{T}$ such that $\operatorname{rk}S_i=d_i$ and all the inclusions $S_i\subset S_j$ are locally direct. Acting as in § 1 of Chapter 1, we introduce in each of the three categories a flag functor of the indicated type by the following definition.

This functor is defined on the category of superspaces over M and it sets up a correspondence between an object $\phi\colon N\to M$ (where ϕ is a morphism of superschemes, supermanifolds or superanalytic spaces appropriate to the type of M) and a set of flags of type $(d_1,\dots,d_k)$ in the sheaf $\mathcal{T}_N=\phi^*(\mathcal{T})$. The extension of the definition to morphisms is clear.

If $d=(d_1,\dots,d_k)$ is a type and $d'=(d'_1,\dots,d'_l)$ is a subtype of it, then just as in the even case a morphism of functors is defined as "projection on subflags of lower type":

$$\pi(d,d')\colon(S_1\subset\cdots\subset S_k)\mapsto(S'_1\subset\cdots\subset S'_k)\,,\quad S'_j=S_{i(j)}\,,$$

where $i(j)$ is defined by the condition $d'_j=d_{i(j)}$.

18. Theorem. *In each of the three categories, the flag functor is represented by a superspace over M which we denote, just as in the purely even case, by*

$F_M(d_1, \ldots, d_l; \mathcal{T})$. *The morphisms* $\pi(d, d')$ *are represented by morphisms of the superspaces over* M.

Sketch of the proof. We can repeat the reasoning carried out in § 1 of Chapter 1 for the purely even case. For $k = 1$, by Theorem 10 one can set $F_M(d_1; \mathcal{T}) = G_M(d_1; \mathcal{T})$. In order to carry out an induction on the length of the flag, we consider the types $d = (d_0, d_1, \ldots, d_k)$ and $d' = (d_1, \ldots, d_k)$. We will assume that $F' = F_M(d'; \mathcal{T})$ has already been constructed. Then on $F_M(d'; \mathcal{T})$ there is a tautological flag $S_1 \subset \cdots \subset S_k$ which corresponds to the identity morphism of F_M to itself. We consider its lowest sheaf S_1 and construct the relative grassmannian:

$$G_{F'}(d_0; S_1) \xrightarrow{\pi_0} F' \xrightarrow{\pi'} M,$$

where π_0 is the structure morphism of this grassmannian and π' is the structure morphism of F'. Then on $G_{F'}$ there is a tautological sheaf $S_0 \subset \pi_0^*(\pi'^*(\mathcal{T}))$. We set $\pi = \pi' \circ \pi_0 \colon G_{F'} \to M$ and consider the flag of type d on $G_{F'}$: $S_0 \subset \pi_0^* S_1 \subset \cdots \subset \pi_0^* S_k \subset \pi^* \mathcal{T}$.

We claim that $G_{F'}$, a superspace over M, along with this flag of type d represents a flag functor of this type.

The construction of the morphism $N \to G_{F'}$ corresponding to the flag of type d on $N \xrightarrow{\phi} M$ reduces to two steps. To begin with, we throw away the lowest component; the remaining part, by the inductive hypothesis, is induced by a morphism $N \to F'$. Then the lowest component is used to lift if to $N \to G_{F'}$.

The same construction allows one to build a superspace morphism $\pi(d^{(1)}, d^{(2)})$ by induction on the length of $d^{(1)}$. The first step ($k^{(1)} = 1$) here is trivial, since for the grassmannian, $\pi(d^{(1)}, d^{(0)})$ is either the identity morphism or the structure projection on M. We leave the remaining details to the reader.

19. Maximal flags. Let $\operatorname{rk} \mathcal{T} = d^0|d^1$. The type $d = (d_1, \ldots, d_k)$ of a flag in $\mathcal{T}$ is said to be maximal if its length is equal to $k = d^0 + d^1 - 1$. An equivalent condition is this: for all $i = 0, \ldots, k$, $d_{i+1} - d_i = 1|0$ or $0|1$, where we assume that $d_0 = 0|0$, $d_{k+1} = d^0|d^1$. Thus, there are $\binom{d^0+d^1}{d^0} = \binom{d^0+d^1}{d^1}$ maximal types; there is trivially more than one if $d_0 d_1 > 0$. In this respect, flag superspaces differ essentially from the purely even case.

In the language of Lie superalgebras, this corresponds to the existence of pairwise nonconjugate Borel subalgebras in $sl(d^0|d^1)$.

20. Picard group and projectivity. As was shown by I. A. Skornyakov, from the computations of subsections 15 and 16 it follows that a grassmannian supermanifold whose underlying space admits a nontrivial decomposition into a direct sum is not projective. In fact, no invertible sheaf on them has a holomorphic section except for the structure sheaf. The simplest proof of the nonprojectivity of $G(a|b;T^{m|n})$, for $0 < a < m$, $0 < b < n$, can be obtained in the manner below.

(a) Let $G = G(1|1;T^{2|2})$, and let $\mathcal{O}(a,b)$ be the standard sheaf on $G_{\text{rd}} = \mathbf{P}^1\times\mathbf{P}^1$. A computation like the one in subsection 15 of the obstruction to extending $\mathcal{O}(a,b)$ to G shows that it is proportional to $a+b$. In particular, the only sheaves which are extendable are the sheaves $\mathcal{O}(a,-a)$; thus G is not projective.

(b) The inclusion $T^{2|2} \subset T^{a+1|b+1}$ induces a closed embedding $G(1|1;T^{2|2}) \subset G(1|1;T^{a+1|b+1})$. The realization of the second grassmannian in dual form gives an inclusion $G(1|1;T^{2|2}) \subset G(a|b;(T^*)^{a+1|b+1})$. Finally, the inclusion $(T^*)^{b+1|a+1} \subset T^{m|n}$ gives a closed embedding $G(1|1;T^{2|2}) \subset G(a|b;T^{m|n})$. Since the smaller grassmannian is not projective, the larger is also not projective.

(c) Supplementing this reasoning with computation of the "continuous part" of the Picard group, I. A. Skornyakov established that Pic $G(a|b;T^{m|n})$ is generated by the class of the sheaf Ber $\mathcal{S}$.

§ 4. The Frobenius Theorem and Connections

1. All the definitions and constructions of §§ 4–6 of Chapter 1 can be carried over word for word to supermanifolds. Here we limit ourselves to a brief survey of the formulations, emphasizing some peculiarities arising from the presence of odd coordinates.

2. Local classification of vector fields. Let X be an even or odd vector field in the neighborhood of a point x_0 of a supermanifold M.

We say that X is rectifiable is there exists a local system of coordinates (x^a, ξ^b) in a neighborhood of x_0 such that in these coordinates X has the following form:

(a) $X = \dfrac{\partial}{\partial x^1}$ ("1|0-rectifiable");

(b) $X = \dfrac{\partial}{\partial \xi^1} + \xi^1\dfrac{\partial}{\partial x^1}$ ("1|1-rectifiable");

(c) $X = \dfrac{\partial}{\partial \xi^1}$ ("0|1-rectifiable").

Possibly the name of the second type deserves some comment. This is the only case when the commutator $[X,X] = 2\partial/\partial x^1$ is different from zero; the minimal Lie algebra containing X has dimension 1|1 over the base field. The following theorem of V. N. Shander allows the characterization of rectifiable vector fields.

3. Theorem. *In order for a vector field X to be rectifiable in a neighborhood of x_0, it is necessary and sufficient that one of the following conditions be fulfilled:*

(a) *X is even and does not vanish at x_0 (i.e., $X \notin {}_{\bullet\, x_0}TM$) (1|0-rectifiability);*

(b) *X is odd and in some neighborhood $U \ni x_0$ the mapping $X\colon \mathcal{O}(U) \to \mathcal{O}(U)$ is surjective ($1|1$-rectifiability);*

(c) *X is odd, does not vanish at x_0 and $X^2 = 0$ ($0|1$-rectifiability).*

Sketch of proof. Let $J = \mathcal{O}_{\bar{1}} + \mathcal{O}_{\bar{1}}^2$. If X is even and is not zero at the point x_0, then in any initial coordinate system (y^a, η^b) we have $X \equiv \sum_a f_a(y)\partial/\partial y^a \bmod J\mathcal{T}M$, so $f_a(y(x_0)) \neq 0$ for some a. By the theorem on rectification of a vector field in even geometry, we can choose a new coordinate system on the underlying space so that $\sum_a f_a(y)\partial/\partial y^a$ takes the form $\partial/\partial x^a$. By Proposition 1.8 this change of coordinates can be extended to a change of coordinates on M with the same properties. We will assume that $X \equiv \partial/\partial x^1 \bmod J^{2k-1}\mathcal{T}M$, $k \geq 1$. This means that $X = (\partial/\partial x^1) + \sum_b \eta^b(\partial/\partial\xi^b) + X'$, where $X' \in J^{2k}\mathcal{T}M$ and (x^a, ξ^b) is a local system of coordinates. We will show that there is a new coordinate system (x^a, ς^b) in which $X = (\partial/\partial x^1) + X''$, where $X'' \in J^{2k}\mathcal{T}M$ and $\varsigma^b \equiv \xi^b \bmod {\diamond}_{x_0}\mathcal{T}M$. For this we set $\varsigma^b = \sum_\alpha g_\alpha^b(x)\xi^\alpha$ and solve these equations for g_α^b:

$$(X - X')\varsigma^b = 0, \quad \varsigma^b = \varsigma^b(x^1, \ldots, x^m), \quad \varsigma^b(0; x^2, \ldots, x^m) = \xi^b.$$

We have reached the point that we have $X \equiv \partial/\partial x^1 \bmod J^{2k}\mathcal{T}$ for $k \geq 1$; the transition from $2k$ to $2k+1$ is carried out in an analogous manner, but with a change of even coordinates. Since $J^{m+1}\mathcal{T}M = 0$, at the end it turns out that $X = \partial/\partial x^1$. This proves assertion (a).

Let X be odd and not be zero at the point x_0. Then $X \equiv \sum_b f_b(y)\partial/\partial\eta^b \bmod J\mathcal{T}M$, and, say, $f_1(x_0) \neq 0$. We rewrite X in the form

$$X = \sum_b g^b(y, \eta_2, \ldots, \eta_n)\frac{\partial}{\partial\eta^b} + \sum_a \varsigma^a(y, \eta_2, \ldots, \eta_n)\frac{\partial}{\partial y^a} + \eta_1 X'.$$

We pass to coordinates (x, ξ): $x^a = y^a - (\eta^1/g^1)\,\varsigma^a$, $\xi^1 = \eta^1/g^1$ and $\xi^b = \eta^b - \eta^1 g^b/g^1$. In these coordinates $X = (\partial/\partial\xi^1) + \xi^1 X''$. We set $X'' = Y + \eta\,\partial/\partial\xi^1$ where $\eta = X''\xi^1$. It is not difficult to see that the even derivation Y is not zero at the point x_0 and so Y is $1|0$-rectifiable. This permits us to conclude that $X^2 = (1+\xi^1\eta)(\partial/\partial x^1) + \eta(\partial/\partial\xi^1)$ in a suitable system of coordinates. Let $X^2 u = 1$ and $\nu = Xu$; supplementing (u, ν) to get a local system of coordinates, we obtain $X = (\partial/\partial\nu) + \nu(\partial/\partial u)$.

We will omit the analysis of case (c). □

4. Frobenius theorem. A distribution on a manifold M is defined to be a locally direct subsheaf $\mathcal{T} \subset \mathcal{T}M$. A distribution is said to be integrable if $\mathcal{T}$

is closed with respect to the supercommutator. The Frobenius theorem for supermanifolds asserts the same thing as the theorem for purely even manifolds: integrability of a distribution is equivalent to the property that locally $\mathcal{T}$ be freely generated by vector fields of the form $(\partial/\partial x^a, \partial/\partial \xi^b)$, where (x^a, ξ^b) can be extended to a local system of coordinates. This is true for the holomorphic and the infinitely differentiable cases.

The proof can be carried out by induction on $p+q$, where $\operatorname{rk} \mathcal{T} = p|q$. The cases $1|0$ and $0|1$ form the content of Theorem 3.

5. Connections on a locally free sheaf. Let $\mathcal{E}$ be a locally free sheaf on a supermanifold M. A connection on $\mathcal{E}$ is defined by a covariant differential $\nabla: \mathcal{E} \to \mathcal{E} \otimes \Omega^1 M$ which satisfies the Leibniz formula

$$\nabla(ae) = (-1)^{(\widetilde{a}+1)\widetilde{e}} e \otimes da + (-1)^{\widetilde{a}} a \nabla e.$$

The signs are adapted to the relations: $\Omega^1 = \Omega^1_{\text{odd}}$ and $\widetilde{\nabla} = 1$. The concepts of the de Rham sequence of the connection, the curvature, and integrability are all defined as in the purely even case. A specifically odd construction is the definition of a right connection; the next subsection is devoted to this.

§ 5. Right Connections and Integral Forms

1. Right connections. We continue to work in the category of differentiable or analytic supermanifolds and set $K = \mathbf{R}$ or $\mathbf{C}$, respectively. We will consider a connection on a locally free sheaf $\mathcal{E}$ on M to be a covariant derivative, i.e., a K-bilinear mapping $\Delta: \mathcal{T}M \otimes_K \mathcal{E} \to \mathcal{E}$. This can be extended to a K-bilinear mapping $\Delta_l: \mathcal{D}_1 \otimes_K \mathcal{E} \to \mathcal{E}$, where $\mathcal{D}_1$ is the sheaf of differentiable operators on M of order ≤ 1, by setting $\Delta_l(f \otimes e) = fe$, where f is a local function viewed as an operator of order zero, and by extending the operator by additivity. The axioms of a connection, defined using Δ_l, take on a simpler form; in addition to the K-bilinearity of Δ_l, we have for any local operator X of order ≤ 1, any function f and any section e:

(a) $\Delta_l(f \otimes e) = fe$,

(b) $\Delta_l(X \otimes fe) = \Delta_l(X \circ f \otimes e)$;

(c) $\Delta_l(fX \otimes e) = f\Delta_l(X \otimes e)$.

The second identity is the Leibniz formula. $X \circ f$ denotes the product of X and f as operators; the circle is written to distinguish the result from the application of X fo f: $X \circ f = (-1)^{\widetilde{f}\widetilde{X}} fX + Xf$. We introduce the idea of a right connection in an analogous way:

2. Definition. A *right connection* on $\mathcal{E}$ is defined to be a K-bilinear mapping $\Delta_r: \mathcal{E} \otimes \mathcal{D}_1 \to \mathcal{E}$ with the following properties:

(a) $\Delta_r(e \otimes f) = ef$;

(b) $\Delta_r(e \otimes X \circ f) = \Delta_r(e \otimes X)f$;

(c) $\Delta_r(e \otimes fX) = \Delta_r(ef \otimes X)$.

In works on physics one writes $e\overleftarrow{X}$ instead of $\Delta_r(e \otimes X)$. If X is a vector field, the operator $\nabla_r(X)$ of right covariant differentiation along X is defined by the formula $\nabla_r(X)e = (-1)^{\tilde{X}\tilde{e}}\Delta_r(ef \otimes X)$.

On $\mathcal{O}_M$ there is an obvious left connection: $\Delta_l(X \otimes f) = Xf$. In turns out that on $\operatorname{Ber} M = \left(\operatorname{Ber} \Omega^1_{\text{odd}} M\right)^*$ there is a canonical right connection.

3. Proposition. *On* $\operatorname{Ber} M$ *there is a unique right connection* Δ_r *satisfying the following condition: for any local system of coordinates* $(x^a) = x$ *we have*

$$\Delta_r\left(D^*(dx) \otimes \frac{\partial}{\partial x^a}\right) = 0 \qquad \text{for all } a,$$

where $D^*(dx)$ *is the local section of* $\operatorname{Ber} M$ *dual to* $D(dx^1, \ldots, dx^{m+n})$.

Proof. The uniqueness of Δ_r follows from the fact that $D^*(dx)$ is a right $\mathcal{O}_M$-basis for $\operatorname{Ber} M$, while the $\partial/\partial x^a$ form a right $\mathcal{O}_M$-basis of $\mathcal{T}M$ (locally). Thus the axioms for a right connection uniquely define the action on all of $\mathcal{D}_1$. Therefore, in the domain of (x^a) we obtain the following formula for Δ_r:

$$\begin{aligned}
&\Delta_r\left(D^*(dx)\, f \otimes \left(\sum_a g^a \frac{\partial}{\partial x^a} + h\right)\right) \\
&\quad = \Delta_r\left(D^*(dx) \otimes \sum_a (-1)^{(\tilde{f}+\tilde{g}^a)\tilde{x}^a}\left[\frac{\partial}{\partial x^a} \circ (fg^a) - \frac{\partial(fg^a)}{\partial x^a}\right] + fh\right) \\
&\quad = -D^*(dx) \otimes \left[\sum_a (-1)^{(\tilde{f}+\tilde{g}^a)\tilde{x}^a} \frac{\partial(fg^a)}{\partial x^a} + fh\right],
\end{aligned}$$

which, it is not difficult to see, satisfies the axioms for a right connection.

The final point to check is that $\Delta_r\left(D^*(dy) \otimes (\partial/\partial y^b)\right) = 0$, where $y^b = y^b(x)$ is another coordinate system. In fact

$$dy^c = \sum_a dx^a \otimes \frac{\partial y^c}{\partial x^a}, \qquad \frac{\partial}{\partial y^b} = \sum_a \frac{\partial x^a}{\partial y^b}\frac{\partial}{\partial x^a},$$

from which, by the preceding formula,

$$\begin{aligned}\Delta_r\left(D^*(y)\otimes\frac{\partial}{\partial y^b}\right)&=\Delta_r\left(D^*(x)\operatorname{Ber}\left(\frac{\partial y}{\partial x}\right)\otimes\sum_a\frac{\partial x^a}{\partial y^b}\frac{\partial}{\partial x^a}\right)\\&=-D^*(dx)\sum_a(-1)^{\widetilde{x^a}(\widetilde{x^a}+\widetilde{y^b})}\frac{\partial}{\partial x^a}\left(\operatorname{Ber}\left(\frac{\partial y}{\partial x}\right)\frac{\partial x^a}{\partial y^b}\right).\end{aligned}$$

It must be shown that the last sum equals zero. Since this is equivalent to saying that the right connections defined in terms of the coordinates x and y are identical, it is sufficient to show this for elementary changes of coordinates, either linear changes or substitutions of only one function, since any change factors into the product of such ones. For linear changes $\operatorname{Ber}(\partial y/\partial x)\,\partial x^a/\partial y^b$ is constant, so everything is trivial. We will consider substitutions of even and odd functions separately. Working in the standard format, we will assume that an even function being replaced is the first one and that an odd one is the last, with index $m+n$.

Even substitution. Set $y^1=F(x^a)$, $y^b=x^b$ for $b>1$; the inverse change is $x^1=G(y^b)$, $x^a=y^a$ for $a>1$. Then we have

$$\operatorname{Ber}\left(\frac{\partial y}{\partial x}\right)=\frac{\partial F}{\partial x^1},\qquad\frac{\partial x^a}{\partial y^b}=\begin{cases}\delta^a_b & \text{if } a\neq 1,\\ \dfrac{\partial G}{\partial y^b} & \text{for } a=1,\end{cases}$$

$$\begin{aligned}&\sum_a(-1)^{\widetilde{x^a}(\widetilde{x^a}+\widetilde{y^b})}\frac{\partial}{\partial x^a}\left(\operatorname{Ber}\left(\frac{\partial y}{\partial x}\right)\frac{\partial x^a}{\partial y^b}\right)\\&\qquad=\begin{cases}\dfrac{\partial}{\partial x^b}\left(\dfrac{\partial F}{\partial x^1}\right)+\dfrac{\partial}{\partial x^1}\left(\dfrac{\partial F}{\partial x^1}\dfrac{\partial G}{\partial y^b}\right) & \text{for } b\neq 1,\\ \dfrac{\partial}{\partial x^1}\left(\dfrac{\partial F}{\partial x^1}\dfrac{\partial G}{\partial y^1}\right) & \text{for } b=1.\end{cases}\end{aligned}$$

Now we consider the identity

$$\begin{aligned}dy^1&=dx^1\frac{\partial F}{\partial x^1}+\sum_{b\geq 2}dx^b\frac{\partial F}{\partial x^b}\\&=\left(dy^1\frac{\partial G}{\partial y^1}+\sum_{b\geq 2}dy^b\frac{\partial G}{\partial y^b}\right)\frac{\partial F}{\partial x^1}+\sum_{b\geq 2}dy^b\frac{\partial F}{\partial x^b},\end{aligned}$$

whence $(\partial G/\partial y^1)(\partial F/\partial x^1)=1$ and $(\partial G/\partial y^b)(\partial F/\partial x^1)+(\partial F/\partial x^b)=0$. Differentiating with respect to x^1 we get the desired result.

Odd substitution. We set $y^{m+n} = F(x^a)$ and $y^b = x^b$ for $b < m+n$; the inverse mapping is $x^{m+n} = G(y^b)$, $x^a = y^a$ for $a < m+n$. Then we have

$$\operatorname{Ber}\left(\frac{\partial y}{\partial x}\right) = \left(\frac{\partial F}{\partial x^{m+n}}\right)^{-1}, \qquad \frac{\partial x^a}{\partial y^b} = \begin{cases} \delta_b^a & \text{for } a \neq m+n, \\ \dfrac{\partial G}{\partial y^{m+n}} & \text{for } a = m+n. \end{cases}$$

$$\sum_a (-1)^{\widetilde{x^a}(\widetilde{x^a}+\widetilde{y^b})} \frac{\partial}{\partial x^a}\left(\operatorname{Ber}\left(\frac{\partial y}{\partial x}\right)\frac{\partial x^a}{\partial y^b}\right)$$

$$= \begin{cases} \dfrac{\partial}{\partial x^b}\left[\left(\dfrac{\partial F}{\partial x^{m+n}}\right)^{-1}\right] - (-1)^{\widetilde{y^b}} \dfrac{\partial}{\partial x^{m+n}}\left[\left(\dfrac{\partial F}{\partial x^{m+n}}\right)^{-1}\dfrac{\partial G}{\partial y^b}\right] & \\ \qquad\qquad\qquad\qquad\qquad\qquad\qquad\qquad \text{for } b \neq m+n, & \\ \dfrac{\partial}{\partial x^{m+n}}\left[\left(\dfrac{\partial F}{\partial x^{m+n}}\right)^{-1}\dfrac{\partial G}{\partial y^{m+n}}\right] \qquad \text{for } b = m+n. & \end{cases}$$

In analogy with the even case, from the identity

$$dy^{m+n} = dx^{m+n}\frac{\partial F}{\partial x^{m+n}} + \sum_{b<m+n} dx^b \frac{\partial F}{\partial x^b}$$

$$= \left(dy^{m+n}\frac{\partial G}{\partial y^{m+n}} + \sum_{b<m+n} dy^b\frac{\partial G}{\partial y^b}\right)\frac{\partial F}{\partial x^{m+n}} + \sum_{b<m+n} dy^b\frac{\partial F}{\partial x^b}$$

we obtain

$$\frac{\partial G}{\partial y^{m+n}} = \left(\frac{\partial F}{\partial x^{m+n}}\right)^{-1}, \qquad \frac{\partial G}{\partial y^b} + \frac{\partial F}{\partial x^b}\left(\frac{\partial F}{\partial x^{m+n}}\right)^{-1} = 0.$$

Since x^{m+n} is an odd variable, $\partial F/\partial x^{m+n} = H(x^1, \ldots, x^{m+n-1})$. Then it follows from the first equality that $\partial/\partial x^{m+n}\left[(\partial F/\partial x^{m+n})^{-1}\,\partial G/\partial y^{m+n}\right] = 0$. Finally, for $b < m+n$, taking into account that $\widetilde{y^b} = \widetilde{x^b}$:

$$\frac{\partial}{\partial x^b}\left[\left(\frac{\partial F}{\partial x^{m+n}}\right)^{-1}\right] - (-1)^{\widetilde{y^b}}\frac{\partial}{\partial x^{m+n}}\left[\left(\frac{\partial F}{\partial x^{m+n}}\right)^{-1}\frac{\partial G}{\partial y^b}\right]$$

$$= -\left(\frac{\partial F}{\partial x^{m+n}}\right)^{-2}\frac{\partial}{\partial x^b}\frac{\partial F}{\partial x^{m+n}} - (-1)^{\widetilde{y^b}}\left(\frac{\partial F}{\partial x^{m+n}}\right)^{-1}\frac{\partial}{\partial x^{m+n}}\frac{\partial G}{\partial y^b}$$

$$= -(-1)^{\widetilde{x^b}}\left(\frac{\partial F}{\partial x^{m+n}}\right)^{-1}\frac{\partial}{\partial x^{m+n}}\left[\frac{\partial G}{\partial y^b} + \left(\frac{\partial F}{\partial x^{m+n}}\right)^{-1}\frac{\partial F}{\partial x^b}\right] = 0.$$

4. Integral forms. Integral forms are defined to be sections of the sheaf

$$\Sigma M = \operatorname{Ber} M \otimes_{\mathcal{O}_M} S(\mathcal{T} M\Pi) = \mathcal{H}om_{\mathcal{O}_M}\left(\Omega^{\cdot}_{\mathrm{odd}} M, \operatorname{Ber} M\right).$$

Here $\mathcal{T}M\Pi$ is an $\mathcal{O}_M$-module with the same left multiplication by $\mathcal{O}_M$ as $\mathcal{T}M$ but with reversed parity. By analogy with the exterior differential on ΩM, one can define a differential $\delta\colon \Sigma M \to \Sigma M$ with $\delta^2 = 0$. To begin with, we introduce the definition in local coordinates. In the domain of the coordinates (x^a) we have $\Sigma M = \operatorname{Ber} M \otimes_K K\left[\frac{\partial}{\partial x^a}\Pi\right]$; we set

$$\delta^x = \sum_a \nabla_r\left(\frac{\partial}{\partial x^a}\right) \otimes_K \frac{\partial}{\partial(\frac{\partial}{\partial x^a}\Pi)}(-1)^{\widetilde{x^a}}.$$

5. Proposition. *The definition $\delta = \delta^x$ does not depend upon the choice of the coordinates (x^a); also $\delta^2 = 0$.*

Proof. To check that $\delta^2 = 0$ we apply the criterion of § 4 of Chapter 3. The operators $\nabla_r\left(\frac{\partial}{\partial x^a}\right)$ and $\partial/\partial\left(\frac{\partial}{\partial x^a}\Pi\right)$ have opposite parity. It remains to show that the first ones also commute pairwise. Let $X = \partial/\partial x^a$ and $Y = \partial/\partial x^b$. Then for $M = m|n$

$$\begin{aligned}\nabla_r(X)(D^*(dx)\, f) &= (-1)^{\widetilde{X}(m+\widetilde{f})}\Delta_r(D^*(dx)\, f \otimes X)\\ &= (-1)^{\widetilde{X}m+1} D^*(dx)\, Xf,\end{aligned}$$

whence $[\nabla_r(X), \nabla_r(Y)] = (-1)^{(\widetilde{X}+\widetilde{Y})m}\nabla_r([X,Y]) = 0$.

To check that the definition is correct, we first show that on $\operatorname{Ber} M \otimes \mathcal{T}M\Pi$ the operator δ^x does not depend on (x^a); more precisely,

$$\delta^x(\omega \otimes X\Pi) = (-1)^{\widetilde{\omega}+\widetilde{X}}\Delta_r(\omega \otimes X).$$

In fact, let $\omega = D^*(dx)\, f$ and $X = \sum g^a\, (\partial/\partial x^a)$. Then

$$\begin{aligned}\delta^x(\omega\otimes X\Pi) &= \left(\sum_b \nabla_r\left(\frac{\partial}{\partial x^b}\right)\otimes_K \frac{\partial}{\partial(\frac{\partial}{\partial x^b}\Pi)}(-1)^{\widetilde{x^b}}\right)\left(\sum_a D^*(dx)\, fg^a \otimes_K \frac{\partial}{\partial x^a}\Pi\right)\\ &= \sum_a (-1)^{\widetilde{x^a}+(\widetilde{x^a}+1)(\widetilde{\omega}+\widetilde{g^a})}\nabla_r\left(\frac{\partial}{\partial x^a}\right)(D^*(dx)\, fg^a)\\ &= \sum_a (-1)^{\widetilde{x^a}+(\widetilde{x^a}+1)(\widetilde{\omega}+\widetilde{g^a})}(-1)^{\widetilde{x^a}(\widetilde{\omega}+\widetilde{g^a})}\Delta_r\left(D^*(dx)\, fg^a\otimes\frac{\partial}{\partial x^a}\right)\\ &= (-1)^{\widetilde{\omega}+\widetilde{X}}\Delta_r(\omega\otimes X).\end{aligned}$$

Finally, we prove that for $F, G \in K\left[\frac{\partial}{\partial x^b}\Pi\right]$ we have the Leibniz formula:

$$\delta^x(\omega \otimes_K FG) = \delta^x(\omega \otimes_K F)G + (-1)^{GF}\delta^x(\omega \otimes_K G)F.$$

From this it will follow by induction on the degree of F, beginning with one, that $\delta^x(\omega \otimes F)$ does not depend on (x^a).

In fact:

$$\delta^x(\omega \otimes FG) = \left[\sum_a \nabla_r\left(\frac{\partial}{\partial x^a}\right) \otimes_K \frac{\partial}{\partial\left(\frac{\partial}{\partial x^a}\Pi\right)}(-1)^{\widetilde{x}^a}\right](\omega \otimes FG)$$

$$= \sum_a \nabla_r\left(\frac{\partial}{\partial x^a}\right)\omega \otimes (-1)^{\widetilde{\omega}(\widetilde{x}^a+1)+\widetilde{x}^a}\left[\frac{\partial F}{\partial\left(\frac{\partial}{\partial x^a}\Pi\right)}G + (-1)^{\widetilde{F}(\widetilde{x}^a+1)}F\frac{\partial G}{\partial\left(\frac{\partial}{\partial x^a}\Pi\right)}\right];$$

the decomposition into two summands inside the brackets corresponds to the right side of the Leibniz formula. □

6. Remarks. (a) As usual, in the purely even case ($n = 0$) there is an isomorphism $\operatorname{Ber} M \to \Omega^m M$ which carries $D^*(dx)$ to $dx^1 \wedge \cdots \wedge dx^m$. (The parity of this isomorphism is $(-1)^m$ if we assume that $\Omega^m M$ has rank $1|0$.) The contraction operation of a vector field with a differentiable form permits us to construct an isomorphism $\Omega^m M \otimes \wedge^i(\mathcal{T}M) \xrightarrow{\sim} \wedge^{m-i}(\mathcal{T}^*M) = \Omega^{m-i}M$. Over the sheaf of purely even rings $\mathcal{O}_M$, one can identify $\wedge^i(\mathcal{T}M)$ with $S^i(\mathcal{T}M\Pi)$. Thus in the purely even case, integral forms essentially are the same as differential forms; in this case the differentials are also identified up to some signs.

(b) If the $\mathbb{Z}$-grading of the module of differential forms is chosen in the usual way, i.e., set $\mathcal{O}_M = \Omega^0$ and take $dx^a \in \Omega^1$, then it is natural to say that $S^i(\mathcal{T}M\Pi)$ has degree $-i$ and that δ increases degree by 1. The degree of $D^*(dx)$ can be decided in several possible ways. Since for even constants t we have $D^*(d(tx)) = t^{m-n}D^*(dx)$, one can consider $\operatorname{Ber} M = \Sigma_{m-n}M$. If one wishes to have the homology of the complex of sections of integral forms vanish in the same interval as the differential forms, then it is natural to consider $\operatorname{Ber} M = \Sigma_m M$:

$$\mathcal{O}_M \to \Omega^1 M \to \cdots \to \Omega^m M \to \Omega^{m+1} M \to \cdots$$
$$\cdots \to \Sigma_{-1}M \to \Sigma_0 M \to \Sigma_1 M \to \cdots \to \operatorname{Ber} M.$$

7. Spencer sequence of a sheaf with right connection. Now let $\mathcal{E}$ be a locally free sheaf with right connection Δ_r on M. We set $\Sigma(\mathcal{E}) = \mathcal{E} \otimes_{\mathcal{O}_M} S(\mathcal{T}M\Pi)$. The preceding construction can be carried out for $\mathcal{E}$ instead of $\operatorname{Ber} M$. More precisely, in the domain of the coordinates (x^a) we set

$$\delta^x(\mathcal{E}, \Delta_r) = \sum_a \nabla_r\left(\frac{\partial}{\partial x^a}\right) \otimes_K \frac{\partial}{\partial\left(\frac{\partial}{\partial x^a}\Pi\right)}(-1)^{\widetilde{x}^a} : \Sigma(\mathcal{E}) \to \Sigma(\mathcal{E}).$$

Then the following facts are true:

8. Proposition. (a) $\nabla_r = \delta^x(\mathcal{E}, \Delta_r)$ *does not depend on the choice of coordinate system* (x^a) *and is determined only by the connection* Δ_r.

(b) $\nabla_r^2 \colon \Sigma(\mathcal{E}) \to \Sigma(\mathcal{E})$ *is an* $\mathcal{O}_M$*-linear operator which is called the curvature of the right connection* Δ_r.

Proof. Assertion (a) can be verified by a literal repetition of the appropriate part of the proof of Proposition 5.

To prove (b) it is convenient to compute ∇_r^2 in the coordinates (x^a). We set $\Phi_{ba} = \left[\nabla_r\left(\frac{\partial}{\partial x^b}\right), \nabla_r\left(\frac{\partial}{\partial x^a}\right)\right] : \mathcal{E} \to \mathcal{E}$. Then a direct computation shows that on $\mathcal{E} \otimes_K S(\mathcal{T}M\Pi)$

$$\nabla_r^2 = \sum_{a,b} (-1)^{\widetilde{x^a}+1} \Phi_{ba} \otimes \frac{\partial}{\partial\left(\frac{\partial}{\partial x^a}\Pi\right)} \frac{\partial}{\partial\left(\frac{\partial}{\partial x^b}\Pi\right)} .$$

From the definition of $\nabla_r(X)$ it follows that $\nabla_r(X)(fe) = (-1)^{\widetilde{X}\widetilde{f}} f \nabla_r(X)(e) +$ (*a term including the derivative* Xf), where f is a function on M and e is a section of $\mathcal{E}$. Therefore, $\Phi_{ba}(fe) = (-1)^{\widetilde{f}(\widetilde{x^a}+\widetilde{x^b})} f \Phi_{ba}(e)$. From this it follows that ∇_r^2 is linear. □

9. Theorem. *The categories of right and left connections are equivalent. More precisely, we consider the mapping B which sets up a correspondence between the pair* $(\mathcal{E}, \Delta_l)$ *and the pair* $(\operatorname{Ber} M \otimes \mathcal{E}, \Delta_r)$, *where for any vector field X, by definition,*

$$\Delta_r(\omega \otimes e \otimes X) = (-1)^{\widetilde{X}\widetilde{e}} \omega \otimes \Delta_l(X \otimes e) + (-1)^{\widetilde{X}\widetilde{e}} \Delta_r(\omega \otimes X) \otimes e .$$

This mapping is well-defined and extends uniquely to a functor on the category of morphisms $(\mathcal{E}, \Delta_l) \to (\mathcal{E}', \Delta_l')$ *which commute with connections. This functor is an equivalence.* □

We omit the proof, which consists of a series of automatic verifications.

We only observe that the curvatures of the corresponding right and left connections can in a natural sense be identified. In the de Rham sequence $\mathcal{E} \otimes \Omega^{\cdot} M$ the curvature ∇_l^2 is multiplication by a $\Phi_l \in \Gamma(\operatorname{End} \mathcal{E} \otimes \Omega^{\cdot} M)$. In the Spencer sequence $\Sigma(\operatorname{Ber} M \otimes \mathcal{E})$, the curvature ∇_r^2 can be interpreted as contraction with the same form Φ_l.

§ 6. The Berezin Integral

1. We recall that the complex of sheaves of differential forms on a purely even differentiable manifold $\mathcal{O}_M \to \Omega^1 M \to \Omega^2 M \to \cdots$ is exact everywhere except at the zero term while the complex of sections with compact support (or finite sections) has as its cohomology $H_c^{\cdot}(M, \mathbf{R})$. Later we will establish the corresponding fact for supermanifolds and the complex of integral forms.

We begin with the right side of this complex, $\Sigma_m M = \operatorname{Ber} M$ and $\Sigma_m^c(M) = \operatorname{Ber}_c(M) = \Gamma_c(M, \operatorname{Ber} M)$; M is assumed to be an oriented, connected differentiable supermanifold, and c indicates compact supports. We recall the role of the compactness condition in the purely even situation. For functions and forms on $M^{1|0} = \mathbf{R}$ there are two exact sequances:

$$0 \to \mathbf{R} \to \mathcal{O}(\mathbf{R}) \xrightarrow{d} \Omega^1(\mathbf{R}) \to 0,$$

$$0 \to \mathcal{O}_c(\mathbf{R}) \xrightarrow{d} \Omega_c^1(\mathbf{R}) \xrightarrow{\int} \mathbf{R} \to 0,$$

since a form $\omega \in \Omega_c^1(\mathbf{R})$ can always be represented in the form $\omega = df$ for some $f \in \mathcal{O}(\mathbf{R})$. But a choice of $f \in \mathcal{O}_c(\mathbf{R})$ is possible only if $\int \omega = 0$. The functional $\omega \mapsto \int_m \omega$ plays an analogous role for general volume forms with compact support. The construction of the Berezin integral $\int : \operatorname{Ber}_c(M) \to \mathbf{R}$ is the topic of this section. We assume that M_{rd} is oriented.

2. Definition. (a) Let (x^a) be a local system of coordinates on M such that the $(x^1, \ldots, x^m)$ are even and define the given orientation on M_{rd} and the $(x^{m+1}, \ldots, x^m)$ are odd. Let $\omega \in \operatorname{Ber}_c(M)$ be a form with compact support in the domain of the coordinates (x^a):

$$\omega = D^*(dx) \otimes \sum_\alpha (x^{m+1})^{\alpha_1} \cdots (x^{m+n})^{\alpha_n} f_\alpha(x^1, \ldots, x^m).$$

In such a case we set

$$\int_M \omega = (-1)^{mn} \int_{M_{\mathrm{rd}}} dx^1 \ldots dx^m f_{1\ldots 1}(x_1, \ldots, x_m).$$

(b) For a general volume form with compact support $\omega \in \operatorname{Ber}_c(M)$ we define $\int_M \omega$ by additivity, using a partition of unity and (a).

3. Theorem. *The integral is well-defined. For forms $\omega \in \operatorname{Ber}_c(M)$ there exists a $\nu \in \Sigma_{m-1}^c(M)$ with $\delta\nu = \omega$ if and only if $\int_M \omega = 0$.*

Proof. In the domain of the coordinates (x^a) we have

$$\delta\left(D^*(dx)\, f \otimes \frac{\partial}{\partial x^a}\Pi\right) = \pm D^*(dx)\frac{\partial f}{\partial x^a}\,.$$

Consequently, the forms $D^*(dx)\left(\sum_{i,a}\partial f_i/\partial x^a\right)$ are in the image of δ, where the f_i have compact support. Specifically, if $\alpha \neq (1,\dots,1)$, there exists an x^{m+j} with $\alpha_j = 0$; then

$$(x^{m+1})^{\alpha_1}\cdots(x^{m+n})^{\alpha_n} f_\alpha(x^1,\dots,x^n) = \pm\frac{\partial}{\partial x^{m+j}}(x^{m+j}x^\alpha f_\alpha).$$

In particular, there is no contribution to the Berezin integral from the forms which are in the image of δ in our coordinate domain. Finally, the image of δ contains $D^*(dx)dx^{m+1}\cdots dx^{m+n}f_{1\dots1}$ only if $f_{1\dots1}$ is a divergence in the usual sense.

The remaining point is to reformulate this reasoning globally and invariantly with respect to coordinate changes applied to the form. We break the reasoning into several steps.

(a) $\operatorname{Ber}_c(M) = \delta\left(\Sigma^c_{m-1}(M)\right) + J^n_M \operatorname{Ber}_c(M)$.

To prove this using a partition of unity, we represent any form $\omega \in \operatorname{Ber}_c(M)$ as a locally finite sum $\omega = \sum \omega^{(i)}$, where the support of each $\omega^{(i)}$ lies in the coordinate neighborhood of $(x^a_{(i)})$ and is compact. We decompose $\omega^{(i)}$ as in the beginning of the proof: $\omega^{(i)} = \omega_1^{(i)} + x^{m+1}_{(i)}\cdots x^{m+n}_{(i)}\omega_2^{(i)}$, where $\omega_1^{(i)}$ contains only monomials in the odd coordinates of degree $< n$. Then $\sum_i \omega_1^{(i)} \in \delta\left(\Sigma^c_{m-1}(M)\right)$ and $\omega - \sum_i \omega_1^{(i)} \in J^n_M \operatorname{Ber}_c(M)$.

(b) There is a natural isomorphism of sheaves of $\mathcal{O}_{M_{\mathrm{rd}}}$-modules

$$b\colon J^n_M \operatorname{Ber}_c(M) \,\tilde{\to}\, \Omega^m_c(M_{\mathrm{rd}}),$$

which in any local coordinate system is defined by the formula

$$b(D^*(dx)x^{m+1}\cdots x^{m+n}f) = dx^1_{\mathrm{rd}}\cdots dx^m_{\mathrm{rd}} f_{\mathrm{rd}}\,, \quad f_{\mathrm{rd}} = f \bmod J_N\,.$$

Actually there is a corresponding sheaf isomorphism $J^n_M \operatorname{Ber} M \to \Omega^n M_{\mathrm{rd}}$. To check this, it is sufficient to be satisfied that the left and right sides agree under change of coordinates to (y^a). The following computation shows this also for analytic manifolds:

$$\begin{aligned} D^*(dy)y^{m+1}\cdots y^{m+n} &= D^*(dx)\operatorname{Ber}\left(\frac{\partial y}{\partial x}\right) y^{m+1}\cdots y^{m+n} \\ &= D^*(dx)\det\left(\frac{\partial y^{\le m}}{\partial x^{\le m}}\right)\det{}^{-1}\left(\frac{\partial y^{>m}}{\partial x^{>m}}\right) y^{m+1}\cdots y^{m+n}; \end{aligned}$$

$$dy^1_{\mathrm{rd}}\cdots dy^m_{\mathrm{rd}} = \det\left(\frac{\partial {y^{\le m}}_{\mathrm{rd}}}{\partial {x^{\le m}}_{\mathrm{rd}}}\right) dx^1_{\mathrm{rd}}\cdots dx^m_{\mathrm{rd}}\,.$$

Since clearly $\det\left(\partial y^{\leq m}/\partial x^{\leq m}\right)_{\mathrm{rd}} = \det\left(\partial y^{\leq m}{}_{\mathrm{rd}}/\partial x^{\leq m}{}_{\mathrm{rd}}\right)$, the last point is to check that $\det^{-1}\left(\partial y^{>m}/\partial x^{>m}\right) y^{m+1}\cdots y^{m+n} = g x^{m+1}\cdots x^{m+n}$, where $g_{\mathrm{rd}} = 1$. Let $y^{m+j} = \sum_i f_i^j x^{m+i} \bmod J_M^2$ where the f_i^j are odd functions. Then $\det\left(\partial y^{>m}/\partial x^{>m}\right)_{\mathrm{rd}} = \det(f_i^j)_{\mathrm{rd}}$ and $y^{m+1}\cdots y^{m+n} = \det(f_i^j) x^{m+1}\cdots x^{m+n}$, which proves what was required.

(c) $b\left(\delta(\Sigma_{m-1}^c(M)) \cap J_M^n \operatorname{Ber}_c(M)\right) = d\left(\Omega_c^{m-1}(M_{\mathrm{rd}})\right)$.

We check both sides of the inclusion. Let $\omega \in J_M^n \operatorname{Ber}_c(M)$, with $\omega = d\nu$ for some $\nu \in \Sigma_{m-1}^c(M)$. We decompose $\nu = \sum \nu^{(i)}$ as in (a). We may assume that $\nu^{(i)} = x_{(i)}^{m+1}\cdots x_{(i)}^{m+n}\mu^{(i)}$ since $\delta^2 = 0$ and since the components $\nu^{(i)}$ without a complete number of odd $x_{(i)}$ are in the image of δ. Then $b(\delta\nu^{(i)}) \in d\left(\Omega_c^{m-1}(M_{\mathrm{rd}})\right)$, and summation with respect to i proves that the left side is contained in the right.

Conversely, let $b(\omega) = d\left(\Omega_c^{m-1}(M)\right)$, where $\omega \in J_M^n \operatorname{Ber}_c(M)$. We set $b(\omega) = d\left(\sum_i \nu_{(i)}\right)$, where the $\nu_{(i)}$ are differential forms with compact support on M_{rd} in the domain of $\left(x_{(i)\mathrm{rd}}^1, \ldots, x_{(i)\mathrm{rd}}^m\right)$. We "lift" the $\nu_{(i)}$ to integral forms $\mu_{(i)}$; in coordinates:

$$\nu_{(i)} = \sum_j dx_{(i)\mathrm{rd}}^1 \cdots \widehat{dx_{(i)\mathrm{rd}}^j} \cdots dx_{(i)\mathrm{rd}}^m f_{(i)}^j\,,$$

$$\mu_{(i)} = \sum_j \pm D^*(dx) x_{(i)}^{m+1}\cdots x_{(i)}^{m+n} f_{(i)}^j \otimes \frac{\partial}{\partial x_{(i)}^j}\Pi\,,$$

where the $\pm$ signs are chosen such that $b(\delta\mu_{(i)}) = \nu_{(i)}$. Then the support of $\mu_{(i)}$ is the same as that of $\nu_{(i)}$ and $b(\omega - \delta\mu) = 0$, where $\mu = \sum_i \mu_{(i)} \in J_M^n \Sigma_{m-1}^c(M)$.

We set $\omega' = \omega - \delta\mu$. We choose a partition of unity $(g_{(i)})$ subordinate to the coordinate domains. Then $b(g_{(i)}\omega') = g_{(i)\mathrm{rd}} b(\omega') = 0$. Let $g_{(i)}\omega' = \omega'_{(i)}$. Since $b\left(\omega'_{(i)}\right) = 0$, the expansion of $\omega'_{(i)}$ with respect to the coordinates $x_{(i)}^a$ does not contain any terms divisible by $x_{(i)}^{m+1}\cdots x_{(i)}^{m+n}$. This means that $\omega'_{(i)} = \delta\mu'_{(i)}$, where the support of $\mu'_{(i)}$ is no greater than the support of $\omega'_{(i)}$. Finally, $\omega = \delta\left(\mu + \sum \mu'_{(i)}\right)$.

Now we can conclude the proof of the theorem. The mapping b induces an isomorphism

$$\operatorname{Ber}_c(M)/\delta\left(\Sigma_{m-1}^c(M)\right) \xrightarrow{\sim} \Omega_c^m(M_{\mathrm{rd}})/d\Omega_c^{m-1}(M_{\mathrm{rd}})\,.$$

On the other hand, the ordinary integral of a form with compact support over M_{rd} induces an isomorphism of the latter space with $\mathbf{R}$. The Berezin integral is a composition of these two isomorphisms; this is clear from its definition. □

4. Remark. If a volume form $\omega \in \Gamma(M, \operatorname{Ber} M)$ does not have compact support, the Berezin integral $\int \omega$ can be viewed as a generalized superfunction on M; that is, the functional on $\Gamma_c(M, \mathcal{O}_M)$ which associates to a function f the integral $\int \omega f$.

§ 7. Densities

1. Let M be a supermanifold with $d \leq \dim M$, its superdimension. The goal of this section is to introduce objects with can be integrated over immersed d-dimensional subsupermanifolds in M. The idea is to define these objects on M so that they behave functorially with respect to immersions $\phi\colon N^d \to M$ and such that the pullback to N^d is a volume form on N^d. Then the Berezin integral of this form will be what is required.

More precisely, we construct the relative grassmannian $G = G_M(d; \mathcal{T}M) \xrightarrow{\pi} M$ and denote by $\mathcal{S}$ the tautological sheaf on it.

2. Definition. A *d-density* on M is defined to be any local section of the sheaf $(\operatorname{Ber} \Pi \mathcal{S}^*)^*$ on G. □

We observe that the domain of definition of the density is an open subsuperspace in G and not in M; from the point of view of M this domain of definition consists of some points of M and some distinguished d-dimensional tangent spaces at these points.

We will denote by Ξ^d the sheaf $(\operatorname{Ber} \Pi \mathcal{S}^*)^*$.

3. Proposition. *Let $\phi\colon N \to M$ be an immersion of supermanifolds, with $\dim N = d$ and τ a density on M. Then on N one can define canonically a volume form $\phi^*(\tau)$, that is, a section of* $\operatorname{Ber} N$ *over a suitable open subset of N.*

Proof. Since ϕ is an immersion, $d\phi\colon \mathcal{T}N \to \phi^*(\mathcal{T}M)$ is a locally direct injection. By the universal property of the grassmannian G there exists a unique morphism $G\phi\colon N \to G$ with the properties $\mathcal{T}N = (G\phi)^*(\mathcal{S})$ and $\phi = \pi \circ G\phi$; this identifies $(G\phi)^*(\mathcal{S})$ naturally with the image of $\mathcal{T}N$ with respect to $d\phi$. Therefore, $\operatorname{Ber} N = (\operatorname{Ber} \Pi \mathcal{T}^* N)^* = (G\phi)^*(\operatorname{Ber} \Pi \mathcal{S}^*)^*$ (the reader is asked to forgive the profusion of stars denoting different things in the same formula).

Now if $\tau \in \Gamma(V, (\operatorname{Ber} \Pi \mathcal{S}^*)^*)$ for an open $V \subset G$, we set

$$\phi^*(\tau) = (G\phi)^*(\tau) \in \Gamma((G\phi)^{-1}(V), \operatorname{Ber} N). \quad \Box$$

4. Remarks. A d-density on M as an object of integration has at least two peculiarities:

(a) $\phi^*(\tau)$ "depends on the first derivatives" of the immersion ϕ, since $d\phi$ is used in the construction of the volume $G\phi$ on the grassmannian.

(b) There are two ways that the domain of definition of τ may not contain a cylinder $\pi^{-1}(U)$, for $U \subset M$, i.e., that for the immersion ϕ there are forbidden directions. We recall that in the purely even complex-analytic geometry on the absolute grassmannian $G(d;T)$ the tautological sheaves S and $\widetilde{S}$ are negative in the sense that the tensor algebra generated by them (without dualizing) has no nonconstant holomorphic sections over the whole grassmannian. In contrast, S^* and $\widetilde{S}^*$ are positive.

Now let $G = G(d_0|d_1; T_0 \oplus T_1)$. As we already know (see § 3), $G_{\mathrm{rd}} = G(d_0;T_0) \times G(d_1;T_1)$ and $S_{\mathrm{rd}} = \pi_0^* S_0 \oplus \pi_1^* \Pi S_1$. Therefore, $\Pi S_{\mathrm{rd}}^* = \pi_0^* \Pi S_0 \oplus \pi_1^* \Pi S_1^*$ and $(\mathrm{Ber}\, \Pi S_{\mathrm{rd}}^*)^*$ will generally be the tensor product of two invertible sheaves coming from G_0 and G_1, one of which is positive and the other negative. In a case when the negative factor is present, there are no holomorphic sections over G of $(\mathrm{Ber}\, \Pi S_{\mathrm{rd}}^*)^*$. In the relative variant, this means that there are no holomorphic sections along the fiber in the cylinder $\pi^{-1}(U)$.

The following theorem shows that in those cases when there are holomorphic sections along the fibers they are provided using differential and integral forms.

5. Theorem. *Let* $\dim M = m|n$. *The sheaf* Ξ^d *of* d*-densities is purely positive along the fibers of* π *only if* $d = p|0$ *or* $d = p|n$ *and purely negative only if* $d = 0|q$ *or* $d = m|q$. *In the first two cases, there are canonical mappings from the forms to the densities:*

(a) $d = p|0$: $\Gamma(U; \Omega^p M) \to \Gamma(\pi^{-1}(U), \Xi^{p|0})$,

(b) $d = p|n$: $\Gamma(U; \Sigma_p M) \to \Gamma(\pi^{-1}(U), \Xi^{p|n})$.

In the remaining two cases we have:

(c) $d = 0|q$: $\Xi^{0|q} = S^q(S)\Pi^q$,

(d) $d = m|q$: $\Xi^{m|q} = \pi^*(\mathrm{Ber}\, M) \otimes \Pi^{n-q} S^{n-q}(\widetilde{S})$.

In particular, an $m|n$*-density is a section of the sheaf* Ber M.

Proof. We observe first of all that for any locally free sheaf $\mathcal{F}$ of purely even or odd rank there are isomorphisms expressing the tensor character of the berezinian (in distinction from the general case):

$\mathrm{rk}\, \mathcal{F} = r|0$:

$$(\mathrm{Ber}\, \Pi \mathcal{F})^* = S^r(\Pi \mathcal{F}) \colon D^*(\Pi f_1, \ldots, \Pi f_r) \mapsto \Pi f_1 \cdots \Pi f_r;$$

$\mathrm{rk}\, \mathcal{F} = 0|r$:

$$(\mathrm{Ber}\, \Pi \mathcal{F}) = \Pi^r S^r(\mathcal{F}) \colon D(\Pi f_1, \ldots, \Pi f_r) \mapsto \Pi^r (f_1 \cdots f_r).$$

From this and from the remarks in the preceding section it follows that if $0 < d_0 < m$ and $0 < d_1 < n$ then the sheaf $\mathrm{Gr}_0\, \Xi^d$ cannot be either purely positive or purely negative. Furthermore, for $d = p|0$ we find $(\mathrm{Ber}\,\Pi S^*)^* = S^p(\Pi S^*)$. After this, from the standard morphism $a\colon \pi^*(\Omega^1_{\mathrm{odd}} M) \to \Pi S^*$ we find the morphism of sheaves

$$S^p(a)\colon \pi^*(\Omega^p M) \to S^p(\Pi S^*) = \Xi^{p|0},$$

from which follow the mappings of (a).

For the construction of the mappings of (b) one must first express $\Xi^{p|n}$ in terms of $\widetilde{S}$ using the exact sequence $0 \to \Pi\widetilde{S} \to \pi^*(\Omega^1_{\mathrm{odd}} M) \to \Pi S^* \to 0$, which gives

$$(\mathrm{Ber}\,\Pi S^*)^* \otimes (\mathrm{Ber}\,\Pi\widetilde{S})^* = \mathrm{Ber}\, M\,,$$

and further,

$$\Xi^{p|n} = \mathrm{Ber}\, M \otimes \mathrm{Ber}\,\Pi\widetilde{S} = \mathrm{Ber}\, M \otimes (S^{m-p}(\Pi\widetilde{S}))^* = \mathrm{Ber}\, M \otimes S^{m-p}(\widetilde{S}\Pi^*)\,.$$

To conclude we construct the symmetric power of the morphism $b\colon \pi^*(\mathcal{T} M\Pi) \to \widetilde{S}^*\Pi$:

$$\mathrm{id} \otimes S^p(b)\colon \pi^*(\mathrm{Ber}\, M \otimes S^{m-p}(\mathcal{T} M\Pi)) = \pi^*(\Sigma_p M) \to \Xi^{p|n}\,.$$

The remaining two assertions of the theorem are verified in an analogous way, (c) directly and (d) by passing to $\widetilde{S}$. □

6. Thus from Theorem 5 it follows that differential forms can be integrated over submanifolds of odd dimension zero and integral forms can be integrated over submanifolds of odd codimension zero. For submanifolds of intermediate odd dimension, at present we know nothing except about densities. It is natural to pose the question whether there exist constructions which by themselves lead to densities. In purely even geometry, for example, from a symplectic form ω on M one constructs the forms $\omega^{\wedge p}$, i.e., $2p$-densities, and from a metric $ds^2 = g_{ab}\, dx^a\, dx^b$ one constructs induced densities on p-dimensional subsupermanifolds defined up to sign such as arc length $\sqrt{ds^2}$ or volume $\sqrt{|\det g|}\, d^m x$.

The algebraic mechanism on which this construction is based carries over into supergeometry.

7. Lemma. *Let A be a supercommutative ring and let S be a free A-module, with $b\colon S \otimes S \to A$ a nondegenerate even bilinear form on S. Then from this there is uniquely defined the square of a "volume form" $w(b) \in (\mathrm{Ber}\, S^*)^{-2}$.*

Proof. We view b as an isomorphism $b\colon S^* \to S$. then $\mathrm{Ber}\, b$ is an isomorphism

$$\mathrm{Ber}\, b\colon \mathrm{Ber}\, S^* \to \mathrm{Ber}\, S = (\mathrm{Ber}\, S^*)^{-1}.$$

Dualizing, we obtain the isomorphism $A \xrightarrow{\sim} (\operatorname{Ber} S^*)^{-2}$; then $w(b)$ is the image of the unit under this mapping.

Globalizing this construction leads to the following result:

8. Proposition. *Let M be a supermanifold with a supermetric ω, that is, a nondegenerate bilinear form on $\mathcal{T}M\Pi$. Then for any superdimension $d \leq \dim M$, from ω one can construct a canonical square of a d-density $w_d(\omega) \in \Gamma(V; (\Xi^d)^2)$, where $V \subset G_M(d; \mathcal{T}M)$ is the maximal open set on which the restriction of the form $\pi^*(\omega)$ to the subsheaf $S\Pi \subset \pi^*(\mathcal{T}M\Pi)$ is nondegenerate.* □

9. Remarks and variants. (a) If we apply this construction to the purely even case and a skew-symmetric form ω, then we obtain $w_d(\omega) = 0$ for odd d since the restriction of a skew-symmetric form to an odd-dimensional space is always degenerate, while $w_{2d}(\omega) = (\omega^{\wedge d})^2$ (in the sense of the identification of forms with densities). The possibility of extracting a square root is related to whether the determinant of the skew-symmetric matrix is the square of its pfaffian. For even (anti)symmetric forms in the even-odd case, where O and Sp are intermingled, the O part always prevents taking a square root.

In the purely even case, an essential role is also played by the condition that ω be closed; from this it follows that $\omega^{\wedge d}$ is closed. In supergeometry the important concept of a closed densitiy can also be introduced. In some measure this compensates for the absence of an exterior differential relating densities of different dimensions.

(b) As a specifically supergeometric variant of a symplectic structure, it has been suggested that one consider a supermanifold M supplied with an odd nondegenerate closed form $\omega \in (\Omega^2_{\text{odd}} M)_1$. The attempt to construct densities from such a form in analogy with Lemma 7 are not successful for the following reason.

In the notation of Lemma 7 we have an even isomorphism b: $S^* \to \Pi S$ (instead of S). Its berezinian sets up an isomorphism of modules which are already canonically isomorphic (and not dual): $\operatorname{Ber} S^* = (\operatorname{Ber} S)^{-1} \to \operatorname{Ber} \Pi S = (\operatorname{Ber} S)^{-1}$. Thus $\operatorname{Ber} b$ can be identified with an element of the ring A_0. the globalization of this construction leads to a family of functions $v_d(\omega)$ on $G_M(d; \mathcal{T}M)$ whose role is not completely clear.

(c) The preceding reasoning shows that if we replace a bilinear form b by an isomorphism p: $S \to \Pi S$ then it is again possible to construct the square of a volume form. Putting a "Π-symmetry" on M, that is, an isomorphism p: $\mathcal{T}M \to \Pi\mathcal{T}M$ with the condition $p^2 = 1$, gives a very interesting supergeometric structure. In the language of G-structures of Cartan, this means reducing the structure group of the tangent bundle to a special analog of the complete linear group in superalgebra. However, not every direct subsheaf $S \subset \mathcal{T}M$ is carried into itself by a Π-symmetry

p. Such subsheaves are classified by the grassmannian $G\Pi_M(d;\mathcal{T}M,p)\xrightarrow{\pi}M$, using which one can introduce the analogs of d-densities $\Xi_p^d = (\mathrm{Ber}\, S_{G\Pi}^*)^*$. These densities turn into volume forms on the immersed supermanifolds whose tangent bundles are p-invariant. This leads to the concept of an integrable p-symmetry and to corresponding characteristic classes.

10. Functorial properties of densities. Again let $\phi: N \to M$ be an immersion of supermanifolds with $d \leq \dim N$. Then, generalizing Proposition 3, it is not difficult to define a pullback mapping of d-densities from M to N. Actually, a d-density τ on M is a system of volume forms defined on an open set of d-dimensional tangent directions on M; one should simply consider the part of these directions which are tangent to $\phi(N)$.

More formally, an immersion $\phi: N \to M$ defines a locally direct inclusion $\mathcal{T}N \subset \phi^*(\mathcal{T}M)$ and so an immersion over M: $\psi: G_N(d;\mathcal{T}N) \to G_M(d;\mathcal{T}M)$. This immersion is defined along with an isomorphism of sheaves $S_N^d = \psi^*(S_M^d)$ which induces the corresponding transfer map.

§ 8. The Stokes Formula and the Cohomology of Integral Forms

1. Superdomain with boundary. Let $M^{m|n}$ be a differentiable oriented supermanifold, and let $U \subset M_{\mathrm{rd}}$ be a connected open set with compact closure, whose boundary ∂U is an immersed oriented compact closed submanifold in M_{rd}. We introduce on U the natural structure of an open subsuperspace in M and call a boundary of the superdomain U any immersion $\phi: N^{m-1|n} \to M^{m|n}$ such that ϕ_{rd} defines a diffeomorphism of N_{rd} with ∂U. All orientations are taken to be consistent in the following way: in a neighborhood of every point of the boundary of U, there is a local system of coordinates on M_{rd}, $(x^0, x^1, \ldots, x^{m-1})$, such that U is described by the inequality $x^0 < 0$, and the (x^a) are consistent with the orientations of M and ∂U.

We observe that the boundary of s superdomain is not defined uniquely (even up to the natural concept of an isomorphism). Locally, the boundary can be given by one even equation $x^0 = 0$ such that x^0 can be completed to a local system of coordinates.

Now let $\sigma \in \Sigma_{m-1}(M)$ be an integral form on M. By Theorem 7.5 we can view this as a density $\sigma \in \Xi^{m-1|n}$ defined everywhere on $G_M(m-1|n;\mathcal{T}M)$ and then can construct with it a volume form $\psi^*(\sigma) \in \mathrm{Ber}(N)$. On the other hand, $\delta\sigma$ is a volume form on M, and the integral $\int_U \delta\sigma$ is defined, since the closure of U is compact.

2. Theorem (Stokes formula). $\int_N \phi^*(\sigma) = \int_U \delta\sigma$.

Proof. Since locally the Berezin integral is computed like the ordinary integral of an appropriate volume form, this Stokes formula naturally is a corollary of the usual Stokes formula. Using the additivity of both sides of the equality in σ and in U, we can reduce everything to the case $M = \mathbf{R}^{m|n}$, where U is described by the inequality $x^0 < 0$ and $\sigma = D^*(dx)\, f \otimes \frac{\partial}{\partial x^a}\Pi$, where $f = \sum (x^m)^{\alpha_1} \cdots (x^{m+n-1})^{\alpha_n} f_\alpha$ is a superfunction with compact support. Then

$$\delta\left[D^*(dx)\, f \otimes \frac{\partial}{\partial x^a}\Pi\right] = -(-1)^{m+\widetilde{x^a}(\tilde{f}+1)} D^*(dx) \frac{\partial f}{\partial x^a}$$

and furthermore

$$\int_U \delta\sigma = \begin{cases} 0 & \text{for } \widetilde{x}^a = 1, \\ (-1)^{m+1} \displaystyle\int_U D^*(dx) \frac{\partial f}{\partial x^a} & \\ \quad = (-1)^{mn+m+1} \displaystyle\int_{x^0<0} dx^0 \cdots dx^{m-1} \frac{\partial f_{1\dots 1}}{\partial x^a} & \text{for } \widetilde{x}^a = 0. \end{cases}$$

Integrating with respect to x^a from $-\infty$ to ∞ for $a > 0$ and with respect to x^0 from $-\infty$ to 0, we obtain finally

$$\int_U \delta\sigma = \begin{cases} 0 & \text{for } a > 0; \\ (-1)^{mn+m+1} \displaystyle\int_{\mathbf{R}^{m-1}} dx^1 \cdots dx^{m-1} f_{1\dots 1}(0; x^1, \dots, x^{m-1}) & \text{for } a = 0. \end{cases}$$

To compute $\phi^*(\sigma)$, where $\phi\colon \mathbf{R}^{m-1|n} \to \mathbf{R}^{m|n}$ is the standard immersion, one must carry out in coordinates the construction of Theorem 7.5. This is the answer:

$$\phi^*\left(D^*(dx)\, f \otimes \frac{\partial}{\partial x^a}\Pi\right)$$
$$= \begin{cases} 0 & \text{for } a > 0, \\ (-1)^{m+n-1} D^*(dx^1, \dots, dx^{m+n-1}) \phi^*(f) & \text{for } a = 0. \end{cases}$$

However,

$$\int_{\partial U} \phi^*(\sigma) = (-1)^{m+n-1+(m-1)n} \int_{\mathbf{R}^{m-1}} dx^1 \cdots dx^{m+n-1} (\phi^*(f))_{1\dots 1}\,.$$

Clearly, $(\phi^*(f))_{1\dots 1} = f_{1\dots 1}(0; x^1, \dots, x^{m-1})$, and this completes the proof. □

3. Theorem. (a) *$\mathcal{H}^i(\Sigma, M) = 0$ for $i \neq 0$, and $\mathcal{H}^0(\Sigma, M) = \mathbf{R}$ (the constant sheaf associated with $\mathbf{R}$).*

(b) *A canonical multiplication $\Sigma_p \otimes \Omega^q M \to \Sigma_{p+q} M$ can be defined for $p+q \leq m$, with the following properties (σ is an integral form and ω is a differential form):*

$$\sigma(\omega_1\omega_2) = (\sigma\omega_1)\omega_2,$$
$$\delta(\sigma\omega) = \delta\sigma \cdot \omega + (-1)^{\tilde{\sigma}}\sigma\, d\omega,$$
$$\text{supp}\,(\sigma\omega) \subset \text{supp}\,\sigma \cap \text{supp}\,\omega.$$

This multiplication defines a nondegenerate pairing $\langle$class of σ, class of $\omega\rangle \mapsto \int \sigma\omega$:

$$H^{m-p}\left(\Gamma_c(\Sigma_{.}M)\right) \times H^p\left(\Gamma(\Omega^{\cdot}M)\right) \to \mathbf{R},$$
$$H^{m-p}\left(\Gamma(\Sigma_{.}M)\right) \times H^p\left(\Gamma_c(\Omega^{\cdot}M)\right) \to \mathbf{R},$$

Sketch of proof. The unit section of $\mathcal{H}^0(\Sigma_{.}M)$ is represented in local coordinates by the closed integral form

$$D^*(dx) \otimes x^m \cdots x^{m+n-1}\left(\frac{\partial}{\partial x^0}\Pi\right) \cdots \left(\frac{\partial}{\partial x^{m-1}}\Pi\right)$$

(with notation as in subsection 1). The exactness of the complex $\Sigma_{.}M$ at the remaining terms (i.e., the Poincaré lemma for integral forms) can be proved by constructing a homotopy as in § 4 of Chapter 5.

The multiplication $\Sigma_{.} \otimes \Omega^{\cdot} \to \Sigma_{.}$ is induced, of course, by the contraction $\mathcal{T}M\Pi \times \Omega^1_{\text{odd}}M \to \mathcal{O}_M$ extended to the symmetric powers such that the associativity condition $(\sigma\omega_1)\omega_2 = \sigma(\omega_1\omega_2)$ is satisfied. The formula for compatibility with the differential can be checked in local coordinates. From this it follows that if σ and ω are cycles, then $\sigma\omega$ is also a cycle; and if either σ or ω is a boundary, then $\sigma\omega$ is also a boundary. Thus $\int \sigma\omega$ in fact induces a pairing of the cohomology of the complexes of sections of $\Sigma_{.}$ and $\Omega^{\cdot}$, with one of the complexes consisting of finite forms. The nondegeneracy of the pairing follows formally from the fact that $H^p\left(\Gamma(\Omega^{\cdot}M)\right)$ and $H^p\left(\Gamma_c(\Omega^{\cdot}M)\right)$ are finite-dimensional and from the truth of the assertion for $\mathbf{R}^{m|n}$, that is, essentially from the Poincaré lemma. □

§ 9. Supermanifolds with Distinguished Volume Form. Pseudodifferential and Pseudointegral Forms

1. Distinguished volume forms. On a supermanifold M provided with some supplementary structures, the structures may define a distinguished volume form v_M or, equivalently, (outside the zeroes of v_M) a trivialization of the invertible sheaf Ber M. Then on M functions can be integrated by setting $\int f = \int f v_M$ (if f is finite or if it decreases sufficiently rapidly); this leads to various duality relations. For example, on the projective space $\mathsf{P}^{m|m+1}$ there is a canonical form of this kind (up to multiplication by a constant).

In purely even geometry the cotangent bundle is provided with the Liouville volume form. We formulate the corresponding facts in supergeometry.

Let $\mathcal{E}$ be a locally free sheaf on a supermanifold M, and let $E \xrightarrow{\pi} M$ be its total space. From the definitions it follows that there is a canonical exact sequence on E:

$$0 \to \pi^*(\mathcal{E}) = \mathcal{T}E/M \to \mathcal{T}E \xrightarrow{d\pi} \pi^*(\mathcal{T}M) \to 0$$

from which, dualizing and applying Π, we find

$$(\mathrm{Ber}\, E)^* = \pi^*(\mathrm{Ber}\, \Pi\mathcal{E}^* \otimes \mathrm{Ber}\, \Pi\mathcal{T}^*M)\,.$$

For $\mathcal{E} = \mathcal{T}M$, or $\mathcal{T}M\Pi$, or $\mathcal{T}^*M = \Omega^1_{\mathrm{ev}}M$ or $\Pi\mathcal{T}^*M = \Omega^1_{\mathrm{odd}}M$, we will denote the corresponding total spaces by the symbols TM, $TM\Pi$, T^*M or ΠT^*M. Applying the preceding remark to these spaces, we obtain the following fact.

2. Proposition. (a) *On T^*M and $TM\Pi$ there are canonical volume forms.*

(b) *On TM and ΠT^*M the sheaf of volume forms can be identified canonically with $\pi^*(\mathrm{Ber}\, M)^2$. In particular, if on M there is a distinguished volume form, then the same is true for TM and ΠTM.* □

Corollary. *There are distinguished volume forms on the supermanifolds $T(T^*M)$, $T(TM\Pi)$, $\Pi T^*(T^*M)$, and $\Pi T^*(TM\Pi)$.* □

3. Pseudodifferential forms. According to the definition of differential forms on M, they can be viewed as functions on $TM\Pi$. We have already established that on $TM\Pi$ there is a canonical volume form v_M; therefore, any differential form ω on M can be put in correspondence with the integral $\int_M \omega = \int_{TM\Pi} \omega v_M$. However, in general this is only a generalized function on M (or $TM\Pi$) since for $\dim M = m|n$, $n > 0$, the fibers of the projection $TM\Pi \to M$ are the noncompact superspaces $\mathsf{R}^{m|n}$, while $\mathrm{supp}\, v_M = TM\Pi$ (this will be evident below from the coordinate computation). The form ω is polynomial along the fibers $\mathsf{R}^{m|n}$, and the coordinates along the fibers are the differentials dx, with $\tilde{x} = 1$. Therefore, the form is non-integrable as a rule. A general function on $TM\Pi$ can depend non-polynomially on the even dx (more precisely, on the $\pi^*(dx)$), like $\exp(-(dx^{m+1})^2 - \ldots - (dx^{m+n})^2)$ for instance, it could then be integrable. This motivates the following definition.

4. Definition. (a) A *pseudodifferential form* on a supermanifold M is defined to be a local section of the sheaf $\mathcal{O}_{TM\Pi}$.

(b) A pseudodifferential form ω on a manifold M is said to be *integrable* if it is defined on all of $TM\Pi$ and the integral $\int \omega v_M$ converges absolutely. □

We have assumed that M is oriented; then one can construct a consistent orientation on $TM\Pi$ in an obvious way.

5. Homological vector field on $TM\Pi$. On $TM\Pi$ there is an odd global vector field $\mathcal{D}$ with the following property: if ω is any differential form on M, viewed as a function on $TM\Pi$, then $\mathcal{D}\omega = d\omega$. Clearly, $\mathcal{D}^2 = 0$.

6. Pseudodifferential forms. Using the canonical volume form w_M on T^*M, in an analogous manner one can integrate functions on T^*M, in particular the sections of the sheaf $S_{\mathcal{O}_M}(\mathcal{T}M)$ on M.

7. Computations in coordinates. Let (x^a) be a local system of coordinates on M. This system is related to trivializations of the sheaves $\mathcal{T}M$ and $\Pi\mathcal{T}^*M$ by means of the sections $\partial/\partial x^a$ and dx^a, respectively, and so to distinguished atlases on T^*M, $TM\Pi$ and also on ΠT^*M and TM. We write explicitly the notation and the transformation rules for the corresponding systems of coordinates under a coordinate change $y^a = y^a(x)$ on M:

$$T^*M : (x^a, X_a), \quad \left(y^a = y^a(x), Y_a = \sum_b \frac{\partial x^b}{\partial y^a} X_b = \sum_b \left(\frac{\partial x}{\partial y}\right)_{ab} X_b\right);$$

$$\Pi T^*M : (x^a, X_a\Pi), \quad \left(y^a = y^a(x), Y_a\Pi = \sum_b \left(\frac{\partial x}{\partial y}\right)_{ab} X_b\Pi\right);$$

$$TM\Pi : (x^a, X^a), \quad \left(y^a = y^a(x), Y^a = \sum_b X^b \left(\frac{\partial y}{\partial x}\right)_{ba}\right);$$

$$TM : (x^a, \Pi X^a), \quad \left(y^a = y^a(x), \Pi Y^a = \sum_b \Pi X^b \left(\frac{\partial y}{\partial x}\right)_{ba}\right).$$

The canonical volume forms are:

$$T^*M : \; w_M = D^*(dx^a; dX_a),$$

$$TM\Pi : \; v_M = D^*(dx^a; dX^a),$$

In fact, just as in the purely even case, on T^*M and ΠT^*M there are canonical "symplectic" forms, even and odd, respectively:

$$T^*M : \; \omega = \sum (-1)^{\widetilde{x^a}} \, dx^a \, dX_a,$$

$$\Pi T^*M : \; \omega = \sum (-1)^{\widetilde{x^a}} \, dx^a \, d(X_a\Pi).$$

We verify coordinate independence for, say, the first of these:

$$\sum_a (-1)^{\widetilde{y^a}}\, dy^a\, dY_a = \sum_a (-1)^{\widetilde{y^a}} \left(\sum_b dx^b \frac{\partial y^a}{\partial x^b}\right) \left(\sum_c d\left(\frac{\partial x^c}{\partial y^a}\right) X_c \right.$$

$$\left. + \sum_c (-1)^{\widetilde{x^a}+\widetilde{x^b}} \frac{\partial x^c}{\partial y^a}\, dX_b \right).$$

Since $\sum_a (\partial y^a/\partial x^b)(\partial x^c/\partial y^a) = \delta_b^c$, the second sum in the parentheses on the right, along with the outside summation, gives

$$\sum_b (-1)^{\widetilde{x^b}}\, dx^b\, dX_b\,.$$

The remaining part equals zero:

$$\sum_a (-1)^{\widetilde{y^a}}\, dy^a\, d\left(\frac{\partial x^c}{\partial y^a}\right) = -d\left(\sum_a dy^a \frac{\partial x^c}{\partial y^a}\right) = -d(dx^c) = 0\,.$$

Proposition 7.8 tells when on T^*M there are canonical squares of densities in all dimensions.

§ 10. Lie Superalgebras of Vector Fields and Finite-dimensional Simple Lie Superalgebras

1. Algebras of vector fields. The goal of this section is to describe without proofs the classification theory of simple finite-dimensional complex Lie superalgebras.

The main class of naturally-arising Lie superalgebras is made up of the vector fields on supermanifolds. Let M be a supermanifold; for concreteness, we suppose it is complex. We set

$$w(M) = H^0(M, \mathcal{T}M)\,.$$

Vector fields act on the sections of natural sheaves on M by means of the Lie derivative, and the commutator is carried to the commutator. Therefore, the set of vector fields which preserve (or send to zero) a fixed section of a natural sheaf is a Lie subalgebra. In this construction a special role is played by the volume forms $v \in H^0(M, \operatorname{Ber} M)$ and by the "metrics" $\omega \in H^0(M, \Omega^1 M \otimes \Omega^1 M)$ (where $\Omega^1 = \Omega^1_{\mathrm{ev}}$ or Ω^1_{odd}) which obey the symmetry conditions which in § 5 of Chapter 3 are denoted as types O Sp, Π Sp and as Π-inversions of them. The corresponding Lie algebras are denoted by the following symbols:

$$s(M,v) = \{ X \in w(M) \mid L_X v = 0 \} ,$$
$$h(M,\omega) = \{ X \in w(M) \mid L_X \omega = 0 \} ,$$

where L_X denotes the Lie derivative with respect to the vector field X.

Generally speaking, these Lie algebras need not be finite-dimensional. Finite-dimensionality can be attained by requiring that M be compact. Another way is to set $M = \mathbf{C}^{m|n}$ and to use the linear structure to define a grading on the polynomial fields in $w(\mathbf{C}^{m|n})$; in coordinates,

$$w_i(\mathbf{C}^{m|n}) = \left\{ \sum_{a=1}^{m+n} f^a \frac{\partial}{\partial x^a} \,\middle|\, f^a \text{ are homogeneous polynomials of degree } i+1 \right\} .$$

It is not difficult to see that $[w_i, w_j] \subset w_{i+j}$; so $w_0(\mathbf{C}^{m|n})$ is a Lie subalgebra. Let V be the superspace of linear functions on $\mathbf{C}^{m|n}$; if $X \in w_0(\mathbf{C}^{m|n})$, then $L_X V \subset V$ and the assignment $X \mapsto L_X$ defines an isomorphism

$$w_0(\mathbf{C}^{m|n}) \cong gl(V) = gl(m|n) .$$

Let the volume form v be chosen to be $D^*(dx^a)$ and $s_0(\mathbf{C}^{m|n}, v) = s(\mathbf{C}^{m|n}, v) \cap w_0$. Then

$$s_0(\mathbf{C}^{m|n}, v) = sl(V) = sl(m|n) ,$$

for, as it is not difficult to see,

$$L_X D^*(dx^a) = 0 \iff \operatorname{str} L_X = 0$$

(the supertrace refers to the operator $L_X: V \to V$). Using supermetrics, Lie superalgebras of classical type are defined:

$$osp(m|2n) = h_0 \left(\mathbf{C}^{2n|m};\ \sum_{i=1}^{n} dx^i\, dx^{i+n} + \sum_{j=1}^{m} (d\xi^j)^2,\ d = d_{\mathrm{odd}} \right) ,$$

$$\pi sp(m|m) = h_0 \left(\mathbf{C}^{m|m};\ \sum_{i=1}^{m} dx^i\, d\xi^i \right) .$$

If on V there is an odd involution $p: V \to V$, $p^2 = 1$, then the vector fields which commute with it form a Lie superalgebra which has no even analog:

$$\pi(m|m) = \{ X \in gl(m|m) \mid [L_X, p] = 0 \} .$$

Finally, for $M = \mathbf{C}^{0|n}$ there is no need to replace w with w_0 in order to obtain a finite-dimensional Lie superalgebra, so we can also consider

$$w(o|n); \quad s(0|n, v); \quad h(0|n, \omega).$$

2. Cartan classification. We recall the result of the Cartan classification of finite-dimensional Lie algebras.

A Lie algebra g is said to be semisimple if it is non-abelian and has no nontrivial solvable ideals; it is simple if it has no nontrivial ideals. Every finite-dimensional semisimple algebra is the direct sum of simple ones.

Every simple Lie algebra is either a member of one of the classical series sl, o or sp or else is isomorphic to one of the exceptional simple Lie algebras G_2, F_4, E_6, E_7 or E_8, whose definition is more complicated and requires one to draw on root techniques or linear algebra over the quaternions and octonions.

The members of the classical series sl, o, sp are simple and pairwise non-isomorphic, with a small number of exceptions. More precisely, the simple Lie algebras are contained one time in the following table, where their alternative notation is indicated:

$sl(n+1)$	$o(2n+1)$	$sp(2n)$	$o(2n)$
A_n, $n \geq 1$	B_n, $n \geq 2$	C_n, $n \geq 3$	D_n, $n \geq 3$

The "accidental" isomorphisms are: $o(3) = sl(2)$; $sp(2) = sl(2)$; $sp(4) = o(5)$; $o(4) = sl(2) \times sl(2)$. The separation of the series o into two parts is defined, as is well-known, by a difference in the root structure: the Dynkin diagram of the D series contains a branch, in contrast with the B series.

3. Classification of simple Lie superalgebras. As was shown by Kac [61], the simple complex finite-dimensional Lie superalgebras admit a classification parallel to that of Cartan.

To be more precise, every such Lie superalgebra is either a simple Lie algebra or else is isomorphic to one of the Lie superalgebras described above, which are in the cartanian series sl, osp, πsp, π, w, s; an algebra h either differs from an element of a series by a simple construction, such as taking the quotient by the center, or else finally belongs to one of the exceptional types, denoted $D(2|1, \alpha)$ (a one-parameter family), $F(4)$, $G(3)$.

Below we will describe in turn all the cartanian Lie superalgebras g in some detail, adhering to the following plan: realization in a fundamental representation, the transition to a simple superalgebra when this is necessary, the structure of g_0 and the representation of g_0 on g_1, $\mathbf{Z}$-grading, "accidental" isomorphisms with superalgebras described earlier, supplementary information.

4. Series A. We set

$$A(m|n) = \begin{cases} sl(m+1|n+1) & \text{for } m \neq n;\ m,n \geq 0; \\ sl(n+1|n+1)/\mathbb{C}E_{n+1|n+1} & \text{for } m = n \geq 1. \end{cases}$$

Matrices which are proportional to the identity in $gl(n+1|n+1)$ have zero trace and so fall in $sl(n+1|n+1)$; moreover, they clearly form an ideal. After taking the quotient with respect to this ideal, the Lie superalgebra becomes simple. Clearly,

$$\dim A(m|n) = \begin{cases} (m+1)^2 + (n+1)^2 - 1|(m+1)(n+1) & \text{for } m \neq n, \\ 2(n+1)^2 - 2|(n+1)^2 & \text{for } m = n. \end{cases}$$

Furthermore:

$$A(m|n)_0 = \begin{cases} A_m \oplus A_n \oplus \mathbb{C} & \text{for } m \neq n, \\ A_n \oplus A_n & \text{for } m = n, \end{cases}$$

while the representation of $A(m|n)_0$ on $A(m|n)_1$ is $f_m \otimes f_n' \otimes \mathbb{C}$ or $f_n \otimes f_n$ respectively. Here and below we denote by the symbol f_n the fundamental n-dimensional representation of the respective summands of the even part of the Lie superalgebra.

The algebras $A(n|n)$ have zero Killing form, $(a,b) = \operatorname{str}(\operatorname{ad} a, \operatorname{ad} b)$; for the remaining algebras it is nondegenerate.

The "accidental" isomorphisms are: $A(m|n) \simeq A(n|m)$. The invariant explanation is the existence of the canonical isomorphism $gl(V) \to gl(\Pi V)$: $a \mapsto a^{\Pi}$.

5. Series B. We set

$$B(m|n) = osp(2m+1|2n), \quad m \geq 0,\ n > 0\,.$$

We have

$$B(m|n)_0 = B_m \oplus C_n\,, \qquad B(m|n)_1 = f_{2m+1} \otimes f_{2n}'\,,$$

$$\dim B(m|n) = 2m^2 + m + 2n^2 + n|4mn + 2n\,.$$

All these Lie superalgebras are simple and are not isomorphic to the superalgebras of series A described above.

6. Series C. We set

$$C(n) = osp(2|2n-2)\,,\ n \geq 2.$$

We have

$$C(n)_0 = C_{n-1} \oplus \mathbb{C}\,, \qquad C(n)_1 = f_{2n-2} \otimes \mathbb{C}\,,$$

$$\dim C(n) = 2n^2 - 3n + 2|4n - 4\,.$$

The accidental isomorphism is:

$$A(1|0) = sl(2|1) = osp(2|2) = C(2).$$

All the superalgebras in the series C are simple.

7. Series D. We set

$$D(m|n) = ops(2m|2n),\ m \geq 2,\ n > 0.$$

We have

$$D(m|n)_0 = D_m \oplus C_n, \qquad D(m|n)_1 = f_{2m} \otimes f'_{2n},$$
$$\dim D(m|n) = 2m^2 - m + 2n^2 - n|4mn.$$

All the superalgebras in the series D are simple; there are no accidental isomorphisms.

We will describe yet another realization of the superalgebras of the series B, C, D (i.e., osp) in terms of the tensor algebra of the fundamental representation. Let $T = T_0 \oplus T_1$, let $\dim T = m|n$, and let $(,)_0$ and $(,)_1$ be nondegenerate symmetric forms on T_0 and on T_1, respectively. In the formulas below, we employ as always the law of signs so that $\wedge^2 V_1 = S^2(\Pi V_1)$, etc. Then there are isomorphisms

$$osp(m|n) = (\wedge^2 T_0 \oplus \wedge^2 T_1) \oplus (T_0 \otimes T_1)$$

with the following law of composition:

$$[ab, c] = (a,c)_0 b - (b,c)_0 a, \quad \text{for } ab \in \wedge^2 T_0,\ c \in T_0;$$
$$[ab, c] = (a,c)_1 b - (b,c)_1 a, \quad \text{for } ab \in \wedge^2 T_1,\ c \in T_1;$$
$$[ab, cd] = [ab, c]d + c[ab, d], \quad \text{for } ab,\ cd \in \wedge^2 T_0 \text{ or } \wedge^2 T_1;$$
$$[a \otimes c, b \otimes d] = (a,b)_0 cd + (c,d)_1 ab, \quad \text{for } a \otimes c,\ b \otimes d \in T_0 \otimes T_1.$$

8. Series P. We set

$$P(n) = \pi sp(n+1|n+1) \cap sl(n+1|n+1), \quad n \geq 2.$$

We have

$$P(n)_0 = A_n, \qquad P(n)_1 = \wedge^2 f^*_{n+1} \otimes S^2 f_{n+1},$$
$$\dim P(n) = n^2 + 2n|(n+1)^2.$$

There are no accidental isomorphisms with the algebras constructed earlier.

9. Series Q. The Lie superalgebra $\pi(n+1|n+1)$ in the natural basis is realized by matrices with the format

$$g = \left(\begin{array}{c|c} \mathrm{a} & \mathrm{b} \\ \hline \mathrm{c} & \mathrm{d} \end{array}\right), \quad a \in gl(n+1|n+1).$$

On such matrices there is defined an operation of the "odd trace," which is equal to zero on commutators:

$$\begin{aligned} \mathrm{otr}\ g &= \mathrm{tr}\, b\,, \\ \mathrm{otr}\ [g, g'] &= \mathrm{tr}[a, b'] - \mathrm{tr}[a', b] = 0\,. \end{aligned}$$

Therefore, the kernel of otr is a Lie superalgebra; moreover, the superalgebra contains a nontrivial ideal C consisting of multiples of the unit matrix. Finally, we set

$$Q(n) = \left\{ \left(\begin{array}{c|c} \mathrm{a} & \mathrm{b} \\ \hline \mathrm{b} & \mathrm{a} \end{array}\right) \middle|\, a \in gl(n+1),\ b \in sl(n+1) \right\} \Big/ C, \quad n \geq 2.$$

We have

$$\begin{aligned} Q(n)_0 = A_n\,, &\qquad Q(n)_1 = \mathrm{ad}\ A_n\,, \\ \dim Q(n) &= n^2 + 2n|n^2 + 2n\,. \end{aligned}$$

All these superalgebras are simple; there are no accidental isomorphisms with the algebras described earlier.

10. Exceptional superalgebras of classical type. All the simple superalgebras which have been described, A–D, P, Q, have the following property: their even part is semisimple and the representation of the even part on the odd part is completely reducible. There are three other types of simple superalgebras with this property, $F(4)$, $G(3)$, and $D(2|1,\alpha)$. We will list some information about them:

$$\begin{aligned} F(4)_0 &= B_3 \oplus A_1\,, \\ F(4)_1 &= \mathrm{spin}(7) \otimes f_2\,, \quad \dim \mathrm{spin}(7) = 8; \\ \dim F(4) &= 24|16; \\ G(3)_0 &= G_2 \oplus A_1\,, \quad G(3)_1 = f_7 \otimes f_3\,; \\ \dim G(3) &= 10|21; \\ D(2|1,\alpha)_0 &= A_1 \oplus A_1 \oplus A_1\,, \quad D(2|1,\alpha)_1 = f_2 \otimes f_2 \otimes f_2\,, \\ \dim D(2|1,\alpha) &= 9|8. \end{aligned}$$

11. Series W. We set

$$W(n) = w(\mathrm{C}^{0|n}) = \left\{ \sum_{a=1}^{n} f^a(\xi^1, \dots, \xi^n) \frac{\partial}{\partial \xi^a} \;\middle|\; f^a \in \mathrm{C}[\xi^1, \dots, \xi^n] \right\},$$

$$\dim W(n) = n2^{n-1} | n2^{n-1}.$$

The algebra $W(n)$ is simple for $n \geq 2$ and is not isomorphic to any described earlier if $n \geq 3$; there is an accidental isomorphism:

$$W(2) = C(2) = A(1|0).$$

It is natural to think of $W(n)$ as a "curved" version of $gl(0|n)$; it acts on $\mathrm{C}^{0|n}$ by infinitesimal automorphisms which do not necessarily preserve the linear structure.

The Lie algebra $W(n)_0$ is not semisimple; its semisimple quotient is $gl(n)$.

12. Series S and $\widetilde{S}$. A nondegenerate volume form on $\mathrm{C}^{0|n}$ can be written in the form $f(\xi^1, \dots, \xi^n) D^*(d\xi^1, \dots, d\xi^n)$, with $f(0) \neq 0$. It can be shown that for odd n, by means of automorphisms of $\mathrm{C}^{0|n}$, such a form can be reduced to the form $D^*(d\xi^1, \dots, d\xi^n) = v$, while for even n it can either be reduced to this form or else to the form $\widetilde{v}_t = (1 + t\xi^1 \cdots \xi^n) \times D^*(d\xi^1, \dots, d\xi^n)$. We set

$$S(n) = \{ X \in W(n) \mid \Delta_r(v \otimes X) = 0 \},$$

$$\widetilde{S}(n) = \{ X \in W(n) \mid \Delta_r(\widetilde{v}_1 \otimes X) = 0 \}, \qquad n \equiv 0 \bmod 2,$$

where the right action $\Delta_r(v \otimes X)$ was defined in § 5. These algebras are simple for $n \geq 3$; the semisimple parts of $S(n)_0$ and $\widetilde{S}(n)_0$ are isomorphic to $sl(n)$. Their dimensions are equal to $(n-1)2^{n-1} | (n-1)2^{n-1} + 1$. The accidental isomorphism is:

$$S(3) = P(2).$$

13. Series H. We set $\omega = (d\xi^1)^2 + \dots + (d\xi^n)^2 \in \Omega^2_{\mathrm{odd}} \mathrm{C}^{0|n}$; and further,

$$\widetilde{H}(n) = \{ X \in W(n),\ L_X \omega = 0 \}, \quad H(n) = [\widetilde{H}(n), \widetilde{H}(n)].$$

The algebra $H(n)$ is simple for $n \geq 4$; its dimension is equal to $2^{n-1} | 2^{n-1} - 2$. The accidental isomorphism is:

$$H(4) = A(1|1).$$

The semisimple part of $H(n)$ is isomorphic to $so(n)$.

The simple finite-dimensional Lie superalgebras over C described in subsections 4–13 exhaust all such algebras up to isomorphism.

References for Chapter 4

The systematic theory of smooth supermanifolds from various points of view is expounded in the large works of Kostant [62] and Leites [68]. Superspaces can be considered in several geometric categories: the definition of analytic superspaces which we have adopted was formulated by Deligne. The Frobenius theorem and differential equations on supermanifolds were studied in the work of Shander [99]. The integration theory of volume forms of Berezin has been generalized by Bernstein and Leites [15], [16]; to them also belongs the Stokes formula in supergeometry. The flag superspaces were constructed here by the standard method of (relative) coverings by big cells. Because of the fact that most of the grassmannians are not projective, it is necessary to generalize to the supercase the technique of proofs, which do not use projectivity. The duality theorem was recently proved by Penkov using the theory of D-modules on supermanifolds [87]. For the geometry of G-structures on superspaces, see the work of Švarc [97].

CHAPTER 5

GEOMETRIC STRUCTURES OF SUPERSYMMETRY AND GRAVITATION

In this chapter we present several basic constructions in the theory of fields which employ the idea of a superspace. In § 1 the Minkowski superspace M is described. If one begins with twistors, then the compact complex version of M is naturally realized as a flag space. The geometry of the flat case corresponding to the minimal number ($N = 1$) of odd coordinates, after a suitable twist, turns into the geometry of simple gravity according to Ogievetskii-Sokachev, to which § 7 is devoted. In § 2, we explain for the simplest example of scalar superfields several fundamental ideas of supersymmetric field theory which are used in the physical literature. In § 3 connections and dynamical equations for Yang-Mills superfields are studied on the basis of an idea of Witten to treat them as equations of integrability along light supergeodesics. In § 4 we present a method for constructing solutions of supersymmetric Yang-Mills equations using the Penrose transform in supergeometry; the coordinate computations relating to this are carried out in § 5. In § 6 we classify the other flag superspaces whose underlying space is the Penrose model. They have several exotic properties, but they can be useful for understanding such constructions as the Fayet-Sohnius multiplet and other questions of extended supersymmetry and supergravity.

§ 1. Supertwistors and Minkowski Superspace

1. Supertwistors. In this section T denotes the complex linear superspace of dimension $4|N$, the supertwistor space. We set $M = F(2|0, 2|N; T)$. This superanalytic manifold is a compact complex model for the N-extended Minkowski superspace. This extended superspace is itself a real supermanifold, the set of fixed points of a big cell of M relative to an appropriate real structure. Another real structure leads to a euclidean version of the superspace.

The goal of this section is to study the geometry of M. Since $M_{\mathrm{rd}} = G(2|0, T_0)$, the underlying space of M is the Plücker quadric analyzed in detail in § 3 of Chapter 1. Therefore, the stress here is placed on the special properties related to the presence of odd coordinates.

2. Right and left superspaces. We set $M_l = G(2|0; T)$ and $M_r = G(2|N; T) = G(2|0; T^*)$. The superspace M has two canonical projections

$$M_l \xleftarrow{\pi_l} M \xrightarrow{\pi_r} M_r\,.$$

Over M_l and M_r the space M is represented in the form of relative grassmannians:

$$M = G_{M_l}\left(0|N; (\widetilde{S}_l^{2|N})^*\right) = G_{M_l}\left(2|0; \widetilde{S}_l^{2|N}\right),$$
$$M = G_{M_r}\left(2|0; S_r^{2|N}\right) = G_{M_r}\left(0|N; (S_r^{2|N})^*\right).$$

Here S_l and S_r are the tautological sheaves on M_l and M_r respectively. See subsection 1.1.18 for the principle of this notation. In particular, π_l and π_r are submersions of relative dimension $2|0 \times 0|N = 0|2N$. Therefore, the complex dimensions of these analytic supermanifolds are these:

$$\dim_{\mathbf{C}} M_l = \dim_{\mathbf{C}} M_r = 2|0 \times 2|N = 4|2N\,; \quad \dim_{\mathbf{C}} M = 4|4N\,.$$

3. Structure of the tangent sheaf. We set

$$\mathcal{T}_l/M = \mathcal{T}M/M_r\,, \qquad \mathcal{T}_rM = \mathcal{T}M/M_l\,.$$

By Theorem 4.3.11 we have canonical isomorphisms

$$\mathcal{T}_lM = \mathcal{H}om(S^{2|0}, S^{2|N}/S^{2|0}) = \mathcal{H}om(\widetilde{S}^{2|N}/\widetilde{S}^{2|0}, (S^{2|0})^*)\,,$$
$$\mathcal{T}_rM = \mathcal{H}om(S^{2|N}/S^{2|0}, (\widetilde{S}^{2|0})^*) = \mathcal{H}om(\widetilde{S}^{2|0}, \widetilde{S}^{2|N}/\widetilde{S}^{2|0})\,,$$

where S and $\widetilde{S}$ are the tautological sheaves on M. In particular, composition of homomorphisms provides a natural bilinear mapping

$$\mathcal{T}_lM \otimes \mathcal{T}_rM \xrightarrow{a} (S^{2|0})^* \otimes (\widetilde{S}^{2|0})^*.$$

For a precise formula for a, see subsection 7 below.

On the other hand, $\mathcal{T}_lM$ and $\mathcal{T}_rM$ are subsheaves of a Lie superalgebra in $\mathcal{T}M$, and the Frobenius form between $\mathcal{T}_lM$ and $\mathcal{T}_rM$, i.e., the supercommutator modulo $\mathcal{T}_lM + \mathcal{T}_rM$, defines a mapping

$$[\ ,\ ]\colon \mathcal{T}_lM \otimes \mathcal{T}_rM \xrightarrow{b} \mathcal{T}M/(\mathcal{T}_lM + \mathcal{T}_rM) = \mathcal{T}_0M\,.$$

4. Proposition. (a) *The sum* $\mathcal{T}_lM + \mathcal{T}_rM$ *in* $\mathcal{T}M$ *is direct, and* $\mathcal{T}_lM \oplus \mathcal{T}_rM$ *is a locally free and locally direct subsheaf of rank* $0|4N$.

(b) *There exists a unique isomorphism* $\mathcal{T}_0M = (S^{2|0})^* \otimes (\widetilde{S})^{2|0})^*$, *with respect to which the forms* a *and* b *defined above coincide. Under reduction of odd coordinates, this isomorphism is carried to the standard isomorphism* $\mathcal{T}M_{\mathrm{rd}} = (S_{\mathrm{rd}}^{2|0})^* \otimes (\widetilde{S}_{\mathrm{rd}}^{2|0})^*$. □

We will establish this assertion by means of computations in standard coordinates on a big cell of M.

5. Coordinates in T and T^*. Adding odd functions to the coordinates in the ordinary twistor spaces from § 3 of Chapter 1, we set

$$T\colon (z_\alpha, z^{\dot\beta}, \varsigma^j), \quad \alpha, \beta = 0, 1, \quad j = 1, \ldots, N;$$
$$T^*\colon (w^\alpha, w_{\dot\beta}, \omega_j).$$

The canonical bilinear form (as an element of $T^* \otimes T$) will be written in the form

$$\langle (z, \varsigma), (w, \omega) \rangle = z_\alpha w^\alpha - z^{\dot\beta} w_{\dot\beta} + 2i\varsigma^j \omega_j$$

(Summation is implied with respect to repeated indices; the factor $2i = 2\sqrt{-1}$ has been inserted for consistency with conventions in the physical literature. Certainly this is important only when considering real structures).

The choice of a big cell in M_{rd} and its identification with $\mathrm{C}^{4|4N}$ defines four subspaces in the T-component of the tautological flag at the origin of coordinates and at the vertex of the cone at infinity; they are defined by the following equations:

$$S(0)^{2|0} : z^{\dot\beta} = 0,\ \varsigma^j = 0; \qquad S(\infty)^{2|N} : z_\alpha = 0;$$
$$S(0)^{2|N} : z^{\dot\beta} = 0; \qquad S(\infty)^{2|0} : z_\alpha = 0,\ \varsigma^j = 0.$$

6. Coordinates on big cells in M_l and M_r. Writing them down in the general format of the matrices Z_I as in § 3 of Chapter 3, we have

$$M_l : \begin{array}{c|cccc|ccc|} s^0 & 1 & 0 & x_l^{0\dot 0} & x_l^{0\dot 1} & \theta_l^{01} & \ldots & \theta_l^{0N} \\ s^1 & 0 & 1 & x_l^{1\dot 0} & x_l^{1\dot 1} & \theta_l^{11} & \ldots & \theta_l^{1N} \\ & z_0 & z_1 & z^{\dot 0} & z^{\dot 1} & \varsigma^1 & \ldots & \varsigma^N \end{array} = (\delta^{\alpha\beta} \mid x_l^{\alpha\dot\beta} \mid \theta_l^{\alpha j})$$

and analogously

$$M_r : \begin{array}{c|cccc|ccc|} s^{\dot 0} & x_r^{\dot 0 0} & x_r^{\dot 0 1} & 1 & 0 & \theta_{r1}^{\dot 0} & \ldots & \theta_{rN}^{\dot 0} \\ s^{\dot 1} & x_r^{\dot 1 0} & x_r^{\dot 1 1} & 0 & 1 & \theta_{r1}^{\dot 1} & \ldots & \theta_{rN}^{\dot 1} \\ & w^0 & w^1 & w_{\dot 0} & w_{\dot 1} & \omega_1 & \ldots & \omega_N \end{array} = (x_r^{\dot\beta\alpha} \mid \delta^{\dot\beta\dot\alpha} \mid \theta_{rj}^{\dot\beta})$$

The order of the indices is adapted to the numbering of the coordinates in T and T^*; the first indices enumerate the rows of the matrix. We recall that, as modules of sections of $\mathcal{O}_{M_l} \otimes T$ and $\mathcal{O}_{M_r} \otimes T$ over the big cell, $S_l(x_l, \theta_l)$ and $S_r(x_r, \theta_r)$ are generated by the lines of the first and second matrices respectively; we have denoted these by s^α and $s^{\dot\beta}$.

7. Coordinates on a big cell in M. A $(2|0, 2|N)$-flag in T is the same thing as a pair of $2|0$-subspaces in T and T^*, orthogonal with respect to the form $\langle\ \rangle$. The condition of orthogonality of $S_l(x_l, \theta_l)$ and $S_r(x_r, \theta_r)$ consists of four relations:

$$\delta^{\alpha\gamma} x_r^{\dot\beta\gamma} - x_l^{\alpha\dot\gamma} \delta^{\dot\beta\dot\gamma} + 2i\theta_l^{\alpha j} \theta_{rj}^{\dot\beta} = 0, \qquad \alpha = 0, 1, \quad \dot\beta = \dot 0, \dot 1\,.$$

We will consider x_l, x_r, θ_l, θ_r as functions on M, having lifted them from M_l and M_r using π_l^* and π_r^*. We set

$$x^{\alpha\dot\beta} = \frac{1}{2}\left(x_l^{\alpha\dot\beta} + x_r^{\dot\beta\alpha}\right).$$

Rewriting the orthogonality relation in the form

$$x_l^{\alpha\dot\beta} - x_r^{\dot\beta\alpha} = 2i\theta_l^{\alpha j} \theta_{rj}^{\dot\beta},$$

we obtain that as a standard system of coordinates on the big cell of M we can take $(x^{\alpha\dot\beta}, \theta_l^{\alpha j}, \theta_{rj}^{\dot\beta})$. The even coordinates of the left and right (chiral, in physical terminology) superspaces have the form

$$x_l^{\alpha\dot\beta} = x^{\alpha\dot\beta} + i\theta_l^{\alpha j} \theta_{rj}^{\dot\beta},$$

$$x_r^{\dot\beta\alpha} = x^{\alpha\dot\beta} - i\theta_l^{\alpha j} \theta_{rj}^{\dot\beta}.$$

8. Bases of $\mathcal{T}_l M$ and $\mathcal{T}_r M$. Sections of $\mathcal{T}_l M$ (respectively $\mathcal{T}_r M$) are superderivations on a big cell of M which take (x_r, θ_r) (respectively (x_l, θ_l)) to zero. It is not difficult to see that a basis of these sections is given by the following family of vector fields, which in the physical literature are sometimes called "covariant derivatives":

$$\mathcal{T}_l M : D_{l\alpha j} = \frac{\partial}{\partial \theta_l^{\alpha j}} + i\theta_{rj}^{\dot\beta} \frac{\partial}{\partial x^{\alpha\dot\beta}}, \quad \alpha = 0, 1, \quad j = 1, \ldots, N,$$

$$\mathcal{T}_r M : D_{r\dot\beta}^{j} = \frac{\partial}{\partial \theta_{rj}^{\dot\beta}} + i\theta_l^{\alpha j} \frac{\partial}{\partial x^{\alpha\dot\beta}}, \quad \dot\beta = \dot 0, \dot 1, \quad j = 1, \ldots, N.$$

From this it is evident that the sum $\mathcal{T}_l M + \mathcal{T}_r M$ is direct and the vector fields $\partial/\partial x^{\alpha\dot\beta}$ form a basis of a sheaf of rank 4|0 extending it to $\mathcal{T}M$. Furthermore, a direct computation of the commutator gives

$$\left[D_{l\alpha j}, D^k_{r\dot\beta}\right] = 2i\frac{\partial}{\partial x^{\alpha\dot\beta}}\delta^k_j .$$

9. Proof of Proposition 4. Our remaining task is to compare the bilinear mappings a and b. Let X_l, X_r, s and t be local sections, respectively, of the sheaves $\mathcal{T}_l M$, $\mathcal{T}_r M$, $S^{2|0}$ and $\widetilde{S}^{2|0}$. We defined a by either of two formulas, whose equivalence will be checked below:

$$a(X_l \otimes X_r)(s \otimes t) = (-1)^{\widetilde{X}_l\widetilde{X}_r}\langle X_r X_l s, t\rangle = (-1)^{\widetilde{s}(\widetilde{X}_l+\widetilde{X}_r)}\langle s, X_l X_r t\rangle .$$

The middle expression is interpreted in this way: $X_l s$ is a section of $S^{2|N}/S^{2|0}$ which is obtained by coordinatewise differentiation of s expressed in terms of a basis of T, after which $X_r X_l s$ is the result of an analogous differentiation of $X_l s$ in a basis lifted from M_r (on a big cell we can again assume that this is the initial basis of T); $X_r X_l s$ is a section of $(\widetilde{S}^{2|0})^*$. We apply this formula to $D_{l\alpha j}$, $D^k_{r\dot\beta}$, s^γ and $s^{\dot\delta}$. We have

$$D^k_{r\dot\beta} D_{l\alpha j} s^\gamma = \left(\frac{\partial}{\partial\theta^{\dot\beta}_{rk}} + i\theta^{\mu k}_l \frac{\partial}{\partial x^{\mu\dot\beta}}\right)\left(\frac{\partial}{\partial\theta^{\alpha j}_l} + i\theta^{\dot\nu}_{rj}\frac{\partial}{\partial x^{\alpha\dot\nu}}\right)\left(\delta^{\gamma\rho}\middle| x^{\gamma\dot\rho} + i\theta^{\gamma c}_l\theta^{\dot\rho}_{rc}\middle|\theta^{\gamma d}_l\right)$$

$$= \left(0\,\middle|\, i\delta_{\gamma\dot\rho,\alpha\dot\nu}\,\delta_{\dot\beta k,\dot\nu j} + i\delta_{\alpha j,\gamma c}\,\delta_{\dot\beta k,\dot\rho c}\,\middle|\,0\right),$$

whence

$$(-1)^{\widetilde{D}_r\widetilde{D}_l}\left\langle D^k_{r\dot\beta} D_{l\alpha j} s^\gamma, s^{\dot\delta}\right\rangle = 2i\delta^{\alpha\dot\beta k}_{\gamma\dot\delta j} .$$

In an analogous way we compute the second scalar product, which it is convenient to represent in the form $(-1)^{\widetilde{s}\widetilde{t}}\langle X_l X_r t, s\rangle$ and obtain

$$D_{l\alpha j} D^k_{r\dot\beta} s^{\dot\delta} = \left(\frac{\partial}{\partial\theta^{\alpha j}_l} + i\theta^{\dot\nu}_{rj}\frac{\partial}{\partial x^{\alpha\dot\nu}}\right)\left(\frac{\partial}{\partial\theta^{\dot\beta}_{rk}} + i\theta^{\mu k}_l\frac{\partial}{\partial x^{\mu\dot\beta}}\right)\left(x^{\rho\dot\delta} - i\theta^{\rho c}_l\theta^{\dot\delta}_{rc} \mid \delta^{\dot\delta\dot\rho} \mid \theta^{\dot\delta}_{rd}\right).$$

Again we have

$$\left\langle D_{l\alpha j} D^k_{r\dot\beta} s^{\dot\delta}, s^\gamma\right\rangle = 2i\delta^{\alpha\dot\beta k}_{\gamma\dot\delta j} .$$

From this it follows that the mapping a is surjective and that it can be identified with the supercommutator with respect to the isomorphism

$$\left(S^{2|0}\right)^* \otimes \left(\widetilde{S}^{2|0}\right)^* \to \mathcal{T}_0 M$$

$$s_\alpha \otimes s_{\dot\beta} \mapsto \frac{\partial}{\partial x^{\alpha\dot\beta}} \bmod (\mathcal{T}_l M \oplus \mathcal{T}_r M) ,$$

where (s_α) and $(s_{\dot\beta}$ are bases for the sections of $\left(S^{2|0}\right)^*$ and $\left(\widetilde{S}^{2|0}\right)^*$ dual to (s^α) and $(s^{\dot\beta})$. A comparison with the case $N = 0$, which corresponds to the structure of M_{rd}, completes the proof.

10. Spinor decompositions on the superspace M. The sheaves $S_M^{2|0} = S$ and $\widetilde{S}_M^{2|0} = \widetilde{S}$ on M are sheaves of two-component spinors. Besides them, on M there are two bundles of rank $0|N$: $S^{2|N}/S^{2|0}$ and $\widetilde{S}^{2|N}/\widetilde{S}^{2|0}$ which are related by the canonical duality induced by the form $\langle\,\rangle$. We pass to the Π-inverted bundles in order to write down the structure of $\Omega^1_{\mathrm{odd}} M = \Pi\mathcal{T}^* M$; we set

$$\mathcal{E}_l = \Pi\left(S^{2|N}/S^{2|0}\right)^* = \Pi\left(\widetilde{S}^{2|N}/\widetilde{S}^{2|0}\right) ,$$

$$\mathcal{E}_r = \Pi\left(\widetilde{S}^{2|N}/\widetilde{S}^{2|0}\right)^* = \Pi\left(S^{2|N}/S^{2|0}\right) .$$

Now we can present the following information about differential and integral forms on M in terms of "superspinor algebra."

(a) The sheaf of 1-forms $\Omega^1 M$ is part of an exact sequence:

$$0 \to \Omega^1_0 M \xrightarrow{a} \Omega^1 M \xrightarrow{(b_l, b_r)} \Omega^1_l M \oplus \Omega^1_r M \to 0 ,$$

$$\Omega^1_l M = S^{2|0} \otimes \mathcal{E}_l , \quad \Omega^1_r M = \widetilde{S}^{2|0} \otimes \mathcal{E}_r , \quad \Omega^1_0 M = \Pi\left(S^{2|0} \otimes \widetilde{S}^{2|0}\right) , \quad \mathcal{E}_l = \mathcal{E}_r^* .$$

In particular, on M there are invariant differential equations $d_l f = 0$ and $d_r f = 0$, where f is a function and $d_{l,r} = b_{l,r} \circ d$. The solutions of these are called chiral scalar fields. Of course $d_l f = 0$ means that $f = \pi_r^* g$, where g is a function on M_r. We can define chiral sections of the sheaf $\mathcal{E}$ analogously by means of the connection ∇ as solutions of the equations $\nabla_{l,r} f = 0$, where $\nabla_{l,r} = (\mathrm{id}_{\mathcal{E}} \otimes b_{l,r}) \circ \nabla$.

(b) On the sheaf of 2-forms there is a canonical filtration

$$\Omega^2_0 M = S^2\left(\Omega^2_0 M\right) \subset \Omega^2_1 M = \Omega^1_0 M \cdot \Omega^1 M \subset \Omega^2_2 M = \Omega^2 M ,$$

with quotients

$$\Omega^2_1/\Omega^2_0 = \Omega^1_0 M \otimes \left(\Omega^1_l M \oplus \Omega^1_r M\right) ,$$

$$\Omega^2_2/\Omega^2_1 = \Omega^2_l M \oplus \Omega^2_r M \oplus \Omega^1_l M \otimes \Omega^1_r M .$$

In particular, there is a canonical odd mapping $c\colon \Omega^2 M \to \Omega^1 M$ which carries the quotient sheaf $\Omega^1_l M \otimes \Omega^1_r M$ to the subsheaf $\Omega^1_0 M$ (see the identification in (a)). With this, one defines part of the specific connections on Yang-Mills superfields, $\nabla\colon \mathcal{E} \to \mathcal{E} \otimes \Omega^1 M$.

(c) The berezinian of M, M_l and M_r can be computed directly from the exact sequences for Ω^1:

$$\begin{aligned}
\operatorname{Ber} M = (\operatorname{Ber} \Omega^1 M)^* &= (\operatorname{Ber} \Omega^1_0 M)^* \otimes (\operatorname{Ber} \Omega^1_l M)^* \otimes (\operatorname{Ber} \Omega^1_r M)^*, \\
(\operatorname{Ber} \Omega^1_0 M)^* &= \operatorname{Ber}(S^{2|0} \otimes \widetilde{S}^{2|0}) = (\operatorname{Ber} S^{2|0})^2 \otimes (\operatorname{Ber} \widetilde{S}^{2|0})^2, \\
(\operatorname{Ber} \Omega^1_l M)^* &= \left[(\operatorname{Ber} S^{2|0})^N \otimes (\operatorname{Ber} \mathcal{E}_l)^2\right]^*, \\
(\operatorname{Ber} \Omega^1_r M)^* &= \left[(\operatorname{Ber} \widetilde{S}^{2|0})^N \otimes (\operatorname{Ber} \mathcal{E}_r)^2\right]^*,
\end{aligned}$$

from which finally

$$\operatorname{Ber} M = (\operatorname{Ber} S^{2|0})^{2-N} \otimes (\operatorname{Ber} \widetilde{S}^{2|0})^{2-N}.$$

In particular, for $N = 2$ on M there is a canonical volume form.

Analogously, we find

$$\begin{aligned}
\pi_l^*(\operatorname{Ber} M_l) &= (\operatorname{Ber} S^{2|0})^{2-N} \otimes (\operatorname{Ber} \widetilde{S}^{2|0})^2 \otimes (\operatorname{Ber} \mathcal{E}_r)^2, \\
\pi_r^*(\operatorname{Ber} M_r) &= (\operatorname{Ber} \widetilde{S}^{2|0})^{2-N} \otimes (\operatorname{Ber} S^{2|0})^2 \otimes (\operatorname{Ber} \mathcal{E}_l)^2.
\end{aligned}$$

11. Real structures: Minkowski signature. In § 3 of Chapter 1 we defined a real structure on $T \times T^*$ ($N = 0$). In turn it induces a real structure on $G(2, T)$; on the set of its fixed points, the standard conformal metric has Minkowski signature. Analogously, on $\mathbf{P}(T) \times \mathbf{P}(T^*)$ a real structure was introduced which reduced to a Riemann metric on the corresponding section of $G(2, T)$. Here we extend these real structures to the case of an arbitrary N. For the general definitions of the type of a real structure, see § 6 of Chapter 3.

(a) We introduce on $T \times T^*$ a real structure ρ of type $(1, 1, 1)$

$$\begin{aligned}
(w^\beta, w_{\dot\alpha}, \omega_j)^\rho &= (z^{\dot\beta}, z_\alpha, \varsigma^j), \\
(z^{\dot\beta}, z_\alpha, \varsigma^j)^\rho &= (w^\beta, w_{\dot\alpha}, \omega_j).
\end{aligned}$$

For this structure we have

$$\langle (z, \varsigma), (w, \omega) \rangle^\rho = w_{\dot\alpha} z^{\dot\alpha} - w^\beta z_\beta - 2i\varsigma^j \omega_j = -\langle (z, \varsigma), (w, \omega) \rangle.$$

Thus the standard scalar product is a pure imaginary element of $T \otimes T^*$.

(b) Using the coordinates from subsection 6, we get an induced real structure on the product of big cells $M_l \times M_r$:

$$\left(x_l^{\alpha\dot\beta}, \theta_l^{\alpha j}; x_r^{\dot\alpha\beta}, \theta_{rj}^{\dot\alpha}\right)^\rho = \left(x_l^{\dot\alpha\beta}, \theta_{rj}^{\dot\alpha}; x_l^{\alpha\dot\beta}, \theta_l^{\alpha j}\right).$$

It has the same type $(1,1,1)$. On $S^{2|0}$ and $\widetilde{S}^{2|0}$ there is induced an extension of this involution of type $(\eta = 1, \widetilde{\rho} = 0)$: $(s^\alpha, s^{\dot\alpha})^\rho = (s^{\dot\alpha}, s^\alpha)$.

(c) Using the coordinates on the big cell M introduced in subsection 7, we finally find a real structure on it:

$$(x^{\alpha\dot\beta})^\rho = (x^{\alpha\dot\beta})^t; \qquad (\theta_l^{\alpha j}, \theta_{rk}^{\dot\beta})^\rho = (\theta_{rj}^{\dot\alpha}, \theta_l^{\beta k}).$$

A generalized real structure of type $(1,1,1)$ is close to an ordinary real structure; if we write $\overline{f}$ instead of f^ρ, these rules hold: $\overline{\overline{f}} = f$ and $\overline{fg} = (-1)^{\widetilde{f}\widetilde{g}}\overline{f}\,\overline{g} = \overline{g}\,\overline{f}$ (here f, g are in a supercommutative C-algebra A with real structure of this type). A real A-point of the big cell M is defined as a homomorphism from the ring of functions to A commuting with ρ, i.e., such that if x is carried to x_0 then x^ρ is carried to $\overline{x}_0$. Therefore, at real points the values of the coordinates (x, θ_l, θ_r) satisfy the following conditions:

$$(x^{\alpha\dot\beta})^t = (\overline{x}^{\alpha\dot\beta}); \qquad (\theta_{rj}^{\dot\alpha}, \theta_l^{\beta k}) = (\overline{\theta}_l^{\alpha j}, \overline{\theta}_{rk}^{\dot\beta}).$$

In particular, the coordinates $x^a = \sigma^a_{\alpha\dot\beta} x^{\alpha\dot\beta}$ are real, just as in the purely even case. The indices l, r are redundant and can be omitted since the position of the indices α, $\dot\beta$ and j for θ already indicate the chirality. The usual choice of coordinates in the physical literature is $(x^a, \theta^{\alpha j}, \theta_{\dot\beta k})$, with relations $\theta_{\dot\beta k} = \overline{\theta^{\beta k}}$ at the real points.

In particular, the set of real points of M is included, by means of π_l (respectively, π_r), in the set of complex points of M_l (respectively, M_r). This gives the possibility, after postulating an analogous structure in the twisted case, of considering the geometry for $N = 1$ of supergravity as an analog of the geometry of real submanifolds of complex manifolds (A. S. Švarc).

12. Real structures: Euclidean signature. On T and T^* one introduces quaternionic structures :

$$\left(z_\alpha, z^{\dot\alpha}, \varsigma^j\right)^\rho = \left(\epsilon^{\alpha\beta} z_\beta, \epsilon^{\dot\alpha\dot\beta} z^{\dot\beta}, \lambda\varsigma^j\right),$$

$$(w^\alpha, w_{\dot\alpha}, \omega_j)^\rho = \left(\epsilon_{\alpha\beta} w^\beta, \epsilon_{\dot\alpha\dot\beta} w_{\dot\beta}, -\overline{\lambda}\omega_j\right).$$

Here $\epsilon^{01} = \epsilon_{01} = 1$, and $\lambda \in \mathrm{C}$, $|\lambda| = 1$, is a fixed complex number. In distinction to the previous passage, where T and T^* were considered as superspaces, i.e., ρ acted on their rings of functions, here T and T^* are considered to be modules over a supercommutative ring. The type of the real structure on the latter is understood to be $(1,1,1)$; the type on T, T^* will be $(\eta = 1, \widetilde{\rho} = 0)$. The scalar product is again purely imaginary:

$$\begin{aligned}\langle (z,\varsigma),(w,\omega)\rangle^{\rho} &= \epsilon^{\alpha\beta} z_{\beta}\epsilon_{\alpha\gamma}w^{\gamma} - \epsilon^{\dot\alpha\dot\beta}z^{\dot\beta}\epsilon_{\dot\alpha\dot\gamma}w_{\dot\gamma} + 2i\omega_j\varsigma^j \\ &= -\langle (z,\varsigma),(w,\omega)\rangle\,.\end{aligned}$$

The induced structures on $\mathsf{P}(T)$ and $\mathsf{P}(T^*)$ are real but without real points.

§ 2. Scalar Superfields and Component Analysis

1. General concepts. (a) Superfields on a Minkowski superspace M are sections of natural bundles but also sections of exterior bundles and connections on them. By natural bundles, we mean the sheaves $\mathcal{S}$, $\widetilde{\mathcal{S}}$, $\mathcal{E}_{l,r}$, $\mathcal{T}M$, as well as elements of the tensor algebra generated by these sheaves and their invariantly defined subquotients.

(b) Natural differential operators act between natural bundles. Two classes of such operators, constraints and dynamical equations, play a special role.

In the flat case which we have been considering, the solution of constraint equations often corresponds to picking out subspaces, irreducible with respect to the Poincaré supergroup, of all the sections (of the given natural bundle). The Poincaré supergroup is the ρ-invariant part of SL (T) which carries the light cone at infinity into itself; this is the cone which compactifies the big cell in $F(2|0,2|N;T)$ which we chose as a model of Minkowski superspace. Its irreducible representations and also their realizations by superfields are called supermultiplets.

In the curved case, without simplifying matters too much, we can say that constraints are defined by the geometric structure of the superspace itself or that a choice of constraint is a choice of the structure.

Dynamical equations, in contrast to pure constraints, impose conditions on the dependence of superfields on the even coordinates of the superspace. Generally speaking, dynamical equations are Euler-Lagrange equations corresponding to an action whose density is a volume form on the superspace M (or its chiral version $M_{l,r}$) depending on superfields. In many cases, however, such a superlagrangian is unknown.

(c) The connection of the theory of superfields to ordinary field theory is supplied by the method which is called component decomposition or component analysis. In the language of coordinates, this consists simply of the fact that a superfunction Φ can be represented in the form $\sum \phi_\alpha(x)\theta^\alpha$; the ϕ_α are viewed as ordinary fields on the ordinary manifold M_{rd}. More invariantly, such an analysis begins with the choice of isomorphisms $M \simeq \operatorname{Gr} M$ and $\mathcal{E} \simeq \operatorname{Gr} \mathcal{E}$ for the natural bundles $\mathcal{E}$ and so forth. After this, the results of the analysis are the homogeneous components of fields, components of constraint equations, dynamic equations, etc., which are studied by ordinary field formalism. (To this we should add that M is usually considered not over $\mathbf{R}$ (or $\mathbf{C}$) but over some unspecified supercommutative algebra from which odd constants in the needed quantity are taken.)

The nonuniqueness in the choice of these isomorphisms can act as a different kind of gauge equivalence. It can turn out that part of the component fields can be set to zero or expressed in terms of the others using constraints, equations of motion and an arbitrary choice of coordinates. The components in this part are called auxiliary fields.

The problem of seeking a superspace formulation of a field theory is often stated as a problem of constructing auxiliary fields. This "component synthesis" as a rule poses more nontrivial questions than component analysis. It has not been worked out for extended supergravity.

In the following subsection we will illustrate the more evident aspects of component analysis by the example of Yang-Mills superfields. Here we will limit ourselves to the simplest scalar superfields.

2. Scalar superfields. A scalar superfield Φ is an even function (section of the structure sheaf) on M. We will assume for definiteness that its domain of definition contains the big cell described in the preceding section. As we have already recalled in § 1.8, the superfield Φ is called left (respectively, right) chiral if $d_r\Phi = 0$ (respectively, $d_l\Phi = 0$) or, what is the same thing, if Φ belongs to $\mathcal{O}_{M_l}$ (respectively, $\mathcal{O}_{M_r}$).

The standard notation for Φ when $N = 1$ can have the form (in the coordinates of § 1)

$$\begin{aligned}\Phi(x^a, \theta^\alpha, \theta_{\dot\beta}) = A(x^a) + \theta^\alpha \psi_\alpha(x^a) + \overline{\phi}^{\dot\beta}(x)\theta_{\dot\beta} + \theta^\alpha\theta^\beta F_{\alpha\beta}(x^a) \\ + \theta_{\dot\alpha}\theta_{\dot\beta}\overline{G}^{\dot\alpha\dot\beta}(x^a) + \sigma_b^{\alpha\dot\beta}\theta^\alpha\theta_{\dot\beta}B^b(x^a) + \theta^\alpha\theta^\beta\overline{\kappa}^{\dot\gamma}_{\alpha\beta}(x^a)\theta_{\dot\gamma} \\ + \theta_{\dot\alpha}\theta_{\dot\beta}\theta^\gamma\lambda_\gamma^{\dot\alpha\dot\beta}(x^a) + \epsilon_{\alpha\beta}\epsilon_{\dot\gamma\dot\delta}\theta^\alpha\theta^\beta\theta_{\dot\gamma}\theta_{\dot\delta}D(x^a)\,.\end{aligned}$$

The fields A, F, G, B and D on M_{rd} are the boson components of the superfield Φ; the fields ψ, ϕ, κ and λ are its fermion components. The fermion components are (depending on x) odd elements of some supercommutative ring of "parameters." (If one does not wish to introduce this ring, it would be possible to remove the limitation to even functions Φ, but the insufficiency of odd constants would manifest itself in another place.) In this ring the bar denotes a real structure of the same type $(1,1,1)$ as in $\mathcal{O}_M$. The notation is adapted so as to write out the condition of reality $\Phi^\rho = \Phi$ in the form $\phi^{\dot\alpha} = \psi_\alpha$, $F_{\alpha\beta} = G^{\dot\beta\dot\alpha}$, $\overline{B}^b = B^b$, $\kappa^{\dot\gamma}_{\alpha\beta} = \lambda^{\dot\beta\dot\alpha}_\gamma$, $\overline{D} = D$.

The condition of left chirality can be written in the form of a system of differential equations $D^{\dot\beta}_\tau \Phi = 0$. It is clear that this can be solved immediately by passing to left coordinates:

$$\Phi_l = A(x_l^{\gamma\dot\delta}) + \theta^\alpha \psi_\alpha(x_l^{\gamma\dot\delta}) + \theta^\alpha\theta^\beta F_{\alpha\beta}(x_l^{\gamma\dot\delta}) .$$

§ 3. Yang-Mills Fields and Integrability Equations along Light Supergeodesics

1. Light supergeodesics. The double fibration $L \leftarrow F \rightarrow M$ used for the construction of the non-autodual Penrose transformation in Chapter 2 for the case of ordinary (compact complex) Minkowski space M can also be constructed over a general Minkowski superspace. It has the following structure:

$$L^{2|2N} = F(1|0,3|N;T) \xleftarrow{\pi_1} F^{6|4N}$$
$$= F(1|0,2|0,2|N,3|N;T) \xrightarrow{\pi_2} M = F(2|0,2|N;T).$$

The dimensions of the superspaces L, F and M are computed, e.g., with the help of the results of § 3 of Chapter 4. The projections π_1 and π_2, which are the standard projections onto a subflag, represent F in the form of relative flag spaces, so locally L and M are direct products. The fibers of F over L of dimension $1|2N$ are called (lifted) light supergeodesics of the superspace M.

In order to better imagine the geometry of this double fibration, it is convenient to use the system of liftings introduced in § 6 of Chapter 1 for ordinary complex geometry. The concepts carry over without change to supergeometry.

(a) A supermanifold F defines a $1|2N$-conical structure on M, i.e., a natural closed imbedding $\phi: F \rightarrow G_M(1|2N; \mathcal{T}M)$ is defined which commutes with the projections on M. It is defined such that the preimage of the tautological sheaf on the last grassmannian is $\mathcal{T}F/L$, which is embedded into $\pi_2^*(\mathcal{T}M)$ by means of $d\phi$. To verify the correctness of this definition, one must establish that $d\phi: \mathcal{T}F/L \rightarrow \pi_2^*(\mathcal{T}M)$ is a locally direct embedding; this will follow from the computation carried below.

(b) The distribution $\mathcal{T}F/L$ is an integrable conical connection, and the fibers of the projection π_1 are its geodesic supermanifolds.

2. Structure of $\Omega^i F/L$. It is not difficult to see that F/L is a relative flag manifold of the following kind: $F = F_L(1|0, 1|N; S_L^{3|N}/S_L^{1|0})$, where the S_L are components of the tautological flag on L in its standard flag realization. As we did for M in § 1, we introduce left and right spaces of F:

$$F_l = G_L\left(1|0; S_L^{3|N} \;/\; S_L^{1|0}\right),$$
$$F_r = G_L\left(1|N; S_L^{3|N} \;/\; S_L^{1|0}\right)$$

and the corresponding relative spaces of differentials $(\Omega^\cdot = \Omega^\cdot_{\text{odd}})$:

$$\Omega^1_l F/L = \Omega^1 F/F_r\,, \qquad \Omega^1_r F/L = \Omega^1 F/F_l\,.$$

Using the theorem about relative tangent sheaves of grassmannians, we find this exact sequence:

$$0 \to \Omega^1_0 F/L \to \Omega^1 F/L \to \Omega^1_l F/L \oplus \Omega^1_r F/L \to 0\,,$$

where

$$\Omega^1_l F/L = \Pi\,\mathcal{H}om\left(S_F^{2|N}/S_F^{2|0}, S_F^{2|0}/S_F^{1|0}\right) = S_F^{2|0}/S_F^{1|0} \otimes \Pi\left(S_F^{2|N}/S_F^{2|0}\right)^*,$$
$$\Omega^1_r F/L = \Pi\,\mathcal{H}om\left(S_F^{3|N}/S_F^{2|N}, S_F^{2|N}/S_F^{2|0}\right) = \Pi S_F^{2|N}/S_F^{2|0} \otimes \tilde{S}_F^{2|0}/\tilde{S}_F^{1|0},$$
$$\Omega^1_0 F/L = \Pi\left(S_F^{2|0}/S_F^{1|0} \otimes \tilde{S}_F^{2|0}/\tilde{S}_F^{1|0}\right).$$

Here $\Omega^1_0 F/L$ is defined as the kernel of the restriction of the sheaf of 1-forms to the direct sum of the left and right relative forms. Its spinor decomposition, described above, is established as the analogous result is established for $\Omega_0 M$ in § 1, by computing the Frobenius form $\mathcal{T}_l F/L \otimes \mathcal{T}_r F/L \to \mathcal{T}_0 F/L$ and the subsequent dualization. We omit the details; however, we recall for comparison information about $\Omega^1 M$ represented in the same form, from § 1:

$$0 \to \Omega^1_0 M \to \Omega^1 M \to \Omega^1_l M \oplus \Omega^1_r M \to 0\,,$$

where

$$\Omega^1_l M = S_M^{2|0} \otimes \Pi\left(S_M^{2|N}/S_M^{2|0}\right)^*,$$
$$\Omega^1_r M = \tilde{S}_M^{2|0} \otimes \Pi\left(S_M^{2|N}/S_M^{2|0}\right),$$
$$\Omega^1_0 M = \Pi\left(S_M^{2|0} \otimes \tilde{S}_M^{2|0}\right).$$

Now we introduce the sheaf $\mathcal{N}$, having defined it by means of the exact sequence

$$0 \to \mathcal{N} \to \pi_2^*(\Omega^1 M) \xrightarrow{\text{res}} \Omega^1 F/L \to 0,$$

where res is the restriction of the forms to the π_1-vertical vector fields. Comparing the exact sequences for $\Omega^1 M$ and $\Omega^1 F/L$, after checking that res is induced by natural mappings of corresponding tautological sheaves, we obtain an analogous decomposition for $\mathcal{N}$ (consider that $S_F^{2|0}$ and $S_F^{2|N}$ and the analogous sheaves with tildes are lifted from M):

$$0 \to \mathcal{N}_0 \to \mathcal{N} \to \mathcal{N}_l \oplus \mathcal{N}_r \to 0,$$

where

$$\mathcal{N}_l = S_F^{1|0} \otimes \Pi\left(S_F^{2|N}/S_F^{2|0}\right)^*,$$

$$\mathcal{N}_r = \widetilde{S}_F^{1|0} \otimes \Pi\left(S_F^{2|N}/S_F^{2|0}\right),$$

$$0 \to \Pi\left(S_F^{1|0} \otimes \widetilde{S}_F^{1|0}\right) \to \Pi\left(S_F^{2|0} \otimes \widetilde{S}_F^{1|0} \oplus S_F^{1|0} \otimes \widetilde{S}_F^{2|0}\right) \to \mathcal{N}_0 \to 0.$$

Using these facts about $\mathcal{N}$, we establish the following result; later it will play a fundamental role in the analysis of constraints and dynamical equations for Yang-Mills superfields on M.

3. Theorem. (a) $\pi_{2*}(\text{res})\colon \Omega^1 M \to \pi_{2*}\Omega^1 F/L$ *is an isomorphism.*

(b) *On 2-forms the mapping $\pi_{2*}(\text{res})$ is surjective. Its kernel $\Omega^2_{\text{con}} M$, which we call the sheaf of 2-forms satisfying constraints, has the following structure:*

$$\Omega^2_{\text{con}} M = R^1\pi_{2*}(\wedge^2 N),$$

$$\Omega^2_{\text{con}} M/\Omega^2_{1,\text{con}} M = \wedge^2\left(S_M^{2|0}\right) \otimes \wedge^2 \mathcal{E}_l \oplus \wedge^2\left(\widetilde{S}_M^{2|0}\right) \otimes \wedge^2 \mathcal{E}_r,$$

$$\Omega^2_{1,\text{con}} M = \Pi\left(\wedge^2\left(S_M^{2|0}\right) \otimes \widetilde{S}_M^{2|0} \otimes \mathcal{E}_l \oplus \wedge^2\left(\widetilde{S}_M^{2|0}\right) \otimes S_M^{2|0} \otimes \mathcal{E}_r\right).$$

Here $\Omega^2_{i,\text{con}}$ is the filtration which is induced by the filtration on $\Omega^2 M$ introduced in subsection 1.8.

Proof. It is clear that $F/M = \mathsf{P}\left(S_M^{2|0}\right) \times_M \mathsf{P}\left(\widetilde{S}_M^{2|0}\right)$ is a relative two-dimensional quadric. From this it follows that $\pi_{2*}\mathcal{O}_F = \mathcal{O}_M$ and more generally that $\pi_{2*}(\pi_2^*\mathcal{E}) = \mathcal{E}$ and $R^i\pi_{2*}(\pi_2^*\mathcal{E}) = 0$ when $i \geq 1$ for a locally free sheaf $\mathcal{E}$ on M. The sheaves $S_F^{1|0}$ and $\widetilde{S}_F^{1|0}$ on this relative quadric in standard notation are

$\mathcal{O}(-1,0)$ and $\mathcal{O}(0,-1)$. All the sheaves $\mathcal{O}(a,b)$ for $a=-1$ and $b=-1$ are acyclic. However, $R^i\pi_{2*}(\pi_2^*\mathcal{E}(a,b))=0$ for all $i\geq 0$, if $a=-1$ or $b=-1$.

Applying this reasoning to the sheaves $\mathcal{N}_0$, $\mathcal{N}_l$, $\mathcal{N}_r$ and $\mathcal{N}$ introduced in the preceding section, we immediately obtain that they are acyclic over M. From this it follows immediately that $\pi_{2*}(\text{res})\colon \Omega^1 M \to \pi_{2*}(\Omega^1 F/L)$ is an isomorphism.

The proof of the second part of the theorem is a bit longer. First of all, from the exact sequence

$$0\to \mathcal{N}\pi_2^*(\Omega^1 M)\to \pi_2^*\Omega^2 M\to \Omega^2 F/L\to 0$$

we find

$$0\to \pi_{2*}\left(\mathcal{N}\pi_2^*(\Omega^1 M)\right)\to \Omega^2 M\to \pi_{2*}\left(\Omega^2 F/L\right)\to R^1\pi_{2*}\left(\mathcal{N}\pi_2^*(\Omega^1 M)\right)\to 0.$$

Since $\mathcal{N}$ is embedded in $\pi_2^*(\Omega^1 M)$ in the form of a locally direct subsheaf and since $\Omega^2 M$ is the symmetric square of $\Omega^1 M$, it is not difficult to see that there is an exact sequence

$$0\to \wedge^2\mathcal{N}\to \mathcal{N}\otimes\pi_2^*(\Omega^1 M)\to \mathcal{N}\pi_2^*(\Omega^1 M)\to 0.$$

The middle sheaf is acyclic, so $R^i\pi_{2*}\left(\mathcal{N}\pi_2^*(\Omega^1 M)\right)=R^{i+1}\pi_{2*}\left(\wedge^2\mathcal{N}\right)$.

We will check that $R^2\pi_{2*}\left(\wedge^2\mathcal{N}\right)=0$. By relative Serre duality (which here reduces to the usual duality on a quadric), this is equivalent to $\pi_{2*}\left(\wedge^2\mathcal{N}^*(-2,-2)\right)=0$. The latter assertion is established again using the results of the preceding section:

(a) $\pi_{2*}\left(\wedge^2\mathcal{N}_0^*(-2,-2)\right)=0$ since $\wedge^2\mathcal{N}_0^*$ is embedded in a sheaf which is a sum of type $\bigoplus_{a+b=2}\pi_2^*(\mathcal{E}_{ab})(a,b)$, with $a,b\geq 0$;

(b) if $\mathcal{K}=\operatorname{Ker}\left(\wedge^2\mathcal{N}^*\to\wedge^2\mathcal{N}_0^*\right)$, then $\pi_{2*}(\mathcal{K}(-2,-2))=0$ since $\mathcal{K}$ is a subsheaf of a filtered sheaf and factors of the form $(\mathcal{N}_l^*\oplus\mathcal{N}_r^*)^{\otimes 2}$, $(\mathcal{N}_l^*\oplus\mathcal{N}_r^*)\otimes\mathcal{N}_0^*$, which after multiplication by $\mathcal{O}(-2,-2)$ become negative along the fiber. Thus we have established that $\Omega^2_{\text{con}}M=R^1\pi_{2*}(\wedge^2\mathcal{N})$ and that the morphism $\Omega^2 M\to \pi_{2*}\Omega^2 F/L$ is surjective.

In 1.8 a three-term filtration of $\Omega^2 M$ was described. Analogous filtrations naturally arise on $\pi_{2*}\Omega^2 F/L$ and $\Omega^2_{\text{con}}M$.

Since $\operatorname{rk}\Omega^1_0 F/L=0|1$, then $S^2\left(\Omega^1_0 F/L\right)=0$, and the filtration becomes two-term on $\Omega^2 F/L$:

$$0\to \Omega^1_0 F/L\otimes\left(\Omega^1_l F/L\oplus\Omega^1_r F/L\right)\to \Omega^2 F/L\to S^2\left(\Omega^1_l F/L\oplus\Omega^1_r F/L\right)\to 0,$$

whence

$$0 \to \pi_{2*}\left[\Omega_0^1 F/L \otimes (\Omega_l^1 F/L \oplus \Omega_r^1 F/L)\right] \to \pi_{2*}\Omega^2 F/L$$
$$\to S^2(S_M^{2|0}) \otimes S^2\mathcal{E}_l \oplus \Omega_l^1 M \otimes \Omega_r^1 M \oplus S^2(\widetilde{S}_M^{2|0}) \otimes S^2\mathcal{E}_r \to 0 .$$

The corresponding two-term filtration on $\Omega^2_{\text{con}}M$, induced from $\Omega^2 M$ is this:

$$0 \to \Omega^2_{1,\text{con}}M \to \Omega^2_{\text{con}}M \to \wedge^2(S_M^{2|0}) \otimes \wedge^2\mathcal{E}_l \oplus \wedge^2(\widetilde{S}_M^{2|0}) \otimes \wedge^2\mathcal{E}_r \to 0 ,$$

where $\Omega^2_{1,\text{con}}M = \Omega_1^2 M \cap \Omega^2_{\text{con}}M$ can also be computed as the kernel of the composition of mappings

$$\Omega_1^2 M \to \Omega_1^2 M/\Omega_0^2 M = \Omega_0^1 M \otimes (\Omega_l^1 M \oplus \Omega_r^1 M)$$
$$\to \pi_{2*}\left[\Omega_0^1 F/L \otimes (\Omega_l^1 F/L \oplus \Omega_r^1 F/L)\right] .$$

The meaning of the second line is evident in tensor algebra. For the left component we have

$$\Omega_0^1 M \otimes \Omega_l^1 M = \Pi\left[S_M^{2|0} \otimes \widetilde{S}_M^{2|0} \otimes S_M^{2|0} \otimes \mathcal{E}_l\right] ,$$

$$\pi_{2*}\left[\Omega_0^1 F/L \otimes \Omega_l^1 F/L\right] = \left[S^2\left(S_M^{2|0}\right) \otimes \widetilde{S}_M^{2|0} \otimes \mathcal{E}_l\right] ,$$

and this morphism reduces to symmetrization with respect to S_M. The right components are constructed analogously.

4. Yang-Mills fields which are integrable along light supergeodesics. Now let $\mathcal{E}$ be a locally free sheaf over a domain $U \subset M$ and $\nabla\colon \mathcal{E} \to \mathcal{E} \otimes \Omega^1 M$ a connection on it. As in purely even geometry, on $\mathcal{E}_F = \pi_2^*(\mathcal{E})$ a connection $\pi_2^*(\nabla)$ is defined as well as its restriction $\nabla_{F/L}\colon \mathcal{E}_F \to \mathcal{E}_F \otimes \Omega^1 F/L$ to the fibers of π_1. We say that a Yang-Mills field $(\mathcal{E}, \nabla)$ is integrable along light geodesics, in short, null-integrable, if it fulfills the following equivalent conditions:

(a) $\nabla^2_{F/L} = 0$;

(b) $\Phi(\nabla) \in \Gamma\left(\mathcal{E}nd\,\mathcal{E} \otimes \Omega^2_{\text{con}}M\right)$, where $\Phi(\nabla)$ is the curvature form of ∇. Their equivalence follows from Theorem 3; if $\Phi(\nabla)$ is the curvature of ∇, then the equation $\nabla^2_{F/L} = 0$ means that the restriction of $\pi_2^*(\Phi(\nabla))$ to $\mathcal{T}F/L$ is equal to zero, i.e., $\Phi(\nabla) \in \operatorname{Ker}\pi_{2*}(\text{res}) = \mathcal{E}nd\,\mathcal{E} \otimes \Omega^2_{\text{con}}M$.

We compare the description given above of the sheaf $\Omega^2_{\text{con}}M$ with the coordinate description of constraints which is usually cited in the journal literature. On a big cell the curvature form is written in terms of the basis of $\Omega^1 M$ which is dual to the basis

$$D_{\alpha j},\quad D^k_{\dot\beta},\quad D_a = \frac{1}{N}\sigma_a^{\alpha\dot\beta}\left[D_{\alpha j}, D^j_{\dot\beta}\right] .$$

The corresponding components have the form Φ_{AB}, where the indices A, B are taken from the set $\{\, a = 0, 1, 2, 3; (\alpha j), \binom{k}{\dot\beta}\,\}$.

In this notation we have the following equations.

(a) $N = 1$. The indices j, k can be omitted; constraints have the form $\Phi_{\alpha\beta} = \Phi_{\dot\alpha\dot\beta} = \Phi_{\alpha\dot\beta} = 0$. In our language this corresponds to the vanishing of the image of Φ in $\mathcal{E}nd\,\mathcal{E} \otimes \Omega^2/\Omega^2_1$. This agrees with the fact that for $N = 1$ we have $\wedge^2 \mathcal{E}_l = \wedge^2 \mathcal{E}_r = 0$, so that by Theorem 3 we find $\Omega^2_{\text{con}} = \Omega^2_{1,\text{con}}$.

After this, it is established that the remaining components of the curvature can be expressed by the prepotentials W_α, $W_{\dot\beta} \in \mathcal{E}nd\,\mathcal{E} \otimes \mathcal{S}^{2|0}$ (or $\widetilde{\mathcal{S}}^{2|0}$). In our language, this is evident: after trivializing the sheaves of rank one, $\wedge^2 \mathcal{S}^{2|0}$, $\wedge^2 \widetilde{\mathcal{S}}^{2|0}$, $\mathcal{E}_l$ and $\mathcal{E}_r$ over a big cell, we can identify $\Omega^2_{1,\text{con}}$ directly with $\mathcal{S}^{2|0}_l \oplus \mathcal{S}^{2|0}_r$; the components of the curvature under this identification are the prepotentials.

A lagrangian of a Yang-Mills field must be a real section of $\operatorname{Ber} M$, i.e., the sheaf $\wedge^2(\mathcal{S}^{2|0}) \otimes \wedge^2(\widetilde{\mathcal{S}}^{2|0})$ (see 1.8). This section can be constructed from $\Phi = \Phi_l \oplus \Phi_r$, having set $\mathcal{L} = \operatorname{Tr}(\Phi_l \oplus \Phi_r)$. Here the trace Tr includes, besides the trace in $\mathcal{E}nd\,\mathcal{E}$, the contraction of $\mathcal{S}_L$ and $\mathcal{S}_r$ by means of the section $\epsilon_{\alpha\beta}$ and the real structure. In coordinate notation, $\mathcal{L} = \operatorname{Tr}\left(W^\alpha W_\alpha + \overline{W}_{\dot\alpha} \overline{W}^{\dot\alpha}\right)$.

(b) $N = 2$. Beginning with $N = 2$, the sheaf Ω^2_{con} no longer coincides with $\Omega^2_{1,\text{con}}$, and the conditions on Φ become meaningful modulo $\Omega^2_1 M$. The coordinate notation for the constraints found in the literature is this:

$$\Phi_{\alpha i, \beta j} = -\Phi_{\beta i, \alpha j}\,, \quad \Phi^{ij}_{\dot\alpha\dot\beta} = -\Phi^{ij}_{\dot\beta\dot\alpha}\,, \quad \Phi_{\alpha i,}{}^{j}{}_{\dot\beta} = 0\,.$$

It is clear that this is equivalent to our formulation, following Theorem 3:

$$\Phi \bmod \Omega^2_{1,\text{con}} M \in \wedge^2(\mathcal{S}^{2|0}) \otimes \wedge^2 \mathcal{E}_l \oplus \wedge^2(\widetilde{\mathcal{S}}^{2|0}) \otimes \wedge^2 \mathcal{E}_r\,.$$

As above, using trivializations of these two invertible sheaves, we can write

$$\Phi \bmod \Omega^2_{1,\text{con}} M = W_l \oplus W_r\,,$$

where the scalar functions W_l and W_r are prepotentials of the Yang-Mills field for $N = 2$.

Here it is already not obvious that any form Φ satisfying the condition

$$\Phi \bmod \Omega^2_{1,\text{con}} M \in \mathcal{E}nd\,\mathcal{E} \otimes \Omega^2_{\text{con}} M / \Omega^2_{1,\text{con}} M$$

automatically belongs to $\mathcal{E}nd\,\mathcal{E} \otimes \Omega^2_{\text{con}} M$. In fact this is not true; however, if Φ is the curvature of a connection, then it satisfies the Bianchi identity. From this it follows first that Φ can be expressed completely by W_l and W_r and also that the form obtained lies in $\mathcal{E}nd\,\mathcal{E} \otimes \Omega^2_{\mathrm{con}} M$. We omit this computation.

A lagrangian of a Yang-Mills field for $N = 2$ is proportional to $\mathrm{Tr}(W_l^2 \otimes W_r^2)$. We recall that $\mathrm{Ber}\, M \simeq \mathcal{O}_M$ in this case, so any scalar function can be considered a volume form.

(c) $N = 3$. Apparently, the lagrangian in this case is unknown, while the constraint equation, written above, coincides with the equations of motion.

(d) $N = 4$. The equations of motion are obtained by adding to the constraints the conditions

$$\epsilon^{\alpha\beta}\Phi_{\alpha i,\beta j} = \frac{1}{2}\epsilon_{ijkl}\epsilon^{\dot\alpha\dot\beta}\Phi^{kl}_{\dot\alpha\dot\beta}\,.$$

This is a peculiar condition of autoduality with respect to $\wedge^2 \mathcal{E}_{l,\,r}$ which can only appear for $N = 4$.

5. Method of constructing null-integrable Yang-Mills fields. Let $U \subset M$ be an open set, and let $L(U) = \pi_1\pi_2^{-1}(U)$. We consider $L(U)$ as an open subsuperspace of L. As in Chapter 2 setting $L(x) = \pi_2\pi_1^{-1}(x)$ for $x \in U$, we have $L(U) = \bigcup_{x\in U} L(x)$. We view the quadric $L(x)$ as a closed sub superspace of L. A locally free sheaf $\mathcal{E}_L$ on $L(U)$ is said to be U-trivial if its restriction to any of the quadrics, $L(x)$, $x \in U$, is a trivial (or free) sheaf, i.e., isomorphic to $\mathcal{O}^{p|q}_{L(x)}$. A locally free sheaf $\mathcal{E}_L$ is called a YM-sheaf if it is defined on some open subset $V \subset L$ such that for a nonempty $U \subset M$ we have $V \supset L(U)$ and $\mathcal{E}_L|_{L(U)}$ is U-trivial.

Let $\mathcal{E}_L$ be a YM-sheaf. The Penrose transformation carrying it to a null-integrable Yang-Mills field on the domain of triviality U is defined just as in the purely even case:

(a) We construct the sheaf $\mathcal{E}_F = \pi_1^*(\mathcal{E}_L)$ on $\pi_1^{-1}(L(U))$ and the relative connection $\nabla_{F/L}\colon \mathcal{E}_F \to \mathcal{E}_F \otimes \Omega^1 F/L$ is uniquely defined such that $\nabla_{F/L}$ is zero on $\pi_1^{-1}(\mathcal{E}_L)$. Clearly, $\nabla^2_{F/L} = 0$.

(b) We restrict $\mathcal{E}_F$ and $\nabla_{F/L}$ to $\pi_2^{-1}(U)$; for conciseness we keep the notation the same. We set $\mathcal{E} = \pi_{2*}(\mathcal{E}_F)$ and $\nabla = \pi_{2*}(\nabla_{F/L})$. From the U-triviality of $\mathcal{E}_L$ it follows that $\mathcal{E}_F = \pi_2^*(\mathcal{E})$ (canonically). By Theorem 3 we can then identify $\pi_{2*}(\mathcal{E}_F \otimes \Omega^1 F/L) = \pi_{2*}(\pi_2^*\mathcal{E} \otimes \Omega^1 F/L)$ with $\mathcal{E} \otimes \Omega^1 M$. It is not difficult to check that the differential operator $\nabla\colon \mathcal{E} \to \mathcal{E} \otimes \Omega^1 M$ is a connection. By the same construction, the Yang-Mills field $(\mathcal{E}, \nabla)$ is null-integrable.

There is an obvious restriction which $(\mathcal{E}, \nabla)$ satisfies: the connection ∇ has trivial monodromy along all nonempty intersections of light geodesics with U. Essentially this is the only restriction on $(\mathcal{E}, \nabla)$ which can be obtained in this manner.

6. Theorem. *Let the nonempty intersections of light geodesics with U be connected. Then the following categories are equivalent:*

(a) *The category of null-integrable Yang-Mills fields $(\mathcal{E}, \nabla)$ on U with trivial monodromy along light geodesics.*

(b) *The category of U-trivial YM-sheaves on $L(U)$.*

The proof is formally the same as in the purely even case. □

7. Supersymmetric Yang-Mills equations. Suppose that a YM-sheaf $\mathcal{E}_L$, which is defined on $L(U)$ and is U-trivial, extends to a locally free sheaf $\mathcal{E}_L^{(m)}$ on the m-th infinitesimal neighborhood $L^{(m)}(U)$ in $\mathsf{P}(T) \times \mathsf{P}(T^*) = \mathsf{P} \times \widehat{\mathsf{P}}$ (certainly $L^{(m)}(U)_{\mathrm{rd}} = L(U)_{\mathrm{rd}}$, the sheaf induced by $\mathcal{O}_{\mathsf{P}\times\widehat{\mathsf{P}}}/I_L^{m+1}$, where I_L is the sheaf of equations of L). We will call such an $\mathcal{E}_L$ m-extendable. The property of m-extendability, as in the purely even case, can be expressed by a supplementary system of differential equations on the Yang-Mills field $(\mathcal{E}, \nabla)$ corresponding to $\mathcal{E}_L$. Following Witten, we will apply this condition to define supersymmetric Yang-Mills equations for $N \leq 3$.

8. Definition. A field $(\mathcal{E}, \nabla)$ is said to be a *solution of supersymmetric Yang-Mills equations* if (locally on M) it corresponds to a $(3 - N)$-extendable YM-sheaf $\mathcal{E}_L$. □

It is known that for $N \leq 2$ these equations can be deduced from the lagrangians which were introduced in subsection 4 and can be defined on fields satisfying constraints. For $N = 3$ and $N = 4$ the equations were written down independently of the Penrose transformation; the fact that they coincide for $N = 3$ with Definition 8 must be checked independently. For $N = 4$ an interpretation in terms of $\mathcal{E}_L$ is not completely understood.

In the next section we will apply the algebro-geometric method of monads to construct solutions of supersymmetric Yang-Mills equations.

9. Cohomological component analysis. In § 2 we explained that component analysis of a scalar superfield on M consists simply of expanding it into powers of θ in a fixed coordinate system and interpreting the coefficients as ordinary (classical) fields on M_{rd}. We can carry out component analysis for a Yang-Mills field analogously; however, here the analysis is complicated by two circumstances: first, the necessity of taking gauge equivalence into account and, second, the necessity taking care of the connection form and the curvature form in parallel.

However, in the framework of the Penrose transformation it is interesting to try to carry out component analysis by cohomological methods directly in terms of the YM-sheaf $\mathcal{E}_L$. Naturally for this we will have to limit ourselves to null-integrable fields and will have to understand how the presence of constraints and dynamical equations is reflected in cohomological language. It turns out that the basic characteristics of this are the same as working in the purely even case with the infinitesimal neighborhoods $L^{(m)}$, $N=0$.

We will repeat briefly, with necessary changes, the formalism of obstruction theory and extensions in the superanalytic case.

10. Extensions and obstructions. Let M be an analytic supermanifold, and let $J = \mathcal{O}_{M,1} + \mathcal{O}_{M,2}{}^2$ and $M^{(n)} = (M, \mathcal{O}_M/J^{n+1})$. Let G be a connected (complex) analytic supergroup, and let $\rho\colon G \to \mathrm{GL}(p|q)$ be a finite-dimensional representation of it. We fix a class in noncommutative cohomology, $e \in H^1(M, G(\mathcal{O}_M)$ is the sheaf of analytic mappings of domains in M to G. The pair (e, ρ) can be considered as a locally free sheaf of rank $p|q$ on M whose structure group is reduced to G with representation ρ. In particular, for $G = \mathrm{GL}(p|q)$ and $\rho = \mathrm{id}$, the elements of $H^1(M, \mathrm{GL}(p|q; \mathcal{O}_M))$ classify the sheaves of rank $p|q$ up to isomorphism.

Now let $e^{(n)} \in H^1(M, G(\mathcal{O}_M/J^{n+1}))$ and let $\mathcal{G}$ be the locally free sheaf on $M^{(0)} = M_{\mathrm{rd}}$ corresponding to the pair $(e^{(0)}, \mathrm{Ad})$, where $e^{(0)}$ is the reduction of $e^{(n)}$ mod J and Ad is the adjoint representation. As in the purely even case, the following properties hold:

(a) A mapping $\omega\colon H^1(M, G(\mathcal{O}_M/J^{n+1})) \to H^2(M, (\mathcal{G} \otimes J^{n+1}/J^{n+2})_0)$ is defined each that $\omega(e^{(n)}) = 0$ if and only if $e^{(n)}$ extends to a class

$$e^{(n+1)} \in H^1(M, G(\mathcal{O}_M/J^{n+2})).$$

This is the construction of ω. There is a commutative diagram of exact triples:

$$\begin{array}{ccccccc}
0 \to J^{n+1}/J^{n+2} \to & & \mathcal{O}_{M^{(n+1)}} & \to & \mathcal{O}_{M^{(n)}} & \to 0 \\
\downarrow \simeq & & \downarrow d\otimes 1 & & \downarrow d & \\
0 \to J^{n+1}/J^{n+2} \to & & \Omega^1 M^{(n+1)} \otimes \mathcal{O}_{M^{(n)}} & \to & \Omega^1 M^{(n)} & \to 0.
\end{array}$$

From this an exact diagram of sheaves is obtained:

$$\begin{array}{ccccccc}
0 \to (\mathcal{G}\otimes J^{n+1}/J^{n+2})_0 & \xrightarrow{\exp} & G(\mathcal{O}_{M^{(n+1)}}) & \to & G(\mathcal{O}_{M^{(n)}}) & \to 1 \\
\| & & \downarrow D\otimes 1 & & \downarrow D & \\
0 \to (\mathcal{G}\otimes J^{n+1}/J^{n+2})_0 & \longrightarrow & (\mathcal{G}\otimes \Omega^1 M^{(n+1)} \otimes \mathcal{O}_{M^{(n)}})_1 & \to & (\mathcal{G}\otimes \Omega^1 M^{(n)})_1 & \to 0,
\end{array}$$

where $Dg = g^{-1}dg$ in a matrix realization. The obstruction mapping ω is represented as a composition:

$$\omega\colon H^1(M, G(\mathcal{O}_{M^{(n+1)}})) \xrightarrow{H^1(D)} H^1(M, (\mathcal{G}\otimes\Omega^1 M^{(n)})_1) \xrightarrow{\delta} H^2(M, (\mathcal{G}\otimes J^{n+1}/J^{n+2})_0).$$

The indices 0, 1 refer of course to the $\mathbb{Z}_2$-grading.

(b) If $\omega(e^{(n)}) = 0$, then the group $H^1(M, (\mathcal{G} \otimes J^{n+1}/J^{n+2})_0)$ acts transitively on the set of extensions $\{ e^{(n+1)} \}$ of the class $e^{(n)}$.

(c) This action is effective if the mapping $H^0(M, \mathcal{G}^{(n+1)}) \to H^0(M, \mathcal{G}^{(n)})$ is a surjection for one of the extensions, where $\mathcal{G}^{(n+1)}$ and $\mathcal{G}^{(n)}$ are the sheaves associated with $(e^{(n+1)}, \mathrm{Ad})$ and $(e^{(n)}, \mathrm{Ad})$, respectively.

We observe that even if this action is not effective, one can still define canonically the difference $e_1^{(n+1)} - e_2^{(n+1)} \in H^1(M, (\mathcal{G} \otimes J^{n+1}/J^{n+2})_0)$ of two extensions e_1, e_2.

11. Component analysis of $\mathcal{E}_L$. From the results of § 3 of Chapter 4 it is clear that on $\mathsf{P} \times \widehat{\mathsf{P}}$ we have $J^{n+1}/J^{n+2} = S^{n+1}(J/J^2) = S^{n+1}(T_1^* \otimes \mathcal{O}(-1,0) \oplus T_1 \otimes \mathcal{O}(0,-1))$, where T_1 is the odd part of the space of supertwistors T. Since L is a supermanifold and its odd dimension is the same as that of $\mathsf{P} \times \widehat{\mathsf{P}}$, i.e., $2N$, the same formula holds for J_L^{n+1}/J_L^{n+2}. The symmetric (i.e., the grassmannian) powers must also be constructed over $\mathcal{O}_{L_{\mathrm{rd}}}$; for $n \geq 2N$, $J_L^{n+1} = 0$.

Let $e \in H^1(L(U), G(\mathcal{O}_L))$, with image $e^{(n)}$ in $H^1(L(U), G(\mathcal{O}_L/J_L^{n+1}))$. We can imagine that e is constructed step-by-step from $e^{(n)}$ to $e^{(n+1)}$. The pair $(e^{(0)}, \rho)$ represents a YM-sheaf $\mathcal{E}_L^{(0)}$ on L_{rd} with structure group reduced to G. (We observe that the structure group can be a supergroup, although L_{rd} is a purely even manifold.) Corresponding to this is an ordinary Yang-Mills field on $U \subset G(2; T_0)$ with group G if $\mathcal{E}_L$ is U-trivial.

If $e^{(n)}$ has already been constructed and $\omega(e^{(n)}) = 0$, we choose some extension $\tilde{e}^{(n+1)}$, and we parametrize the other extensions by elements $h^{(n+1)}$ of the corresponding cohomology groups and write the equation $\omega(\tilde{e}^{(n+1)} + h^{(n+1)}) = 0$. The elements $h^{(n+1)}$ are turned into fields on U_{rd} by the Radon-Penrose transformation just as are the elements of the cohomology groups in which the obstructions $\omega(e^{(n)})$ lie. The non-uniqueness in the choice of the initial extension $\tilde{e}^{(n+1)}$ under the condition $\omega(e^{(n)}) = 0$ can be viewed as a remnant of the theory of noncommutative cohomology (nonabelian Yang-Mills fields) which is not reducible to the purely abelian case. The type of the operator ω defines the type of the equation of null-integrability and, as L passes to $L^{(m)}$ via intermediate neighborhoods, the type of the supersymmetric Yang-Mills equations.

In order not to complicate the exposition and to limit it to the three cohomology groups already computed in Chapter 2, we present in the table below information relating only to the case $N = 3$, $m = 0$.

It is assumed that U is Stein and has connected intersection with the light lines. In the box indexed by (n, H^i) is the sheaf $\mathcal{F}(n, i)$ on U such that

$$H^i\left(L(U), (\mathcal{G}_L \otimes J^n/J^{n+1})_0\right) = H^0(U, \mathcal{F}(n,i)).$$

This is notation used in the table: $(\mathcal{G}, \nabla)$ is a sheaf with connection on U which is the Penrose transform of $(e^{(0)}, \mathrm{Ad} \circ \rho)$ on $L(U)_{\mathrm{rd}}$; $T^{(i)} = S^i(T_1)$ for $i \geq 0$ and $S^{-i}(T_1^*)$ for $i < 0$. D_∇ is the Dirac operator based on ∇; the tautological sheaves $S_\pm$ are constructed on $G(2; T_0) = M_{\mathrm{rd}}$.

The basic observation about the table is that, as we move upward through the levels $L(U)^{(n)}$, we first collect fields (up to $n \leq 3$) and then equations ($3 \leq n \leq 6$). The last obstruction is the current for the ordinary Yang-Mills field which is obtained by putting odd coordinates equal to zero.

§ 4. Monads on Superspaces and YM-sheaves

1. Combined YM-sheaves. The goal of this section is to construct a class of YM-sheaves on the space L and its infinitesimal neighborhoods $L^{(m)} \subset \mathsf{P}\times\widehat{\mathsf{P}}$, especially solutions of supersymmetric Yang-Mills equations, which by Definition 3.8 correspond to YM-sheaves on $L_N^{(3-N)}$ for $0 \leq N \leq 3$.

Let $\pi: L \to \mathsf{P}$ and $\widehat{\pi}: L \to \widehat{\mathsf{P}}$ be the standard projections. We say that a YM-sheaf is selfdual if $\mathcal{E}_L = \pi^*(\mathcal{E}_\mathsf{P})$, where $\mathcal{E}_\mathsf{P}$ is some sheaf on P or on an open subset of P containing $\mathsf{P}(U) = \pi_1\pi_2^{-1}(U)$. Analogously, we say that $\mathcal{E}_L$ is anti-selfdual if $\mathcal{E}_L = \widehat{\pi}^*(\mathcal{E}_{\widehat{\mathsf{P}}})$. Finally, we call $\mathcal{E}_L$ combined if it is contained in the tensor algebra generated by the selfdual and the anti-selfdual sheaves, i.e., it is isomorphic to a sum of the form $\bigoplus_i \pi^*(\mathcal{E}_\mathsf{P}^{(i)}) \otimes \widehat{\pi}^*(\mathcal{E}_{\widehat{\mathsf{P}}}^{(i)})$.

It is clear that combined YM-sheaves are defined on domains of the form $\mathsf{P}(U) \times \widehat{\mathsf{P}}(U)$ and in particular on any neighborhoods of $L(U)$.

It is easier to construct selfdual and anti-selfdual sheaves than general YM-sheaves since it is easier to take care of the triviality of the restriction of $\mathcal{E}_\mathsf{P}$ to a general line $\mathsf{P}^1(x) \subset \mathsf{P}(U)$. In fact, by a theorem of Grothendieck, the class of such a restriction is described by a collection of integers: $\mathcal{E}_\mathsf{P}|\mathsf{P}^1(x) \simeq \sum_i \mathcal{O}^{p_i|q_i}(m_i)$. Thus any method of constructing $\mathcal{E}_\mathsf{P}$ allows one to control the values of m_i. The locally free sheaves on the quadric $\mathsf{P}(x) \times \widehat{\mathsf{P}}(x)$ depend on an arbitrarily large number of continuous parameters, so it is much more difficult to produce triviality.

The construction of this section is based on two observations.

First, it turns out that combined sheaves can admit nontrivial deformations on a finite neighborhood $L^{(m)}$ which do not extend to higher neighborhoods. On the level of infinitesimal deformations, this is a cohomological phenomenon. We will begin with a description of this in order to make it clearer which cohomology groups, by their nontriviality, are responsible for it.

Cohomology Table ($N = 3$)

n	H^1	H^2
1	$\mathcal{G}_1 \otimes \left(S_+ \otimes \wedge^2 S_- \otimes T^{(1)} \oplus S_- \otimes \wedge^2 S_+ \otimes T^{(-1)} \right)$	0
2	$\mathcal{G}_1 \otimes \left(\wedge^2 S_+ \otimes T^{(-2)} \oplus \wedge^2 S_- \otimes T^{(2)} \oplus \wedge^2 S_+ \otimes \wedge^2 S_- \otimes T^{(1)} \otimes T^{(-1)} \right)$	0
3	Ker $D_{\nabla,1}$	Coker $D_{\nabla,1}$
	where $D_{\nabla,1}\colon \mathcal{G}_1 \otimes \left(S_+ \otimes \wedge^2 S_+ \otimes T^{(-3)} \oplus S_- \otimes \wedge^2 S_- \otimes T^{(3)} \right) \to \mathcal{G}_1 \otimes \left(S_- \otimes (\wedge^2 S_+)^2 \otimes T^{(-3)} \oplus S_+ \otimes (\wedge^2 S_-)^2 \otimes T^{(3)} \right)$	
4	0	$\mathcal{G}_0 \otimes \left((\wedge^2 S_+)^2 \otimes \wedge^2 S_- \otimes T^{(-3)} \otimes T^{(1)} \oplus \wedge^2 S_+ \otimes (\wedge^2 S_-)^2 \otimes T^{(3)} \otimes T^{(-1)} \oplus \wedge^2 S_+ \otimes \wedge^2 S_- \otimes T^{(2)} \otimes T^{(-2)} \right)$
5	0	$\mathcal{G}_1 \otimes \left(S_+ \otimes \wedge^2 S_+ \otimes \wedge^2 S_- \otimes T^{(-3)} \otimes T^{(2)} \oplus S_- \otimes \wedge^2 S_- \otimes \wedge^2 S_+ \otimes T^{(3)} \otimes T^{(-2)} \right)$
6	0	$\mathrm{Ker}_0\, \nabla_3$ where $\nabla_3\colon \mathcal{G} \otimes \Omega^3 U \to \mathcal{G} \otimes \Omega^4 U$ (covariant differential).

Second, it has been established that the problem of integrating these infinitesimal deformations to get global ones can sometimes be circumvented by constructing the deformed sheaves directly by the method of monads, as instanton sheaves were constructed in Chapter 2.

2. Extensions, deformations and cohomology. Our initial observation is that the existence of nontrivial deformations of combined sheaves on $L^{(m)}$ is related to the nonvanishing of some cohomology groups of selfdual and antiselfdual sheaves. The group $H^1\left(L(U)^{(m)}, \pi^*\mathcal{E}_{\mathbf{P}} \otimes \widehat{\pi}^*\mathcal{E}_{\widehat{\mathbf{P}}}\right)$ plays an obvious role (a more standard notation for this would be $H^1\left(L(U), j_m^*(\pi^*\mathcal{E}_{\mathbf{P}} \otimes \widehat{\pi}^*\mathcal{E}_{\widehat{\mathbf{P}}})\right)$, where $j_m: L(U)^{(m)} \to \mathbf{P}(U) \times \widehat{\mathbf{P}}(U)$ is the closed inclusion). It appears in the following contexts.

(a) Its elements classify the extensions $0 \to \widehat{\pi}^*\mathcal{E}_{\widehat{\mathbf{P}}_1} \to \mathcal{E}_L \to \pi^*\mathcal{E}_{\mathbf{P}}^* \to 0$. (In the supercase the even elements of H^1 classify these extensions while the odd ones classify the extensions of $\pi^*\Pi\mathcal{E}_{\mathbf{P}}^*$ by $\widehat{\pi}^*\mathcal{E}_{\widehat{\mathbf{P}}}$.) The classification up to isomorphism of the sheaves $\mathcal{E}_L$ corresponding to different elements of H^1 is of course a separate problem, but at least for $H^1 \neq 0$ there exist sheaves $\mathcal{E}_L$ which are not isomorphic to $\pi^*\mathcal{E}_{\mathbf{P}}^* \oplus \widehat{\pi}^*\mathcal{E}_{\widehat{\mathbf{P}}_1}$. They are U-trivial YM-sheaves for any U for which $\mathcal{E}_{\mathbf{P}}$ and $\mathcal{E}_{\widehat{\mathbf{P}}}$ are U-trivial, since on $L(x) = \mathbf{P}^1(x) \times \widehat{\mathbf{P}}^1(x)$ any extension of a free sheaf is free.

(b) The elements of the group $H^1\left(L(U)^{(m)}, \pi^*\mathcal{E}_{\mathbf{P}}^* \otimes \widehat{\pi}^*\mathcal{E}_{\widehat{\mathbf{P}}} \oplus \pi^*\mathcal{E}_{\mathbf{P}} \otimes \widehat{\pi}^*\mathcal{E}_{\widehat{\mathbf{P}}}^*\right)$ classify the infinitesimal deformations of first order of the direct sum $\pi^*\mathcal{E}_{\mathbf{P}} \oplus \widehat{\pi}^*\mathcal{E}_{\widehat{\mathbf{P}}}$ on $L^{(m)}(U)$ for which $\pi^*\mathcal{E}_{\mathbf{P}}$ and $\widehat{\pi}^*\mathcal{E}_{\widehat{\mathbf{P}}}$ do not deform. Actually, the infinitesimal deformations of a sheaf $\mathcal{F}$ are classified by the group $H^1(\mathcal{F}^* \otimes \mathcal{F})$; in the super case deformations with even parameters correspond to even cohomology classes and odd parameters to odd classes. Here it is necessary to resolve separately the question of whether one can extend an infinitesimal deformation to a global one in the case of an even parameter. As was shown in the preceding paragraph, mixed directions in such deformations correspond in a way to the superposition of an extension of $\pi^*\mathcal{E}_{\mathbf{P}}$ by $\widehat{\pi}^*\mathcal{E}_{\widehat{\mathbf{P}}}$ and an extension of $\widehat{\pi}^*\mathcal{E}_{\widehat{\mathbf{P}}}$ by $\pi^*\mathcal{E}_{\mathbf{P}}$. The odd constants squared are equal to zero; therefore, for them there is no extension problem.

(c) Infinitesimal deformations of the combined sheaf $\pi^*\mathcal{E}_{\mathbf{P}} \otimes \widehat{\pi}^*\mathcal{E}_{\widehat{\mathbf{P}}}$ are classified by the 1-cohomology of $\pi^*\,\mathcal{E}nd\,\mathcal{E}_{\mathbf{P}} \otimes \widehat{\pi}^*\,\mathcal{E}nd\,\mathcal{E}_{\widehat{\mathbf{P}}}$. The sheaves of endomorphisms of (anti)selfdual sheaves themselves belong to the same class.

The second remark is that $H^1(\mathbf{P}(U), \mathcal{E}_{\mathbf{P}}(-m-1))$ and $H^1(\widehat{\mathbf{P}}(U), \mathcal{E}_{\widehat{\mathbf{P}}}(-m-1))$ make a fundamental contribution to the group $H^1(L(U)^{(m)}, \pi^*\mathcal{E}\mathbf{P} \otimes \widehat{\pi}^*\mathcal{E}_{\widehat{\mathbf{P}}})$. Here is the simplest situation. Let us suppose that $L(U) = L \cap (\mathbf{P}(U) \times \widehat{\mathbf{P}}(U))$.

3. Proposition. *Let us suppose that* $H^0(\mathsf{P}(U), \mathcal{E}_{\mathsf{P}}) = H^1(\mathsf{P}(U), \mathcal{E}_{\mathsf{P}}) = 0$ *and analogously for* $\mathcal{E}_{\widehat{\mathsf{P}}}$. *Then*

$$H^1(L(U)^{(m)}, \pi^*\mathcal{E}_{\mathsf{P}} \otimes \widehat{\pi}^*\mathcal{E}_{\widehat{\mathsf{P}}}) = H^1(\mathsf{P}(U), \mathcal{E}_{\mathsf{P}}(-m-1)) \otimes H^1(\widehat{\mathsf{P}}(U), \mathcal{E}_{\widehat{\mathsf{P}}}(-m-1)).$$

Proof. We denote by $s \in H^0(\mathsf{P} \times \widehat{\mathsf{P}}, \mathcal{O}(1,1))$ the equation of L and write on $\mathsf{P}(U) \times \widehat{\mathsf{P}}(U)$ the exact sequence

$$0 \to \mathcal{O}_{\mathsf{P}\times\widehat{\mathsf{P}}}(-m-1, -m-1) \xrightarrow{s^{m+1}} \mathcal{O}_{\mathsf{P}\times\widehat{\mathsf{P}}} \to j^*_m\mathcal{O}_{L^{(m)}} \to 0$$

and take its tensor product with $\pi^*\mathcal{E}_{\mathsf{P}} \otimes \widehat{\pi}^*\mathcal{E}_{\widehat{\mathsf{P}}}$. The boundary differential $\delta\colon H^1 \to H^2$ gives the desired isomorphism by using the Künneth formula. (For noncompact $\mathsf{P}(U)$ and $\widehat{\mathsf{P}}(U)$ the tensor products that occur are completed products if we are working in the analytic category.) □

We recall some facts about the cohomology of selfdual sheaves for $N = 0$. We consider two cases separately; when U is sufficiently small and when U is so large that $\mathsf{P}(U) = \mathsf{P}$.

4. Proposition. *Let* $U \subset M$ *be a connected Stein manifold, and let any nonempty intersection of* U *with a complex light ray be connected and simply-connected. In addition, let* $N = 0$ *and let* $\mathcal{E}_{\mathsf{P}}$ *be a* U*-trivial sheaf on* $\mathsf{P}(U)$ *with corresponding Yang-Mills field* $(\mathcal{E}, \nabla)$ *on* U. *Let* $S_{\pm}$ *be the tautological (spinor) bundles on* U. *Then the following isomorphisms exist:*

(a) $H^0(\mathsf{P}(U), \mathcal{E}_{\mathsf{P}}) = H^0(U, \operatorname{Ker}\nabla)$, *and* $H^1(\mathsf{P}(U), \mathcal{E}_{\mathsf{P}})$ *is isomorphic to the (middle) homology group of the complex*

$$H^0(U, \mathcal{E}) \xrightarrow{\nabla} H^0(U, \mathcal{E} \otimes \Omega^1 U) \xrightarrow{\nabla_+} H^0(U, \mathcal{E} \otimes S^2 S_+ \otimes \wedge^2 S_-),$$

where ∇_+ *is the composition of* ∇ *and the projection* $\Omega^2 \to \Omega^2_+$.

(b) $H^1(\mathsf{P}(U), \mathcal{E}_{\mathsf{P}}(-i)) = \operatorname{Ker} H^0(D_i)$, *where* D_i *is the following differential operator on* U:

$$D_1\colon \mathcal{E} \otimes S_- \otimes \wedge^2 S_+ \to \mathcal{E} \otimes S_+ \otimes (\wedge^2 S_+)^2 \otimes \wedge^2 S_-$$

(the Dirac operator on the background $(\mathcal{E}, \nabla)$*);*

$$D_2\colon \mathcal{E} \otimes \wedge^2 S_+ \to \mathcal{E} \otimes (\wedge^2 S_+)^2 \otimes \wedge^2 S_-$$

(the Klein-Gordan operator on the background $(\mathcal{E}, \nabla)$); it is of second order); for $i \geq 3$:

$$D_i: \mathcal{E} \otimes S^{i-2} S_+ \otimes \wedge^2 S_+ \to \mathcal{E} \otimes S^{i-3} S_+ \otimes S_- \otimes (\wedge^2 S_+)^2$$

(the operator of massless fields of spin $(i-2)/2$ on the background $(\mathcal{E}, \nabla)$)).

Analogous results are true for $\mathcal{E}_{\widehat{\mathsf{P}}}$ when $S_\pm$ is replaced by $S_\mp$. □

This result shows that the elements of the group $H^2(L(U)^{(i-1)}, \pi^* \mathcal{E}_{\mathsf{P}} \otimes \widehat{\pi}^* \mathcal{E}_{\widehat{\mathsf{P}}})$ can be constructed by solving the equations $D_i \psi = 0$ and $\widehat{D}_i \widehat{\psi} = 0$. If $H^1(\mathcal{E}_{\mathsf{P}}) \neq 0$, then one must compute $\operatorname{Ker}(H^2(s^{i+1}))$ in $\operatorname{Ker} D_i \otimes \operatorname{Ker} \widehat{D}_i$ (here and later, if $\alpha: \mathcal{F} \to \mathcal{G}$ is a morphism of sheaves, then $H^i(\alpha): H^i(\mathcal{F}) \to H^i(\mathcal{G})$ is the corresponding morphism on cohomology).

Now we take up the case $\mathsf{P}(U) = \mathsf{P}$ and the possibilities for purely algebraic constructions. From now on we assume that $N \geq 0$.

5. Cohomology of projective superspace. The structure of projective superspace $\mathsf{P} = \mathsf{P}(T_0 \oplus T_1)$, i.e., the grassmannian of $1|0$-subspaces in T, is this: $\mathsf{P}_{\mathrm{rd}} = \mathsf{P}(T_0)$ and $\mathcal{O}_{\mathsf{P}} = S(T_1^* \otimes \mathcal{O}_{\mathsf{P}(T_0)}(-1))$ (since T_1 is odd, $\mathcal{O}_{\mathsf{P}}$ is a sheaf of Grassmann algebras). The following facts are a slight generalization of standard ones (we assume that $\dim T_0 = 4$).

(a) The dualizing sheaf $\operatorname{Ber} \Omega^1 \mathsf{P}$ is isomorphic to the sheaf $\Pi^N \mathcal{O}_{\mathsf{P}}(-4 + N)$ (see [88]).

(b) $H^0(\mathsf{P}, \mathcal{O}_{\mathsf{P}}(n)) = S^n(T^*)$ for $n \geq 0$ and is 0 for $n < 0$; $H^i(\mathsf{P}, \mathcal{O}_{\mathsf{P}}(n)) = 0$ for all n and $i = 1, 2$; $H^3(\mathsf{P}, \mathcal{O}_{\mathsf{P}}(n)) = S^{-n+N-4}(T) \otimes \operatorname{Ber} T^*$ for $n \leq N - 4$ and equals 0 for $n > N - 4$.

(c) For any locally free sheaf $\mathcal{E} = \mathcal{E}_{\mathsf{P}}$ on P the behavior of $H^i(\mathsf{P}, \mathcal{E}(n))$ is qualitatively the same as for $\mathcal{O}_{\mathsf{P}}$: the groups $H^0(\mathsf{P}, \mathcal{E}(n))$ are different from zero only for $n \geq n_0$, $H^3(\mathsf{P}, \mathcal{E}(n)) \neq 0$ for $n \leq n_3$, and $H^{1,2}(\mathsf{P}, \mathcal{E}(n)) \neq 0$ for $n_{1,2}^- \leq n \leq n_{1,2}^+$.

(d) The invertible sheaves on P of rank $1|0$ (respectively $0|1$) are exhausted up to isomorphism by the sheaves $\mathcal{O}(n)$ (respectively $\Pi\mathcal{O}(n)$).

6. Monads. A *monad* (on an arbitrary superspace) is defined to be a complex of locally free sheaves $\mathcal{F}_\cdot$: $\mathcal{F}_{-1} \xrightarrow{\alpha} \mathcal{F}_0 \xrightarrow{\beta} \mathcal{F}_1$, where α is a locally direct injection and β is a surjection, with $\beta \circ \alpha = 0$. We denote by $\mathcal{E}(\mathcal{F}_\cdot) = \operatorname{Ker} \beta / \operatorname{Im} \alpha$ the sheaf of zero cohomology of the monad. We will construct U-trivial sheaves on the type $\mathcal{E}(\mathcal{F}_\cdot)$ where all the $\mathcal{F}_i$ are direct sums of invertible sheaves. The instantons and superinstantons are contained as a partial case among the fields corresponding to these sheaves. We will write $\alpha(i)$ instead of $\alpha \otimes \mathrm{id}_{\mathcal{O}(i)}: \mathcal{F}_{-1}(i) \to \mathcal{F}_0(i)$, etc. The corresponding morphism of cohomology groups will be denoted by $H^k(\alpha(i))$.

7. Proposition. *Let $\mathcal{F}_{\cdot}$ be a monad consisting of direct sums of invertible sheaves on $\mathbf{P}$, and let $\mathcal{E} = \mathcal{E}(\mathcal{F}_{\cdot})$. Then for all i we have:*

$$H^0(\mathcal{E}(i)) = \operatorname{Ker} H^0(\beta(i))/\operatorname{Im} H^0(\alpha(i));$$
$$H^1(\mathcal{E}(i)) = \operatorname{Coker} H^0(\beta(i));$$
$$H^2(\mathcal{E}(i)) = \operatorname{Ker} H^3(\alpha'(i)), \quad \alpha' : \mathcal{F}_{-1} \to \operatorname{Ker}\beta;$$
$$H^3(\mathcal{E}(i)) = \operatorname{Ker} H^3(\beta(i))/\operatorname{Im} H^3(\alpha'(i)).$$

Proof. We set $\mathcal{K}(i) = \operatorname{Ker}\beta(i)$ and write down two exact sequences:

$$0 \to \mathcal{K}(i) \longrightarrow \mathcal{F}_0(i) \xrightarrow{\beta(i)} \mathcal{F}_1(i) \to 0,$$
$$0 \to \mathcal{F}_{-1}(i) \xrightarrow{\alpha'(i)} \mathcal{K}(i) \longrightarrow \mathcal{E}(i) \to 0.$$

Now we take into account that $H^{1,2}(\mathcal{F}_\alpha(i)) = 0$. In fact, $\mathcal{F}_a$ is a direct sum of sheaves $\mathcal{O}(i)$ and $\Pi\mathcal{O}(i)$, and so the same assertion if true for $\mathcal{F}_a(i)$ as an $\mathcal{O}_{\mathbf{P}_{rd}}$-module. Thus we can use the classical theorem of Serre. This allows us to compute the cohomology $\mathcal{K}(i)$ using the first exact triple:

$$H^0(\mathcal{K}(i)) = \operatorname{Ker} H^0(\beta(i)),$$
$$H^1(\mathcal{K}(i)) = \operatorname{Coker} H^0(\beta(i)),$$
$$H^2(\mathcal{K}(i)) = 0, \qquad H^3(\mathcal{K}(i)) = \operatorname{Ker} H^3(\beta(i)).$$

Putting these results into the cohomology sequence arising from the second exact triple, we obtained the desired result. □

To verify the U-triviality of the sheaf given by the monad, we will use the following criterion.

8. Proposition. *Let $\mathcal{F}_{\cdot}$ be a monad on $\mathbf{P}^1$ with the following properties: $\mathcal{F}_0$ is free and $H^0(\mathcal{F}_{-1}) = H^0(\mathcal{F}_1^*) = 0$. Then a mapping is defined, $\epsilon = \epsilon(\mathcal{F}_{\cdot})\colon H^1(\mathcal{F}_{-1}(-1)) \to H^0(\mathcal{F}_{-1}(-1))$, which is an isomorphism if and only if $\mathcal{E}(\mathcal{F}_{\cdot})$ is free on $\mathbf{P}^1$.*

Proof. As above, we set $\mathcal{K} = \operatorname{Ker}\beta$. From $0 \to \mathcal{K}(-1) \to \mathcal{F}_0(-1) \to \mathcal{F}_1(-1) \to 0$ we find that $H^0(\mathcal{F}_1(-1)) = H^1(\mathcal{K}(-1))$ and $H^0(\mathcal{K}(-1)) = 0$. From $0 \to \mathcal{F}_{-1}(-1) \to \mathcal{K}(-1) \to \mathcal{E}(-1) \to 0$ we obtain a morphism $H^1(\mathcal{F}_{-1}(-1)) \to H^1(\mathcal{K}(-1)) \xrightarrow{\sim} H^0(\mathcal{F}_1(-1))$, which by definition is ϵ. Its kernel and cokernel are, respectively, $H^0(\mathcal{E}(-1))$ and $H^1\mathcal{E}(-1))$. But these two groups vanish if and only if $\mathcal{E}$ is free. □

9. m-monads on P. We define an *m-monad* ($m \geq 0$) to be a complex of the form

$$\mathcal{F}_{\cdot}: F_{-1} \otimes \mathcal{O}_{\mathsf{P}}(-m-1) \xrightarrow{\alpha} F_0 \otimes \mathcal{O}_{\mathsf{P}} \xrightarrow{\beta} F_1 \otimes \mathcal{O}_{\mathsf{P}}(m+1),$$

where the F_i are linear (super)spaces. We set $\mathcal{E}_{\mathsf{P}} = \mathcal{E}(\mathcal{F}_{\cdot})$. From Proposition 6 we find for $0 \leq i \leq m$:

$$H^1(\mathsf{P}, \mathcal{E}_{\mathsf{P}}(-i-1)) = F_1 \otimes S^{m-i}(T^*).$$

Thus we have a supply of one-dimensional cohomologies for constructing nontrivial extensions and deformations. As will be shown below, the sheaf $\mathcal{E}(\mathcal{F}_{\cdot})$ can be U-trivial for nonempty U.

The conditions of Proposition 3 are verified by a monad such that:

$$\left.\begin{aligned} H^0(\mathsf{P}, \mathcal{E}_{\mathsf{P}}) = 0 &\iff \operatorname{Ker} H^0(\beta) = 0, \\ H^1(\mathsf{P}, \mathcal{E}_{\mathsf{P}}) = 0 &\iff \operatorname{Coker} H^0(\beta) = 0 \end{aligned}\right\} \iff F_0 \xrightarrow[\sim]{H^0(\beta)} F_1 \otimes S^{m+1}(T^*).$$

These same conditions are fulfilled for the sheaf $\mathcal{E}_{\mathsf{P}}^*$, which is defined by the dual monad, if the isomorphism is also the mapping $F_0^* \xrightarrow[\sim]{H^0(\alpha^*)} F_1^* \otimes S^{m+1}(T^*)$. A monad with these two properties is said to be *maximal.*

10. m-monads on $L^{(l)}$. We consider a diagram of sheaves on $\mathsf{P} \times \widehat{\mathsf{P}}$ of the following type:

$$\begin{aligned} \mathcal{F}_{\cdot} \;:\; & F_{-1}^+ \otimes \mathcal{O}(-m-1, 0) \oplus F_{-1}^- \otimes \mathcal{O}(0, -m-1) \xrightarrow{\alpha} (F_0^+ \oplus F_0^-) \otimes \mathcal{O} \\ & \xrightarrow{\beta} F_1^+ \otimes \mathcal{O}(m+1, 0) \oplus F_1^- \otimes \mathcal{O}(0, m+1), \end{aligned}$$

where $\alpha = \begin{pmatrix} \alpha^+ & 0 \\ 0 & \alpha^- \end{pmatrix}$ and $\beta = \begin{pmatrix} \beta^+ & \gamma^+ \\ \gamma^- & \beta^- \end{pmatrix}$. Here and below, when we write a morphism in the form of a block matrix, we mean that it multiplies a column of components on the left. In particular,

$$\begin{aligned} \gamma^+ &: F_0^- \otimes \mathcal{O} \to F_1^+ \otimes \mathcal{O}(m+1, 0), \\ \gamma^- &: F_0^+ \otimes \mathcal{O} \to F_1^- \otimes \mathcal{O}(0, m+1). \end{aligned}$$

We assume that the following conditions are fulfilled:

$$\begin{aligned} \beta^+\alpha^+ &= 0, & \beta^-\alpha^- &= 0, \\ \gamma^+\alpha^- &= s^{i+1}\gamma', & \gamma^-\alpha^+ &= s^{i+1}\gamma''. \end{aligned}$$

In addition to this, let α be a locally direct injection and let β be a surjection. Then the restriction $\mathcal{F}_\cdot|L^{(i)}$ is a monad and therefore defines a sheaf on $L^{(i)}$. We will call such monads *standard*.

The monad $\mathcal{F}_\cdot$ is a deformation of a direct sum of monads lifted from $\mathbf{P}$ and $\widehat{\mathbf{P}}$, respectively. Now let $\mathcal{F}_\cdot(t_+, t_-)$ be the complex obtained from $\mathcal{F}_\cdot$ by replacing $\gamma^\pm$ by $t_\pm\gamma^\pm$. For $t_\pm = 0$, this is a direct sum: $\mathcal{F}_\cdot(0,0) = \mathcal{F}_\cdot^+ \oplus \mathcal{F}_\cdot^-$. If, say, $t_- \neq 0$ and $t_+ = 0$, then there is a natural inclusion of complexes $\mathcal{F}_\cdot^- \subset \mathcal{F}_\cdot(0, t_-)$, with a corresponding extension $0 \to \mathcal{E}(\mathcal{F}_\cdot^-) \to \mathcal{E}(\mathcal{F}_\cdot(0,t_-)) \to \mathcal{E}(\mathcal{F}_\cdot^+) \to 0$. The class $e(t_-)$ of this extension, which depends on t_-, lies in the group

$$H^1\left(L^{(i)}, \mathcal{E}(\mathcal{F}_\cdot^+) \otimes \mathcal{E}(\mathcal{F}_\cdot^-)\right) = H^1\left(\mathbf{P}, \mathcal{E}^*_{\mathbf{P}}(-i-1)\right) \otimes H^1\left(\widehat{\mathbf{P}}, \mathcal{E}_{\widehat{\mathbf{P}}}(-i-1)\right)$$
$$= (F_{-1}^+)^* \otimes S^{m-i}(T^*) \otimes F_1^- \otimes S^{m-i}(T)$$

(we assume that the monads $\mathcal{F}_\cdot^\pm$ are maximal). On the other hand, on $\mathbf{P} \times \widehat{\mathbf{P}}$ we have a morphism $t_-\gamma^-\alpha^+ = t_- s^{i+1}\gamma'' \colon F_{-1}^+ \otimes \mathcal{O}(-m-1, 0) \to F_1^- \otimes \mathcal{O}(0, m+1)$. It is clear that $t_-\gamma''$ can be considered to be an element of the same space $\mathrm{Hom}(F_{-1}^+, F_1^-) \otimes H^0\left(\mathbf{P} \times \widehat{\mathbf{P}}, \mathcal{O}(m-i, m-i)\right)$ which contains $e(t_-)$. It is not difficult to conjecture, and it can be verified, that $e(t_-) = \mathrm{const} \cdot t_-\gamma''$ under this identification.

Therefore, to verify that the monads $\mathcal{F}_\cdot$, for all possible $t_\pm\gamma^\pm$ with $t_\pm^2 = 0$, realize all the infinitesimal deformations of $\mathcal{E}(\mathcal{F}_\cdot^+ \oplus \mathcal{F}_\cdot^-)$, it is sufficient to see that the corresponding pairs (γ', γ'') can run through all possible values. We restrict ourselves to the cases when $\mathcal{F}_\cdot^\pm$ is maximal. Then the monad $\mathcal{F}_\cdot$ can be written in the following form, which we will also call standard:

$$\begin{array}{c}
F_{-1}^+ \otimes \mathcal{O}(-m-1, 0) \oplus F_{-1}^- \otimes \mathcal{O}(0, -m-1) \\
\Big\downarrow \begin{pmatrix} a^+ & 0 \\ 0 & a^- \end{pmatrix} \\
\mathcal{O} \otimes \left[F_{-1}^+ \otimes S^{m+1}(T) \oplus F_{-1}^- \otimes S^{m+1}(T^*)\right] \\
\downarrow \ B \\
\mathcal{O} \otimes \left[F_1^+ \otimes S^{m+1}(T^*) \oplus F_1^- \otimes S^{m+1}(T)\right] \\
\Big\downarrow \begin{pmatrix} b^+ & 0 \\ 0 & b^- \end{pmatrix} \\
F_1^+ \otimes \mathcal{O}(m+1, 0) \oplus F_1^- \otimes \mathcal{O}(0, m+1)\,.
\end{array}$$

In this notation the morphisms $a^\pm$ and $b^\pm$ are defined once and for all, and the parameters of the monad are the elements of the invertible constant operator $B =$

$\begin{pmatrix} B^+ & C^+ \\ C^- & B^- \end{pmatrix}$. More precisely,

$$a^+(f^+ \otimes t) = f^+ \otimes h(ts^{m+1}),$$

where $f^+ \in F^+_{-1}$, t is a section of $\mathcal{O}(-m-1,0)$ and $h\colon \mathcal{O}(0,m+1) \to S^{m+1}(T)\otimes\mathcal{O}$ is the canonical morphism; a^- is defined analogously. Furthermore, b^+ is defined such that $H^0(b^+)$ is the identity mapping of $F_1^+ \otimes S^{m+1}(T^*)$. The matrix B describes the identity isomorphism between the two natural realizations of the sheaf $\mathcal{F}_0$ of the monad in which α and β respectively take on canonical form.

In order to see that a monad exists with a given γ' and γ'', one must use the freedom of choice of $C^\pm$ (the $B^\pm$ are defined by the monads of $\mathcal{F}^\pm_\cdot$). Let us say that C^- can be any element in $\mathrm{Hom}(F^+_{-1}, F^-_1) \otimes S^{m+1}(T^*) \otimes S^{m+1}(T)$, for example, $\gamma' s^{m+1}$, $i \le m$.

The following result shows that the monads in some natural class are standard.

11. Theorem. *Let the monad $\mathcal{F}_\cdot$ have the form*

$$\mathcal{F}_\cdot\colon \bigoplus_{a+b=m+1} F_{-1}^{(a,b)} \otimes \mathcal{O}(-a,-b) \xrightarrow{\alpha} F_0 \otimes \mathcal{O} \xrightarrow{\beta} \bigoplus_{a+b=m+1} F_1^{(a,b)} \otimes \mathcal{O}(a,b)$$

where the $F(a,b)_j$ $(a,b \ge 0)$ are linear superspaces. Suppose that its cohomology sheaf is defined over $L^{(i)}$ and is U-free for nonempty U. Then the following assertions are true:

(a) *The monad $\mathcal{F}_\cdot$ is isomorphic to a standard one; in particular, it defines two monads $\mathcal{F}^\pm_\cdot$ lifted from P and $\widehat{\mathsf{P}}$ respectively.*

(b) *The sheaf $\mathcal{E}(\mathcal{F}_\cdot)|L(x)$ is free if and only if $\mathcal{E}(\mathcal{F}^+_\cdot)|\mathsf{P}^1(x)$ and $\mathcal{E}(\mathcal{F}^-_\cdot)|\widehat{\mathsf{P}}^1(x)$ are free.*

Proof. First of all, $F(-a,-b)_{-1} = \{0\}$ for $ab > 0$. To prove this we establish the following: if $F(-a,-b)_{-1} \ne \{0\}$ for some pair (a,b) with $ab > 0$, then $H^1(\mathcal{E}(\mathcal{F}_\cdot)(-1,-1)|L(x)) \ne 0$ for any quadric $L(x) \subset L$, so $\mathcal{E}(\mathcal{F}_\cdot)|L(x)$ cannot be free. We will write $\mathcal{F}_i(x)$ instead of $\mathcal{F}_i|L(x)$, etc. It follows from $0 \to \mathcal{K}(x) \to \mathcal{F}_0(x) \to \mathcal{F}_1(x) \to 0$ that $H^2(\mathcal{K}(x)(-1,-1)) = 0$. From $0 \to \mathcal{F}_{-1}(x) \to \mathcal{K}(x) \to \mathcal{E}(x) \to 0$ it is evident that $\delta\colon H^1(\mathcal{E}(x)(-1,-1)) \to H^2(\mathcal{F}_{-1}(-1,-1))$ is surjective. Finally, $H^2(\mathcal{F}_{-1} \otimes \mathcal{O}_{L(x)}(-a-1,-b-1)) \ne 0$ for $a,b > 0$.

Applying this result to the dual monad, we find that the U-triviality of $\mathcal{E}(\mathcal{F}^*_\cdot)$ implies that $F_1^{(a,b)} = \{0\}$ for $a,b > 0$. Thus $\mathcal{F}_\cdot$ on $L^{(i)}$ has the form

$$F^+_{-1} \otimes \mathcal{O}(-m-1,0) \oplus F^-_{-1} \otimes \mathcal{O}(0,-m-1)$$

$$\xrightarrow{(\alpha^+,\alpha^-)} F_0 \otimes \mathcal{O} \xrightarrow{\binom{\beta^+}{\beta^-}} F_1^+ \otimes \mathcal{O}(m+1,0) \oplus F_1^- \otimes \mathcal{O}(0,m+1).$$

We have $\beta^+\alpha^+ + \beta^+\alpha^- = 0$ and $\beta^-\alpha^+ + \beta^-\alpha^- = 0$. But $\beta^+\alpha^+$ is multiplication by a section of the sheaf $\mathrm{Hom}(F_{-1}^+, F_1^+) \otimes \mathcal{O}(2m+2, 0)$, and $\beta^+\alpha^-$ is multiplication by a section of the sheaf $\mathrm{Hom}(F_{-1}^+, F_1^-) \otimes \mathcal{O}(m+1, m+1)$. Consequently, $\beta^+\alpha^+ = 0$ and $\beta^+\alpha^- = 0$ on $L^{(i)}$; analogously, $\beta^-\alpha^+ = 0$ and $\beta^-\alpha^- = 0$.

Let F_0^+ (respectively F_0^-) be a minimal subspace in F_0 such that the image of α^+ (respectively α^-) is contained in $F_0^+ \otimes \mathcal{O}$ (respectively $F_0^- \otimes \mathcal{O}$). Since (α^+, α^-) is a locally direct inclusion, we have $F_0^+ \cap F_0^- = \{0\}$. We may assume that $F_0 = F_0^+ \oplus F_0^-$ (if this is not true, we can redefine F_0^-, say, by adding to it a missing complement of F_0). Then the monad acquires standard form:

$$F_0 = F_0^+ \oplus F_0^-, \qquad \alpha = \begin{pmatrix} \alpha^+ & 0 \\ 0 & \alpha^- \end{pmatrix}, \qquad \beta = \begin{pmatrix} \beta^+ & \gamma^+ \\ \gamma^- & \beta^- \end{pmatrix}.$$

All these sheaves and morphisms extend uniquely from $L^{(i)}$ to $\mathbf{P} \times \widehat{\mathbf{P}}$, but the relations such as $\gamma^+\alpha^- = 0$ on $L^{(i)}$ become $\gamma^+\alpha^- = s^{i+1}\gamma'$ on $\mathbf{P} \times \widehat{\mathbf{P}}$, since an extension of $\gamma^+\alpha^-$ is multiplication by a section of the sheaf $\mathrm{Hom}(F_{-1}^+, F_1^+) \otimes \mathcal{O}(m+1, m+1)$.

We will now determine for which $L(x)$ the sheaf $\mathcal{E}(x)$ is free. We will use the following criterion for freeness: $H^i(\mathcal{E}(x)(\epsilon, \eta)) = 0$ for all i and $(\epsilon, \eta) = (-1, 0)$, $(0, -1)$ or $(-1, -1)$. Its necessity is obvious. The sufficiency is obtained by writing the Koszul resolution of the geometric fiber $\mathcal{E}(x)$ at any point $(u, v) \in L(x)$ and checking that it reduces to the canonical trivialization of $\mathcal{E}(x)$ which establishes an isomorphism of $H^0(\mathcal{E}(x))$ with this fiber.

Now, as usual, we write out the monad $\mathcal{F}_{\cdot}(\epsilon, \eta)$ restricted to $L(x)$:

$$0 \to F_{-1}^+ \otimes \mathcal{O}_{L(x)}(-m-1+\epsilon, \eta) \oplus F_{-1}^- \otimes \mathcal{O}_{L(x)}(\epsilon, -m-1+\eta) \xrightarrow{\tilde{\alpha}(\epsilon,\eta)} \mathcal{K}(x)(\epsilon, \eta) \to \mathcal{E}(x)(\epsilon, \eta) \to 0,$$

$$0 \to \mathcal{K}(x)(\epsilon, \eta) \to (F_0^+ \oplus F_0^-) \otimes \mathcal{O}_{L(x)}(\epsilon, \eta) \to F_1^+ \otimes \mathcal{O}_{L(x)}(m+1+\epsilon, \eta) \oplus F_1^- \otimes \mathcal{O}_{L(x)}(\epsilon, m+1+\eta) \to 0.$$

From the second triple we find the isomorphisms:

$$\delta^+(\beta) \colon H^1(\mathcal{K})(x)(-1, 0)) \xrightarrow{\sim} F_1^+ \otimes H^0(\mathcal{O}_{L(x)}(m, 0)),$$
$$\delta^-(\beta) \colon H^1(\mathcal{K})(x)(0, -1)) \xrightarrow{\sim} F_1^- \otimes H^0(\mathcal{O}_{L(x)}(0, m)).$$

All the remaining groups $H^i(\mathcal{K}(x)(\epsilon,\eta))$ are zero, so $H^i(\mathcal{E}(x)(-1,-1)) = 0$ for all i while the vanishing of the remaining groups is equivalent to having the following mappings be isomorphisms:

$$\delta^+(\beta)\circ H^1(\widetilde{\alpha}(-1,0)) : F^+_{-1}\otimes H^1(\mathcal{O}_{L(x)}(-m-2,0) \to F^+_1 \otimes H^0(\mathcal{O}_{L(x)}(m,0)),$$
$$\delta^-(\beta)\circ H^1(\widetilde{\alpha}(0,-1)) : F^-_{-1}\otimes H^1(\mathcal{O}_{L(x)}(0,-m-2) \to F^-_1 \otimes H^0(\mathcal{O}_{L(x)}(0,m)).$$

For $\gamma^\pm = 0$, that is, for $\beta = \begin{pmatrix}\beta^+ & 0\\ 0 & \beta^-\end{pmatrix}$, the two last mappings are clearly $\epsilon(\mathcal{F}^+_{\cdot}(x))$ and $\epsilon(\mathcal{F}^-_{\cdot}(x))$ in the notation of Proposition 8. We will now show that they actually do not depend on $\gamma^\pm$, so the freeness of $\mathcal{E}(\mathcal{F}_{\cdot}(x))$ is equivalent to the simultaneous freeness of $\mathcal{E}(\mathcal{F}^+_{\cdot}(x))$ and $\mathcal{E}\mathcal{F}^-_{\cdot}(x))$.

Now take a cohomology class in $H^1(\mathcal{F}_{-1}(x)(-1,0))$; suppose that it is a cocycle in the usual two-element covering (lifted from $\mathbf{P}^1$ to $\mathbf{P}^1(x)\times\widehat{\mathbf{P}}^1(x)$). In order to apply $\delta^+(\beta)\circ H^1(\widetilde{\alpha}(-1,0))$ to it, we must split this cocycle in $\mathcal{F}_0(x)(-1,0)$ and then apply the morphism $H^0(\beta(x)(-1,0))$ to one of the two terms. But all that is dependent on $\gamma^\pm$ is the component of the morphism $\beta(x)(-1,0)$ which lies in $F^-_1\otimes\mathcal{O}_{L(x)}(-1,m+1)$, and H^0 of this component is zero. Compare this with the computation in subsection 2 of § 5 below.

12. Remarks. (a) It is essential that the deformation of combined sheaves which we constructed on $L^{(i)}$ remains nontrivial after restriction to $L^{(0)} = L$. In fact, from the proof of Proposition 3 and the identification of subsection 10, it is clear that the restriction mapping from $L^{(i)}$ to L on H^2 reduces to multiplication by s^i:

$$S^{m-i}(T\oplus T^*) \to S^m(T\oplus T^*).$$

(b) Most likely, a sheaf $\mathcal{E}_L$ on $L^{(m)}$ given by an m-monad with maximal $\mathcal{F}^\pm$ and $(\gamma',\gamma'')\neq(0,0)$ does not extend to $L^{(m+1)}$. To check this one must compute the obstruction to the extension, which lies in $H^2(L,\mathcal{E}nd\,\mathcal{E}^{(0)}_L(-m-1,-m-1))$.

§ 5. Some Coordinate Computations

1. The simplest m-monads on P in coordinates. We will change slightly the notation of § 1. Let $(z_1,\dots,z_4;\varsigma_1,\dots,\varsigma_N)$ be a basis of T^*, and let (Z^a) be the basis of $S^{m+1}(T^*)$ consisting of monomials of degree $m+1$: $Z^a = z_1^{a_1}\cdots z_4^{a_4}\varsigma_1^{\alpha_1}\cdots\varsigma_N^{\alpha_N}$, where the $a_i\geq 0$, $\alpha_i = 0$ or 1, and $\sum a_i + \sum\alpha_j = m+1$. We set $A(Z) = (Z^a)^t$ (a column of monomials) and $B(Z) = (\sum_b \beta^a_b Z^b)$ (a row of polynomials), where the constants β^a_b, besides what is required in superalgebra by the parity of B(Z), satisfy two conditions: (a) $\sum_{a+b=c} \beta^a_b = 0$ for any c, and (b) the elements of $B(Z)$ generate $S^{m+1}(T^*)$. Then the diagram

$$\mathcal{F}_\cdot : \mathcal{O}_{\mathbf{P}}(-m-1) \xrightarrow{A(Z)} \mathcal{O}_{\mathbf{P}}^{r(m)|s(m)} \xrightarrow{B(Z)} \mathcal{O}_{\mathbf{P}}(m+1),$$

where $r(m)|s(m) = \dim S^{m+1}(T^*)$, is a maximal monad on $\mathbf{P}$; all maximal monads with $\dim F_{-1} = 1$ are obtained in this way.

After replacing the constants β^a_b with the matrices of the mappings $F^+_{-1} \to F^+_1$, we obtain the general case.

2. A criterion for U-triviality. We consider the line $\mathbf{P}^1 = \{ z_3 = z_4 = 0;\ \zeta_1 = \ldots = \zeta_N = 0 \}$ and express in terms of the (β^a_b) the condition that the restriction of $\mathcal{E}(\mathcal{F}_\cdot)$ to this line is free. With this aim we will apply Proposition 4.8, but first we will simplify $\mathcal{F}_\cdot|\mathbf{P}^1$. We have $\mathcal{F}_0 = \mathcal{F}'_0 \oplus \mathcal{F}''_0$, where $\mathcal{F}'_0$ is generated by the monomials $z_1^{a_1} z_2^{a_2}$, with $a_1 + a_2 = m+1$ and $\mathcal{F}''_0$ is generated by the remaining monomials. We have respectively $A(Z) = \binom{A'(Z)}{A''(Z)}$ and $B(Z) = (B'(Z)|B''(Z))$, so $A''(Z)|\mathbf{P}^1 = 0$ and $B''(Z)|\mathbf{P}^1 = 0$. Therefore, $\mathcal{E}(\mathcal{F}_\cdot|\mathbf{P}^1)$ contains the sheaf $\mathcal{F}''_0|\mathbf{P}^1$ as a direct summand, and it is sufficient to determine when the cohomology sheaf of a monad on $\mathbf{P}^1$ of the following form is free:

$$\mathcal{G}_\cdot : \mathcal{O}_{\mathbf{P}^1}(-m-1) \xrightarrow{A'(Z)} \mathcal{O}_{\mathbf{P}^1}^{m+2} \xrightarrow{B'(Z)} \mathcal{O}_{\mathbf{P}}(m+1)\,.$$

This depends only on $(\beta^{a_1 a_2}_{b_1 b_2})$, where $a_1 + a_2 = b_1 + b_2 = m+1$. We compute the mapping $\epsilon(\mathcal{G}_\cdot)\colon H^1(\mathcal{G}_{-1}(-1)) \to H^0(\mathcal{G}_1(-1))$ from Proposition 4.8. In the complex of the Čech covering $U_i = \{ z_i \neq 0 \}$, a basis of $H^1(\mathcal{G}_{-1}(-1)) = H^1(\mathcal{O}(-m-2))$ is represented by classes of cocycles

$$h_i(U_{12}) = z_1^{-i} z_2^{i-m-2}, \qquad i = 1, \ldots, m+1\,.$$

We should view these as cochains with values in $\mathcal{K}(-1)$ and explicitly decompose their images in $\mathcal{F}_0(-1)$ after applying $A'(Z)$. The decomposition is like this:

$$z_1^{-i} z_2^{i-m-2} \overset{A'(Z)}{\longmapsto} \left(z_1^{a_1 - i} z_2^{i - a_1 - 1} \right)^t_{a_1 = 0,\ldots,m+1}$$

$$= \left(\ldots z_1^{a_1-i} z_2^{i-a_1-1} \ldots \mid 0 \ldots 0 \right)^t_{a_1 \leq i-1} + \left(0 \ldots 0 \mid \ldots z_1^{a_1-i} z_2^{i-a_1-1} \ldots \right)^t_{a_1 \geq i}\,.$$

The first term on the right is regular on U_1; the second is regular on U_2. Finally,

$$\epsilon(h_i) = B'(z)h_i^2 = -B'(Z)h_i^1$$
$$= -\sum_{a_1=0}^{i-1}\sum_{b_1=0}^{m+1} \beta^{a_1,m+1-a_1}_{b_1,m+1-b_1} z_1^{a_1+b_1-i} z_2^{m+i-a_1-b_1}$$

The sheaf $\mathcal{E}\mathcal{F}_{\cdot}|\mathbf{P}^1$ is free if and only if all the polynomials $\epsilon(h_i)$, $i = 1,\ldots,m+1$, are linearly independent. The set of such β is a Zariski-open set. In order to see that it is nonempty, we impose supplementary relations: $\beta^{a_1a_2}_{b_1b_2} = 0$ if $(a_1 + b_1, a_2 + b_2) \neq (m+1, m+1)$ and $\beta^{a_1a_2}_{b_1b_2} = \beta(a_1) - \gamma$ otherwise. In addition, let $\sum_{a_1=0}^{m+1} \beta(a_1) = (m+2)\gamma$; this gives the condition $\sum_{a+b=c} \beta^a_b = 0$ for all a, b of the form $(a_1, a_2, 0, \ldots, 0)$. The condition of freeness is expressed as the system of inequalities $\sum_{a_1=0}^{i-1}(\beta(a_1) - \gamma) \neq 0$ for all $i \leq m+1$, since

$$\epsilon(h_i) = -\left(\sum_{a_1=0}^{i-1}(\beta(a_1) - \gamma)\right) z_1^{m+1-i} z_2^{i-1}, \quad i = 1,\ldots,m+1.$$

In order to build up a submatrix (β^a_b) to the complete matrix of a monad, one can set $\beta^a_b = 0$ if a (or b) has the form $(a_1, a_2, 0, \ldots, 0)$ while b (or a) does not have this form. Then the elements of $B(Z)$ will be polynomials either in the variables z_1, z_2 only or else only in the remaining coordinates, so the requirements on the second part of the polynomials can be satisfied independently.

3. The simplest m-monads on $L^{(m)}$ in coordinates. Let (W^a) be the dual basis of monomials on $S^{m+1}(T)$ in the elements $(w_1,\ldots,w_4,\nu_1,\ldots,\nu_N)$. Then the equation of L is $s = \sum z_a w_a + \sum \varsigma_\alpha \nu_\alpha = 0$. In $S(T \oplus T^*)$ we set $s^{m+1} = \sum_a C(a,m)Z^aW^a$. Further, let $A_+(Z) = (Z^a_{\cdot})^t$, $A_-(W) = (W^a)^t$, $B_+(Z) = (\sum_b \beta^a_{b+}Z^b)$ and $B_-(W) = (\sum_b \beta^a_{b-}W^b)$. Finally,

$$\Gamma_+(Z) = t_+\left(C(a,m)Z^a\right),$$
$$\Gamma_-(W) = t_-\left((-1)^{\widetilde{a}}C(a,m)W^a\right),$$

where the $t_\pm$ are constants and $\widetilde{a}$ is the parity of W^a, i.e., $\sum \alpha_i$. If $B_\pm$ satisfy the conditions of 1, then the diagram on $\mathbf{P} \times \widehat{\mathbf{P}}$ is

$$\mathcal{F}_{\cdot} : \mathcal{O}(-m-1,0) \oplus \mathcal{O}(0,-m-1) \xrightarrow{\alpha} \mathcal{O} \otimes \left[S^{m+1}(T) \oplus S^{m+1}(T^*)\right]$$
$$\xrightarrow{\beta} \mathcal{O}(m+1,0) \oplus \mathcal{O}(0,m+1),$$

where

$$\alpha = \begin{pmatrix} A_+(Z) & 0 \\ 0 & A_-(W) \end{pmatrix}, \qquad \beta = \begin{pmatrix} B_+(Z) & \Gamma_+(Z) \\ \Gamma_-(W) & B_-(W) \end{pmatrix},$$

defines a monad on $L^{(m)}$ since $\Gamma_+ A_- = t_+ s^{m+1}$ and $\Gamma_- A_+ = t_- s^{m+1}$ (except for exceptional values of $t_\pm$ for which β is possibly not surjective). The U-triviality of such a monad guarantees the preceding point and Theorem 4.11.

4. Ranks of the simplest monads. The simplest solution of the vacuum Yang-Mills equations from the point of view of the structure of a monad is obtained for $m = 3$, $N = 0$. Since $\dim S^4(\mathbf{C}^4) = 35$, the rank of $\mathcal{E}_l$ is equal to 66.

The gauge group which is most tractable in size is obtained for the supersymmetric case: $m = 0$, $N = 3$. This is the group $\mathrm{GL}(4|6; \mathbf{C})$.

Actually, we can get rather precise information about the sheaf $\mathcal{E} = \pi_{2*}\pi_1^*\mathcal{E}_L$ itself and not only about its rank, where $\mathcal{E}_L$ is the cohomology sheaf of the standard m-monad, by using the idea of the computation in subsection 2. We will write $S_+ = S^{2|0}$ and $S_- = \widetilde{S}^{2|0}$ as in the purely even case; then $S_+^\perp = \widetilde{S}^{2|N}$ and $S_-^\perp = S^{2|N}$.

We set $\pi_1^*(\mathcal{F}_i) = \mathcal{G}_i$ and will write α, β instead of $\pi_1^*(\alpha)$, $\pi_1^*(\beta)$. Then $\mathcal{E} = \pi_{2*}\mathcal{E}_F$, where $\mathcal{E}_f = \mathcal{E}(\mathcal{G}_.)$.

We denote by $\mathcal{G}_0' \subset \mathcal{G}_0$ the minimal locally free sheaf containing the image of α. If α is realized in standard form (subsection 4.10), then $\mathcal{G}_0' = F_{-1}^+ \otimes S^{m+1}(S_{+F} \oplus F_{-1}^- \otimes S^{m+1}(S_{-F})$, where $S_{\pm F} = \pi_2^*(S_\pm)$ and the inclusion $\mathcal{G}_0' \subset \mathcal{G}_0$ is induced by the inclusions $S_{+F} \subset \mathcal{O}_F \otimes T$ and $S_{-F} \subset \mathcal{O}_F \otimes T^*$. We set $\beta' = \beta | \mathcal{G}_0'$.

5. Proposition. *A Yang-Mills sheaf $\mathcal{E}$ on M, the support of the connection ∇, has the following structure. There is defined an exact sequence $0 \to \mathcal{E}'' \to \mathcal{E} \to \mathcal{E}' \to 0$, where*

$$\mathcal{E}' = F_{-1}^+ \otimes S^{m-1}(S_+) \otimes \wedge^2 S_+ \bigoplus F_{-1}^- \otimes S^{m-1}(S_-) \otimes \wedge^2(S_-) \quad \text{for } m \geq 1;$$

$$\mathcal{E}' = 0 \quad \text{for } m = 0;$$

$$\mathcal{E}'' = F_1^+ \otimes S_+^\perp \cdot S^m(T^* \otimes \mathcal{O}_M) \bigoplus F_1^- \otimes S_-^\perp \cdot S^m(T \otimes \mathcal{O}_M).$$

Further, we assume that over an open subset $\pi_2^{-1}(U)$, where $U \subset M$, the morphism $\beta' \colon \mathcal{G}_0' \to \mathcal{G}_1$ is surjective. Then on U the canonical sequence above splits: $\mathcal{E}|U = \mathcal{E}'|U \oplus \mathcal{E}''|U$.

Particular case. The simplest non-selfdual solution of the supersymmetric ($N = 3$) Yang-Mills equation is a connection on the sheaf $S_+^\perp \oplus S_-^\perp$, the "interacting" sum of superinstantons and superanti-instantons.

Proof. From the exact sequence $0 \to \mathcal{K} = \operatorname{Ker}\beta \to \mathcal{G}_0 \xrightarrow{\beta} \mathcal{G}_1 \to 0$ on F, we find $\pi_{2*}\mathcal{K} = \operatorname{Ker}\pi_{2*}(\beta)$ and $\mathbf{R}^1\pi_{2*}\mathcal{K} = 0$. The kernel of the morphism $\pi_{2*}(\beta)$ can be identified naturally with $\mathcal{E}''$ if one writes $\mathcal{F}_.$ in standard form and respectively realizes $\mathcal{G}_0$ as $F_1^+ \otimes S^{m+1}(T^*) \otimes \mathcal{O}_F \oplus F_1^- \otimes S^{m+1}(T) \otimes \mathcal{O}_F$ and β as a morphism for which $H^0(\beta)$ on L is the identity map.

After this, from the exact sequence $0 \to \mathcal{G}_{-1} \to \mathcal{K} \to \mathcal{E}_F \to 0$ we obtain

$$0 \to \pi_{2*}\mathcal{K} = \mathcal{E}'' \to \mathcal{E} \to \mathbf{R}^1\pi_{2*}\mathcal{G}_{-1} \to 0\,.$$

The standard isomorphism $\mathbf{R}^1\pi_{2*}\mathcal{O}_F(-m-1,0) = S^{m-1}(\mathcal{S}_+)\otimes\wedge^2\mathcal{S}_+$ and the analogous one for $\mathcal{O}(0,-m-1)$ allow one to identify $\mathbf{R}^1\pi^{2*}\mathcal{G}_{-1}$ with $\mathcal{E}'$. All that remains is to indicate the canonical splitting of this exact sequence over U.

Over $\pi_2^{-1}(U)$ we consider the monad $\mathcal{G}'_. : 0 \to \mathcal{G}_{-1} \xrightarrow{\alpha} \mathcal{G}'_0 \xrightarrow{\beta'} \mathcal{G}_1 \to 0$. We set $\mathcal{K}' = \operatorname{Ker}\beta'$. Computing as before, we find $\pi_{2*}\mathcal{K}' = 0$ and $\pi_{2*}(\mathcal{E}(\mathcal{G}'_.)) = \mathbf{R}^1\pi_{2*}\mathcal{G}_{-1} = \mathcal{E}'$. The inclusion of the monad $\mathcal{G}'_. \subset \mathcal{G}_.$ defines an inclusion $\mathcal{E}(\mathcal{G}'_.) \subset \mathcal{E}(\mathcal{G}_.)$ and so $\mathcal{E}' = \pi_{2*}(\mathcal{E}(\mathcal{G}')) \subset \pi_{2*}(\mathcal{E}(\mathcal{G})) = \mathcal{E}$, which splits our exact triple.

§ 6. Flag Superspaces of Classical Type and Exotic Minkowski Superspaces

1. Statement of the problem. We have been considering fields on the Minkowski superspace $M = G(2|0,2|N;T^{4|N})$. This space has the following two properties:

(a) M is a compact homogeneous complex superspace; the supergroup $\mathrm{GL}(T)$ acts on it transitively;

(b) $M_{\mathrm{rd}} = G(2,\mathbf{C}^4)$ is the Penrose model.

It is natural to pose the general problem of classifying such spaces. In this section we will limit ourselves to enumerating those spaces M which can be realized in the form of flag manifolds for supergroups of classical type. These flag manifolds are described in subsections 2–6. Theorem 7, a list of exotic superspaces of Minkowski flag type, is the fundamental result of this section.

2. Π-symmetry and isotropic flags. The definitions below are parallel to those of § 3 of Chapter 4. The new flag functors which we shall define are represented by the corresponding flag superspaces in any of three standard categories: superschemes, differentiable supermanifolds and analytic supermanifolds. Sometimes for brevity we speak of superschemes. Below we will give a survey of the basic properties of flag manifolds but will often not check details.

Let M be a superspace and let $\mathcal{T}$ be a locally free sheaf of finite rank on it. We recall that a flag of length k in $\mathcal{T}$ is defined to be a sequence of locally direct subsheaves $\mathcal{T}_1 \subset \mathcal{T}_2 \subset \ldots \subset \mathcal{T}_{k+1} = \mathcal{T}$. The type of the flag is the sequence of ranks $\operatorname{rk}\mathcal{T}_i$. Let $\phi: N \to M$ be a morphism; then ϕ^* carries the flags of a given type in $\mathcal{T}$ into flags of the same type in $\phi^*(\mathcal{T}) = \mathcal{T}_N$.

We will also consider sheaves $\mathcal{T}$ furnished with one of the following structures: (a) a Π-symmetry, i.e., an isomorphism $p: \mathcal{T} \to \Pi\mathcal{T}$ with $p^2 = \mathrm{id}$ (sometimes we will consider p to be an odd involution of $\mathcal{T}$); or (b) a nondegenerate form b of one of the types $\mathrm{O\,Sp}$, $\mathrm{Sp\,O}$, $\Pi\,\mathrm{Sp}$ or $\Pi\,\mathrm{O}$. In this case the pullback sheaves $\mathcal{T}_N$ inherit the same structure. The presence of a structure permits one to distinguish subclasses of Π-symmetric and b-isotropic flags in $\mathcal{T}$ and $\mathcal{T}_N$.

Corresponding to a pair $(M, \mathcal{T})$, a structure on $\mathcal{T}$ and the type of a flag, are the flag functors in the category of superspaces over M. We introduce the notation which we will use for them:

$F_M(\text{type of flag}, \mathcal{T})$:

$$(N, \phi) \mapsto \{\,\text{flags of the given type in } \phi^*(\mathcal{T})\,\},$$

$F\Pi_M(\text{type of flag}, \mathcal{T})$:

$$(N, \phi) \mapsto \{\,\Pi\text{-symmetric flags of the given type in } \phi^*(\mathcal{T})\,\},$$

$FI_M(\text{type of flag}, \mathcal{T}; \text{type of form})$:

$$(N, \phi) \mapsto \{\,\text{isotropic flags of the given type in } \phi^*(\mathcal{T})\,\},$$

The flag functors of length one are the grassmannians; we will often denote them by G, $G\Pi$, GI.

All the flag functors are representable by spaces of relatively finite type over M. The natural inclusions $F\Pi_M$, $FI_M \subset F_M$ are represented by closed inclusions; the natural projections "onto the subflags of lower type" are represented by morphisms which are themselves flag spaces. The flag superspaces over $\mathbb{C}$ are smooth; from this it follws that the projections $F_M \to M$ are smooth morphisms.

Let $F = F_M$ be one of the flag spaces for $(M, \mathcal{T})$, and let $\pi: F_M \to M$ be the canonical projection. As usual, we denote a tautological flag in $\pi^*(\mathcal{T})$ by indicating the ranks of its components: $S_F^{d_1} \subset S_F^{d_2} \subset \cdots \subset \pi^*(\mathcal{T}) = \mathcal{T}_F^d$. On F there is also an orthogonal flag whose components are denoted with a tilde: $\tilde{S}_F^{d-d_k} \subset \cdots \subset \tilde{S}_F^{d-d_2} \subset \tilde{S}_F^{d-d_1} \subset \pi^*(\mathcal{T}^*)$, where $\tilde{S}_F^{d-d_i} = (S_F^{d_i})^{\perp}$.

Let $\pi: G \to M$ be one of the grassmannian subsheaves in $\mathcal{T}$ and let $S \subset \pi^*(\mathcal{T})$ be the tautological sheaf on it. We consider a local vertical vector field X on G (i.e., a section of $\mathcal{T}G/M$) and define a natural action of X on $\pi^*(\mathcal{T})$ (the lifted sections are horizontal). The Leibniz formula shows that the mapping $\overline{X}: S \to \pi^*(\mathcal{T})/S$, where $\overline{X}s = Xs \bmod S$ is linear in s. Moreover, the mapping $X \mapsto \overline{X}$ is linear in X. Therefore, we obtain a morphism of $\mathcal{O}_G$-modules $\mathcal{T}G/M \to \mathcal{H}om(\mathcal{S}, \pi^*(\mathcal{T})/\mathcal{S}) = \mathcal{S}^* \otimes \pi^*(\mathcal{T})/\mathcal{S} = \mathcal{S}^* \otimes \widetilde{\mathcal{S}}^*$. For $G = G\Pi$ or GI the image of this morphism does not coincide with all of $\mathcal{S}^* \otimes \widetilde{\mathcal{S}}^*$. In fact, we realize $G\Pi(d; \mathcal{T})$ as a closed subspace of $G(d; \mathcal{T})$ which is invariant with respect to the involution induced by p, which we also denote by p. This involution also acts on the tangent sheaf, and $\mathcal{T}G\Pi/M$ is the invariant part of $\mathcal{T}G/M$ restricted to $G\Pi$. It is not difficult to relate this action to the action of p on $\mathcal{S}$ and on $\widetilde{\mathcal{S}}$, but we will not need a precise formula. Furthermore, for isotropic grassmannians,the form b_G on $\pi^*(\mathcal{T})$ allows one to construct a mapping λ: $\widetilde{\mathcal{S}}^* \to \mathcal{S}^*$ (for even b) or $\widetilde{\mathcal{S}}^* \to \Pi\mathcal{S}^*$ (for odd b). Below we describe the image of $\mathcal{T}G/M$ in $\mathcal{S}^* \otimes \mathcal{S}^*$ or in $\Pi(\mathcal{S}^* \otimes \mathcal{S}^* = \mathcal{S}^* \otimes \Pi\mathcal{S}^*$ relative to $\mathrm{id} \otimes \lambda$. The following theorem holds; it can be deduced from the coordinate description on the big cells of grassmannians which is given below.

3. Theorem. *The relative tangent sheaves on the grassmannians of various type are described by the following isomorphisms and exact sequences:*

$$\mathcal{T}G_M(d; \mathcal{T})/M = \mathcal{S}^* \otimes \widetilde{\mathcal{S}}^*\,;$$

$$\mathcal{T}G\Pi_M(d; \mathcal{T})/M = \mathcal{H}om^p(\mathcal{S}, \widetilde{\mathcal{S}}^*) = (\mathcal{S}^* \otimes \widetilde{\mathcal{S}}^*)^p\,;$$

$$0 \to \mathcal{S}^* \otimes (\mathcal{S}^{\perp}_{\pi^*(b)}/\mathcal{S}) \to \mathcal{T}GI_M(d; \mathcal{T}, b)/M \to \mathcal{R} \to 0\,,$$

where $\mathcal{R} = \wedge^2(\mathcal{S}^*)$, $S^2(\mathcal{S}^*)$, $\Pi S^2(\mathcal{S}^*)$ or $\Pi \wedge^2 (\mathcal{S}^*)$ respectively for forms b of type O Sp, Sp O, Π Sp or Π O. □

4. Π-symmetric grassmannians. As in § 3 of Chapter 4, we will work locally on M. This means that we replace M with a supercommutatative ring A and the sheaf $\mathcal{T}$ with a free A-module T. After choosing a basis for T, we can assume that $T = A^{d_0+c_0} \oplus (\Pi A)^{d_0+c_0}$. If $p: T \to T$ is a Π-symmetry, then we may assume that the basis of T has the form $(e_1, \dots, e_{c_0+d_0}; pe_1, \dots, pe_{c_0+d_0})$, where the e_i are even elements.

With respect to such a basis a submodule $S \subset T$ is Π-symmetric if and only if for any element of the form $x^i e_i + \xi^j pe_j$ that it contains it also contains the element of the form $-\xi^j e_j + x^i pe_i$. This means that the grassmannian $G\Pi_M \subset G_M$, $M = \operatorname{Spec} A$, is covered by the spectra of rings $A[x_I, \xi_I]$, where the x_I, ξ_I fill the free places in matrices of the form

$$Z_I^{\Pi} = \begin{array}{c} \\ d_0 \\ d_0 \end{array}\!\!\begin{array}{cccc} c_0 & d_0 & d_0 & c_0 \\ \hline x_I & E_{d_0} & \xi_I & 0 \\ -\xi_I & 0 & x_I & E_{d_0} \\ \hline & \underbrace{\phantom{E_{d_0}}}_{I_0} & & \underbrace{\phantom{E_{d_0}}}_{I_1} \end{array}$$

There is a more invariant reasoning which shows that $G\Pi_M$ is closed inside G_M. On G_M we consider sheaves $\mathcal{S}$ with $p(\mathcal{S}) \subset \pi^*(\mathcal{T})$. The N-point $\phi: N \to G_M$ lies in ΠG_M if and only if $\phi^*(\mathcal{S}) = \phi^*(p(\mathcal{S}))$, i.e., if the lift to N of the sheaf homomorphism $\mathcal{S} \to \pi^*(\mathcal{T})/p(\mathcal{S})$ is equal to zero. Since $\pi^*(\mathcal{T})/p(\mathcal{S})$ is locally free, this condition can be written in terms of local equations which generate the sheaf of ideals defining ΠG_M.

The representability of the functor of Π-symmetric flags is established by induction on the length of the flag just as for ordinary flags. Leaving the details to the reader, we now will point out several additional properties of Π-symmetry.

a) Let $p: S \to S$ be a Π-symmetry of an A-module, that is, an odd homomorphism with $p^2 = \mathrm{id}$, and let T be any A-module. Then on $S \otimes T$, $T \otimes S$, $\mathrm{Hom}(S,T)$ and $\mathrm{Hom}(T,S)$, Π-symmetries can be defined by the formulas $p(s \otimes t) = p(s) \otimes t$, $p(t \otimes s) = (-1)^{\tilde{t}} t \otimes p(s)$, $p(f)(s) = (-1)^{\tilde{f}+1} f(p(s))$ and $p(f)(t) = p(f(t))$, respectively. If a Π-symmetry is also given on T, then an analogous construction can be carried out using T. The product of two such symmetries is then an even automorphism with square equal to $-\mathrm{id}$, for $q(s \otimes t) = (-1)^s p(s) \otimes p(t)$ and $q^2(s \otimes t) = -s \otimes t$.

b) Let T be an A-module of rank $1|1$, and let $p: T \to T$ be a Π-symmetry with $p^2 = -\mathrm{id}$. Then p defines a subset $Q \subset T_0$ of the t such that $T = At \oplus Ap(t)$. On Q the multiplicative group A^* acts according to the formula $(a_0 + a_1) \circ t = a_0 t + a_1 p(t)$. It is not difficult to see that Q forms a principal homogeneous space over A^*. This allows one to associate a cohomology class in the set $H^1(M, \mathcal{O}_M{}^*)$ to any pair $(\mathcal{L}, p)$, where $\mathcal{L}$ is a sheaf of rank $1|1$ on a superscheme M with Π-symmetry p, with $p^2 = \mathrm{id}$. This cohomology set is a specific version of the Picard functor for superschemes.

5. Isotropic flags. We will study isotropic flags in more detail in the case when the form b on the sheaf $\mathcal{T}$ is split. The definition of split forms was given in § 5 of Chapter 3.

Now we will introduce their invariant characterization.

Proposition. *Let b be a nondegenerate form on the sheaf $\mathcal{T}$ of type* O Sp *or* Π Sp. *The following conditions are equivalent:*

(a) *b is split;*

(b) *in a neighborhood of each point $\mathcal{T}$ has a locally direct isotropic subsheaf of maximal rank $r|s$ (for $\mathrm{O\,Sp}(2r|2s)$ or $\mathrm{O\,Sp}(2r+1|2s)$) or of one of the maximal ranks $r|s$ (for $\Pi\,\mathrm{Sp}(r+s|r+s)$), or any of the maximal ranks.*

If these conditions are fulfilled, then any direct isotropic subsheaf $\mathcal{S} \subset \mathcal{T}$ locally embeds into a direct isotropic subsheaf of any maximal rank greater than the rank of $\mathcal{S}$ and also admits a local basis of sections which is part of a standard local basis in $\mathcal{T}$ (for the form b).

Proof. (a) $\Longrightarrow$ (b). If b is split, then isotropic direct subsheaves of maximal rank are generated by parts of standard bases.

(b) $\Longrightarrow$ (a). We will prove this implication and the last assertion by induction on the rank of $\mathcal{T}$. For the smallest rank $1|0$ the assertion is trivial; the rank $0|1$ is impossible. Let the rank of $\mathcal{T} \geq 1|1$, and let $\mathcal{S} \subset \mathcal{T}$ be a direct isotropic subsheave of nonzero rank in a neighborhood of $x \in M$. Working locally, we choose a subsheaf $\mathcal{S}_0 \subset \mathcal{S}$ of rank $1|0$ or $0|1$. By the nondegeneracy of b there exists a locally direct subsheaf $\mathcal{S}_0' \subset \mathcal{T}$ such that b induces a nondegenerate pairing between $\mathcal{S}_0$ and $\mathcal{S}_0'$ (this is possible in a small neighborhood). Classical reasoning (about "hyperbolic space") shows that the sum $\mathcal{S}_0 + \mathcal{S}_0'$ is direct and that the restriction of b to it is nondegenerate and admits a standard basis. Moreover, $\mathcal{T} = \mathcal{S}_0 \oplus \mathcal{S}_0' \oplus (\mathcal{S}_0 \oplus \mathcal{S}_0')_b^{\perp}$ (all locally). We set $\mathcal{T}' = (\mathcal{S}_0 \oplus \mathcal{S}_0')_b^{\perp}$ and $\mathcal{S}' = \mathcal{S} \cap \mathcal{T}'$. Then $\mathcal{T}'$ is a sheaf of lower rank than $\mathcal{T}$ with a nondegenerate form of the same type, and $\mathcal{S}' \subset \mathcal{T}'$ is an isotropic direct subsheaf. If $\mathcal{S}$ were of maximal rank, then $\mathcal{S}'$ would also be of maximal rank; by the inductive hypothesis, b is split on $\mathcal{T}'$ and thus b is split on $\mathcal{T}$. If $\mathcal{S}$ and $\mathcal{S}'$ are not of maximal rank, then by the inductive hypothesis a suitable local basis of $\mathcal{S}'$ can be completed to a standard basis in $\mathcal{T}'$, and so this is true for $\mathcal{T}$. □

6. Isotropic grassmannians. First of all, it is easy to establish that the morphism of functors $GI_M \to G_M$ can always be represented by a closed embedding if $b\colon \mathcal{T} \to \mathcal{T}^*$ (or $\Pi\mathcal{T}^*$) is a direct form without any conditions at all of nondegeneracy or symmetry. In fact, the sheaf $\mathcal{S}_b^{\perp} \subset \mathcal{T}$ is then locally direct and so $\pi^*\mathcal{T}/\mathcal{S}_b^{\perp}$ is locally free. Therefore, the subfunctor of G_M corresponding to the morphisms $N \xrightarrow{\phi} M$ for which $\phi^*(\mathcal{S}) \to \phi^*(\pi^*\mathcal{T}/\mathcal{S}_b^{\perp})$ is the zero homomorphism is closed (the same reasoning as in subsection 4). But this subfunctor is GI_M.

The goal of the last computation is to show that if b is nondegenerate and split, then GI_M can be covered by relative affine subspaces when the rank of the isotropic subsheaves is maximal. We simply write equations in the lines of the matrix Z_I which signify the isotropy and show than they are explicitly solvable. The implicit local basis of $\mathcal{T}$ is assumed to be standard, and the choice of identity submatrix in Z_I is made such that for the zero values of the remaining elements of Z_I the basis of the corresponding isotropic subsheaf (the lines of Z_I) are part of a standard basis of $\mathcal{T}$. By Proposition 5 the affine spaces which we obtain thus in fact cover GI_M.

Let B be the Gram matrix of the form. The equations of isotropy for $\mathrm{O\,Sp}$ have the following form. (The matrices are broken up into blocks so that it is convenient to multiply them block-wise; $\mathrm{O\,Sp}(2r+1|2s)$ and $\mathrm{O\,Sp}(2r|2s)$ are denoted together: the part separated by the dotted line belongs to $2r+1$; the Latin blocks consist of even elements and the Greek ones consist of odd elements.)

$$
\begin{array}{c|c|c|c|c|c|}
 & 1 & r & r & s & s \\ \hline
r & Y & A & E_r & 0 & \Gamma \\ \hline
s & \Xi & \Lambda & 0 & E_s & B \\ \hline
\end{array}
\;
\left(\begin{array}{c|cccc}
1 & & 0 & & \\ \hline
 & 0 & E_r & & \\
 & E_r & 0 & \multicolumn{2}{c}{0} \\
0 & & & 0 & E_s \\
 & \multicolumn{2}{c}{0} & E_s & 0
\end{array}\right)
\;
\begin{array}{|c|c|}
\hline
Y^{\mathrm t} & \Xi^{\mathrm t} \\ \hline
A^{\mathrm t} & \Lambda^{\mathrm t} \\ \hline
E_r & 0 \\ \hline
0 & E_s \\ \hline
-\Gamma^{\mathrm t} & B^{\mathrm t} \\ \hline
\end{array}
= 0.
$$

Computing, we find in the case of $\mathrm{O\,Sp}(2r|2s)$ the conditions: $A+A^{\mathrm t}=0$, $B-B^{\mathrm t}=0$, $\Gamma=\Lambda^{\mathrm t}$. The condition for $\mathrm{O\,Sp}(2r+1|2s)$ is somewhat more complicated:

$$
A+A^{\mathrm t}+YY^{\mathrm t}=0, \qquad \Lambda^{\mathrm t}-\Gamma+Y\Xi^{\mathrm t}=0, \qquad B^{\mathrm t}-B+\Xi\Xi^{\mathrm t}=0.
$$

Here we can take as independent coordinates the elements of Y, Ξ, Γ, the elements of A strictly beneath the diagonal and the elements of B below and on the diagonal.

For the group $\Pi\,\mathrm{Sp}(r+s|r+s)$ the equations of isotropy for a subsheaf of rank $r|s$ have the form

$$
\begin{array}{c|c|c|c|c|}
 & s & r & s & r \\ \hline
r & A & E_r & 0 & \Gamma \\ \hline
s & -\Delta & 0 & -E_s & -B \\ \hline
\end{array}
\;
\begin{array}{|c|c|}
\hline
0 & E_{r+s} \\ \hline
E_{r+s} & 0 \\ \hline
\end{array}
\;
\begin{array}{|c|c|}
\hline
A^{\mathrm t} & \Delta^{\mathrm t} \\ \hline
E_r & 0 \\ \hline
0 & E_s \\ \hline
-\Gamma^{\mathrm t} & B^{\mathrm t} \\ \hline
\end{array}
= 0.
$$

We make a remark about signs: we have in mind that the odd part of a basis of the isotropic submodule under consideration is generated by the lines of the matrix $(\Delta|0|E_S|B)$; the minus in front of them reflects the $(-1)^{\widetilde{bt}}$ in the right side of the formula. It does not influence the final form of the equation:

$$
\Gamma-\Gamma^{\mathrm t}=0, \qquad A+B^{\mathrm t}=0, \qquad \Delta+\Delta^{\mathrm t}=0.
$$

Consequently, the corresponding open subset of $GI_M(r|s;\mathcal{T},b)$ is represented by a relative affine subspace of dimension $\big(rs|\frac{1}{2}r(r+1)+\frac{1}{2}s(s-1)\big)$.

Let GI_M be one of the maximal isotropic grassmannians that have been constructed, with π the structure morphism and $\mathcal{S}$ the tautological sheaf. As in the two preceding cases, we can construct a morphism $t\colon \mathcal{T}GI_M|_M \to \mathcal{H}om(\mathcal{S},\pi^*(\mathcal{T})/\mathcal{S})$. Now, using the form $\pi^*(b)$ on $\pi^*(\mathcal{T})$, for $\mathrm{O\,Sp}(2r|2s)$ we can identify $\pi^*(\mathcal{T})/\mathcal{S}$ with $\mathcal{S}^*$ or with $\Pi\mathcal{S}^*$ for $\Pi\,\mathrm{Sp}$. In the case of $\mathrm{O\,Sp}(2r+1|s)$, there is $\beta\colon \pi^*(\mathcal{T})/\mathcal{S}\to\mathcal{S}^*$, a surjection with a locally direct kernel of rank $1|0$. Therefore, we have three possible types of morphisms: $t\colon \mathcal{T}GI_M/M \to \mathcal{S}^*\otimes\mathcal{S}^*$, or $(1\otimes\beta)\circ t\colon \mathcal{T}GI_M/M\to\mathcal{S}^*\otimes\mathcal{S}^*$

(O Sp type) or $t\colon \mathcal{T}GI_M/M \to S^* \otimes \Pi S^* = \Pi(S^* \otimes S^*)$ (represent ΠS^* in the form $\Pi\,\mathrm{O} \otimes S^*$ and make $\Pi\,\mathrm{O}$ an exterior factor). Direct computations show that the images are $\wedge^2(S^*)$ and $\Pi S^2(S^*)$, respectively. In coordinates it is evident that one obtains isomorphisms $\mathcal{T}GI_M = \wedge^2(S^*)$, $(1\otimes\beta)^{-1}\wedge^2(S^*)$ or $\Pi S^2(S^*)$, respectively.

The isotropic grassmannians of nonmaximal type admit a useful covering which is constructed first by passing to the flags of length two for which the submodule is maximal and then by applying the morphism which forgets the big submodule. Superschemes of isotropic flags are constructed by the same induction as in the previous cases.

We now give a list of flag spaces whose underlying space is the Penrose model. Naturally, the list begins with the class which we have just been considering. It is assumed that the base of the flag space is a point.

7. Theorem. *All flag spaces M with $M_{\mathrm{rd}} = G(2;\mathbb{C}^4)$ are contained in the following list (the dimensions of T and M are indicated):*

(a) $F(2|0, 2|N; T^{4|N})$, $4|4N$.

(b) $F(2|0, 4|0; T^{4|N})$ *and* $F(0|N, 2|N; T^{N|4})$, $4|4N$.

(c) $G(2|0, T^{4|N})$ *and* $G(2|N, T^{4|N})$, $4|2N$.

Moreover, there are cases (Πa)–(Πc) *obtained from* (a)–(c) *by replacing T with $T' = \Pi T$, for example,* (Πa): $F(0|2, N|2; T^{N|4})$.

(d) $G\Pi(2|2; T^{4|4}$, $4|4$.

(e) $GI(1|0; T^{6|2N}, \mathrm{O\,Sp})$ *and* $GI(0|1; T^{2N|6}, \mathrm{Sp\,O})$, $4|2N$.

(f) $GI(2|2; T^{4|4}, \Pi\,\mathrm{Sp})$ *and* $GI(2|2; T^{4|4}, \Pi\,\mathrm{O})$, $4|4$.

(g) $GI(2|0; T^{4|4}, \Pi\,\mathrm{Sp})$, $4|7$ *and* $GI(0|2; T^{4|4}, \Pi\,\mathrm{O})$, $4|5$.

(h) $FI(2|0, 2|2; T^{4|4}, \Pi\,\mathrm{Sp})$ *and* $FI(0|2, 2|2; T^{4|4}, \Pi\,\mathrm{O})$, $4|8$.

(i) $FI(2|0, 4|0; T^{4|4}, \Pi\,\mathrm{Sp})$, $4|10$ *and* $FI(0|2, 0|4; T^{4|4}, \Pi\,\mathrm{O})$, $4|6$.

Proof. This is the principle for enumerating all the spaces that we need. Let F be a flag manifold with $F_{\mathrm{rd}} = G(2|0; T^{4|0})$. Let $F \to G$ be the projection of F onto the grassmannian of top subspaces in the flag. Then either $G_{\mathrm{rd}} = G(2|0; T^{4|0})$ or G_{rd} is a point. The manifolds G_{rd} are computed using information obtained earlier; the choice of either of the two possibilities for the top subspaces sharply limits the remaining possibilities. The discussions below in (a)–(d) are devoted to one of the type of flag spaces.

(a) First of all,

$$G(d_0|d_1; T^{d_0+c_0|d_1+c_1})_{\mathrm{rd}} = G(d_0|0; T^{d_0+c_0|0}) \times G(0|d_1; T^{0|d_1+c_1}).$$

Therefore, G_{rd} is a point if and only if $c_0 = d_1 = 0$ or $c_1 = d_0 = 0$ (except for the trivial cases $c_0 = c_1 = 0$ or $d_0 = d_1 = 0$). Thus single-point grassmannians for the top subspaces have either the form $G(d|0;T^{d|c})$ or $G(0|d;T^{c|d})$; the latter are obtained from the former by replacing T with ΠT. Let $F \to G(d|0;T^{d|c})$ be the projection on the top flag and let $F_{\mathrm{rd}} = G(2|0;T^{4|0})$; then F has a projection on $F(d_1|0,d|0;T^{d|c})$ for some $d_1 < d$. This is only possible for $d_1 = 2$, $d = 4$ and c arbitrary. In the remaining cases the top grassmannian for F can have only one of the following forms: $G(2|N;T^{4|N})$, $G(2|0;T^{4|N})$ and the cases Π-symmetric to these. Without changing G_{rd}, one can lengthen the flags of first type by adding subspaces of dimension either $2|0$ or $2|N$. This settles possibilities (a)–(c) and (Πa)–(Πc) in Theorem 7.

(b) Turning to the Π-symmetric flags, we observe first of all that if on T there is a Π-symmetry p, then $\operatorname{rk} T = r|r$ and that all pairs (T,p) of the same rank (over $\mathbb{C}$) are isomorphic. Further, $G\Pi(d|d;T^{c+d|c+d})_{\mathrm{rd}} = G(d|0;T^{c+d|0})$. Therefore, if $F\Pi_{\mathrm{rd}} = G(2|0;T^{4|0})$, then $F\Pi = G\Pi(2|2;T^{4|4})$.

(c) For $\mathrm{O\,Sp}$-isotropic flags we have

$$GI(r|s;T^{m|n},\mathrm{O\,Sp})_{\mathrm{rd}} = GI(r|0;T^{m|0},\mathrm{O}) \times GI(s|0;T^{n|0},\mathrm{Sp}).$$

Thus the grassmannian of top spaces cannot have even dimension zero. Since $G(2|0;T^{4|0})$ is indecomposable, only the cases $s=0$ or $r=0$ are possible for it. In the first case the equation for even dimension $r(m-2r)+\frac{1}{2}r(r-1)=4$ has the unique solution $r=1$, $m=6$. In the second case the equation $s(2n-s)+\frac{1}{2}s(s+1)=4$ also has a unique solution $s=1$, $n=5$, but an $\mathrm{O\,Sp}$-form on the space $T^{m|5}$ is degenerate.

The fact that $GI(1|0;T^{6|2N},\mathrm{O\,Sp})_{\mathrm{rd}} \simeq G(2;\mathbb{C}^4)$ follows from the Plücker realization of this grassmannian in the form of the quadric of decomposable bivectors in $\mathbf{P}(\wedge^2\mathbb{C}^4)$.

(d) Finally, we consider $\Pi\,\mathrm{Sp}$- and $\Pi\,\mathrm{O}$-isotropic flags. If there is a nondegenerate form of this type on T, then $\operatorname{rk} T = m|m$; in this case the even and odd subspaces in T are dual. Therefore, $GI(r|s;T^{m|m},\Pi\,\mathrm{Sp}\ \mathit{or}\ \Pi\,\mathrm{O})_{\mathrm{rd}}$ is isomorphic to the relative grassmannian $G_H(s|0,\widetilde{S}^{m-r|0})$ over the grassmannian $H = G(r|0;T^{m|0})$. Even dimension equal to zero is obtained for $GI(m|0;T^{m|m})$ or $GI(0|m;T^{m|m})$; then completion of flags leads to cases (i) of Theorem 7. The remaining cases correspond to the cases when the grassmannian of the top subspaces is isomorphic after reduction to $G(2|0;T^{4|0})$. □

8. Penrose diagrams for exotic models. Since we have a list of Minkowski flag superspaces, it is natural to pose the question of which ones can be completed to diagrams $L \xleftarrow{\pi_1} F \xrightarrow{\pi_2} M$ with the property that after reduction of odd coordinates this diagram becomes isomorphic to $F(1,3;T^{4|0}) \xleftarrow{\pi_1} F(1,2,3;T^{4|0}) \xrightarrow{\pi_2} M_{\mathrm{rd}}$. One also asks what properties of the Penrose transforms are related to these diagrams. In particular, it is important to know when the conditions of integrability along the fibers of π_1 are nontrivial, that is, when does their dimension have the form $1|a$ with $a > 0$? Also important is the "quantity of nilpotents" on L, since on this depend the restrictions to a Yang-Mills field which can be obtained from YM-sheaves on L.

Without studying these questions exhaustively, we will limit ourselves to some preliminary information. Here are some of the diagrams for the spaces on our list:

For A_N:

$$\begin{array}{ccccccc} L^{5|2N} & = & F(1|0,3|N;T^{4|N}) & \xleftarrow{\pi_1} & F^{6|4N} & = & F(1|0,2|0,2|N,3|N;T) \\ & & & & \downarrow \pi_2 & & \\ & & & & M^{4|4N} & = & F(2|0,2|N;T) \end{array}$$

For P:

$$\begin{array}{ccccccc} L^{5|5} & = & GI(1|1;T^{4|4},\Pi\,\mathrm{Sp}) & \xleftarrow{\pi_1} & F^{6|6} & = & FI(1|1,2|2;T^{4|4},\Pi\,\mathrm{Sp}) \\ & & & & \downarrow \pi_2 & & \\ & & & & M^{4|4} & = & GI(2|2,T^{4|4},\Pi\,\mathrm{Sp}) \end{array}$$

For Q:

$$\begin{array}{ccccccc} L^{5|5} & = & F\Pi(1|1,3|3;T^{4|4}) & \xleftarrow{\pi_1} & F^{6|6} & = & F\Pi(1|1,2|2,3|3;T^{4|4}) \\ & & & & \downarrow \pi_2 & & \\ & & & & M^{4|4} & = & G\Pi(2|2;T^{4|4}) \end{array}$$

For R:

$$\begin{array}{ccccccc} L^{5|4N} & = & GI(2|0;T^{6|2N},\mathrm{O\,Sp}) & \xleftarrow{\pi_1} & F^{6|4N} & = & FI(1|0,2|0;T^{6|2N},\mathrm{O\,Sp}) \\ & & & & \downarrow \pi_2 & & \\ & & & & M^{4|2N} & = & GL(1|0;T,\mathrm{O\,Sp}) \end{array}$$

We considered the diagram A_N in the previous paragraphs. The diagram R does not satisfy the condition that superlight geodesics have nonzero odd dimension. Here, on the contrary, odd coordinates appear on "celestial spheres" $L(x)$.

With respect to the quantity of nilpotents on L, we will prove the following result. Let $L_0^{(n)}$ be the n-th infinitesimal neighborhood of L in the purely even case.

9. Proposition. *For the diagrams A_N, P, Q there exist surjective mappings $L \to L_0^{(N)}$, $L \to L_0^{(2)}$, $L \to L_0^{(2)}$, respectively.*

Proof. (a) The case A_N. Let z_i, ξ_j ($i = 1,2,3,4;;\ j = 1,\dots,N$) be a basis of $(T^{4|N})^*$ and let z^i, ξ^j be the dual basis of $T^{4|N}$. These elements can be interpreted as sections of the sheaves $(S^{1|0})^*$ on $\mathsf{P}(T^{4|N})$ and $\mathsf{P}(T^{4|N*})$ respectively. The subspace $L(A_N) \subset \mathsf{P}(T) \times \mathsf{P}(T^*)$ is given by the incidence equation $\sum x_i \otimes x^i + \sum \xi_j \otimes \xi^j = 0$ (the left-hand side is a section of the invertible sheaf $\mathcal{O}(1,1)$ on $\mathsf{P} \times \widehat{\mathsf{P}}$). On the other hand, the subspace $L_0^{(N)}$ can be given by the equations $\xi_j = \xi^j = 0$ and $(\sum_{i=4}^4 z_i \otimes z^i)^{N+1} = 0$. Since $(\sum_{j=1}^N \xi_j \otimes \xi^j)^{N+1} = 0$, it follows from this that the identification of the underlying space $L(A_N)_{\text{rd}} = L_0$ can extend to a morphism $L(A_N) \to L_0^{(N)}$. It is not difficult to see that it is surjective; for example, we can pass to the standard affine cover and use the fact that $(\sum_{j=1}^N \xi_j \otimes \xi^j)^N \neq 0$.

(b) The case P. We choose a basis in $T^{4|4}$ for which a form of $\Pi\,$Sp-type has the standard Gram matrix $\left(\begin{smallmatrix} 0 & E \\ E & 0 \end{smallmatrix}\right)$. The full grassmannian $G = G(1|1; T^{4|4})$ can be given "homogeneous coordinates" written in the form of a matrix

$$Z = \begin{pmatrix} x_1 & x_2 & x_3 & x_4 & \xi_1 & \xi_2 & \xi_3 & \xi_4 \\ \eta_1 & \eta_2 & \eta_3 & \eta_4 & y_1 & y_2 & y_3 & y_4 \end{pmatrix}.$$

It is covered by affine supermanifolds G_{ij}, $i,j = 1,2,3,4$ which are defined in this way. Let $Z_{ij} = \left(\begin{smallmatrix} x_i & \xi_j \\ \eta_i & y_j \end{smallmatrix}\right)$. Then G_{ij} is the spectrum of the ring generated by the "inhomogeneous coordinates," the elements of the matrix $Z_{ij}^{-1}Z$. On G_{ij} the sheaf $S_G^{1|1}$ is given in these coordinates along with its trivialization; the first row of $Z_{ij}^{-1}Z$ is its even section and the second row is its odd section. These sections are written in the coordinates related to the chosen basis of T. The transition functions for G and $S^{1|1}$ are clear from this description. The subspace $L(P) \subset G$ is distinguished by the homogeneous equations which describe the isotropy of $S^{1|1}$: $\sum_{i=1}^4 x_i y_i + \sum_{i=1}^4 \xi_i \eta_i = 0$ and $\sum_{i=1}^4 y_i \eta_i = 0$. This differs from the case of A_3 in that, because of the pasting rules and the second (odd) equation, now $(\sum_{i=1}^4 x_i y_i)^3 = 0$ and only $(\sum_{i=1}^4 x_i y_i)^2 \neq 0$. Now we consider as an example the open set G_{11}, the spectrum of the ring generated by elements of a matrix of the form

$$\begin{pmatrix} 1 & X_2 & X_3 & X_4 & 0 & \Xi_2 & \Xi_3 & \Xi_4 \\ 0 & \mathrm{H}_2 & \mathrm{H}_3 & \mathrm{H}_4 & 1 & Y_2 & Y_3 & Y_4 \end{pmatrix}.$$

On this $L(P)$ is determined by the equations

$$1 + \sum_{i=2}^4 X_i Y_i + \sum_{i=2}^4 \Xi_i \mathrm{H}_i = 0, \qquad \sum_{i=2}^4 Y_i \mathrm{H}_i = 0.$$

From the first equation it follows that at any point one of the Y_i coordinates is invertible; the second shows that the corresponding H_i can be expressed linearly in terms of the two remaining ones. Therefore, $\mathrm{H}_2\mathrm{H}_3\mathrm{H}_4 = 0$ and $(\sum_{i=2}^4 \Xi_i \mathrm{H}_i)^3 = 0$.

(c) The case Q. Here we consider the closed embedding in the form of an "incidence quadric" $L(Q) \subset G\Pi(1|1,T) \times G\Pi(1|1,T^*)$. The basis of $T^{4|4}$ is chosen in the form (e_i, pe_i), where p is a Π-symmetry. The grassmannian $G\Pi(1|1;T)$ is covered by open sets G_{ii} with coordinates

$$\begin{pmatrix} x_i & \xi_i \\ -\xi_i & x_i \end{pmatrix}^{-1} \begin{pmatrix} x_1 & x_2 & x_3 & x_4 & \xi_1 & \xi_2 & \xi_3 & \xi_4 \\ -\xi_1 & -\xi_2 & -\xi_3 & -\xi_4 & x_1 & x_2 & x_3 & x_4 \end{pmatrix}$$

We denote the analogous coordinates on the second grassmannian by upper indices. The incidence equations are $\sum_{i=1}^4 x_i x^i + \sum_{j=1}^4 \xi_j \xi^j = 0$ and $-\sum_{i=1}^4 x_i \xi^i + \sum_{j=1}^4 x^j \xi_j = 0$. Passing to homogeneous coordinates, we see that $(\sum_{i=1}^4 x_i x^i)^3 = 0$ and $(\sum_{i=1}^4 x_i x^i)^2 \neq 0$ as in the previous case. This completes the proof.

§ 7. Geometry of Simple Supergravity

1. Basic structures. We will define a complex superspace of simple gravity ($N = 1$) to be a comples supermanifold of dimension $4|4$ on which are given structures from the following list.

(a) Two integrable complex distributions $\mathcal{T}_l M$, $\mathcal{T}_r M \subset \mathcal{T} M$ of rank $0|2$ whose sum in $\mathcal{T} M$ is direct. They should staisfy the following condition. Let $\mathcal{T}_0 M = \mathcal{T} M/(\mathcal{T}_l M \oplus \mathcal{T}_r M)$. Then the Frobenius form

$$\phi\colon \mathcal{T}_l M \otimes \mathcal{T}_r M \to \mathcal{T}_0 M$$
$$\phi(X \otimes Y) = [X, Y] \bmod (\mathcal{T}_l M \oplus \mathcal{T}_r M)$$

is an isomorphism.

According to the general Frobenius theorem, $\mathcal{T}_l M$ and $\mathcal{T}_r M$ are integrable to fibrations. There are no topological obstructions to this, for the leaves of the corresponding fibers have dimension $0|2$. More precisely, we set $\mathcal{O}_{M_l}$ (respectively, $\mathcal{O}_{M_r}$) to be the subsheaf of $\mathcal{O}_M$ annihilated by all vector fields in $\mathcal{T}_r M$ (respectively, $\mathcal{T}_l M$). Then $M_l = (M, \mathcal{O}_{M_l})$ and $M_r = (M, \mathcal{O}_{M_r})$ are supermanifolds of dimension $4|2$; the inclusions $\mathcal{O}_{M_{l,r}} \subset \mathcal{O}_M$ define canonical projections $\pi_{l,r}\colon M \to M_{l,r}$ which are the identity on the subsheaf. Finally, $\mathcal{T}_l M = \mathcal{T} M/M_r$ and $\mathcal{T}_r M = \mathcal{T} M/M_l$.

(b) A real structure ρ on M of type $(1,1,1)$ which has a four-dimensional real manifold of fixed points on M_{rd} and which exchanges $\mathcal{T}_l M$ with $\mathcal{T}_r M$.

The data of (a) define a superconformal structure on M. We observe that by no means is a superconformal structure given by a conformal class of supermetrics.

The last data violate the conformal structure roughly in the same way as the choice of a concrete metric in a conformal class.

(c) Two even nondegenerate volume forms $v_{l,r} \in H^0(M_{l,r}, \mathrm{Ber}\, M_{l,r})$ with the condition $v_l^\rho = v_r$.

Now we will describe several derived structures. The most important of these is the lagrangian, a distinguished volume form on M. Its construction is based on the following fact.

2. Proposition. *There is a canonical isomorphism of sheaves:*

$$(\mathrm{Ber}\, M)^3 = \pi_l^* \mathrm{Ber}\, M_l \otimes \pi_r^* \mathrm{Ber}\, M_r \,.$$

Proof. We set $\Omega^1_{l,r} M = \Pi(\mathcal{T}_{l,r} M)^*$ and $\Omega^1_0 M = \Pi(\mathcal{T}_0 M)^* \subset \Omega^1 M$. According to the definitions, there is an exact sequence

$$0 \to \Omega^1_0 M \to \Omega^1 M \to \Omega^1_l M \oplus \Omega^1_r M \to 0 \,.$$

On the other hand, by dualizing the Frobenius form, we obtain the identification $\Omega^1_0 M = \Pi(\Omega^1_l M \otimes \Omega^1_r M)$, whence

$$\begin{aligned}(\mathrm{Ber}\, \Omega^1_0 M)^* &= \mathrm{Ber}(\Omega^1_l M \otimes \Omega^1_r M) \\ &= (\mathrm{Ber}\, \Omega^1_l M)^2 \otimes (\mathrm{Ber}\, \Omega^1_r M)^2 .\end{aligned} \tag{1}$$

Combining these data, we obtain as in § 1.10

$$\begin{aligned}\mathrm{Ber}\, M = (\mathrm{Ber}\, \Omega^1 M)^* &= (\mathrm{Ber}\, \Omega^1_0 M)^* \otimes (\mathrm{Ber}\, \Omega^1_l M)^* \otimes (\mathrm{Ber}\, \Omega^1_r M)^* \\ &= \mathrm{Ber}\, \Omega^1_l M \otimes \mathrm{Ber}\, \Omega^1_r M \,.\end{aligned}$$

We will now consider the subsheaf $\pi_l^* \Omega^1 M_l \subset \Omega^1 M$. In subsection 11 we will check using local coordinates that under the mapping $\Omega^1 M \to \Omega^1_l M$ this subsheaf is projected onto all of $\Omega^1_l M$ while the kernel of the projection coincides with $\Omega^1_0 M$; the analogous statement is true for right 1-forms. Thus there are two exact sequences

$$0 \to \Omega^1_0 M \to \pi_{l,r}^*(\Omega^1 M_{l,r}) \to \Omega^1_{l,r} M \to 0 \,,$$

from which, as above, we obtain

$$\begin{aligned}\pi_l^* \mathrm{Ber}\, M_l &= \mathrm{Ber}(\pi_l^* \Omega^1 M_l)^* = \mathrm{Ber}\, \Omega^1_l M \otimes (\mathrm{Ber}\, \Omega^1_r M)^2 \,, \\ \pi_r^* \mathrm{Ber}\, M_r &= \mathrm{Ber}(\pi_r^* \Omega^1 M_r)^* = (\mathrm{Ber}\, \Omega^1_l M)^2 \otimes \mathrm{Ber}\, \Omega^1_r M \,.\end{aligned} \tag{2}$$

The comparison of (1) and (2) completes the proof. □

3. Lagrangian. A lagrangian of a space of simple supergravity M is defined to be a real volume form on it,

$$w = (\pi_l^* v_l \otimes \pi_r^* v_r)^{1/3} \,.$$

Of course in this formula the identification of sheaves described in Proposition 2 is implicit.

Now there are two problems before us.

We must explain the relation between simple supergravity and ordinary gravity. This "component analysis" includes in particular an explanation of the structures induced on $M_{\rm rd}$; we will show the simplest of these in subsection 4–6.

Next we must learn how to describe by means of superfields the geometric structures introduced above. This is necessary if for no other reason than to learn how to write variational equations in terms of the lagrangian; up to now we have had nothing to vary. There are many ways to introduce superfield descriptions and to remove redundant degrees of freedom with constraints; we will choose the most natural one in our context, the formalism of Ogievetskii and Sokachev. This will be the subject of the reminder of this section, beginning with subsection 7.

4. Spinor structure on $M_{\rm rd}$. We recall that a spinor structure on $M_{\rm rd}$ is an isomorphism $S \otimes \widetilde{S} \xrightarrow{\sim} \Omega^1 M_{\rm rd}$, with $\operatorname{rk} S = \operatorname{rk} \widetilde{S} = 2$. On $M_{\rm rd}$ a canonical spinor structure is defined by the following data: $S = \Pi(\Omega_l^1 M)_{\rm rd}$ and $\widetilde{S} = \Pi(\Omega_r^1 M)_{\rm rd}$. The isomorphism is constructed in this way. First of all, dualizing the Frobenius form gives an isomorphism $\Pi\Omega_l^1 M \otimes \Pi\Omega_r^1 M \xrightarrow{\sim} \Pi\Omega_0^1 M$. Further, the composition of the inclusion $\Omega_0^1 M \subset \Omega^1 M$ and the reduction of the odd coordinates produces a mapping $\Pi(\Omega_0^1 M)_{\rm rd} \to \Pi\Omega^1 M_{\rm rd}$. In subsection 12 below, we will verify that this is an isomorphism. (The extra Π appeared because in purely even geometry we used by tradition $\Omega_{\rm ev}^1$ but in supergeometry used $\Omega_{\rm odd}^1$).

In particular, on $M_{\rm rd}$ there is a holomorphic conformal metric that justifies the name "superconformal structure" in applying it to the data (a) of subsection 1. From the volume forms v_l and v_r one can construce two spinor metrics on $M_{\rm rd}$.

5. Spinor metrics. We set

$$\epsilon_l = (\pi_l^* v_l)^{1/3} \otimes (\pi_r^* v_r)^{-2/3}, \qquad \epsilon_r = (\pi_l^* v_l)^{-2/3} \otimes (\pi_r^* v_r)^{1/3}.$$

From the identifications (2) in the proof of Proposition 2 we obtain

$$\epsilon_{l,r} \in (\operatorname{Ber} \Omega_{l,r}^1 M)^{-1} = \operatorname{Ber} \Pi\Omega_{l,r}^1 M.$$

Therefore, after reducing the odd coordinates, the $\epsilon_{l,r}$ reduce to spinor metrics ϵ, $\widetilde{\epsilon}$ on $M_{\rm rd}$, sections of $\wedge^2 S$ and $\wedge^2 \widetilde{S}$, and $g = \epsilon \otimes \widetilde{\epsilon}$ becomes a holomorphic metric on $M_{\rm rd}$.

6. Real structure on $M_{\rm rd}$. This is induced by ρ and clearly is compatible with the spinor structure and the spinor metrics in the sense of § 1 of Chapter 2. On the real points of $M_{\rm rd}$ the metric $\epsilon \otimes \widetilde{\epsilon}$ has Lorentz signature.

Thus we have described all the structures on $M_{\rm rd}$ which turn this manifold into a complex space-time with the exception of the spinor connections. They can also be constructed after inducing the canonical superconnection on M; we will omit this construction.

7. Adapted coordinate systems. Let (x_l^a, θ_l^α) be a local coordinate system on M_l. We will assume that the following conditions satisfy:

(a) The functions $(x_l^a)_{\rm rd}$ on $M_{\rm rd}$ are real (i.e., ρ-invariant).

(b) The functions $(x^a = \frac{1}{2}(x_l^a + x_r^a), \theta_l^\alpha, \theta_r^{\dot\alpha})$, where $x_r^a = (x_l^a)^\rho$ and $\theta_r^{\dot\alpha} = (\theta_l^\alpha)^\rho$, form a local system of coordinates on M.

We will call systems of coordinates such as (x_l, θ_l) on M_l, (x_r, θ_r) on M_r and $(x_l, \theta_l, \theta_r)$ on M adapted coordinates. Their existence follows directly from the description of the basic structures: the x_l^a exist because the set of real points on $M_{\rm rd}$ is four-dimensional; for the θ_l^α we can take two odd functions which locally rectify $\mathcal{T}_l M$; condition (b) essentially follows from the fact that the sum of $\mathcal{T}_l M$ and $\mathcal{T}_r M$ is direct.

Later, we will carry out a sequence of computations in adapted coordinates.

8. Description of superspaces by superfields. We set

$$H^a = \frac{1}{2i}(x_l^a - x_r^a).$$

These are four functions on M; they are real, i.e., $(H^a)^\rho = H^a$, and nilpotent, since $(x_r^a)_{\rm rd} = (x_l^a)_{\rm rd}$. The 4-tuple of functions (H^a) is called an Ogievetskii-Sokachev prepotential. A change of adapted coordinate system changes the prepotential (H^a); such transformations are called gauge transformations. Following Siegel and Gates, it would be more consistent to admit any system of coordinates (y_l, η_l) on M_l (or only to require the condition of reality of $y_{l,\rm rd}$) and to define an eight-component prepotential $H^a = \frac{1}{2i}(y_l^a - (y_l^a)^\rho)$, $H^\alpha = \frac{1}{2i}(\eta_l^\alpha - (\eta_l^\alpha)^\rho)$. The choice of Ogievetskii-Sokachev already assumes a partial fixing of the gauge.

We set further

$$v_l = \Phi_l^3 D^*(d\theta_l^\alpha, dx_l^a), \qquad v_r = \Phi_r^3 D^*(d\theta_r^{\dot\alpha}, dx_r^a).$$

The cubes are to reduce the quantity of fractional powers in the upcoming formula for the lagrangian.

The prepotential (H^a) defines the structures (a) and (b) of subsection 1 on a general 4|4-manifold $(X^a, \theta^\alpha, \theta^{\dot\alpha})$: it suffices to set

$$(X^a)^\rho = X^a + 2iH^a, \qquad (\theta^{\dot\alpha})^\rho = \theta^\alpha;$$

$\mathcal{O}_l$ consists of the functions in X^a, θ^α; $\mathcal{O}_r$ the functions in $(X^a)^\rho$, $\theta^{\dot\alpha}$.

Two even superfunctions $\Phi_{l,r}$ finish this description. Below, we will compute all the structures that have been introduced, including the lagrangian, in terms of H and Φ. By means of them a variation can be introduced for deducing dynamical equations. The functions $\Phi_{l,r}$ also depend on coordinates; often the gauge is chosen such that $\Phi_l = \Phi_r = 1$.

We set $\partial_a = \partial/\partial x^a$, $\partial_\alpha = \partial/\partial\theta_l^{\dot\alpha}$, and $\partial_{\dot\alpha} = \partial/\partial\theta_r^{\dot\alpha}$ to be the vector fields on M described in a "central" coordinate system. We introduce the following notation: $(\partial H/\partial x)_b^a = (\partial_l H^a)$ (a matrix), and further

$$X_\alpha^a = i\left[\left(1 - i\frac{\partial H}{\partial x}\right)^{-1}\right]_b^a \partial_\alpha H^b;$$

$$X_{\dot\alpha}^a = -i\left[\left(1 + i\frac{\partial H}{\partial x}\right)^{-1}\right]_b^a \partial_{\dot\alpha} H^b.$$

These are odd and thus are nilpotent functions on M.

9. Lemma. *The derivations*

$$\Delta_\alpha = \partial_\alpha + X_\alpha^a \partial_a, \qquad \Delta_{\dot\alpha} = -\partial_{\dot\alpha} - X_{\dot\alpha}^a \partial_a \tag{3}$$

form local bases of $\mathcal{T}_l M$ and $\mathcal{T}_r M$ respectively.

Proof. In order to belong to $\mathcal{T}_l M$ (respectively, $\mathcal{T}_r M$), the Δ_α (respectively, $\Delta_{\dot\alpha}$) must annihilate $\theta^{\dot\alpha}$ and $x_r^a = x^a - iH^a$ (respectively θ^α and $x_l^a = x^a + iH^a$). The coefficients X_α^a (respectively $X_{\dot\alpha}^a$) are found from these conditions:

$$(\partial_\alpha + X_\alpha^a \partial_a)(x^b - iH^b) = 0 \iff (\delta_a^b - i\partial_a H^b) X_\alpha^a = i\partial_\alpha H^b,$$

and analogously for $\Delta_{\dot\alpha}$. Further, let $D = A^a\partial_a + B^\alpha\partial_\alpha + C^{\dot\alpha}\partial_{\dot\alpha}$ lie in $\mathcal{T}_l M$. Then, subtracting the combination $B^\alpha\Delta_\alpha$ from D, we can assume that $B^\alpha = 0$. Applying D to $\theta^{\dot\alpha}$, we obtain $C^{\dot\alpha} = 0$; and applying it to x_r^b, we get $A^a(\delta_b^a - i\partial_a H^b) = 0$. From this we get $A^a = 0$ by the invertibility of the matrix $1 - (\partial H/\partial x)$. □

10. Corollary. *The forms*

$$\omega^a = dx^a - X_\alpha^a\, d\theta^\alpha - X_{\dot\alpha}^a\, d\theta^{\dot\alpha}$$

form a basis for the sheaf $\Omega_0^1 M$.

Proof. It is sufficient to check that $\Delta_\alpha \lrcorner\, \omega^a = 0 = \Delta_{\dot\alpha} \lrcorner\, \omega^a$. □

11. Corollary. *There are two exact sequences*

$$0 \to \Omega_0^1 M \to \pi_{l,r}^* \Omega^1 M_{l,r} \to \Omega_{l,r}^1 M \to 0,$$

used in the proof of Proposition 2. The forms (dx_l^b) *and* (dx_r^b) *form bases of* $\Omega_0^1 M$ *in* $\pi_{l,r}^* \Omega^1 M_{l,r}$ *and are expressed in terms of the* ω^a *by the formulas*

$$dx_l^a = (\delta_b^a + i\partial_b H^a)\omega^b, \qquad dx_r^a = (\delta_b^a - i\partial_b H^a)\omega^b.$$

Proof. The subsheaf $\pi_l^* \Omega^1 M_l \subset \Omega^1 M$ is freely generated by the differentials dx_l^a, $d\theta_l^\alpha$. Since $\Delta_\alpha \rfloor d\theta^\beta = \delta_\alpha^\beta$, the mapping $\pi_l^* \Omega^1 M_l \to \Omega_l^1 M$ is surjective. Clearly its kernel is freely generated by the dx_l^a. From the formula for ω^a it is not difficult to see that the difference $dx_l^a - (\delta_b^a + i\partial_b H^a)\omega^b$, expressed in terms of the central basis $(dx^a, d\theta_l^\alpha, d\theta_r^{\dot\alpha})$, is a linear combination of $d\theta^\alpha$ and $d\theta^{\dot\alpha}$. On the other hand, it lies in $\Omega_0^1 M$ and is therefore equal to zero. The formula for the right basis can be verified analogously. □

12. Corollary. *The restriction* $(\Omega_0^1 M)_{\mathrm{rd}} \to \Omega^1(M_{\mathrm{rd}})$ *is an isomorphism.*

Proof. Under the restriction, the form ω^a is carried to dx_{rd}^a since X_α^a and $X_{\dot\alpha}^a$ are nilpotent. □

We used this situation in subsection 4 to construct a spinor structure on M_{rd}.

13. Computation of the Frobenius form. We denote by $(\mathcal{D}_a)$ the basis of $\mathcal{T}_0 M$ dual to the basis (ω^a) of subsection 10: $(\mathcal{D}_a, \omega^b) = \delta_a^b$. We set

$$\phi(\Delta_\alpha \otimes \Delta_{\dot\beta}) = \phi^a_{\alpha\dot\beta} \mathcal{D}_a ,$$

where ϕ is the Frobenius form introduced in subsection 1. To compute the coefficients of $\phi^a_{\alpha\dot\beta}$ we use the definition:

$$\left(\phi(\Delta_\alpha \otimes \Delta_{\dot\beta}), \omega^b\right) = [\Delta_\alpha, \Delta_{\dot\beta}] \rfloor \omega^b = \phi^a_{\alpha\dot\beta}\, \mathcal{D}_a \rfloor \omega^b = \phi^b_{\alpha\dot\beta} ,$$

whence

$$\phi^a_{\alpha\dot\beta} = [\Delta_\alpha, \Delta_{\dot\beta}] \rfloor \omega^a = [\Delta_\alpha, \Delta_{\dot\beta}] \rfloor \left(dx^a - d\theta^\gamma X_\gamma^a - d\theta^{\dot\delta} X_{\dot\delta}^a\right) .$$

The formulas for Δ_α and $\Delta_{\dot\beta}$ from Lemma 9 show that $[\Delta_\alpha, \Delta_{\dot\beta}]$ is a linear combination of the ∂_a, so that the definition of $\phi^a_{\alpha\dot\beta}$ can be rewritten in the form

$$[\Delta_\alpha, \Delta_{\dot\beta}] = \phi^a_{\alpha\dot\beta}\, \partial_a .$$

It is convenient to seek the expression in terms of H^a thus: $\Delta_a(x^b - iH^b) = 0$, whence $\Delta_\alpha x^b = i\Delta_\alpha H^b$ and $\Delta_{\dot\beta}\Delta_\alpha x^b = i\Delta_{\dot\beta}\Delta_\alpha H^b$. Analogously, $\Delta_{\dot\beta}(x^b+iH^b) = 0$ and also $\Delta_\alpha\Delta_{\dot\beta}x^b = -i\Delta_\alpha\Delta_{\dot\beta}H^b$. Finally,

$$\phi^a_{\alpha\dot\beta} = -i\{\Delta_\alpha, \Delta_{\dot\beta}\}H^a . \tag{4}$$

We recall that $\{\ ,\ \}$ is the supercommutator; when applied to the odd derivations Δ_α and $\Delta_{\dot\beta}$ it looks like a commutator.

The fundamental axiom of a superconformal structure is the maximal nondegeneracy of the Frobenius form, i.e., the invertibility of the matrix of second spinor derivatives of the H^a. (Sometimes, instead of $\phi^a_{\alpha\dot\beta}$ one passes to $\phi^a_b = \sigma_b^{\alpha\dot\beta}\phi^a_{\alpha\dot\beta}$; in this matrix the lines and columns are enumerated the same.)

14. Computation of $\epsilon_{l,r}$. We will carry out in dual form, for $\mathcal{T}$ instead of Ω^1, the basic computations for the isomorphisms described in Proposition 2 which are related to the berezinians. This will give a certain economy in Π and in dualizations.

From Corollary 11 we find exact sequences

$$0 \to \mathcal{T}_{l,r}M \to \pi^*_{l,r}\mathcal{T}M_{l,r} \to \mathcal{T}_0 M \to 0 .$$

We set

$$\widetilde{\mathcal{D}}_{l,a} = l^b_a\pi^*_l\left(\frac{\partial}{\partial x^b_l}\right), \qquad \widetilde{\mathcal{D}}_{r,a} = r^b_a\pi^*_r\left(\frac{\partial}{\partial x^b_r}\right),$$

where

$$l^b_a = \delta^b_a + i\partial_a H^b , \qquad r^b_a = \delta - i\partial_a H^b . \tag{5}$$

We claim that the images of $\widetilde{\mathcal{D}}_{l,a}$ and $\widetilde{\mathcal{D}}_{r,a}$ in $\mathcal{T}_0M$ coincide with $\mathcal{D}_a$. Now for this it is necessary to check that $\widetilde{\mathcal{D}}_{l,a} \lrcorner\, \omega^c = \delta^c_a = \widetilde{\mathcal{D}}_{r,a} \lrcorner\, \omega^c$; this follows from the expression for ω^c in terms of dx^a_l and dx^a_r given in Corollary 11.

Now in $\pi^*_{l,r}\mathcal{T}M_{l,r}$ we have pairs of bases and transition matrices between them:

$$\begin{pmatrix} \widetilde{\mathcal{D}}_{l,a} \\ \Delta_\alpha \end{pmatrix} = \left(\begin{array}{c|c} l^b_a & 0 \\ \hline X^b_\alpha & \delta^\beta_\alpha \end{array}\right) \begin{pmatrix} \pi^*_l(\partial/\partial x^b_l) \\ \pi^*_l(\partial/\partial\theta^\beta_l) \end{pmatrix},$$

and analogously for $\mathcal{T}M_r$. From this

$$\begin{aligned} D(\widetilde{\mathcal{D}}_{l,a}, \Delta_\alpha) &= \det|l^b_a|\pi^*_l\, D\left(\frac{\partial}{\partial x^a_l}, \frac{\partial}{\partial\theta^\alpha}\right), \\ D(\widetilde{\mathcal{D}}_{r,a}, \Delta_{\dot\alpha}) &= \det|r^b_a|\pi^*_r\, D\left(\frac{\partial}{\partial x^a_r}, \frac{\partial}{\partial\theta^{\dot\alpha}}\right), \end{aligned} \tag{6}$$

We can write in coordinates the fundamental identity from Proposition 2:

$$\pi_l^* \operatorname{Ber} \mathcal{T} M_l = (\operatorname{Ber} \mathcal{T}_l M)^{-1} \otimes (\operatorname{Ber} \mathcal{T}_r M)^{-2};$$

$$D(\widetilde{\mathcal{D}}_{l,a}, \Delta_\alpha) = D\left((\phi^{-1})_a^{\alpha\dot\beta} \Delta_\alpha \otimes \Delta_{\dot\beta}, \Delta_\alpha\right) = (\det \phi)^{-1}\, D(\Delta_\alpha)^{-1} D(\Delta_{\dot\beta})^{-1},$$

where ϕ is the Frobenius form computed in the preceding passage; analogously,

$$D(\widetilde{\mathcal{D}}_{r,a}, \Delta_{\dot\alpha}) = (\det \phi)^{-1}\, D(\Delta_\alpha)^{-2} D(\Delta_{\dot\beta})^{-1}.$$

Substituting this equation into (6) we obtain

$$\pi_l^* D\left(\frac{\partial}{\partial x_l^a}, \frac{\partial}{\partial \theta_l^\alpha}\right) = (\det l_a^b)^{-1} (\det \phi)^{-1}\, D(\Delta_\alpha)^{-1}\, D(\Delta_{\dot\beta})^{-2},$$

$$\pi_r^* D\left(\frac{\partial}{\partial x_r^a}, \frac{\partial}{\partial \theta_r^\alpha}\right) = (\det r_a^b)^{-1} (\det \phi)^{-1}\, D(\Delta_\alpha)^{-2}\, D(\Delta_{\dot\beta})^{-1}.$$

We identify $D(\partial/\partial x_l^a, \partial/\partial \theta_l^\alpha)$ with the section of $\operatorname{Ber} \mathcal{T} M_l$ dual to the section $D^*(d\theta_l^\alpha, dx_l^a)$ of the sheaf $\operatorname{Ber}^* \Omega^1 M_l$, and analogously for M_r. Then the formulas for the $v_{l,r}$ take the form

$$\begin{aligned} \pi_l^*(v_l) &= \Phi_l^3 \pi_l^* D^*(d\theta_l^\alpha, dx_l^a) \\ &= \Phi_l^3 \det(l_a^b) \det \phi\, D(\Delta_\alpha)\, D(\Delta_{\dot\beta})^2, \\ \pi_r^*(v_r) &= \Phi_r^3 \pi_r^* D^*(d\theta_r^{\dot\alpha}, dx_r^a) \\ &= \Phi_r^3 \det(r_a^b) \det \phi\, D(\Delta_\alpha)^2\, D(\Delta_{\dot\beta}), \end{aligned} \tag{7}$$

from which finally

$$\begin{aligned} \epsilon_l &= (\pi_l^* v_l)^{1/3} \otimes (\pi_r^* v_r)^{-2/3} \\ &= \Phi_l \Phi_r^{-2} (\det l_a^b)^{1/3} (\det r_a^b)^{-2/3} (\det \phi)^{-1/3}\, D(\Delta_\alpha)^{-1}, \\ \epsilon_r &= (\pi_l^* v_l)^{-2/3} \otimes (\pi_r^* v_r)^{1/3} \\ &= \Phi_l^{-2} \Phi_r (\det l_a^b)^{-2/3} (\det r_a^b)^{1/3} (\det \phi)^{-1/3}\, D(\Delta_{\dot\alpha})^{-1}. \end{aligned} \tag{8}$$

15. Structure frames. We call a structure frame of a superspace M any local basis of vector fields on M of the form

$$\left(\widetilde{\Delta}_\alpha, \widetilde{\Delta}_{\dot\alpha}, \frac{i}{2}\left[\widetilde{\Delta}_\alpha, \widetilde{\Delta}_{\dot\alpha}\right]\right), \quad \text{where } D(\widetilde{\Delta}_\alpha) = \epsilon_l^{-1}, \quad D(\widetilde{\Delta}_{\dot\alpha}) = \epsilon_r^{-1}.$$

To every adapted coordinate system corresponds a structure frame for which

$$\widetilde{\Delta}_\alpha = F\Delta_\alpha, \qquad \widetilde{\Delta}_{\dot\alpha} = F^\rho \Delta_{\dot\alpha} .$$

The functions F and F^ρ are found from formula (8):

$$F = \Phi_l^{1/2}\Phi_r^{-1}(\det l_a^b)^{1/6}(\det r_a^b)^{-1/3}(\det \phi)^{-1/6},$$
$$F^\rho = \Phi_l^{-1}\Phi_r^{1/2}(\det l_a^b)^{-1/3}(\det r_a^b)^{1/6}(\det \phi)^{-1/6}.$$

16. The lagrangian. According to the definition and the formulas (6) we have

$$\begin{aligned} w &= [\pi_l^*(v_l) \otimes \pi_r^*(v_r)]^{1/3} \\ &= \Phi_l\Phi_r(\det\phi)^{2/3}\det\left(1+\left(\frac{\partial H}{\partial x}\right)^2\right)^{1/3} D(\Delta_\alpha)\, D(\Delta_{\dot\beta}) . \end{aligned}$$

Here $\det(1+(\partial H/\partial x)^2) = \det(l_a^b)\det(r_a^b)$ by the definitions of (5).

This formula for the lagrangian is not yet final: we must express $D(\Delta_\alpha)\,D(\Delta_{\dot\beta})$ in terms of $D^*(d\theta^\alpha, d\theta^{\dot\alpha}, dx^a)$ in order to be able to use directly the formula for the Berezin integral defining the action to write the Euler-Lagrange equations, etc.

Doing what was done at the beginning of subsection 14, we consider the exact sequence $0 \to \mathcal{T}_l M \oplus \mathcal{T}_r M \to \mathcal{T}M \to \mathcal{T}_0 M \to 0$. From the definitions it is clear that ∂_a is mapped to $\mathcal{D}_a$, for $\partial_a \lrcorner\, \omega^b = \delta_a^b$. Therefore, from the enumeration matrices of the two bases of $\mathcal{T}M$

$$\begin{pmatrix} \partial_a \\ \Delta_\alpha \\ \Delta_{\dot\alpha} \end{pmatrix} = \begin{pmatrix} E & 0 & 0 \\ X_\alpha^a & E & 0 \\ X_{\dot\alpha}^a & 0 & E \end{pmatrix} \begin{pmatrix} \partial_a \\ \partial_\alpha \\ \partial_{\dot\alpha} \end{pmatrix}$$

we find

$$\begin{aligned} D(\partial_a, \partial_\alpha, \partial_{\dot\alpha}) &= D(\partial_a, \Delta_\alpha, \Delta_{\dot\alpha}) = D\left((\phi^{-1})_a^{\alpha\dot\alpha}(\Delta_\alpha \otimes \Delta_{\dot\alpha}), \Delta_\alpha, \Delta_{\dot\alpha}\right) \\ &= (\det\phi)^{-1}\, D(\Delta_\alpha)^{-1}\, D(\Delta_{\dot\alpha})^{-1}, \end{aligned}$$

or $D(\Delta_\alpha)\,D(\Delta_{\dot\alpha}) = (\det\phi)^{-1} D^*(d\theta^\alpha, d\theta^{\dot\alpha}, dx^a)$, whence

$$w = \Phi_l\Phi_r(\det\phi)^{-1/3}\det\left(1+\left(\frac{\partial H}{\partial x}\right)^2\right)^{1/3} D^*(d\theta^\alpha, d\theta^{\dot\alpha}, dx^a) .$$

Let L be the density of the action, i.e., the coefficient for D^* in this expression. Further let E_B^A be the transition matrix from the holomorphic frame $(\partial_a, \partial_\alpha, \partial_{\dot\alpha})$

to the structure frame $\left(\frac{i}{2}[\widetilde{\Delta}_\alpha, \widetilde{\Delta}_{\dot\alpha}], \widetilde{\Delta}_\alpha, \widetilde{\Delta}_{\dot\alpha}\right)$. Then from the given formulas we obtain without difficulty the Wess-Zumino formula:

$$L = \frac{1}{8}\,\mathrm{Ber}(E_B^A)\,.$$

17. Wess-Zumino gauge. The Wess-Zumino gauge is defined by an adapted coordinate system in which $\Phi_l = 1$, $\Phi_r = 1$ and the potential H^a has the following form:

$$H^a(x, \theta^\alpha, \theta^{\dot\alpha}) = \theta^\alpha\theta^{\dot\alpha}e^a_{\alpha\dot\alpha} + \epsilon_{\dot\alpha\dot\beta}\theta^{\dot\alpha}\theta^{\dot\beta}\theta^\gamma\psi^a_\gamma + \epsilon_{\alpha\beta}\theta^\alpha\theta^\beta\theta^{\dot\gamma}\psi^a_{\dot\gamma} + \epsilon_{\alpha\beta}\epsilon_{\dot\alpha\dot\beta}\theta^\alpha\theta^\beta\theta^{\dot\alpha}\theta^{\dot\beta}A^a\,.$$

According to component analysis in this gauge, the structure of the superspace M is defined by the following collection of classical fields on M_{rd}: (the 16 components of the tetrad $e^a_{\alpha\dot\alpha}$) + (the 16 components of the fields ψ^a_γ, $\psi^a_{\dot\gamma}$ of spin 3/2) + (the auxiliary field A). The A^a falls out by the equations of motion; the superlagrangian w is a linear combination of the Hilbert-Einstein and Rarita-Schwinger lagrangians multiplied by $\theta^0\theta^1\theta^{\dot 0}\theta^{\dot 1}$ and divergences.

References for Chapter 5

The works of Gol'fand-Lichtman [41] and Volkov [106] were the first works on the theory of supergeometry. A survey of the physical motivation and results up to 1975 was given in the paper of Ogievetskii and Mezincescu [81]; see also the popularization [37]. The physical literature on supersymmetry and supergravity already comprises a thousand works and is growing rapidly. We point out one of the recent surveys [79] and the collections [80], [45], which will help the reader go on further. In Russian, see [73], which contains in particular a translation of [115] and also [100].

Our exposition of the theory of supersymmetric Yang-Mills equations was based on the second part of the paper of Witten [118]. Supertwistors were introduced by Ferber [34]. the material of §§ 4–6 was taken from the author's works [71] and [72], where it was first shown that the method of monads permits one to construct non-selfdual solutions of the Yang-Mills equations, both ordinary and supersymmetric. For the subject of supersymmetric instantons see [106] and [98].

The model of simple supergravity described in § 7 was based on the work of Ogievetskii and Sokachev [82] in a form which was proposed by A. A. Beilinson (oral communication). See [73], [80], [45], [97] for alternative versions.

The geometry of supergravity, especially generalized supergravity, is one of the most fascinating mathematical discoveries accomplished by physicists in the last decade.

BIBLIOGRAPHY

Atiyah, M. F.

1. Geometry of Yang-Mills fields, Lezioni Fermiani, Pisa 1979
2. Green's functions for self-dual four manifolds, Adv. in Math. **7A** (1981) 130–158

Atiyah, M. F., Drin'feld, V.G., Hitchin, N. J.

3. Construction of instantons, Phys. Lett. **65A** (1978) 185–187

Atiyah, M F., Hitchin, N. J., Singer, I. M.

4. Self-duality in four-dimensional Riemannian geometry, Proc. Roy. Soc. London **362** (1978), 425-461

Atiyah, M. F., Jones, J. S.

5. Topological aspects of Yang-Mills theory, Comm. Math. Phys. **61** (1978), 97–118

Atiyah, M. F., Ward, R. S.

6. Instantons and algebraic geometry, Comm. Math. Phys. **55** (1977), 117–124

Barth, W.

7. Moduli of vector bundles on the projective plane, Invent. Math **42** (1977), 63–91

Barth, W., Hulek, K.

8. Monads and moduli of vector bundles, manuscripta math. **25** (1978), 323–347

Beilinson, A. A.

9. Coherent sheaves on $\mathbb{P}^n$ and problems in linear algebra, Funct. Anal. Appl. **12** (1978), 214–216

Beilinson, A. A., Gel'fand, S. I, Manin, Yu. I.

10. An instanton is determined by its complex singularities, Funct. Anal. Appl. **14** (1980), 118–119

Belavin, A. A., Polyakov, A. M., Schwartz, A. S., Tyupkin, Yu.

11. Pseudo-particle solutions of the Yang-Mills equations, Phys. Lett. **59B** (1975), 85–87

Belavin, A. A., Zakharov, V. E.

12. Multidimensional method of the inverse scattering problem and duality equations for the Yang-Mills field, JETP Letters **25** (1977), 567–570

Berezin, F. A.

13. The mathematical basis of supersymmetric field theories, Sov. J. of Nuclear Phys. **29** (1979), 857–866

Bernshtein, I. N., Gel'fand, I. M., Gel'fand, S. I.

14. Algebraic bundles over $\mathbb{P}^n$ and problems of linear algebra, Funct. Anal. Appl. **12** (1978), 212–214

Bernshtein, I. N., Leites, D. A.

15. Integral forms and the Stokes formula on supermanifolds, Funct. Anal. Appl. **11** (1977), 45–47
16. Integration of differential forms on supermanifolds, Funct. Anal. Appl. **11** (1977), 219–221

Bott, R.

17. Homogeneous vector bundles, Annals of Math. **66** (1957), 203–246

Buchdahl, N. P.

18. Analysis on analytic spaces and non-self-dual Yang-Mills fields, Preprint, Oxford, 1982
19. On the relative de Rham sequence, Preprint, Oxford, 1982

Corrigan, E., Goddard, P.

20. An n-monopole solution with $4n - 1$ degrees of freedom, Comm. Math. Phys. **80** (1981), 575–587

Corvin, L., Ne'eman, J., Sternberg, S.

21. Graded Lie algebras in mathematics and physics, Rev. Modern Phys. **47** (1975), 573–604

Deligne, P.

22. Équations différentielles à points singuliers réguliers, Springer Lecture Notes in Mathematics **163**, 1970

Demazure, M.

23. A very simple proof of Bott's theorem, Invent. Math **33** (1976), 271–725

Donaldson, S. K

24. Self-dual connections and the topology of smooth 4-manifolds, Preprint, 1982

Douady, A., Verdier, J.L.

25. Les équations de Yang-Mills, Séminaire ENS 1977–78, Astérisque, 71–72, 1980

Drin'feld, V. G., Manin, Yu. I.

26. A description of instantons, II (Russian), Trudi mezhd. sem. po fizike vysokikh ènergiĭ, Serpukhov, 1978, 71–92
27. A description of instantons, Comm. Math. Phys. **63** (1978), 177–182
28. Instantons and bundles on $\mathbb{CP}^3$, Funct. Anal. Appl. **13** (1979), 124–134
29. Yang-Mills fields, instantons, tensor products of instantons, Sov. J. of Nuclear Phys. **29** (1979), 845–849

Dubrovin, A. T., Novikov, S. P., Fomenko, S. P.

30. Modern geometry – methods and applications, Springer-Verlag, New York, 1984

Eastwood, M. G., Penrose, R., Wells, R. O.

31. Cohomology and massless fields, Comm. Math. Phys. **78** (1981), 305–351

Eguchi, T., Gilkey, P. B., Hanson, A.J.

32. Gravitation, gauge theory and differential geometry, Phys. Rep. **66** (1980), 213–393

Eguchi, T., Hanson, A. J.

33. Asymptotically flat solutions to Euclidian gravity, Phys. Lett. **74B** (1978), 249–251

Ferber, A.

34. Supertwistors and conformal supersymmetry, Nuclear Phys. **B132** (1978), 55–64

Fischer, G.

35. Complex analytic geometry, Springer Lecture Notes in Mathematics **538**, 1976

Flaherty, E. J.

36. Hermitian and Kählerian geometry in relativity, Springer Lecture Notes in Physics **46**, 1976

Freedman, D. Z., Nieuwenhuizen, van, P.

37. Supergravity and the unification of the laws of physics, Scientific American **238** (1978), pp. 126–143

Friedrich, T., ed.

38. Self-dual Riemannian geometry and instantons, Teubner, Leipzig, 1981

Gel'fand, I. M., Gindikin, S. G., Graev, M. I.

39. Integral geometry in affine and projective spaces, J. Soviet Math. **18** (1982), 39–167

Gindikin, S. G., Henkin, G.M.

40. Penrose transformation and complex integral geometry, J. Soviet Math. **21** (1983), 508–550

Gol'fand, Yu. A., Likhtman, E. P.

41. Extension of the algebra of Poincaré group generators and the violation of P invariance, JETP Letters **13** (1971), 323–326

Grauert, H., Kerner, H.

42. Deformationen von Singularitäten komplexer Räume, Math. Ann. **153** (1964), 236–260

Griffiths, P.A.

43. The extension problem in complex analysis, II, Amer. J. of Math. **88** (1966), 366–466

Hansen, R. O., Newman, E. T., Penrose, R., Tod, K. P.

44. The metric and curvature properties of $\mathcal{H}$-space, Proc. Roy. Soc. London A **363** (1978), 445–468

Hawking, S. W., Roček, M., ed.

45. Superspace and supergravity, Cambridge University Press, Cambridge, 1981

Helgason, S.

46. The Radon transform, Birkhäuser, Boston, 1980

Henkin, G. M.

47. Representation of solutions of the 4-equation in the form of holomorphic bundles over twistor space, Sov. Math. Dokl. **24** (1981), 415–419
48. Yang-Mills-Higgs fields as holomorphic vector bundles, Sov. Math. Dokl. **26** (1982), 224–228

Henkin, G. M., Manin, Yu. I.

49. Twistor description of classical Yang-Mills-Dirac fields, Phys. Lett. **95B** (1980), 405–408
50. On the cohomology of twistor flag spaces, Compositio Math. **44** (1981), 103–111

Hitchin, N. J.

51. Polygons and gravitons, Math. Proc. Cambridge Philos. Soc. **85** (1979), 465–476
52. Linear field equations on self-dual spaces, Proc. Roy. Soc. London A **370** (1980), 173–191
53. Kählerian twistor spaces, Proc. London Math. Soc. **43** (1981), 133–150
54. Monopoles and geodesics, Comm. Math. Phys. **83** (1982), 579–602
55. On the construction of monopoles, Preprint, Oxford, 1982

Hoàng Lê Minh (Xoang Le Minh)

56. On the twistor interpretation of the Green's function for a non-self dual Yang-Mills field, Russian Math. Surveys **38** (1983), 166–167

Hughston, L. P., Ward, R. S., ed.
57. Advances in twistor theory, Pitman, London, 1979

Illusie, L.
58. Complexe cotangente et déformations I, Springer Lecture Notes in Mathematics **239**, 1971

Isenberg, J., Yasskin, P. B., Green, P. S.
59. Non-self-dual gauge fields, Phys. Lett. **78B** (1978), 464–468

Jaffe, A., Taubes, C.
60. Vortices and monopoles, Birkhäuser, Boston, 1980

Kac, V. G.
61. Lie superalgebras, Adv. in Math. **26** (1977), 8–96

Kostant, B.
62. Graded manifolds, graded Lie theory and prequantization, Springer Lecture Notes in Mathematics **570** (1977), 177–306

Le Brun, C.R.
63. Spaces of complex geodesics and related structures, Thesis, Oxford, 1980
64. The first formal neighborhood of ambitwistor space for curved space time, Lett. Math. Phys. **6** (1982), 345–354
65. Spaces of complex null geodesic in complex-riemannian geometry, Preprint, IHES, 1982
66. $\mathcal{H}$-space with a cosmological constant, Proc. Roy. Soc. London A **380** (1982), 171–185

Leites, D. A.
67. Spectra of graded-commutative rings (Russian), Uspekhi Mat. Nauk **29** (1974), 209–210
68. Introduction to the theory of supermanifolds, Russian Math. Surveys **35** (1980), 3–57

Lerner, D. E., Sommers, P. D., ed.
69. Complex manifold techniques in theoretical physics, Pitman, London, 1979

Manin, Yu. I.
70. Gauge fields and holomorphic geometry, J. Soviet Math. **2** (1983), 465–507
71. Flag superspaces and supersymmetric Yang-Mills equations, Arithmetic and Geometry, vol. II, Birkhäuser, Boston, 1983
72. New exact solutions and cohomology analysis of ordinary and supersymmetric Yang-Mills equations, Proc. Steklov Inst. of Math. **165** (1984), 107–124

Manin, Yu. I., ed.

73. Geometric ideas in physics (Russian translation of papers in English, see Math. Rev., 85g:83003 for contents), Mir, Moscow, 1980

Manin, Yu. I., Henkin, G. M.

74. Yang-Mills equations as Cauchy-Riemann equations in twistor space, Sov. J. of Nuclear Phys. **35** (1982), 941–950

Manin, Yu. I., Penkov, I. B.

75. Null-geodesics of complex Einstein spaces, Funct. Anal. Appl. **16** (1982), 64–66
76. The formalism of left and right connections on supermanifolds, Proc. Summer School in Phys., Varna, 1982

Nahm, W.

77. All selfdual multimonopoles for arbitrary gauge groups, Preprint, CERN TH.3172, 1981
78. The algebraic geometry of multimonopoles, Preprint, Bonn University, 1982

van Nieuwenhuizen, P.

79. Supergravity, Phys. Rep. **68** (1981), 189–398

van Nieuwenhuizen, P., Freedman, D. Z., ed.

80. Supergravity, North-Holland, Amsterdam, 1979

Ogievetskiĭ, V. I., Mezinchesku, L.

81. Boson-fermion symmetries and superfields, Soviet Phys. Uspekhi **18** (1975), 960–982

Ogievetskiĭ, V. I., Sokachev, È. S.

82. The simplest Einstein supersymmetry group, Sov. J. of Nuclear Phys. **31** (1980), 140–148
83. The gravitational axial-vector superfield and the formalism of differential geometry, Sov. J. of Nuclear Phys. **31** (1980), 424–433
84. The normal gauge in supergravity, Sov. J. of Nuclear Phys. **32** (1980), 443–447
85. Torsion and curvature in terms of the axial-superfield, Sov. J. of Nuclear Phys. **32** (1980), 447–452

Okonek, C., Schneider, M., Spindler, H.

86. Vector bundles on complex projective spaces, Birkhäuser, Boston, 1980

Penkov, I. B.

87. The Penrose transform on general grassmannians, C.R. Acad. Bulgare de Sci. **3** (1980), 1439–1442

88. $\mathcal{D}$-modules on supermanifolds, Invent. Math **71** (1983), 501–512
89. Linear differential operators and the cohomology of analytic spaces, Russian Math. Surveys **37** (1982), 131–132

Penrose, R.
90. Solutions of zero-rest-mass equations, J. Math. Phys. **10** (1969), 38–39
91. Twistor theory, its aims and achievements, Quantum Gravity, ed., C. J. Isham et al., Oxford University Press, Oxford, 1975
92. Non-linear gravitons and curved twistor theory, Gen. Rel. Grav. **7** (1976), 31–52
93. The twistor program, Rep. Math. Phys. **12** (1977), 65–76

Reiffen, H. S.
94. Das Lemma von Poincaré für holomorphe Differentialformen auf komplexen Räumen, Math. Z. **101** (1967), 269–284

Scheunert, M.
95. The theory of Lie superalgebras, Springer Lecture Notes in Mathematics **716**, 1979

Schwarz (Švarts), A. S.
96. Instantons and fermions in the field of instanton, Comm. Math. Phys. **64** (1979), 233–268
97. Supergravity, complex geometry and G-structures, Comm. Math. Phys. **87** (1982), 37–63

Semikhatov, A. M.
98. Supersymmetric instanton, JETP Letters **35** (1982), 560–563

Shander, V. N.
99. Vector fields and differential equations on supermanifolds, Funct. Anal. Appl. **14** (1980), 160–161

Slavnov, A. A.
100. Supersymmetric gauge theories and their possible applications to the weak and electromagnetic interactions, Soviet Phys. Uspekhi **21** (1978), 240–251

Sternberg, S.
101. On the role of field theories in our physical conception of geometry, Springer Lecture Notes in Mathematics **676** (1978), 1–80

Taubes, C. H.
102. Selfdual Yang-Mills connection on non-self-dual 4-manifolds, J. Differential Geom. **17** (1982), 139–170

Todorov, I. T.

103. Conformal description of spinning particles, Preprint, ISAS, Trieste, 1981

Ueno, K., Nakamura, Y.

104. Transformation theory for anti-self-dual equations, Preprint, RIMS **413**, Kyoto, 1982

Vainshtein, A. I., Zakharov, V. I., Novikov, V. A., Shifman, M.A.

105. ABC of instantons, Soviet Phys. Uspekhi **25** (1982), 195–215

Volkov, D. V., Akulov, V. P.

106. Possible universal neutrino interaction, JETP Letters **16** (1972), 438–440

Ward, R. S.

107. On self-dual gauge fields, Phys. Lett. **61A** (1977), 81–82
108. A class of self-dual solutions of Einstein's equations, Proc. Roy. Soc. London **A363** (1978), 289–295
109. Self-dual space-times with cosmological constant, Comm. Math. Phys. **78** (1980), 1–17
110. Ansätze for self-dual Yang-Mills fields, Comm. Math. Phys. **80** (1981), 563–574

Wells, R. O., Jr.

111. Differential analysis on complex manifolds, Springer-Verlag, New York, 1980
112. Complex manifolds and mathematical physics, Bull. Amer. Math. Soc. **1** (1979), 296–336
113. Hyperfunction solutions of the zero-mass field equations, Comm. Math. Phys. **78** (1981), 567–600
114. Complex geometry in mathematical physics, Presses de l'Université de Montréal, Montréal, 1982

Wess, J.

115. Supersymmetry – supergravity, Springer Lect. Notes in Math. **77** (1978), 81–125

Weyl, H.

116. Gravitation und Elektrizität, Sitzungsberichte König. Preuss. Akad. Wiss. Berlin (1918), 465–480

Witten, E.

117. Introduction to supersymmetry, Preprint, 1982
118. An interpretation of classical Yang-Mills theory, Phys. Lett. **78B** (1978), 394–398

Yang, C. N., Mills, R. L.

119. Conservation of isotopic spin and isotopic gauge invariance, Phys. Rev. **96** (1954), 191–195

RECENT DEVELOPMENTS

More than a decade has passed since the first publication of the book "Gauge Field Theory and Complex Geometry"; an epoch, in the time scale of theoretical physics.

Overcome by superstring models, gauge field theory has left the proscenium of grand unification physics ... to find a prominent place in pure mathematics. Using gauge field self-duality equations in 4 dimensions, Donaldson produced a series of new invariants for 4-manifolds which showed that the differentiable classification of smooth 4-manifolds is much richer than their topological classification; in particular, he proved the existence of exotic differentiable structures in $\mathbb{R}^4$. Quite recently, Seiberg and Witten offered a new system of gauge field equations which resulted not only in a dramatic simplification of original arguments of Donaldson but also in many new applications in topology and geometry. Many of the recently obtained low-dimensional invariants — Donaldson and Seiberg-Witten invariants for 4-manifolds, the Jones polynomial for knots, the Floer homologies of 3-manifolds — admit a nice interpretation in the (quantum) gauge field theory framework.

On the other hand, the methods of complex geometry have become one of the basic tools of study of real (not necessarily analytic) differential geometric structures. Recent spectacular achievements in the geometry of quaternionic, quaternionic Kähler and hyper-Kähler manifolds, Riemannian 4-manifolds, compact Kähler surfaces, integrable systems, exotic holonomies, harmonic maps, extended supersymmetry have roots in twistor theory and, more generally, in the theory of conic structures. The latter has also found an unexpected application in the canonical quantization of (super)strings and (super)membranes – it turns out that in order to get a better treatment of the Dirac constraints, one has to lift the quantization problem from the space-time background to an appropriate conic structure over space-time (rather misleadingly, these models are, in physics jargon, termed twistor-like string and twistor-like membrane).

Like the book itself, this Addendum splits into two parts, A (twistor theory) and B (supersymmetric geometry).

For a long time twistor theory existed as a number of fragments, each being, at least at the technical level, a more or less independent story. There is, however, a general theorem which unites many of the fragments into one picture. It is chosen as a leading theme of Part A; as a means to gather the threads of the twistor theory and to browse over a number of the results mentioned in the preceding paragraph.

Part B, on the contrary, is *not* intended to give a summary of new developments in the area of supermanifolds. An illuminating exposition of the most important results can be found in the recent book by Yu.I. Manin, "Topics in Noncommutative geometry", Princeton University Press, Princeton NJ, 1991. So, in Part B we just continue the final paragraph of the final Chapter of "Gauge Field Theory and Complex Geometry" with two more particular examples, Riemannian supermanifolds in $3|2$ dimensions and quaternionic supermanifolds in $4k|2k+2$ dimensions, aiming to give more illustrations of the general machinery developed in this book, and to show how different, even at the level of basic notions, the supersymmetrizations of 3- and 4-dimensional Riemannian structures can be. It looks as if supersymmetry removes the degeneracy of the classical category of Riemannian manifolds in the dimension parameter.

SERGEI A. MERKULOV

CHAPTER A

NEW DEVELOPMENTS IN TWISTOR THEORY

§ 1. A Simple Group-Theoretic Explanation of the Twistor Transform

0. Introduction. Let $X \hookrightarrow Z$ be a rational curve $X = \mathbf{C}P^1$ embedded into a complex 3-fold Z with normal bundle $N = \mathcal{O}(1) \oplus \mathcal{O}(1)$. As shown in Chapter 2, §2, the moduli space of all rational curves obtained by holomorphic deformations of X inside Z is a complex 4-dimensional manifold M which comes equipped with a canonically induced self-dual conformal structure; moreover, *any* local conformal self-dual structure arises in this way. Since the pioneering work of Penrose (1976), several other manifestations of this strange phenomenon have been observed when complex analytic data of the form $(X \hookrightarrow Z, N)$ give rise to a category of local geometric structures C_{geo} (more precisely, it is a successful choice of a pair (X, N) consisting of a complex homogeneous manifold X and a homogeneous vector bundle N on X which uniquely specifies C_{geo}, the choice of a particular ambient manifold Z corresponding to the choice of a particular object in C_{geo}).

The questions then arise: How general is this phenomenon? What is the guiding principle for making a *successful* choice of (X, N)?

In the subsections below we elucidate these questions by unveiling strong links between the (curved, or non-linear) twistor approach to differential geometry and the Borel-Weil approach to representation theory (in this context, see also the works of Baston and Eastwood on applications of the Bott-Borel-Weil technique to the linear twistor transform between flatmodels).

1. Complex G-structures. Let M be an n-dimensional complex manifold and V a fixed n-dimensional complex vector space (typically, $V = \mathbf{C}^n$). Let $\pi : \mathcal{L}^*M \to M$ be the holomorphic bundle of V-valued coframes, whose fibres $\pi^{-1}(t)$ consist, by definition, of all $\mathbf{C}$-linear isomorphisms $e : T_tM \to V$, where T_tM is the tangent space at $t \in M$. The space $\mathcal{L}^*M$ is a principal right $\mathrm{GL}(V)$-bundle with the right action given by $R_g(e) = g^{-1} \circ e$. If G is a closed complex subgroup of $\mathrm{GL}(V)$, then a complex G-structure on M is a principal holomorphic subbundle $\mathcal{G}$ of $\mathcal{L}^*M$ with structure group G. It is clear that there is a one-to-one correspondence between the set of G-structures on M and the set of holomorphic sections σ of the quotient bundle $\tilde{\pi} : \mathcal{L}^*M/G \to M$ whose typical fibre is isomorphic to $\mathrm{GL}(V)/G$. A G-structure on M is called *locally flat* if there is a coordinate patch in the neighbourhood of each point $t \in M$ such that in the associated trivialization of $\mathcal{L}^*M/G$ over this patch, the section σ is represented by a constant $GL(m, \mathbf{C})/G$-valued function. A G-structure is called *k-flat* if, for each $t \in M$, the k-jet of the associated section σ of $\mathcal{L}^*M/G$ at t is isomorphic to the k-jet of some locally flat section of $\mathcal{L}^*M/G$. It is not difficult to show that a G-structure admits a torsion-free affine connection if and only if it is 1-flat. A G-structure on M is called *irreducible* if the action of G on V leaves no non-zero proper subspaces invariant.

The notion of G-structure is a unifying idea for a variety of popular themes in differential geometry. For example, (i) if $G \subset \mathrm{GL}(n, \mathbf{C})$ is the special orthogonal group $\mathrm{SO}(n, \mathbf{C})$, then the associated G-structure is nothing but a complex Riemannian structure; (ii) a $\mathrm{CO}(n, \mathbf{C})$-structure on an n-dimensional manifold coincides with a complex conformal structure; (iii) a $\mathrm{GL}(m, \mathbf{C})\mathrm{GL}(n, \mathbf{C})$-structure on an mn-dimensional manifold is locally identical to a Grassmanian spinor structure defined in Chapter 1, §7. In the first two examples, G-structures are always 1-flat, whereas in the third example this is not the case — 1-flat $\mathrm{GL}(2, \mathbf{C})\mathrm{GL}(n, \mathbf{C})$-structures with $n \geq 3$ are called complexified quaternionic structures.

The notion of G-structure has proven to be very useful in the study of affine connections, especially in the context of Berger's problem of classifying the irreducibly acting holonomies of torsion-free affine connections (see, e.g., the works of Bryant). Given an affine connection ∇ on a connected simply connected complex manifold M, then the set $\mathcal{G}_u$ of all points in the bundle of V-valued coframes $\mathcal{L}^*M$ which can be connected to a fixed point $u \in \mathcal{L}^*M$ by a horizontal curve is a principal right subbundle of $\mathcal{L}^*M$ whose structure group G_u is a Lie subgroup of $\mathrm{GL}(V)$, called the (restricted) *holonomy group of* ∇ *at* u. The conjugacy class of G_u in $\mathrm{GL}(V)$ does, in fact, not depend on the choice of u, and any fixed subgroup $G \subseteq \mathrm{GL}(V)$ in this conjugacy class is called, by abuse of language, the *holonomy group of* ∇. According to Hano and Ozeki (1956), any closed subgroup of a general linear group can be realized as the holonomy group of some affine connection which, in general, has non-zero torsion. Which reductive Lie subgroups G of $\mathrm{GL}(V)$ can be realized as the holonomy group of a torsion-free affine connection? According to Berger (1955), the list of such G must be very restricted, though the full classification is far from being complete.

The basic question about the local geometry of G-structures — what is the obstruction for a k-flat G-structure $\mathcal{G} \to M$ to be $(k+1)$-flat? — has been answered by Guillemin & Singer and Sternberg. The obstruction is given by a section of the associated vector bundle $\mathcal{G} \times_G \mathrm{H}^{k,2}(\mathfrak{g})$, where G-modules $\mathrm{H}^{k,2}(\mathfrak{g})$ are defined in the next paragraph.

2. Spencer cohomology. Let V be a vector space and $\mathfrak{g}$ a Lie subalgebra of $gl(V) \simeq V \otimes V^*$. Define recursively the $\mathfrak{g}$-modules

$$\begin{aligned} \mathfrak{g}^{(-1)} &= V \\ \mathfrak{g}^{(0)} &= \mathfrak{g} \\ \mathfrak{g}^{(k)} &= [\mathfrak{g}^{(k-1)} \otimes V^*] \cap [V \otimes S^{k+1}(V^*)]\ , \ k = 1, 2, \ldots, \end{aligned}$$

and define the map

$$\partial : \mathfrak{g}^{(k)} \otimes \Lambda^{l-1}(V^*) \longrightarrow \mathfrak{g}^{(k-1)} \otimes \Lambda^l(V^*)$$

as the antisymmetrisation over the last l indices.

Since $\partial^2 = 0$, there is a complex

$$\mathfrak{g}^{(k)} \otimes \Lambda^{l-1}(V^*) \xrightarrow{\partial} \mathfrak{g}^{(k-1)} \otimes \Lambda^{l}(V^*) \xrightarrow{\partial} \mathfrak{g}^{(k-2)} \otimes \Lambda^{l+1}(V^*)$$

whose cohomology at the centre term is denoted by $\mathrm{H}^{k,l}(\mathfrak{g})$ and is called the (k,l) *Spencer cohomology group*. In particular,

$$\begin{aligned} \mathrm{H}^{k,1}(\mathfrak{g}) &= 0 \\ \mathrm{H}^{k,2}(\mathfrak{g}) &= \frac{\mathrm{Ker} : \mathfrak{g}^{(k-1)} \otimes \Lambda^2(V^*) \xrightarrow{\partial} \mathfrak{g}^{(k-2)} \otimes \Lambda^3(V^*)}{\mathrm{Image} : \mathfrak{g}^{(k)} \otimes V^* \xrightarrow{\partial} \mathfrak{g}^{(k-1)} \otimes \Lambda^2 V^*} . \end{aligned} \tag{1}$$

In addition to $\mathrm{H}^{k,2}(\mathfrak{g})$, the $\mathfrak{g}$-module $\mathfrak{g}^{(1)}$ has a clear geometric meaning as well. If a G-structure $\mathcal{G} \to M$ is 1-flat then the set of all torsion-free affine connections in $\mathcal{G}$ is an affine space modelled on the vector space $\mathrm{H}^0(M, \mathcal{G} \times_G \mathfrak{g}^{(1)})$. In particular, if $G \subseteq \mathrm{GL}(V)$ is such that $\mathfrak{g}^{(1)} = 0$, then any G-structure admits at most one torsion-free affine connection. If $K(\mathfrak{g})$ denotes the $\mathfrak{g}$-module of formal curvature tensors of torsion-free affine connections with holonomy in $\mathfrak{g}$,

$$K(\mathfrak{g}) = [\mathfrak{g} \otimes \Lambda^2 V^*] \cap [\mathrm{Ker} : V \otimes V^* \otimes \Lambda^2(V^*) \to V \otimes \Lambda^3(V^*)] ,$$

then

$$\mathrm{H}^{1,2}(\mathfrak{g}) = \frac{K(\mathfrak{g})}{\partial(\mathfrak{g}^{(1)} \otimes V^*)}$$

i.e., the cohomology group $\mathrm{H}^{1,2}(\mathfrak{g})$ represents the part of $K(\mathfrak{g})$ which is invariant under $\mathfrak{g}^{(1)}$-valued shifts in a formal torsion-free affine connection with holonomy in $\mathfrak{g}$. For example, if $(G,V) = (\mathrm{CO}(n,\mathbb{C}), \mathbb{C}^n)$, then $\mathfrak{g}^{(1)} = V^*$ and $\mathrm{H}^{1,2}(\mathfrak{g})$ is the vector space of formal Weyl tensors.

If $\mathfrak{g}^{(1)} = 0$, then $\mathrm{H}^{1,2}(\mathfrak{g})$ is exactly $K(\mathfrak{g})$, the $\mathfrak{g}$-module which plays a key role in the theory of torsion-free affine connections with holonomy in $\mathfrak{g}$. The case $\mathfrak{g}^{(1)} = 0$ is generic — there are very few irreducibly acting Lie subgroups $\mathfrak{g} \subset gl(V)$ which have $\mathfrak{g}^{(1)} \neq 0$ and they are all known by now. As of this writing, the list of all irreducibly acting $\mathfrak{g} \subset gl(V)$ which have $K(\mathfrak{g}) \neq 0$ is not known.

3. Twistor formulae for Spencer cohomology. Let V be a finite dimensional complex vector space and $G \subseteq \mathrm{GL}(V)$ an irreducible representation of a reductive complex Lie group in V. Then G also acts irreducibly in V^* via the dual representation. Let $\tilde{X}$ be the G-orbit of a highest weight vector in $V^* \setminus 0$. Then the quotient $X := \tilde{X}/\mathbb{C}^*$ is a compact complex homogeneous-rational manifold canonically embedded into $P(V^*)$, and there is a commutative diagram

$$\begin{array}{ccc} \tilde{X} & \hookrightarrow & V^* \setminus 0 \\ \downarrow & & \downarrow \\ X & \hookrightarrow & P(V^*) \end{array}$$

In fact, $X \simeq G_s/P$, where G_s is the semisimple part of G and P is the parabolic subgroup of G_s leaving a highest weight vector in V^* invariant up to a scalar. Let L be the restriction of the hyperplane section bundle $\mathcal{O}(1)$ on $P(V^*)$ to the submanifold X. Clearly, L is an ample homogeneous line bundle on X.

In summary, there is a natural map

$$(G, V) \longrightarrow (X, L)$$

associating with an irreducibly acting reductive Lie group $G \subsetneq \mathrm{GL}(V)$ a pair (X, L) consisting of a compact complex homogeneous-rational manifold X and an ample line bundle L on X.

Can this map be reversed? According to Borel-Weil, the representation space V can be reconstructed very easily:

$$V = \mathrm{H}^0(X, L)\,.$$

What about G? If G is simple, then, according to Onishchik (1962), the Lie algebra of G is a subalgebra of $\mathrm{H}^0(X, TX)$; moreover, it coincides with $\mathrm{H}^0(X, TX)$ unless one of the following holds: (i) G is the representation of $\mathrm{Sp}(n, \mathbb{C})$ in $\mathbb{C}^{2n}$, in which case $\mathrm{H}^0(X, TX) \simeq sl(n, \mathbb{C})$; (ii) G is the representation of G_2 in $\mathbb{C}^7$, in which case $\mathrm{H}^0(X, TX) \simeq so(7, \mathbb{C})$); (iii) G is the (fundamental) spinor representation of $\mathrm{Spin}(2n+1, \mathbb{C})$ in which case $\mathrm{H}^0(X, TX) \simeq so(2n+2,\ \mathbb{C})$.

Therefore, if $G \subset GL(V)$ is semisimple then, with a few exceptions, G can be reconstructed from (X, L). However, it is often undesirable to restrict oneself to semisimple groups only (especially in the context of the Berger holonomy classification problem). There is a natural central extension of the Lie algebra $\mathrm{H}^0(X, TX)$ which is canonically associated with the pair (X, L).

Fact 1 *For any* (X, L), $\mathfrak{g} := \mathrm{H}^0(X, L \otimes (J^1L)^*)$ *is a reductive Lie algebra canonically represented in* $\mathrm{H}^0(X, L)$.

This fact is easy to explain — $\mathrm{H}^0(X, L \otimes (J^1L)^*)$ is exactly the Lie algebra of the Lie group G of all global biholomorphisms of the line bundle L which commute with the projection $L \to X$.

In conclusion, with a given irreducible representation $G \subsetneq \mathrm{GL}(V)$ there is canonically associated a pair (X, L) consisting of a compact complex homogenous-rational manifold X and a very ample line bundle on X such that much of the original information about G can be restored from (X, L). In the twistor theory context, the crucial observation is that the $\mathfrak{g}$-modules $\mathfrak{g}^{(k)}$ and $\mathrm{H}^{k,2}(\mathfrak{g})$ also admit a nice description in terms of (X, L).

Theorem 2 *For a compact complex manifold* X *and a very ample line bundle* L *on* X*, there is an isomorphism*

$$\mathfrak{g}^{(k)} = \mathrm{H}^0\left(X, L \otimes S^{k+1}(N^*)\right), \quad k = 0, 1, 2, \ldots$$

and an exact sequence of $\mathfrak{g}$*-modules,*

$$0 \longrightarrow \mathrm{H}^{k,2}(\mathfrak{g}) \longrightarrow \mathrm{H}^1\left(X, L \otimes S^{k+2}(N^*)\right) \longrightarrow \mathrm{H}^1\left(X, L \otimes S^{k+1}(N^*)\right) \otimes V^*, \quad k = 1, 2, \ldots$$

where $\mathfrak{g} := \mathrm{H}^0(X, L \otimes (J^1L)^*)$, $N := J^1L$, *and* $\mathrm{H}^{k,2}(\mathfrak{g})$ *are the Spencer cohomology groups associated with the canonical representation of* $\mathfrak{g}$ *in the vector space* $V := \mathrm{H}^0(X, L)$.

Proof. Since L is very ample, there is a natural "evaluationëpimorhism $V \otimes \mathcal{O}_X \to J^1L \to 0$ whose dualization gives rise to the canonical monomorphism $0 \to N^* \to V^* \otimes \mathcal{O}_X$. Then one may construct the following sequences of locally free sheaves,

$$0 \longrightarrow L \otimes S^{k+1}(N^*) \longrightarrow L \otimes S^k(N^*) \otimes V^* \longrightarrow L \otimes S^{k-1}(N^*) \otimes \Lambda^2(V^*) \qquad (2)$$

and

$$0 \longrightarrow L \otimes S^{k+2}(N^*) \longrightarrow L \otimes S^{k+1}(N^*) \otimes V^* \longrightarrow L \otimes S^k(N^*) \otimes \Lambda^2(V^*) \longrightarrow L \otimes S^{k-1}(N^*) \otimes \Lambda^3(V^*), \qquad (3)$$

and notice that they both are exact. [Hint: for any vector space W one has $W \otimes \Lambda^2 W \bmod \Lambda^3 W \simeq W \otimes S^2(W) \bmod S^3(W)$.]

Then computing $\mathrm{H}^0(X, \ldots)$ of (2) and using the inductive definition of $\mathfrak{g}^{(k)}$ one immediately obtains the first statement of the Theorem.

The second statement follows from (3) and the definition (1) of $\mathrm{H}^{k,2}(\mathfrak{g})$. Indeed, define E_k by the exact sequence

$$0 \longrightarrow L \otimes S^{k+2}(N^*) \longrightarrow L \otimes S^{k+1}(N^*) \otimes V^* \longrightarrow E_k \longrightarrow 0.$$

The associated long exact sequence implies the following *exact* sequence of vector spaces

$$0 \longrightarrow \mathrm{H}^0(X, E_k)/\partial[\mathfrak{g}^{(k)} \otimes V^*] \longrightarrow \mathrm{H}^1\left(X, L \otimes S^{k+2}(N^*)\right) \longrightarrow \mathrm{H}^1\left(X, L \otimes S^{k+1}(N^*)\right) \otimes V^*.$$

On the other hand, the exact sequence

$$0 \longrightarrow E_k \longrightarrow L \otimes S^k(N^*) \otimes \Lambda^2(V^*) \longrightarrow L \otimes S^{k-1}(N^*) \otimes \Lambda^3 V^*$$

implies

$$\mathrm{H}^0(X, E_k) = \ker : \mathfrak{g}^{(k-1)} \otimes \Lambda^2(V^*) \xrightarrow{\partial} \mathfrak{g}^{(k-2)} \otimes \Lambda^3(V^*),$$

which in turn implies

$$\mathrm{H}^{k,2}(\mathfrak{g}) = \mathrm{H}^0(X, E_k)/\partial[\mathfrak{g}^{(k)} \otimes V^*].$$

This completes the proof of the second part of the Theorem. □

The above result shows that vector spaces of formal curvature tensors fit nicely into the Borel-Weil paradigm; and explains why twistorial data (X, N), consisting of a homogeneous-rational manifold X and a holomorphic vector bundle N on X, can, in principle, be used as a building block for basic differential-geometric objects. If $\mathrm{rank} N \geq 2$, then, following a common practice in complex analysis, one should replace the pair (X, N) by an equivalent one $(\hat{X} = P(N^*), L = \mathcal{O}(1))$ and apply Theorem 2 to find out which geometric category C_{geo} may correspond to the twistorial data (X, N). Applying this procedure, e.g., to the pair $(\mathbb{C}P^1, \mathbb{C}^{2k} \otimes \mathcal{O}(1))$, $k > 1$, one immediately concludes that C_{geo} can only be the category of complexified quaternionic manifolds. Also, this purely group-theoretic result suggests that there should exist a universal twistor construction for *all* torsion-free geometries. Details of this construction are explained in §3 below.

§ 2. Geometry of Kodaira Moduli Spaces

1. Families of compact complex submanifolds. Let Z and M be complex manifolds and let $\pi_1 : Z \times M \longrightarrow Z$ and $\pi_2 : Z \times M \longrightarrow M$ be the natural projections. An *analytic family of compact submanifolds of the complex manifold Z with the moduli space M* is a complex submanifold $F \hookrightarrow Z \times M$ such that the restriction of the projection π_2 on F is a proper regular map (regularity means that the rank of the differential of $\nu \equiv \pi_2 |_F : F \longrightarrow M$ is equal at every point to $\dim M$). Thus the family F has the structure of a double fibration

$$Z \xleftarrow{\mu} F \xrightarrow{\nu} M$$

where $\mu \equiv \pi_1 |_F$. For each $t \in M$, there is an associated compact complex submanifold X_t in Z which is said to belong to the family F. On the other hand, for each $z \in Z' \equiv \cup_{t \in M} X_t$, there is an associated analytic subspace $\nu \circ \mu^{-1}(z)$ in M. The set of all regular points in $\nu \circ \mu^{-1}(z)$ is denoted by α_z and is called an *alpha subspace* of M. Sometimes we use a more explicit notation $\{X_t \hookrightarrow Z \mid t \in M\}$ to denote an analytic family F of compact submanifolds.

If $F \hookrightarrow Z \times M$ is an analytic family of compact submanifolds, then, for any $t \in M$, there is a canonical linear map

$$k_t : T_t M \longrightarrow \mathrm{H}^0(X_t, N_t)$$

from the tangent space at t to the vector space of global holomorphic sections of the normal bundle, $N_t = TZ|_{X_t} / TX_t$, of $X_t \hookrightarrow Z$ which can be described as follows. First note that the normal bundle of the embedding $\nu^{-1}(t) \hookrightarrow F$ is trivial and thus there is a canonical map $\rho_t : T_t M \longrightarrow \mathrm{H}^0(\nu^{-1}(t), N_{\nu^{-1}(t)|F})$. Then the composition $d\nu \circ \rho_t$ gives the desired map k_t, since the differential of ν maps global sections of $N_{\nu^{-1}(t)|F}$ to global sections of $N_{X_t|Z}$. Here the symbol $N_{A|B}$ stands for the normal bundle of a complex submanifold $A \hookrightarrow B$.

Therefore, there is a sheaf isomorphism $k : TM \to \nu_*^0(N_F)$, where N_F is the normal bundle of $F \hookrightarrow Z \times M$.

In the twistor theory context we shall be interested in two kinds of maximal families of compact complex submanifolds. The first one has been introduced by Kodaira in 1962 and is studied in the subsections below, while the definition/construction of the second one is postponed until §3.

2. Kodaira moduli spaces. An analytic family $F \hookrightarrow Z \times M$ of compact submanifolds is called *complete* if the canonical map $k_t : T_tM \to \mathrm{H}^0(X_t, N_t)$ is an isomorphism at each $t \in M$. It is *maximal* if for any other analytic family $\tilde{F} \hookrightarrow Z \times \tilde{M}$ such that $\mu \circ \nu^{-1}(t) = \tilde{\mu} \circ \tilde{\nu}^{-1}(\tilde{t})$ for some points $t \in M$ and $\tilde{t} \in \tilde{M}$, there is a neighbourhood $\tilde{U} \subset \tilde{M}$ of the point $\tilde{t}$ and a holomorphic map $f : \tilde{U} \to M$ such that $\tilde{\mu} \circ \tilde{\nu}^{-1}(\tilde{t}') = \mu \circ \nu^{-1}\left(f(\tilde{t}')\right)$ for every $\tilde{t}' \in \tilde{U}$.

Theorem 3 *If $X \hookrightarrow Z$ is a compact complex submanifold with normal bundle N such that $\mathrm{H}^1(X, N) = 0$, then X belongs to a complete analytic family $F \hookrightarrow Z \times M$ of compact submanifolds. This family is maximal, and its moduli space is of complex dimension $h^0(X, N)$.*

This theorem was proved by Kodaira in 1962 by covering a neighbourhood of X in Z by a finite number of specially adapted coordinate charts $\{U_i \subset Z\}$ and showing that the infinite series of obstructions to agreements on overlaps $\{U_i \cap U_j\}$ of the formal powers series deformations of $X \cap U_i$ inside U_i all lie in $\mathrm{H}^1(X, N)$.

The moduli space M in the above theorem is called a *Kodaira moduli space.* For example, if $X = \mathbb{C}P^1$ is an intersection of two hyperplanes in $Z = \mathbb{C}P^3$, then the Kodaira moduli space M is exactly the Grassmanian $G(2; \mathbb{C}^4)$, while the complete analytic family $F \hookrightarrow M \times Z$ coincides with the flag manifold $F(1, 2; \mathbb{C}^4)$.

Any Kodaira moduli space M comes equipped canonically with two $\mathcal{O}_M$-modules $\Phi_1 \subset TM \otimes S^2(\Omega^1 M)$ and $\Phi_2 \subset TM \otimes \Omega^1 M$ which are defined as follows. First note that, since $k : TM \to \nu_*^0(N_F)$ is an isomorphism, there is a natural morphism of $\mathcal{O}_M$-modules

$$\begin{array}{rccc} \phi_1 : & \nu_*^0\left(N_F \otimes S^2(N_F)^*\right) & \longrightarrow & TM \otimes S^2(\Omega^1 M) \\ & n & \longrightarrow & \phi(n) \end{array}$$

where $\phi(n)$ is the composition

$$S^2(TM) \xrightarrow{S^2(k)} S^2\left(\nu_*^0(N_F)\right) \longrightarrow \nu_*^0\left(S^2(N_F)\right) \xrightarrow{n} \nu_*^0(N_F) \xrightarrow{k^{-1}} TM.$$

Then $\Phi_1 := \mathrm{Im}\phi_1$.

Analogously, one defines Φ_2 as the image of the natural morphism

$$\phi_2 : \nu_*^0\left(N_F \otimes N_F^*\right) \longrightarrow TM \otimes \Omega^1 M.$$

3. Vector bundles on Z and affine connections on M. Let $\{X_t \hookrightarrow Z \mid t \in M\}$ be a complete analytic family of compact submanifolds. In Chapter 2, §§2 and 3, it has been shown that a holomorphic vector bundle $\mathcal{E}$ on Z which is trivial on submanifolds X_t for all $t \in M$ often gives rise to a pair $(\mathcal{E}_M, \nabla)$ consisting of a holomorphic bundle $\mathcal{E}_M$ on M and a linear connection $\nabla : \mathcal{E}_M \to \mathcal{E}_M \otimes \Omega^1 M$ such that ∇ is integrable on alpha subspaces. In this subsection we explain the geometric meaning of a holomorphic vector bundle $\mathcal{E}$ on Z which, when restricted to X_t, is canonically isomorphic to the normal bundle of $X_t \hookrightarrow Z$, i.e. $\mathcal{E}|_{X_t} = N_t$.

Theorem 4 *Let $\{X_t \hookrightarrow Z \mid t \in M\}$ be a complete analytic family of compact submanifolds with $\mathrm{H}^0\left(X_t, N_t \otimes S^2(N_t^*)\right) = \mathrm{H}^1\left(X_t, N_t \otimes S^2(N_t^*)\right) = 0$ for each $t \in M$. Then a holomorphic vector bundle $\mathcal{E}$ on Z such that $\mathcal{E}|_{X_t} = N_t$ for all $t \in M$ gives rise to an induced torsion-free affine connection ∇ on M relative to which all alpha subspaces are totally geodesic.*

Let $(X, \mathcal{O}_X)$ be an analytic subspace of a complex manifold $(Z, \mathcal{O}_Z)$ defined by the sheaf of ideals $J \subset \mathcal{O}_Z$. The mth-order infinitesimal neighbourhood of X in Z is the ringed space $X^{(m)} = (X, \mathcal{O}_X^{(m)})$ with the structure sheaf $\mathcal{O}_X^{(m)} = \mathcal{O}_Z/J^{m+1}$. With the $(m+1)$th-order infinitesimal neighbourhood there is naturally associated an *mth-order conormal sheaf* of $\mathcal{O}_X^{(m)}$-modules $\mathcal{O}_X^{(m)}(N^*) := J/J^{m+2}$. By construction, $\mathcal{O}_X^{(0)}(N^*)$ is the usual conormal sheaf N^*, while $\mathcal{O}_X^{(1)}(N^*)$ fits into the exact sequence of $\mathcal{O}_X^{(1)}$-modules

$$0 \longrightarrow S^2(N^*) \longrightarrow \mathcal{O}_X^{(1)}(N^*) \longrightarrow N^* \longrightarrow 0.$$

When X is a point in Z, then the dual of $\mathcal{O}_X^{(0)}(N^*)$ is identical to the tangent space $T_X Z$ at $X \in Z$, while the dual of $\mathcal{O}_X^{(m)}(N^*)$ is often denoted by $T_X^{[m]} Z$ and is called mth-order tangent space.

A sketch of the proof of Theorem 4. By assumption, there is a holomorphic vector bundle on Z such that $\mathcal{E}|_{X_t} = N_t$ for any given $t \in M$. The restriction of its dual to the first order infinitesimal neighbourhood of X_t in Z is an $\mathcal{O}_{X_t}^{(1)}$-module which fits into the exact sequence (see Chapter 2, §6)

$$0 \longrightarrow N_t^* \otimes \mathcal{E}^*|_{X_t} \xrightarrow{\ i\ } \mathcal{E}^*|_{X_t^{(1)}} \longrightarrow \mathcal{E}^*|_{X_t} \longrightarrow 0.$$

which implies the following extension

$$0 \longrightarrow S^2(N_t^*) \longrightarrow \mathcal{E}^*|_{X_t^{(1)}} / i(\wedge^2 N^*) \longrightarrow N_t^* \longrightarrow 0.$$

It is straightforward to show (using, e.g., the local coordinate representations of all the objects involved), that the difference

$$\left[\mathcal{O}_{X_t}^{(1)}(N^*) - \mathcal{E}^*|_{X_t^{(1)}} / i(\wedge^2 N^*)\right] \in \mathrm{Ext}_{\mathcal{O}_X^{(1)}}\left(N^*, S^2(N^*)\right)$$

is a locally free $\mathcal{O}_{X_t}$-module which compares second order jets of the embeddings, $X_t \hookrightarrow Z$ and $X_t \hookrightarrow \mathcal{E}|_{X_t}$, which are equivalent at zeroth and first orders. We denote its dual sheaf by $\Delta_t^{[2]}$. The crucial property of the latter is that there is a commutative diagram,

$$\begin{array}{ccccccc} 0\longrightarrow & \mathcal{T}_tM & \longrightarrow & \mathcal{T}_t^{[2]}M & \longrightarrow & S^2(\mathcal{T}_tM) & \longrightarrow 0 \\ & \downarrow & & \downarrow & & \downarrow & \\ 0\longrightarrow & \mathrm{H}^0(X_t, N_t) & \longrightarrow & \mathrm{H}^0\left(X_t, \Delta_t^{[2]}\right) & \longrightarrow & \mathrm{H}^0\left(X_t, S^2(N_t)\right) & \end{array} \tag{4}$$

which relates second levels of the towers of infinitesimal neighbourhoods of two associated embeddings $\{t\} \hookrightarrow M$ and $X_t \hookrightarrow Z$. Again, the verification is straightforward when one uses local coordinates. Since $\mathrm{H}^1\left(X_t, N_t \otimes S^2(N^*)\right) = 0$ by assumption, the exact sequence

$$0\longrightarrow N_t \longrightarrow \Delta_t^{[2]} \longrightarrow S^2(N_t) \longrightarrow 0 \tag{5}$$

splits. Moreover, its splitting is unique, for the group $\mathrm{H}^0\left(X_t, N_t \otimes S^2(N_t^*)\right)$ is also assumed to vanish. The commutative diagram (4) then implies a canonical splitting of the exact sequence for $\mathcal{T}_t^{[2]}M$ which is equivalent to a torsion-free affine connection at t. This shows that the Kodaira moduli space M comes equipped canonically with an induced torsion-free affine connection. The fact that this connection is integrable on alpha surfaces is an easy exercise. □

Given a relative deformation problem $X \hookrightarrow Z$, the most obvious candidates for the class of holomorphic vector bundles $\mathcal{E} \to Z$ which have the property required by Theorem 4, are the holomorphic distributions $\mathcal{E} \subset TZ$ which are transverse to X and have rank equal to $\operatorname{codim} X$. Indeed, if at least one such distribution $\mathcal{E}$ exists, then, for all t in a sufficiently small neighbourhood M_0 of the point $t_0 \in M$ corresponding to X, the submanifolds $X_t \hookrightarrow Z$ remain transverse to $\mathcal{E}$, and one has a *canonical* isomorphism $\mathcal{E}|_{X_t} = N_t$.

EXAMPLE 1 (*Penrose's non-linear graviton*). This construction of Penrose (1976) establishes a one-to-one correspondence between sufficiently small 4-dimensional complex Riemannian manifolds M with self-dual Riemann tensor and 3-dimensional complex manifolds Z equipped with the following data: (i) a submanifold $X \simeq \mathbb{C}P^1$ with normal bundle $N \simeq \mathcal{O}(1) \oplus \mathcal{O}(1)$; (ii) a fibration $\pi : Z \longrightarrow X$; (iii) a global non-vanishing section of the "twisted" determinant bundle $\det(V_\pi) \otimes \pi^*(\mathcal{O}(2))$, where V_π is the vector bundle of π-vertical vector fields. Let us look at this data from the point of view of Theorem 4. The distribution V_π is clearly transversal to X, and one easily checks that $\mathrm{H}^0\left(X_t, N_t \otimes S^2(N_t^*)\right) = \mathrm{H}^1\left(X_t, N_t \otimes S^2(N_t^*)\right) = 0$ for all $t \in M$. Therefore, by Theorem 4, the data (i) and (ii) imply that there is a torsion-free affine connection ∇ induced on M_0 which is integrable on alpha surfaces. What is the role of datum (iii)? It will be shown in subsection 5 below that this part of data puts a severe restriction on the possible holonomy group of ∇ which ensures that ∇ is precisely the Levi-Civita connection of a holomorphic metric with self-dual Riemann curvature tensor.

EXAMPLE 2 (*Quaternionic Kähler manifolds*). Let Z be a complex $(2n+1)$-dimensional manifold equipped with a holomorphic contact structure which is a maximally non-degenerate rank $2n$ holomorphic distribution $D \subset TZ$. Let X be a rational curve embedded into Z transversely to D and with normal bundle $N = \mathbb{C}^{2n} \otimes \mathcal{O}(1)$. Then Theorem 4 says that the $4n$-dimensional Kodaira moduli space M comes equipped with the induced torsion-free affine connection satisfying natural integrability conditions. This is in accordance with well-known results in twistor theory. Ward (1981) showed that in the case $n = 1$ the Kodaira moduli space M has an induced complex Riemannian metric satisfying the self-dual Einstein equation with non-zero scalar curvature. The case $n \geq 2$ has been investigated by LeBrun (1989), Pedersen and Poon (1989), and Bailey and Eastwood (1991) who proved that the Kodaira moduli space M comes equipped canonically with a torsion-free connection compatible with the induced complexified quaternionic-Kähler structure on M. This inverts the construction of Salamon (1982) who associated a contact $(2n+1)$-dimensional manifold Z to any quaternionic-Kähler $4n$-fold M. Nitta and Takeuchi (1987) established a correspondence between contact automorphisms of the twistor space Z and isometries in M. Deforming the standard contact structure in the neighbourhood of a rational curve in $\mathbb{C}P^{2n+1}$, LeBrun (1991) proved that the moduli space of complete quaternionic-Kähler structures on $\mathbb{R}^{4n}$ is infinite dimensional. LeBrun (1989) also found a correspondence between twistor spaces Z of quaternionic-Kähler manifolds and spaces of complex null geodesics of conformal manifolds and used it to show that for every real-analytic conformal pseudo-Riemannian metric of signature $(3, n-1)$ there is a canonically associated quaternionic-Kähler $4n$-fold and, moreover, the original conformal data may be read off from the asymptotics of the quaternionic-Kähler geometry. Using Wisniewski classification results on complex Fano manifolds, LeBrun and Salamon (1994) established that for any n, there are, up to isometries and rescalings, only finitely many compact quaternionic-Kähler $4n$-manifolds of positive scalar curvature; moreover, if M is one of these, then $\pi_1(M) = 0$ and

$$\pi_2(M) = \begin{cases} 0 & \text{iff } M \text{ is } \mathbb{HP}^n \\ \mathbb{Z} & \text{iff } M \text{ is } G(2; \mathbb{C}^{n+2}) \\ \text{finite with 2-torsion} & \text{otherwise.} \end{cases}$$

4. Families of affine connections on moduli space. What happens if all conditions of Theorem 4 are satisfied except the vanishing of $\mathrm{H}^0\left(X_t, N_t \otimes S^2(N_t^*)\right)$? In this case, as was noted in subsection 2 above, the locally free $\mathcal{O}_M$-module $TM \otimes S^2(\Omega^1 M)$ comes equipped with a non-zero $\mathcal{O}_M$-submodule Φ_1. If $\nabla_1 : TM \longrightarrow TM \otimes \Omega^1 M$ and $\nabla_2 : TM \longrightarrow TM \otimes \Omega^1 M$ are two torsion-free affine connections on M, then their difference, $\nabla_1 - \nabla_2$, is a global section of $TM \otimes S^2(\Omega^1 M)$.

We say that ∇_1 and ∇_2 are Φ_1-*equivalent*, if

$$\nabla_1 - \nabla_2 \in \mathrm{H}^0(M, \Phi_1).$$

We define a Φ_1-*connection* on M as a section of the quotient of the $\mathcal{T}M \otimes S^2(\Omega^1 M)$-bundle of torsion-free affine connections on M modulo the action of Φ_1. On an open covering $\{U_i \mid i \in I\}$ of M a Φ_1-connection is represented as a collection of ordinary torsion-free affine connections $\{\nabla_i \mid i \in I\}$, which, on overlaps $U_{ij} = U_i \cap U_j$, have their differences in $\mathrm{H}^0(U_{ij}, \Phi_1)$. Locally, a Φ_1-connection is the same thing as a Φ_1-equivalence class of torsion-free affine connections, but globally they are different — the obstruction for the existence of a Φ_1-connection on M lies in $\mathrm{H}^1\left(M, \mathcal{T}M \otimes S^2(\Omega^1 M)/\Phi_1\right)$, while the obstruction for existence of a Φ_1-equivalence class of torsion-free affine connections is an element of $\mathrm{H}^1\left(M, \mathcal{T}M \otimes S^2(\Omega^1 M)\right)$. A submanifold of M is said to be totally geodesic relative to a Φ_1-connection if it is totally geodesic relative to each of its local representatives ∇_i.

Theorem 5 *Let $\{X_t \hookrightarrow Z \mid t \in M\}$ be a complete analytic family of compact submanifolds with $\mathrm{H}^1\left(X_t, N_t \otimes S^2(N_t^*)\right) = 0$ for each $t \in M$. Then a holomorphic vector bundle $\mathcal{E}$ on Z such that $\mathcal{E}|_{X_t} = N_t$ for all $t \in M$ gives rise to an induced Φ_1-connection ∇ on M relative to which all alpha subspaces are totally geodesic.*

Proof. Since $\mathrm{H}^1\left(X_t, N_t \otimes S^2(N_t^*)\right) = 0$, the exact sequence (5) splits, and any its splitting induces via the commutative diagram (4) an associated splitting of the exact sequence for the second-order tangent space $\mathcal{T}_t^{[2]}M$ which in turn induces a torsion-free affine connection at $t \in M$. The set of all splittings of (5) is a principal homogeneous space for the group $\mathrm{H}^0\left(X_t, N_t \otimes S^2(N_t^*)\right)$. Therefore, the set of induced torsion-free affine connections at t is a principal homogeneous space for the fibre of Φ_1 at t. □

Let sym denote the natural projection

$$\mathrm{sym} : \mathcal{T}M \otimes \Omega^1 M \otimes \Omega^1 M \to \mathcal{T}M \otimes S^2(\Omega^1 M).$$

Associated with a subsheaf $\Phi_2 \subset \mathcal{T}M \otimes \Omega^1 M$ there is a concept of Φ_2-*connection* on M which is, by definition, a collection of ordinary torsion-free affine connections $\{\nabla_i \mid i \in I\}$ on an open covering $\{U_i \mid i \in I\}$ of M which, on overlaps $U_{ij} = U_i \cap U_j$, have their differences in $\mathrm{H}^0\left(U_{ij}, \mathrm{sym}(\Phi_2 \otimes \Omega^1 M)\right)$. If Φ_2 happens to be the structure sheaf $\mathcal{O}_M$ embedded diagonally into $\mathcal{T}M \otimes \Omega^1 M$, then a Φ_2-connection is nothing but a torsion-free projective connection on M.

Theorem 6 *Let $\{X_t \hookrightarrow Z \mid t \in M\}$ be a complete analytic family of compact submanifolds with $\mathrm{H}^1\left(X_t, N_t \otimes N_t^*\right) = 0$ for each $t \in M$. Then a holomorphic vector bundle $\mathcal{E}$ on Z such that $\mathcal{E}|_{X_t} = N_t$ for all $t \in M$ gives rise to an induced Φ_2-connection ∇ on M relative to which all alpha subspaces are totally geodesic.*

The proof of this result is only a slight modification of the construction used in the proof of Theorem 4. Note that the condition $\mathrm{H}^1(X_t, N_t \otimes N_t^*) = 0$ only says that N_t is a rigid vector bundle on X_t.

There is an interesting case when the requirement $\mathcal{E}|_{X_t} = N_t$ can safely be replaced by a weaker condition that vector bundles $\mathcal{E}|_{X_t}$ and N_t are only equivalent to each other, $\mathcal{E}|_{X_t} \simeq N_t$, which means that they define the same cohomology class in $\mathrm{H}^1(X_t, GL(p, \mathcal{O}_{X_t}))$, $p = \operatorname{codim} X$.

Proposition 7 *Let $\{X_t \hookrightarrow Z \mid t \in M\}$ be a complete analytic family of compact submanifolds with $\operatorname{codim} X_t = 1$ and $\mathrm{H}^1(X_t, \mathcal{O}_{X_t}) = 0$ for each $t \in M$. Then a holomorphic vector bundle $\mathcal{E}$ on Z such that $\mathcal{E}|_{X_t} \simeq N_t$ for all $t \in M$ gives rise to an induced projective torsion-free affine connection ∇ on M relative to which all alpha subspaces are totally geodesic.*

Proof. Fix any isomorphism $i : \mathcal{E}|_{X_t} \longrightarrow N_t$ and apply Theorem 7 to construct a torsion-free projective connection ∇ on M. Due to compactness of X_t, the isomorphism i is defined up to multiplication by a non-zero constant. The point is that the construction of ∇ sketched in the proof of Theorem 4 is obviously invariant under such a transformation which shows that ∇ is actually independent on a particular choice of the isomorphism i used in its construction. □

Corollary 8 *Let $\{X_t \hookrightarrow Z \mid t \in M\}$ be a complete analytic family of 1-codimensional compact submanifolds. If $\mathrm{H}^1\left(X_{t_0}, \mathcal{O}_{X_{t_0}}\right) = 0$ for some $t_0 \in M$, then a sufficiently small open neighbourhood M_0 of t_0 in M comes equipped with a canonically induced projective torsion-free affine connection ∇ relative to which all alpha subspaces in M_0 are totally geodesic. The projective Weyl tensor of ∇ is represented by a global section of $\nu_1^*(N_F^*) \otimes \Omega^1 M$ over M_0.*

Proof. By the semi-continuity principle, there is an open neighbourhood U of t_0 such that $H^1(X_t, \mathcal{O}_{X_t}) = 0$ for each $t \in U$. If L is the divisor line bundle of X_{t_0}, then $L^*|_{X_t} \simeq N_t$ for all t in a possibly smaller neighbourhood $M_0 \subseteq U$. This fact combined with Proposition 7 implies the first statement of the Corollary. The second statement about the projective Weyl tensor can be established by comparing the third order infinitesimal neighbourhoods of the embeddings $t \hookrightarrow M$ and $X_t \hookrightarrow Z$. □

EXAMPLE 3 (*Einstein-Weyl geometry in three dimensions*). Let $X \hookrightarrow Z$ be a rational curve $X = \mathbf{C}P^1$ embedded into a complex 2-fold Z with normal bundle $N = \mathcal{O}(m)$, $m \geq 2$. Then the associated Kodaira moduli space M is an $(m+1)$-dimensional complex manifold whose tangent bundle factors locally into the symmetric tensor product $TM = \mathcal{S}^m(\mathcal{S})$, $\mathcal{S}$ being a rank 2 vector bundle. According to Corollary 8, M has a canonically induced torsion-free projective connection whose projective Weyl tensor is a section of $\mathcal{S}^{m-2}(\mathcal{S}^*) \otimes \mathcal{S}^m(\mathcal{S}^*) \otimes \Lambda^2(\mathcal{S}^*)$.

If $m = 2$, the factorization $\mathcal{T}M = S^2(\mathcal{S})$ implies that M has a complex conformal structure; the induced projective connection on the 3-fold M (viewed as an equivalence class of torsion-free affine connections) is easily shown to contain a unique affine connection ∇_0 which comes from a linear connection on the vector bundle $\mathcal{S}$. The structure of the projective Weyl tensor then immediately implies that this ∇_0 solves the 3-dimensional Einstein-Weyl equations. This is a variant of the Hitchin (1982) construction which establishes a 1-1 correspondence between the set of local solutions of 3-dimensional Einstein-Weyl equations and the set of embedding data $(\mathbb{C}P^1 \hookrightarrow Z, N = \mathcal{O}(2))$. Pedersen (1986) and Pedersen and Tod (1993) used this correspondence to find many explicit examples of Einstein-Weyl spaces. Jones and Tod (1985) related the Hitchin construction with the Penrose non-linear graviton construction and showed how, given a conformally anti-self-dual 4-manifold with a conformal isometry, to define an Einstein-Weyl structure on the space of the trajectories of the conformal isometry, and conversely, given an Einstein-Weyl space M, how to construct conformally anti-self-dual 4-manifolds from solutions of a monopole equation on M. Tod (1992) classified all Riemannian-signature Einstein-Weyl structures which can exist on compact real 3-manifolds.

EXAMPLE 4 (*Compact complex homogeneous hypersurfaces*). Let $X \hookrightarrow Z$ be a compact complex homogeneous-rational manifold X with $\dim X \geq 2$ embedded as a hypersurface in a complex manifold Y with positive normal bundle N. By the Kodaira vanishing theorem $\mathrm{H}^1(X, N^*) = 0$, which implies that the projective connection canonically induced on the associated Kodaira moduli space M is flat. This in turn implies that the germ of Y at X is biholomorphic to the germ of the vector bundle $N \to X$ at its zero section. Put another way, any complex $(n+1)$-fold Z containing an n-dimensional compact complex homogeneous-rational manifold X with positive normal bundle is necessarily holomorphically "trivialïn the neighbourhood of X unless $X = \mathbb{C}P^1$.

5. Fibred complex manifolds. Suppose that the ambient manifold Z has the structure of a holomorphic fibration over its compact submanifold X, i.e. there is a submersive holomorphic map $\pi : Z \longrightarrow X$. If $\mathrm{H}^1(X, N) = 0$, then, by Kodaira's theorem, X belongs to the complete family $\{X_t \hookrightarrow Z \mid t \in M\}$ of compact submanifolds. Since X is transverse to the distribution V_π of π-vertical vector fields on Z, so is X_t for every t in some neighbourhood M_0 of the point $t_0 \in M$ which corresponds to X. Thus Z has the structure of a holomorphic fibration $Z \longrightarrow X_t$ for $t \in M_0$. It is not hard to show in this case that for each $t \in M_0$ the mth order conormal sheaf $\mathcal{O}^{(m)}_{X_t}(N_t^*)$ is locally free for any $m \in \mathbb{N}$, and there is a canonical linear map

$$k^{(m)} : \mathcal{T}^{(m)}_t M \longrightarrow \mathrm{H}^0\left(X_t, (\mathcal{O}^{(m)}_{X_t}(N_t^*))^*\right)$$

which for $k = 1$ coincides with the Kodaira map. If also $\mathrm{H}^1\left(X_t, N_t \otimes S^2(N_t^*)\right) = 0$, then, by Theorem 5, the moduli space M comes equipped with a family of induced torsion-free affine connections. Holonomy groups of these induced connections can be estimated with surprising ease.

Theorem 9 *Let ∇ be an induced connection on the moduli space M_0. If the function $f : t \longrightarrow h^0(X_t, N_t \otimes N_t^*)$ is constant on M_0, then the holonomy algebra of ∇ is a subalgebra of the finite dimensional Lie algebra $\mathrm{H}^0(X, N \otimes N^*)$. If in addition there is a holomorphic line bundle L on X such that the bundle $\pi^*(L) \otimes \det V_\pi^*$ admits a nowhere vanishing holomorphic section, then the holonomy algebra of ∇ is a subalgebra of $H^0(X, N \otimes_0 N^*)$, where $\otimes_0$ denotes trace-free tensor product.*

This result can be proved with the help of the remarkably simple correspondence $k^{(3)}$ between the third order infinitesimal neighbourhoods of embeddings $t \hookrightarrow M_0$ and $X_t \hookrightarrow Z$.

EXAMPLE 5 (*Einstein-Weyl geometry in four dimensions*). Let $X = \mathbb{C}P^1$ be the projective line embedded into a 3-dimensional complex manifold Z with normal bundle $N = \mathcal{O}(1) + \mathcal{O}(1)$. If Z has the structure of a holomorphic fibration over X, then Theorem 4 says that there is an induced connection ∇ on the moduli space M_0. By Theorem 9, the holonomy algebra of ∇ is contained in $\mathrm{H}^0(X, N \otimes N^*) = gl(2, \mathbb{C}) \subset co(4, \mathbb{C}) \simeq sl(2, \mathbb{C}) + sl(2, \mathbb{C}) + \mathbb{C}$. Since ∇ is torsion-free, this fact implies that ∇ is a complex Weyl connection on the 4-dimensional complex conformal manifold M_0 which has the anti-self-dual parts of the Weyl tensor and the antisymmetrized Ricci tensor vanishing and satisfies the Einstein-Weyl equations. According to Boyer (1988) and Pedersen and Swann (1994), any such connection arises locally in this way. Madsen et al (1995) have classified compact 4-dimensional Einstein-Weyl manifolds with an at least 4-dimensional symmetry group.

EXAMPLE 6 (*Non-linear graviton again*). Let the pair $X \hookrightarrow Z$ be the same as in Example 1 and assume that the bundle $\det V_\pi^* \otimes \pi^* (\mathcal{O}(2))$ admits a nowhere vanishing holomorphic section. By Theorem 9, the holonomy algebra of the induced connection ∇ on the 4-dimensional parameter space M_0 is a subalgebra of $sl(2, \mathbb{C}) \subset so(4, \mathbb{C}) \simeq sl(2, C) + sl(2, \mathbb{C})$. This means that ∇ is the Levi-Civita connection of a complex Riemannian metric on M_0 which is Ricci-flat and has the anti-self-dual part of the Weyl tensor vanishing. According to Penrose (1976), any such connection arises locally in this way.

EXAMPLE 7 (*Obata connections*). Let $X = \mathbb{C}P^1$ be the projective line embedded into a $(2k+1)$-dimensional complex manifold Z with normal bundle $N \simeq \mathbb{C}^{2k} \otimes \mathcal{O}(1)$, $k \geq 2$. The Kodaira moduli space M is then a $4k$-dimensional complex manifold possessing a complexified almost quaternionic structure. If Z has the structure of a holomorphic fibration over X, then, by Theorems 4 and 9, there is an induced torsion-free affine connection ∇ on $M_0 \subseteq M$ with holonomy in $\mathrm{GL}(2k, \mathbb{C})$ which implies that ∇ is a complexified Obata connection. It is not difficult to show, using results of Bailey and Eastwood (1991), that any local complexified

Obata connection can be constructed in this way. A twistor characterization of real analytic Obata connections can be obtained from this holomorphic picture in the usual way (cf. Chapter 2, §2).

§ 3. Geometry of Legendre Moduli Spaces

1. Complex contact manifolds. Let Y be a complex $(2n+1)$-dimensional manifold. A complex contact structure on Y is a rank $2n$ holomorphic subbundle $D \subset TY$ of the holomorphic tangent bundle to Y such that the Frobenius form

$$\begin{array}{rcl} \Phi : D \times D & \longrightarrow & TY/D \\ (v, w) & \longrightarrow & [v, w] \bmod D \end{array}$$

is non-degenerate. Define the contact line bundle L by the exact sequence

$$0 \longrightarrow D \longrightarrow TY \xrightarrow{\theta} L \longrightarrow 0.$$

One can easily verify that maximal non-degeneracy of the distribution D is equivalent to the fact that the above defined "twisted" 1-form $\theta \in H^0(Y, L \otimes \Omega^1 M)$ satisfies the condition $\theta \wedge (d\theta)^n \neq 0$. An n-dimensional complex submanifold $X \hookrightarrow Y$ is called *Legendre* if $TX \subset D$.

Lemma 10 *The normal bundle $N_{X|Y}$ of a complex Legendre submanifold X of a complex contact manifold (Y, L) is isomorphic to $J^1 L_X$, where $L_X := L|_X$.*

Proof. It is well known that the total space of the fibration $\pi : L^* \setminus 0_{L^*} \to Y$ has a symplectic structure such that the inverse image of a Legendre submanifold $X \hookrightarrow Y$ is a Lagrange submanifold $\pi^{-1}(X) \hookrightarrow L^* \setminus 0_{L^*}$. Since $\pi^{-1}(X)$ is Lagrange, its normal bundle is isomorphic to its cotangent bundle. Then restricting the canonical isomorphism $J^1 L = \Omega^1 L^* / \mathbb{C}^*$ to $\pi^{-1}(X)$, one obtains the desired result. □

Therefore, $N_{X|Y}$ fits into the exact sequence

$$0 \longrightarrow \Omega^1 X \otimes L_X \longrightarrow N_{X|Y} \xrightarrow{pr} L_X \longrightarrow 0. \tag{6}$$

2. Existence theorem. Let Y be a complex contact manifold. An analytic family $F \hookrightarrow Y \times M$ of compact submanifolds of Y is called an *analytic family of compact Legendre submanifolds* if, for any point $t \in M$, the corresponding subset $X_t := \mu \circ \nu^{-1}(t) \hookrightarrow Y$ is a Legendre submanifold. The parameter space M is called a *Legendre moduli space.*

Let $F \hookrightarrow Y \times M$ be an analytic family of compact Legendre submanifolds. According to Kodaira, there is a natural linear map $k_t : T_t M \longrightarrow \mathrm{H}^0(X_t, N_{X_t|Y})$. We say that the family F is *complete at a point* $t \in M$ if the composition

$$s_t : T_t M \xrightarrow{k_t} \mathrm{H}^0(X_t, N_{X_t|Y}) \xrightarrow{pr} \mathrm{H}^0(X_t, L_{X_t})$$

is an isomorphism. One of the motivations behind this definition is the fact that an analytic family of compact Legendre submanifolds $\{X_t \hookrightarrow Y \mid t \in M\}$ which is complete at a point $t_0 \in M$ is also *maximal* at t_0 in the sense that, for any other analytic family of compact Legendre submanifolds $\{X_{\tilde{t}} \hookrightarrow Y \mid \tilde{t} \in \tilde{M}\}$ such that $X_{t_0} = X_{\tilde{t}_0}$ for a point $\tilde{t}_0 \in \tilde{M}$, there exists a neighbourhood $\tilde{U} \subset \tilde{M}$ of $\tilde{t}_0$ and a holomorphic map $f : \tilde{U} \longrightarrow M$ such that $f(\tilde{t}_0) = t_0$ and $X_{f(\tilde{t}')} = X_{\tilde{t}'}$ for each $\tilde{t}' \in \tilde{U}$. An analytic family $F \hookrightarrow Y \times M$ is called *complete* if it is complete at each point of the Legendre moduli space M. In this case M is also called complete.

Theorem 11 *Let X be a compact complex Legendre submanifold of a complex contact manifold (Y, L). If $\mathrm{H}^1(X, L_X) = 0$, then there exists a complete analytic family of compact Legendre submanifolds $F \hookrightarrow Y \times M$ containing X. This family is maximal and* $\dim M = h^0(X, L_X)$.

This theorem can be proved by working in local Darboux coordinates and expanding the defining functions of nearby compact Legendre submanifolds in terms of local coordinates on the moduli space M. This is very much in the spirit of the original proof of Kodaira's theorem; the only essential difference is that the infinite sequence of obstructions to agreements on overlaps of formal Legendre power series deformations of X lies now in $\mathrm{H}^1(X, L_X)$ rather than in $\mathrm{H}^1(X, N_{X|Y})$.

Let X be a complex manifold and L_X a line bundle on X. There is a natural "evaluation" map $H^0(X, L_X) \otimes \mathcal{O}_X \longrightarrow J^1 L_X$ whose dualization gives rise to the canonical map

$$L_X \otimes S^{k+1}(J^1 L_X)^* \longrightarrow L_X \otimes S^k(J^1 L_X)^* \otimes \left[\mathrm{H}^0(X, L_X)\right]^*$$

which in turn gives rise to the map of cohomology groups

$$\mathrm{H}^1\left(X, L_X \otimes S^{k+1}(J^1 L_X)^*\right) \xrightarrow{\phi} \mathrm{H}^1\left(X, L_X \otimes S^k(J^1 L_X)^*\right) \otimes \left[\mathrm{H}^0(X, L_X)\right]^*.$$

For future reference, we define the vector subspace

$$\tilde{\mathrm{H}}^1\left(X, L_X \otimes S^{k+1}(J^1 L_X)^*\right) := \ker \phi \subseteq \mathrm{H}^1\left(X, L_X \otimes S^{k+1}(J^1 L_X)^*\right).$$

3. Twistor theory of G-structures. Recall that any compact complex homogeneous-rational manifold X can be identified with the quotient space G/P for some complex semisimple Lie group G and a fixed parabolic subgroup $P \subset G$.

Theorem 12 *Let X be a compact complex homogeneous-rational manifold embedded as a Legendre submanifold into a complex contact manifold Y with contact line bundle L such that L_X is very ample on X. Then*

(i) There exists a complete analytic family $F \hookrightarrow Y \times M$ of compact Legendre submanifolds with moduli space M being an $h^0(X, L_X)$-dimensional complex manifold. For each $t \in M$, the associated Legendre submanifold X_t is isomorphic to X.

(ii) The Legendre moduli space M comes equipped with an induced irreducible G-structure, $\mathcal{G}_{ind} \to M$, with G isomorphic to the identity component of the group of all global biholomorphisms $\phi : L_X \to L_X$ which commute with the projection $\pi : L_X \to X$. The Lie algebra of G is isomorphic to $\mathrm{H}^0(X, L_X \otimes (J^1 L_X)^*)$.

(iii) If $\mathcal{G}_{ind}$ is k-flat, $k \geq 0$, then the obstruction for $\mathcal{G}_{ind}$ to be $(k+1)$-flat is given by a tensor field on M whose value at each $t \in M$ is represented by a cohomology class $\rho_t^{[k+1]} \in \tilde{\mathrm{H}}^1\left(X_t, L_{X_t} \otimes S^{k+2}(J^1 L_{X_t})^*\right)$.

(iv) If $\mathcal{G}_{ind}$ is 1-flat, then the bundle of all torsion-free connections in $\mathcal{G}_{ind}$ has as the typical fibre the affine space modelled on $\mathrm{H}^0\left(X, L_X \otimes S^2(J^1 L_X)^*\right)$.

A sketch of the proof. If X is a compact complex homogeneous-rational manifold and L_X a very ample line bundle on X, then $h^0(X, L_X) > 0$ and hence $\mathrm{H}^1(X, L_X) = 0$ by the Bott-Borel-Weil theorem and the fact that any holomorphic line bundle on X is homogeneous. Since $\mathrm{H}^1(X, TX) = 0$, X is rigid which implies that every submanifold X_t of the family is isomorphic to X.

In view of the canonical isomorphism $\mathcal{T}_t M \to \mathrm{H}^0(X_t, L_{X_t})$, the item (ii) is not a surprise. Using the fact that L_{X_t} is very ample on X_t, one can easily show that $F := \{(y,t) \in Y \times M \mid y \in X_t\}$ can be realized as a subbundle of the projectivized conormal bundle $P_M(\Omega^1 M)$. Fibrewise, this construction is just the well-known projective realization of a compact complex homogeneous-rational manifold X in $\mathbb{C}P^{m-1} \simeq P\left(\mathrm{H}^0(X, L_X)^*\right)$, identifying a point $x \in X$ with the hyperplane in $\mathrm{H}^0(X, L_X)$ consisting of all global sections of $L_X \to X$ which vanish at $x \in X$. The subgroup $G \subset \mathrm{GL}(m, \mathbb{C})$ which leaves $X \subset \mathbb{C}P^{m-1}$ invariant is exactly the one described in (ii).

To prove (iii), we have to explore the towers of infinitesimal neighbourhoods of two embeddings of analytic spaces, $X_t \hookrightarrow Y$ and $t \hookrightarrow M$. At the first floors of these towers, we have an isomorphism $\mathcal{T}_t M = \mathrm{H}^0(X_t, L_{X_t})$ which is in the basis of the above conclusion about the induced G-structure on M. The second floors of these two towers are related to each other as follows. If N_t denotes the normal bundle of $X_t \hookrightarrow Y$, then, by (6), there is a natural epimorphism $\mathcal{O}_X^{(1)}(N_t^*) \longrightarrow TX_t \otimes L_{X_t}^*$, whose kernel, denoted by $\mathcal{L}_{X_t}^{(1)}(N_t^*)$, fits into an exact sequence of $\mathcal{O}_{X_t}^{(1)}$-modules

$$0 \longrightarrow S^2(N_t^*) \longrightarrow \mathcal{L}_{X_t}^{(1)}(N_t^*) \longrightarrow L_{X_t}^* \longrightarrow 0.$$

Next, consider the restriction of the dual of the contact line bundle L to the first order infinitesimal neighbourhood of X_t in Y, $L_{X_t}^{*(1)} \equiv L^*|_{X_t^{(1)}}$, which, as an $\mathcal{O}_{X_t}^{(1)}$-module, has the following extension structure

$$0 \longrightarrow N_t^* \otimes L_{X_t}^* \xrightarrow{i_1} L_{X_t}^{*(1)} \longrightarrow L_{X_t}^* \longrightarrow 0.$$

Let $\mathbb{I}$ be the canonical global section of $N \otimes N^* \simeq \mathrm{End}(N)$ whose value at each point of X is the multiplicative identity in the corresponding stalk of endomorphism groups. Then the global section $\tilde{\mathbb{I}} \equiv pr \otimes id\,(\mathbb{I}) \in \mathrm{H}^0(X_t, L_{X_t} \otimes N_t^*)$, where pr is the projection $N_t \longrightarrow L_{X_t}$, defines a monomorphism of $\mathcal{O}_{X_t}^{(1)}$-modules

$$\begin{array}{rcccl} i_2: & N_t^* \otimes L_{X_t}^* & \longrightarrow & (N_t^* \otimes L_{X_t}^*) \otimes (L_{X_t} \otimes N_t^*) & \cong \; N_t^* \otimes N_t^* \\ & f & \longrightarrow & f \otimes \tilde{\mathbb{I}}. & \end{array}$$

Consider the composition

$$i_3 : N_t^* \otimes L_{X_t}^* \xrightarrow{j} N_t^* \otimes L_{X_t}^* \oplus N_t^* \otimes L_{X_t}^* \xrightarrow{i_1 \oplus i_2} L_{X_t}^{*(1)} \oplus N_t^* \otimes N_t^*,$$

where

$$\begin{array}{rcl} j : N_t^* \otimes L_{X_t}^* & \longrightarrow & N_t^* \otimes L_{X_t}^* \oplus N_t^* \otimes L_{X_t}^* \\ f & \longrightarrow & f \oplus (-f). \end{array}$$

The quotient sheaf of $\mathcal{O}_{X_t}^{(1)}$-modules,

$$\tilde{L}_{X_t}^{*(1)} := \left(L_{X_t}^{*(1)} \oplus N_t^* \otimes N_t^*\right) / i_3\left(N_t^* \otimes L_{X_t}^*\right)$$

fits into the commutative diagram

$$\begin{array}{ccccccccc} & & 0 & & 0 & & & & \\ & & \downarrow & & \downarrow & & & & \\ 0 & \longrightarrow & N_t^* \otimes L_{X_t}^* & \xrightarrow{i_1} & L_{X_t}^{*(1)} & \longrightarrow & L_{X_t}^* & \longrightarrow & 0 \\ & & i_2\downarrow & & \downarrow & & \| & & \\ 0 & \longrightarrow & N_t^* \otimes N_t^* & \xrightarrow{i_4} & \tilde{L}_{X_t}^{*(1)} & \longrightarrow & L_{X_t}^* & \longrightarrow & 0. \end{array}$$

Then the sheaf of $\mathcal{O}_{X_t}^{(1)}$-modules, $\mathcal{L}_X^{(1)}(L^*) \equiv \tilde{L}_X^{*(1)}/i_4\left(\wedge^2 N^*\right)$, fits into the exact sequence of $\mathcal{O}_{X_t}^{(1)}$-modules

$$0 \longrightarrow S^2(N_t^*) \longrightarrow \mathcal{L}_X^{(1)}(L^*) \longrightarrow L_X^* \longrightarrow 0.$$

The upshot of all these constructions is that the $\mathcal{O}_{X_t}^{(1)}$-module

$$\left[\mathcal{L}_{X_t}^{(1)}(N_t^*) - \mathcal{L}_{X_t}^{(1)}(L^*)\right] \in \mathrm{Ext}^1_{\mathcal{O}_{X_t}^{(1)}}\left(L_{X_t}^*, S^2(N_t^*)\right)$$

is actually a locally free $\mathcal{O}_X$-module whose dual, denoted by $\Delta_{X_t}^{[2]}$, fits into the exact sequence of locally free sheaves

$$0 \longrightarrow L_{X_t} \xrightarrow{\alpha} \Delta_{X_t}^{[2]} \longrightarrow S^2(J^1 L_{X_t}) \longrightarrow 0 \tag{7}$$

and, as is not hard to show, into a commutative diagram of vector spaces

$$\begin{array}{ccccccccc} 0 & \longrightarrow & T_t M & \longrightarrow & T_t^{[2]} M & \longrightarrow & S^2(T_t M) & \longrightarrow & 0 \\ & & \downarrow & & \downarrow & & \downarrow & & \\ 0 & \longrightarrow & H^0(X_t, L_{X_t}) & \longrightarrow & H^0\left(X_t, \Delta_{X_t}^{[2]}\right) & \longrightarrow & H^0\left(X_t, S^2(J^1 L_{X_t})\right) & \longrightarrow & 0 \end{array} \tag{8}$$

which extends the canonical isomorphism $\mathcal{T}_t M \to H^0(X_t, L_{X_t})$ to *second* order infinitesimal neighbourhoods of $t \hookrightarrow M$ and $X_t \hookrightarrow Y$.

Let $\rho_t^{[1]} \in \mathrm{H}^1\left(X_t, L_X \otimes S^2(J^1 L_{X_t})^*\right)$ be the obstruction for a global splitting of (7). If $\rho_t^{[1]} = 0$, then any splitting of (7), i.e. a morphism $\beta : \Delta_{X_t}^{[2]} \to L_{X_t}$ such that $\beta \circ \alpha = id$, induces via the commutative diagram (8) an associated splitting of the exact sequence for $T_t^{[2]} M$ which is equivalent to a torsion-free affine connection at $t \in M$. Thus vanishing of $\rho_t^{[1]}$ implies that the induced G-structure is 1-flat at $t \in M$. Conversely, given a Legendre moduli space M with the induced G-structure $\mathcal{G}_{ind}$ as in Theorem 12(iv), one can use again the commutative diagram (8) to show that that a torsion-free connection on $\mathcal{G}_{ind}$ at t induces canonically a splitting of the exact sequence (7), which implies $\rho_t^{[1]} = 0$. This proves the statement (iii) for $k = 1$. For $k \geq 2$ this statement follows from Theorem 2.

Finally, if $\mathcal{G}_{ind}$ is 1-flat at $t \in M$, then extension (7) splits, and the set of all its splittings, which is the same as the set of all torsion-free affine connections at t, is an affine space modelled on $\mathrm{H}^0\left(X_t, L_{X_t} \otimes S^2(J^1 L_{X_t})^*\right) \simeq \mathrm{H}^0\left(X, L_X \otimes S^2(J^1 L_X)^*\right)$.

□

Remark. Theorem 12 is actually valid for a larger class of compact complex manifolds X than the class of compact complex homogeneous-rational manifolds — the only vital assumptions are that X is rigid and the cohomology groups $\mathrm{H}^1(X, \mathcal{O}_X)$ and $\mathrm{H}^1(X, L_X)$ vanish.

The geometric meaning of cohomology classes $\rho_t^{[k]} \in \tilde{\mathrm{H}}^1(X_t, L_{X_t} \otimes S^{k+1}(J^1 L_{X_t})^*)$ of Theorem 12(iii) is very simple — they compare the germ of the Legendre embedding $X_t \hookrightarrow Y$ with the "flatmodel, $X_t \hookrightarrow J^1 L_{X_t}$, up to $(k+1)$th order, where the ambient contact manifold is just the total space of the vector bundle $J^1 L_{X_t}$, together with its canonical contact structure, and the Legendre submanifold X_t is realised in $J^1 L_{X_t}$ as its zero section. Therefore, the cohomology class $\rho_t^{[k]}$ can be called the kth Legendre jet of X_t in Y. Then it is natural to call a Legendre submanifold $X_t \hookrightarrow Y$ *k-flat* if $\rho_t^{[k]} = 0$. With this terminology, the item (iii) of Theorem 12 acquires a rather symmetric form: *the induced G-structure on the moduli space M of a complete analytic family of compact Legendre submanifolds is k-flat if and only if the family consists of k-flat Legendre submanifolds.*

The next result shows that in the category of irreducible 1-flat G-structures the construction of Theorem 12 is almost universal.

Theorem 13 *(i) Let $H \subset \mathrm{GL}(k, \mathbb{C})$ be one of the following subgroups:*
(a) $\mathrm{Spin}(2n+1, \mathbb{C})$ *when $k = 2^n \geq 8$; (b)* $\mathrm{Sp}(2n+2, \mathbb{C})$ *when $k = 2n+2 \geq 4$; (c) G_2 when $k = 7$. Suppose that $G \subset \mathrm{GL}(m, \mathbb{C})$ is a connected semisimple Lie subgroup whose decomposition into a locally direct product of simple groups contains H. If $\mathcal{G}$ is any irreducible 1-flat $G \cdot \mathbb{C}^*$-structure on an m-dimensional manifold M, then there exists a complex contact manifold (Y, L) and a Legendre*

submanifold $X \hookrightarrow Y$ with $X = G/P$ for some parabolic subgroup $P \subset G$ such that L_X is very ample and, at least locally, M is canonically isomorphic to the associated Legendre moduli space and $\mathcal{G} \subset \mathcal{G}_{ind}$. In particular, when $G = H$ one has in the case (a) $X = \mathrm{SO}(2n+2,\mathbb{C})/\mathrm{U}(n+1)$ and $\mathcal{G}_{ind}$ is a $\mathbb{C}^*\cdot\mathrm{Spin}(2n+2,\mathbb{C})$-structure; in the case (b) $X = \mathbb{C}P^{2n+1}$ and $\mathcal{G}_{ind}$ is a $\mathrm{GL}(2n+2,\mathbb{C})$-structure; and in the case (c) $X = Q_5$ and $\mathcal{G}_{ind}$ is a $\mathrm{CO}(7,\mathbb{C})$-structure.

(ii) Let $G \subset GL(m,\mathbb{C})$ be an arbitrary connected semisimple Lie subgroup whose decomposition into a locally direct product of simple groups does not contain any of the groups H considered in (i). If $\mathcal{G}$ is any irreducible 1-flat $G\cdot\mathbb{C}^*$-structure on an m-dimensional manifold M, then there exists a complex contact manifold (Y,L) and a Legendre submanifold $X \hookrightarrow Y$ with $X = G/P$ for some parabolic subgroup $P \subset G$ such that L_X is very ample and, at least locally, M is canonically isomorphic to the associated Legendre moduli space and $\mathcal{G} = \mathcal{G}_{ind}$.

A sketch of the proof. Consider a complex m-dimensional manifold M and an irreducible G- or $G\cdot\mathbb{C}^*$-structure $\mathcal{G} \to M$, where $G \subset \mathrm{GL}(m,\mathbb{C})$ is a semisimple Lie subgroup (any irreducible H-structure with reductive H must be one of these). Since $\mathcal{G}$ is irreducible, there is a naturally associated subbundle $\tilde{F} \subset \Omega^1 M \setminus 0_{\Omega^1 M}$ whose typical fibre is the cone in $\mathbb{C}^m \setminus 0$ defined as the G-orbit of the line spanned by a highest weight vector. The quotient bundle $F := \tilde{F}/\mathbb{C}^* \xrightarrow{\nu} M$ is then a subbundle of the projectivized cotangent bundle $P(\Omega^1 M)$ whose fibres $\nu^{-1}(t)$ are isomorphic to the compact complex homogeneous-rational manifold G/P, where P is the parabolic subgroup of G which preserves the highest weight vector in $\mathbb{C}^m \setminus 0$ up to a scalar. Let $\dim G/P = n$.

The total space of the cotangent bundle $\Omega^1 M \setminus 0_{\Omega^1 M}$ has a canonical holomorphic symplectic 2-form ω. Then the sheaf of holomorphic functions on $\Omega^1 M$ is a sheaf of Lie algebras relative to the Poisson bracket $\{f,g\} = \omega^{-1}(df,dg)$. The pullback, $i^*\omega$, of the symplectic form ω from $\Omega^1 M \setminus 0_{\Omega^1 M}$ to its submanifold $i : \tilde{\mathcal{F}} \longrightarrow \Omega^1 M \setminus 0_{\Omega^1 M}$ defines a distribution $\tilde{D} \subset T\tilde{F}$ as the kernel of the natural "lowering of indicesmap $T\tilde{F} \xrightarrow{\lrcorner\omega} \Omega^1\tilde{F}$, i.e. $\tilde{D}_e = \left\{V \in T_e\tilde{F} : V \lrcorner i^*\omega = 0\right\}$ at each point $e \in \tilde{F}$. Using the fact that $d(i^*\omega) = i^*d\omega = 0$, one can show that this distribution is integrable and thus defines a foliation of $\tilde{\mathcal{F}}$ by holomorphic leaves. We shall assume that the space of leaves, $\tilde{Y}$, is a complex manifold. This assumption imposes no restrictions on the local structure of M. Since

$$L_V(i^*\omega) = V \lrcorner i * d\omega + d(V \lrcorner i^*\omega) = 0$$

for any vector field V tangent to the leaves, $i^*\omega$ is the pullback, $\tilde{\mu}^*(\tilde{\omega})$, of some closed 2-form $\tilde{\omega}$ on $\tilde{Y}$, where $\tilde{\mu}$ is the projection $\tilde{F} \to \tilde{Y}$. It is easy to check that $\tilde{\omega}$ is non-degenerate which means that the quotient manifold $(\tilde{Y},\tilde{\omega})$ is symplectic.

Let e be a point in $\tilde{F} \subset \Omega^1 M \setminus 0_{\Omega^1 M}$. Restricting the "lowering of indicesmap $T_e(\Omega^1 M) \xrightarrow{\lrcorner\omega} \Omega^1_e(\Omega^1 M)$ to the subspace $\tilde{D}_e$, one obtains an injective map

$$0 \longrightarrow \tilde{D}_e \xrightarrow{\lrcorner\omega} \tilde{N}^*_e,$$

where $\tilde{N}_e^*$ denotes the fibre of the conormal bundle of $\tilde{F} \hookrightarrow \Omega^1 M \setminus 0_{\Omega^1 M}$. Therefore, the rank of the distribution $\tilde{D}$ is equal at most to rank $\tilde{N}^* = m - n - 1$. It is easy to show that rank $\tilde{D}$ assumes its maximal value if and only if $\tilde{F}$ is a coisotropic submanifold in $(\Omega^1 M \setminus 0_{\Omega^1 M}, \omega)$ in which case the G-structure $\mathcal{G}$ is said to be *involutive* (note that $\tilde{F}$ is coisotropic if and only if the ideal sheaf of $\tilde{F} \hookrightarrow \Omega^1 M \setminus 0_{\Omega^1 M}$ is the subsheaf of Lie subalgebras of $(\mathcal{O}_{\Omega^1 M}, \{\ ,\ \})$). It is easy to check that a locally flat G-structure is involutive. Since $\{\ ,\ \}$ is a first order differential operation, this immediately implies that any irreducible 1-flat G-structure is also involutive.

Our next task is to show that if G is a reductive Lie group, then every complex m-manifold M with an involutive G-structure is canonically isomorphic, at least locally, to a Legendre moduli space. Indeed, there is an integrable distribution $\tilde{D}$ of rank $m-n-1$ on the bundle $\tilde{\mathcal{F}} \to M$. Assuming that M is sufficiently "small", the quotient $(\tilde{Y}, \tilde{\omega})$ is a symplectic manifold with $\dim \tilde{Y} = (m+n+1) - (m-n-1) = 2n+2$. There is a double fibration

$$\tilde{Y} \stackrel{\tilde{\mu}}{\longleftarrow} \tilde{F} \stackrel{\tilde{\nu}}{\longrightarrow} M$$

with fibres of $\tilde{\mu}$ being leaves of the integrable distribution $\tilde{D}$ and fibres of $\tilde{\nu}$ being $(n+1)$-dimensional cones $\tilde{\nu}^{-1}(t) \subset \Omega_t^1 M \setminus 0$ identified with G-orbits of highest weight vectors. It is clear that for each $t \in M$, the submanifold $\tilde{X}_t := \tilde{\mu} \circ \tilde{\nu}^{-1}(t) \subset \tilde{Y}$ is a Lagrange submanifold relative to the induced symplectic form $\tilde{\omega}$.

There is a natural action of $\mathbf{C}^*$ on $\tilde{F}$ which leaves $\tilde{D}$ invariant and thus induces the action of $\mathbf{C}^*$ on $\tilde{Y}$. The quotient $Y = \tilde{Y}/\mathbf{C}^*$ is a $(2n+1)$-dimensional complex manifold which has a double fibration structure

$$Y \stackrel{\mu}{\longleftarrow} F = \tilde{F}/\mathbf{C}^* \stackrel{\nu}{\longrightarrow} M$$

and thus contains a family of compact n-dimensional embedded manifolds, $\{X_t := \mu \circ \nu^{-1}(t) \hookrightarrow Y \mid t \in M\}$, with $X_t = \tilde{X}_t/\mathbf{C}^*$. Inverting a well-known procedure of symplectivization of a contact manifold, it is not hard to show that Y has a complex contact structure such that the family $\{X_t \hookrightarrow Y \mid t \in M\}$ is a family of compact Legendre submanifolds. The contact line bundle L on Y is just the quotient $L = \tilde{F} \times \mathbf{C}/\mathbf{C}^*$ relative to the natural multiplication map $\tilde{F} \times \mathbf{C} \longrightarrow \tilde{F} \times \mathbf{C}$, $(p, c) \to (\lambda p, \lambda c)$, where $\lambda \in \mathbf{C}^*$. Then $L|_{X_t}$ is isomorphic to the restriction of the hyperplane section bundle on $P(\Omega_t^1 M)$ to its submanifold $\nu^{-1}(t) \simeq X_t$ and hence is very ample on X_t. Therefore, $h^0\left(X_t, L|_{X_t}\right) = m$ which implies that the Legendre family $\{X_t \hookrightarrow Y \mid t \in M\}$ is complete.

Therefore we have proven that if $G \subset \mathrm{GL}(m, \mathbf{C})$ is a semisimple Lie subgroup and $\mathcal{G}$ any irreducible involutive (in particular 1-flat) $G \cdot \mathbf{C}^*$-structure on an m-dimensional manifold M, then there exists a complex contact manifold (Y, L) and a Legendre submanifold $X \hookrightarrow Y$ with $X = G/P$ for some parabolic subgroup $P \subset G$ such that L_X is very ample and, at least locally, M is canonically isomorphic to the associated Legendre moduli space. The rest of Theorem 13 follows immediately from the result of Onishchik (1962) mentioned in §1, subsect. 3. □

EXAMPLE 8 (*Generic G-structure*). The twistor theory of a torsion-free $GL(m, \mathbb{C})$- or $SL(m, \mathbb{C})$-structure on a complex m-dimensional manifold M is the simplest possible: the associated Legendre deformation problem "$X \hookrightarrow Y$ consists of the projectivized cotangent bundle $Y = P(\Omega^1 M)$ with its natural contact structure and the Legendre submanifold $X = \mathbb{C}P^{m-1}$ which is just a fibre of the projection $P(\Omega^1 M) \to M$. The corresponding complete family $\{X_t \hookrightarrow Y \mid t \in M\}$ is the set of all fibres of this fibration. Since $L_X = \mathcal{O}(1)$ and $J^1 L_X = \mathbb{C}^m \otimes \mathcal{O}_X$, we have $H^1(X, L_X \otimes S^{k+2}(J^1 L_X)^*) = 0$ for all $k \geq 0$, while $H^0(X, L_X \otimes S^2(J^1 L_X)^*) \neq 0$ (the latter group is isomorphic to the typical fibre of the vector bundle $TM \otimes S^2(\Omega^1 M)$). Then Theorem 12(iii) implies that all $GL(m, \mathbb{C})$- or $SL(m, \mathbb{C})$-structures on an m-dimensional manifold are locally flat.

EXAMPLE 9 (*Conformal geometry*). Let $X \hookrightarrow Y$ be a pair consisting of an n-quadric Q_n embedded into a $(2n+1)$-dimensional contact manifold (Y, L) with $L|_X \simeq i^* \mathcal{O}_{\mathbb{C}P^{n+1}}(1)$, $i : Q_n \to \mathbb{C}P^{n+1}$ being a standard projective realization of Q_n. Using the extension

$$0 \longrightarrow L_X^* \longrightarrow H^0(X, L_X) \otimes_{\mathbb{C}} \mathcal{O}_X \longrightarrow J^1 L_X \longrightarrow 0,$$

it is easy to check that in this case $H^0(X, L_X \otimes (J^1 L_X)^*)$ is precisely the conformal algebra, implying that the associated $(n+2)$-dimensional Legendre moduli space M comes equipped canonically with a conformal structure. Since $H^1(X, L_X \otimes S^2(J^1 L_X)^*) = 0$, the induced conformal structure must be torsion-free, in agreement with the classical result of differential geometry. Simple calculations show that the vector space $H^1(X, L_X \otimes S^3(J^1 L_X)^*)$ is exactly the subspace of $TM \otimes \Omega^1 M \otimes \Omega^2 M$ consisting of tensors with Weyl curvature symmetries. Thus, Theorem 12(iii) implies the well-known Schouten conformal flatness criterion. Since $H^0(X, L_X \otimes S^2(J^1 L_X)^*)$ is isomorphic to the typical fibre of $\Omega^1 M$, the set of all torsion-free affine connections preserving the induced conformal structure is, by Theorem 12(iv), an affine space modelled on $H^0(M, \Omega^1 M)$, again in agreement with the classical result.

If $Q_2 \hookrightarrow Y_0$ is a pair consisting of the 2-quadric $Q_2 = \mathbb{C}P^1 \times \mathbb{C}P^1$ standardly embedded as a Legendre submanifold into $Y_0 = F(1, 3; \mathbb{C}^4)$ with its natural contact structure, then the associated Legendre moduli space M, which in this case is the same as the Kodaira moduli space, is exactly the Grassmanian $G(2; \mathbb{C}^4)$ with its canonical conformal structure (see Chapter 2, §7). Let J be the ideal of the natural embedding $F(1, 3; \mathbb{C}^4) \hookrightarrow \mathbb{C}P^3 \times \mathbb{C}P^3$ and let $Y_0^{(m)} := (Y_0, \mathcal{O}_{\mathbb{C}P^3 \times \mathbb{C}P^3}/J^{m+1})$ be the associated tower of infinitesimal neighbourhoods. Analyzing infinitesimal contact deformations of Y_0, Baston and Mason (1987) provided a strong evidence for the conjecture that if Y is the contact 5-manifold associated with a curved conformal space-time $(M, [g])$, then curved analogues of $Y_0^{(4)}$ should always exist, while the obstruction for the existence of an analogue of $Y_0^{(5)}$ should be the Bach tensor of $[g]$; moreover, for a generic conformal structure $[g]$ the existence of an analogue of $Y_0^{(6)}$ should imply that the conformal class $[g]$ contains a metric satisfying the Einstein equations. Later this conjecture has been proved by LeBrun (1991).

EXAMPLE 10 (*Exotic G_3-structures*). Let $X \hookrightarrow Y$ be a pair consisting of a rational Legendre curve $X = \mathbb{C}P^1$ embedded into a complex contact 3-fold (Y, L) as a Legendre submanifold such that $L_X = \mathcal{O}(3)$. Since $J^1\mathcal{O}(3) = \mathbb{C}^2 \otimes \mathcal{O}(2)$, an easy calculation shows that $\mathrm{H}^1\left(X, L_X \otimes S^2(J^1L_X)^*\right) = 0$, while $\mathrm{H}^0(X, L_X \otimes (J^1L_X)^*)$ represented on $\mathrm{H}^0(X, L_X)$ is exactly $gl(2, \mathbb{C})$ represented on $S^3(\mathbb{C}^2)$. Therefore, by Theorem 12, the induced G-structure on the associated 4-dimensional Legendre moduli space M is precisely a torsion-free exotic G_3-structure which has been studied by Bryant (1992) in his search for irreducibly acting holonomy groups of torsion-free affine connections which are missing in the list of Berger (1955). Since $H^0\left(X, L_X \otimes S^2(J^1L_X)^*\right) = 0$, $\mathcal{G}_{ind}$ admits a unique torsion-free affine connection ∇. The cohomology class $\rho_t^{[2]} \in H^1\left(X, L_X \otimes S^3(J^1L_X)^*\right) \simeq \mathbb{C}^2 \otimes S^3(\mathbb{C}^2)$ from Theorem 12(iii) is exactly the curvature tensor of ∇.

EXAMPLE 11 (*Grassmanian structure*). The pair (X, L_X) associated by Theorem 13 to a Grassmanian $\mathrm{GL}(p+1, \mathbb{C}) \cdot \mathrm{GL}(q+1, \mathbb{C})$-structure on a complex manifold M is $(\mathbb{C}P^p \times \mathbb{C}P^q, \mathcal{O}(1,1))$, where

$$\mathcal{O}(1,1) := \pi_l^*\left(\mathcal{O}_{\mathbb{C}P^p}(1)\right) \otimes \pi_r^*\left(\mathcal{O}_{\mathbb{C}P^q}(1)\right)$$

and $\pi_l : \mathbb{C}P^p \times \mathbb{C}P^q \to \mathbb{C}P^p$ and $\pi_r : \mathbb{C}P^p \times \mathbb{C}P^q \to \mathbb{C}P^q$ are natural projections. One may use the extension

$$0 \longrightarrow J^1\mathcal{O}(1,1) \longrightarrow \mathbb{C}^{p+1} \otimes \mathcal{O}(0,1) \oplus \mathbb{C}^{q+1} \otimes \mathcal{O}(1,0) \longrightarrow \mathcal{O}(1,1) \longrightarrow 0$$

to compute the cohomology groups

$$\tilde{\mathrm{H}}^1\left(X, \mathcal{O}(1,1) \otimes S^{k+2}\left(J^1\mathcal{O}(1,1)\right)^*\right). \tag{9}$$

- For $p > 1, q > 1$, all the cohomology groups (9) vanish. By Theorem 12(iii), this means that torsion-free Grassmanian spinor structures for $p > 1, q > 1$ are locally flat (cf. Chapter 1, §7).
- For $p = 1, q > 1$, all the cohomology groups (9) vanish except for $k = 1$ in which case it is isomorphic to the subspace of $\tilde{S} \otimes S^3(\tilde{S}^*)$ consisting of trace-free tensors, where $\tilde{S} = \mathrm{H}^0(X, \mathcal{O}(0,1))$. Therefore, the induced $\mathrm{GL}(2, \mathbb{C}) \cdot \mathrm{GL}(q+1, \mathbb{C})$-structure $\mathcal{G}_{ind}$ on the Legendre moduli space associated to $X \hookrightarrow Y$ is automatically torsion-free, and its obstruction to local flatness is given by a section of $\mathcal{G}_{ind} \times_{GL(q+1,\mathbb{C})} \tilde{S} \otimes_0 S^3(\tilde{S}^*)$. Since $\mathrm{H}^0(X, L_X \otimes S^2(N^*)) = S^* \otimes \tilde{S}^*$, Theorem 12(iv) implies that the typical fibre of the bundle of quaternionic torsion-free affine connections is an affine space modelled on $S^* \otimes \tilde{S}^*$.

4. Twistor theory of torsion-free affine connections. Let $\{X_t \hookrightarrow Y \mid t \in M\}$ be a complete family of compact Legendre submanifolds. A torsion-free connection on M which arises at each $t \in M$ from a splitting of the extension (7) is called an *induced* connection. The arguments used in the proofs of Theorems 12 and 13 imply also the following

Theorem 14 *Let ∇ be a holomorphic torsion-free affine connection on a complex manifold M with irreducibly acting reductive holonomy group G. Then there exists a complex contact manifold (Y, L) and a Legendre submanifold $X \hookrightarrow Y$ with $X = G_s/P$ for some parabolic subgroup P of the semisimple part G_s of G such that L_X is very ample and, at least locally, M is canonically isomorphic to the associated Legendre moduli space and ∇ coincides with an induced torsion-free affine connection in $\mathcal{G}_{ind}$.*

Which irreducible representations of a complex reductive Lie group G can occur as the holonomy of a torsion-free affine connection? Theorems 12 and 14 offer the following necessary condition: *if (X, L_X) is the pair associated to an irreducible representation of G* (as in Chapter A, §1, subsect.4), *then at least one of the following two inequalities must be satisfied*

(C_1) $\mathrm{H}^0\left(X, L_X \otimes S^2(J^1 L_X)^*\right) \neq 0$

(C_2) $\tilde{\mathrm{H}}^1\left(X, L_X \otimes S^3(J^1 L_X)^*\right) \neq 0.$

If we are interested only in non locally symmetric connections, the condition (C_2) must be replaced by a stronger one

(C_2') $\dim \tilde{\mathrm{H}}^1\left(X, L_X \otimes S^3(J^1 L_X)^*\right) > 1.$

EXAMPLE 12 (*All torsion-free holonomy representations of* $\mathrm{GL}(2, \mathbb{C})$ *and* $\mathrm{SL}(2, \mathbb{C})$). The only compact complex homogeneous-rational manifold associated to irreducible representations of $\mathrm{SL}(2, \mathbb{C})$ or $\mathrm{GL}(2, \mathbb{C})$ is $\mathbb{C}P^1$. Any ample line bundle L_X on $X = \mathbb{C}P^1$ is of the form $\mathcal{O}(k)$ for $k \geq 1$. Since $J^1(\mathcal{O}(k)) = \mathbb{C}^2 \otimes \mathcal{O}(k-1)$ for $k \geq 1$, it is not hard to show that the condition (C_1) holds only for $k = 1, 2$, while condition (C_2) is satisfied only for $k = 3, 4$. In the case $k = 4$, however, the condition (C_2') is violated which means that if there exists a torsion-free affine connection with holonomy H_4, the $k = 4$ representation of $\mathrm{SL}(2, \mathbb{C})$, it must be locally symmetric (using the isomorphism $K(\mathfrak{g}) = \tilde{\mathrm{H}}^1\left(X, L_X \otimes S^3(J^1 L_X)^*\right)$ and the Ambrose-Singer theorem, it is not hard to show that there are no torsion-free affine connections with holonomy G_4, the $k = 4$ representation of $\mathrm{GL}(2, \mathbb{C})$). Thus the cases $k = 1, 2, 3$ exhaust all possible non locally symmetric holonomy representations of $\mathrm{SL}(2, \mathbb{C})$ and $\mathrm{GL}(2, \mathbb{C})$. The case $k = 1$ corresponds to the "generictorsion-free affine connections in dimensions 2, the case $k = 2$ corresponds to 3-dimensional Riemannian and Weyl affine connections, while the existence of torsion-free affine connections in dimensions 4 with holonomy representations of $\mathrm{SL}(2, \mathbb{C})$ and $\mathrm{GL}(2, \mathbb{C})$ corresponding to $k = 3$ has been proved by Bryant (1992) (see also the work of Schwachhöfer (1994)).

EXAMPLE 13 (*All torsion-free holonomy representations of* $\mathrm{SL}(2, \mathbb{C})\mathrm{SL}(2, \mathbb{C})$ *and* $\mathrm{GL}(2, \mathbb{C})\mathrm{SL}(2, \mathbb{C})$). There is only one compact complex homogeneous-rational manifold associated to irreducible representations of these groups — this is $X = \mathbb{C}P^1 \times \mathbb{C}P^1$. Any ample line bundle L_X on X is of the form $\mathcal{O}(m,n)$ for $m,n \geq 1$. Standard cohomology computations (cf. Chapter 1, §1) imply that condition (C_1) holds only for $m = n = 1$, while (C_2) holds only for the pairs $(m = 1, n = 1)$, $(m = 1, n = 2)$, $(m = 1, n = 3)$ and $(m = 2, n = 2)$. However the last two pairs violate (C_2') which means that the associated holonomy representations can be only locally symmetric. Next, using the isomorphism $K(\mathfrak{g}) = \tilde{\mathrm{H}}^1\left(X, L_X \otimes S^3(J^1 L_X)^*\right)$ and the Ambrose-Singer theorem, it is not hard to show that only $\mathrm{SL}(2,\mathbb{C})\mathrm{SL}(2,\mathbb{C})$ may have the holonomy representation $(m = 1, n = 2)$. The case $(m = 1, n = 1)$ corresponds to 4-dimensional Riemannian and Weyl connections, while the existence of torsion-free affine connections with exotic holonomy $(m = 1, n = 2)$ has been proved by Chi and Schwachhöfer (1994).

5. Interconnections between Kodaira and Legendre moduli spaces. If $X \hookrightarrow Y$ is a complex submanifold, there is an exact sequence of vector bundles

$$0 \longrightarrow N^*_{X|Y} \longrightarrow \Omega^1 Y\big|_X \longrightarrow \Omega^1 X \longrightarrow 0,$$

which induces a natural embedding, $P(N^*_{X|Y}) \hookrightarrow P(\Omega^1 Y)$, of the total spaces of the associated projectivized bundles. The manifold $\hat{Y} = P(\Omega^1 Y)$ carries a natural contact structure such that the embedding $\hat{X} = P(N^*_{X|Y}) \hookrightarrow \hat{Y}$ is Legendre. Indeed, the contact distribution $D \subset T\hat{Y}$ at each point $\hat{y} \in \hat{Y}$ consists of those tangent vectors $V_{\hat{y}} \in T_{\hat{y}}\hat{Y}$ which satisfy the equation $\langle \hat{y}, \tau_*(V_{\hat{y}}) \rangle = 0$, where $\tau : \hat{Y} \longrightarrow Y$ is a natural projection and the angular brackets denote the pairing of 1-forms and vectors at $\tau(\hat{y}) \in Y$. Since the submanifold $\hat{X} \subset \hat{Y}$ consists precisely of those projective classes of 1-forms in $\Omega^1 Y\big|_X$ which vanish when restricted to TX, we conclude that $T\hat{X} \subset D|_{\hat{X}}$. One may check that the association

$$\begin{array}{ccc} \text{Kodaira moduli space} & \longrightarrow & \text{Legendre moduli space} \\ \{X_t \hookrightarrow Y \mid t \in M\} & \longrightarrow & \left\{\hat{X}_t := P(N^*_{X_t|Y}) \hookrightarrow \hat{Y} := P(\Omega^1 Y) \;\mid t \in M\right\} \end{array}$$

preserves completeness while changing its meaning, i.e. a *complete* Kodaira family of compact complex submanifolds is mapped into a *complete* family of compact complex Legendre submanifolds (which is usually not complete in the Kodaira sense).

The contact line bundle L on $\hat{Y}$ is just the dual of the tautological line bundle $\mathcal{O}_{\hat{Y}}(-1)$. Simplifying notations, $N := N_{X|Y}$ and $\hat{N} := N_{\hat{X}|\hat{Y}}$, we obtain a commutative diagram which explains how $\hat{N}$ is related to $\rho^*(N)$ and L

$$\begin{array}{ccccccccc}
 & & 0 & & 0 & & & & \\
 & & \downarrow & & \downarrow & & & & \\
 & & \rho^*(\Omega^1 X)\otimes L_{\hat{X}} & = & \rho^*(\Omega^1 X)\otimes L_{\hat{X}} & & & & \\
 & & \downarrow & & \downarrow & & & & \\
0 & \longrightarrow & \Omega^1\hat{X}\otimes L_{\hat{X}} & \longrightarrow & \hat{N} & \longrightarrow & L_{\hat{X}} & \longrightarrow & 0 \\
 & & \downarrow & & \downarrow & & \| & & \\
0 & \longrightarrow & \Omega^1_\rho\otimes L_{\hat{X}} & \longrightarrow & \rho^*(N) & \longrightarrow & L_{\hat{X}} & \longrightarrow & 0 \\
 & & \downarrow & & \downarrow & & & & \\
 & & 0 & & 0 & & & &
\end{array}$$

Here, $L_{\hat{X}} = L|_{\hat{X}}$, ρ is the natural projection $\hat{X} \to X$, and Ω^1_ρ is the bundle of ρ-vertical 1-forms, i.e. the dual of $\mathcal{T}_\rho = \ker : T\hat{X} \to TX$. Using this diagram it is not hard to show that there are exact sequences of vector spaces

$$0\longrightarrow \mathrm{H}^0(X, N\otimes N^*)\longrightarrow \mathrm{H}^0(\hat{X}, L_{\hat{X}}\otimes \hat{N}^*)\longrightarrow \mathrm{H}^0(X, \mathcal{T}X)\longrightarrow 0,$$

$$\begin{aligned}
0 \to \mathrm{H}^0\left(X, N\otimes S^2(N^*)\right) &\to \mathrm{H}^0\left(\hat{X}, L_{\hat{X}}\otimes S^2(\hat{N}^*)\right) \to \mathrm{H}^0(X, N^*\otimes \mathcal{T}X) \to \\
\to \mathrm{H}^1\left(X, N\otimes S^2(N^*)\right) &\to \mathrm{H}^1\left(\hat{X}, L_{\hat{X}}\otimes S^2(\hat{N}^*)\right) \to \mathrm{H}^1(X, N^*\otimes \mathcal{T}X) \to \dots
\end{aligned}$$

and

$$\begin{aligned}
0 \to \mathrm{H}^0\left(X, N\otimes S^3(N^*)\right) &\to \mathrm{H}^0\left(\hat{X}, L_{\hat{X}}\otimes S^3(\hat{N}^*)\right) \to \mathrm{H}^0\left(X, S^2(N)^*\otimes \mathcal{T}X)\right) \xrightarrow{i} \\
\to \mathrm{H}^1\left(X, N\otimes S^2(N^*)\right) &\to \mathrm{H}^1\left(\hat{X}, L_{\hat{X}}\otimes S^3(\hat{N}^*)\right) \to \mathrm{H}^1\left(X, S^2(N)^*\otimes \mathcal{T}X\right) \to \dots
\end{aligned}$$

These together with Theorem 12 immediately imply

Proposition 15 *Let $X \hookrightarrow Y$ be a compact complex rigid submanifold with rigid normal bundle N such that $\mathrm{H}^1(X, N) = 0$ and let M be the associated Kodaira moduli space. If*

$$\mathrm{H}^1\left(X, N\otimes S^2(N^*)\right) = \mathrm{H}^1\left(X, N^*\otimes \mathcal{T}X\right) = 0,$$

then the associated Kodaira moduli space M comes equipped with an induced 1-flat G-structure $\mathcal{G}_{ind}$ with the Lie algebra $\mathfrak{g}$ of G being characterized by the following exact sequence of Lie algebras

$$0\longrightarrow \mathrm{H}^0(X, N\otimes N^*)\longrightarrow \mathfrak{g}\longrightarrow \mathrm{H}^0(X, \mathcal{T}X)\longrightarrow 0$$

If $\mathrm{H}^1\left(X, S^2(N^*)\otimes \mathcal{T}X\right) = 0$ as well, the obstruction to 2-flatness of $\mathcal{G}_{ind}$ is given by a section of the associated vector bundle

$$\mathcal{G}_{ind}\times_G \left[\mathrm{H}^1(X, N\otimes S^3(N^*))/i\left(\mathrm{H}^0(X, S^2(N^*)\otimes \mathcal{T}X)\right)\right].$$

EXAMPLE 14 (*Geometry of self-dual 4-manifolds*). Let X be a rational curve embedded into a complex 3-fold Z with normal bundle $N = \mathbb{C}^2 \otimes \mathcal{O}(1)$. Then all vanishing cohomology conditions of Proposition 15 are satisfied. Since $\mathrm{H}^0(X, N \otimes N^*) + \mathrm{H}^0(X, TX) \simeq co(4, \mathbb{C})$, we conclude that the associated Kodaira moduli space M comes equipped with the induced conformal structure. Since

$$\left[\mathrm{H}^1(X, N \otimes S^3(N^*))/i\left(\mathrm{H}^0(X, S^2(N^*) \otimes TX)\right)\right] \simeq S^4(\mathbb{C}^2),$$

this induced conformal structure must be conformally half-flat (say, self-dual). If there is an antiholomorphic involution in Z which fixes X, then the complex Kodaira moduli space M is locally a complexification of a real analytic 4-fold M_0 with real self-dual conformal structure. In fact, Z fibers over M_0 with fibers $\mathbb{C}P^1$ which are often called twistor lines.

This is a variant of the construction of Chapter 2, §2, where one may find also its reversion which associates to a given real self-dual conformal 4-manifold M the *complex* manifold $Z = P_M(S_+)$ such that all fibres of the natural projection $\pi : P_M(S_+) \to M$ are rational curves *holomorphically* embedded into Z with normal bundle $N = \mathbb{C}^2 \otimes \mathcal{O}(1)$. Z is called the twistor space of M.

Given two oriented 4-manifolds M_1 and M_2, by removing a small ball from each and then identifying the resulting boundary spheres by a reflection, one obtains a new oriented 4-manifold which is called the connected sum of M_1 and M_2 and is denoted $M_1 \# M_2$. The projective plane $\mathbb{C}P^2$ with the Fubini-Study metric is a self-dual Riemannian manifold. Using twistor methods, Poon (1986) constructed families of self-dual structures on $\mathbb{C}P^2 \# \mathbb{C}P^2$. Motivated by this example, Donaldson and Friedman (1989) studied general conditions under which the connected sum $M_1 \# M_2$ of two self-dual Riemannian 4-manifolds admits a self-dual structure. One of their main results can be described as follows. Let Z_1 and Z_2 be the twistor spaces of M_1 and M_2 respectively. After blowing up arbitrary twistor lines l_1 and l_2 in Z_1 and Z_2 respectively, one can glue the resulting manifolds $\tilde{Z}_1$ and $\tilde{Z}_2$ along their exceptional divisors Q in such a way that the resulting complex space $Z = \tilde{Z}_1 \cup_Q \tilde{Z}_2$ comes equipped with an antiholomorphic involution σ_0 and only mild singularities. Donaldson and Friedman found the following condition for Z to admit "smoothing".

Theorem 16 (Donaldson and Friedman) *If* $\mathrm{H}^1(Z_i, TZ_i)=0$ *for* $i = 1, 2$, *then* Z *admits a standard deformation, that is a complex family* $\varpi : \mathcal{Z} \to S$ *where* $\mathcal{Z}$ *is smooth,* ϖ *is proper, and* S *is a small open ball in* $\mathbb{C}^n$ *centered at the origin such that:*

(1) $\mathcal{Z}$ *has an antiholomorphic involution* σ *compatible with complex conjugation in* S.

(2) The central fibre $\varpi^{-1}(0)$ *is the singular complex space* Z *whose antiholomorphic involution* σ_0 *coincides with the induced one from* σ.

(3) Let $S \cap \text{Fix}(\sigma) = \{(r_1, r_2, \ldots, r_n) \mid r_i \in \mathbb{R}^n, i = 1, \ldots, n\}$. *For any* $\vec{t} \in S \setminus \{t_1 = 0\}$, $\varpi^{-1}(\vec{t})$ *is a smooth complex manifold, and for any real vector* $\vec{r} = (r_1, r_2, \ldots, r_n)$ *with* $r_1 \neq 0$ *the fibre* $\varpi^{-1}(\vec{r})$ *is a twistor space of a self-dual metric on the connected sum* $M_1 \# M_2$.

One of the immediate corollaries of this theorem is that for any $n > 0$, the connected sum

$$n\mathbb{C}P^2 = \underbrace{\mathbb{C}P^2\# \ldots \#\mathbb{C}P^2}_{n}$$

admits a self-dual Riemannian metric. Explicit descriptions of such metrics have been obtained by LeBrun (1991). Pedersen and Poon (1994) proved that if the twistor space Z of a compact simply connected self-dual Riemannian manifold M is such that $(\Lambda^3 TZ)^{1/2}$ admits a non-trivial real holomorphic section, then M is diffeomorphic to $\tau\mathbb{C}P^2$, where τ is the signature of M. It follows, in particular, that if the 4-manifold is known to be homeomorphic to $4\mathbb{C}P^2$, then it is also diffeomorphic to $4\mathbb{C}P^2$.

It is well known that the signature of an oriented compact 4-manifold M can be written as

$$\tau(M) = \frac{1}{12\pi^2} \int_M (\|W_+\|^2 - \|W_-\|^2)\text{Vol}_g,$$

where W_+ and W_- are self-dual and anti-self-dual parts of the Weyl tensor of any Riemannian g metric on M. So there must exist lots of 4-manifolds which do *not* admit self-dual metrics. Nevertheless, the obstruction to the existence of self-dual metrics is rather mild:

Theorem 17 (Taubes) *For any smooth compact oriented 4-dimensional manifold* M*, the connected sum* $M \# n\mathbb{C}P^2$ *admits a self-dual Riemannian metric for all sufficiently large* n.

Using this theorem, LeBrun (1995) gave a very simple argument proving that, for any finitely-presented group Γ, there exists a compact complex manifold Z with $\pi_1(Z) = \Gamma$. Indeed, by performing elementary surgeries on S^4 one may first construct a compact oriented 4-manifold M with $\pi_1(M) = \Gamma$. According to Taubes' result, there is a natural number n such that $\tilde{M} = M \# n\mathbb{C}P^2$ admits a self-dual metric g, and $\pi_1(\tilde{M}) = \pi_1(M) = \Gamma$ by the Seifert-van Kampen theorem. The twistor space Z of $(\tilde{M}, g)$ is a compact complex manifold which fibers over $\tilde{M}$ with fibres S^2. Then the exact homotopy sequence of this fibration, $\pi_1(S^2) \to \pi_1(Z) \to \pi_1(\tilde{M}) \to 0$, implies that $\pi_1(Z) = \Gamma$.

Example 15 (*Compact Hermitian and Kähler surfaces*). Let (M, g) be a 4-dimensional Kähler manifold. There are two natural decompositions of $\Omega^2 M$,

$$\Omega^2 M = \Omega^2_+ M + \Omega^2_- M$$

which is associated with the conformal class of g, and

$$\Omega^2 M = \mathbb{C}\omega + \Omega_0^{1,1} M + \Omega^{0,2} M + \Omega^{2,0} M$$

which is associated with the complex structure and the Kähler form ω on M. Here $\Omega^2_\pm M$ is the eigenspace of the Hodge operator $*$ with eigenvalue ± 1, and $\Omega_0^{1,1} M := \{\phi \in \Omega^{1,1} M \mid \omega \wedge \phi = 0\}$ is the bundle of primitive 1-forms. It is not hard to check that these decompositions are related to each other as follows

$$\begin{aligned} \Omega^2_+ &= \mathbb{C}\omega + \Omega^{0,2} M + \Omega^{2,0} M \\ \Omega^2_- &= \Omega_0^{1,1} M. \end{aligned}$$

Gauduchon used this fact to show that a Kähler metric g on a complex surface M is anti-self-dual if and only if its Ricci scalar vanishes. Thus, a scalar-flat Kähler surface has an associated twistor space Z.

In the other direction, let Z be a twistor space of an oriented anti-self-dual Riemannian manifold (M, g). The argument given in Chapter 2, §2, subsect. 9, shows that, as a S^2-fibration over M, the twistor space Z can be identified with the bundle of almost complex structures on M which are compatible with g and induce the given orientation on M. So, if (M, g, J) is anti-self-dual Hermitian, then the complex structure J defines a cross section of the fibration $Z \to M$ whose image we denote by Σ. Analogously, $-J$ defines a hypersurface $\bar{\Sigma} \subset Z$. The real structure σ on Z interchanges Σ and $\bar{\Sigma}$. Pontecorvo (1992) proved that (M, g, J) is conformally Kähler if and only if the divisor line bundle of $[\Sigma + \bar{\Sigma}]$ is isomorphic to $(\Lambda^3 TZ)^{1/2}$. This implies the following result, originally obtained by Boyer (1986) with the help of completely different methods: *Let $\pi : Z \to M$ be the twistor fibration of a compact conformal anti-self-dual 4-manifold $(M, [g])$. Let $\Sigma \subset Z$ be a hypersurface which meets each fibre of π in exactly one point. If $b_1(M)$ is even, then the conformal class $[g]$ contains a scalar-flat metric; if $b_1(M)$ is odd, $[g]$ is conformally equivalent to a Kähler metric with strictly positive scalar curvature almost everywhere.* Using this twistor construction, Boyer and Pontecorvo obtained a classification of compact conformally flat Hermitian surfaces.

Using an extension of the Donaldson-Friedman method, Kim, Lebrun and Pontecorvo (1994) proved the following Kähler analogue of Taub's existence theorem: *Let (M, J) be a compact complex 2-manifold which admits a Kähler metric for which the integral of the scalar curvature tensor is non-negative. Then precisely one of the following holds:*

- *(M, J) admits a Ricci-flat Kähler metric;*
- *any blow-up of (M, J) has blow-ups which admit scalar flat Kähler metrics.*

In particular there exist scalar flat Kähler metrics on the following surfaces

- *$\mathbb{C}P^1 \times \mathbb{C}P^1$ blown up at 13 suitably chosen points;*
- *$E \times \mathbb{C}P^1$ blown up at 6 suitably chosen points, where E is any elliptic curve.*

EXAMPLE 16 (*Quaternionic manifolds*). Let X be a rational curve embedded into a complex $(2k+1)$-manifold Z with normal bundle $N = \mathbb{C}^{2k} \otimes \mathcal{O}(1)$. The first half of Proposition 15 implies that the associated Kodaira moduli space M comes equipped with the induced complexified quaternionic structure, while the second half of this proposition gives a local flatness criterion which was originally found by Bailey and Eastwood (1991) with the help of very different arguments. That any quaternionic manifold M can be constructed in this way is shown by Salamon (1986).

CHAPTER B

GEOMETRY ON SUPERMANIFOLDS

§ 1. Simple Supergravity in Dimension Three

1. Basic structures. Let M be a $(3|2)$-dimensional complex supermanifold. *A superconformal structure* on M is, by definition, a rank $(0|2)$ locally direct locally free subsheaf $\mathcal{T}_1 M \subset \mathcal{T}M$ such that the Frobenius map

$$\begin{array}{rccc} \phi: & \Lambda^2(\mathcal{T}_1 M) & \longrightarrow & \mathcal{T}_0 M := \mathcal{T}M/\mathcal{T}_1 M \\ & X \otimes Y & \longrightarrow & [X,Y] \bmod \mathcal{T}_1 M \end{array}$$

is an isomorphism. Defining the rank $(2|0)$ *spinor* bundle $\mathcal{S} := \Pi \mathcal{T}_1 M$ and noting that $\Lambda^2 \mathcal{T}_1 M \simeq S^2(\mathcal{S})$, we may write

$$0 \longrightarrow \Pi\mathcal{S} \xrightarrow{\ i\ } \mathcal{T}M \longrightarrow S^2(\mathcal{S}) \longrightarrow 0 \tag{10}$$

which implies

$$\mathrm{Ber} M = \mathrm{Ber}\left(S^2(\mathcal{S})\right)^* \otimes \mathrm{Ber}(\Pi\mathcal{S})^* = (\mathrm{Ber}\mathcal{S})^{-3} \otimes \mathrm{Ber}\mathcal{S} = (\mathrm{Ber}\mathcal{S})^{-2} = (\Lambda^2\mathcal{S}^*)^2.$$

A complex superconformal $(3|2)$-fold equipped with a superconformal structure is called an $(N=1, D=3)$ *superconformal supermanifold.* It is not hard to check that the isotropic super-Grassmanian $M = GI(0|2; \mathbb{C}^{4|1}; b)$, $b \in \mathrm{Osp}(1|4)$, together with the rank $(0|2)$ tautological vector bundle $\mathcal{T}_1 M$ is superconformal.

A *scale* on an $N=1, D=3$ superconformal supermanifold M is, by definition, a nowhere vanishing section ε of $\Lambda^2\mathcal{S}^*$. A choice of scale on M determines an associated volume form $\omega = \varepsilon^2$ and thus provides us with a well-defined integration theory.

A scaled superconformal supermanifold M is called *Lorentzian* (resp. *Euclidean*), if there is a real structure ρ on M of type (1,1,1) (resp. (-1,1,1)) which has a 3-dimensional manifold of fixed points in M_{rd} and satisfies the following two conditions : $\rho_*(\mathcal{S}) = \overline{\mathcal{S}}$ (resp. $\rho_*(\mathcal{S}) = \mathcal{S}$) and $\rho_*(\epsilon) = \epsilon$.

Clearly, the Frobenius form induces a factorization of the tangent bundle of the 3-manifold M_{rd} underlying M as a symmetric tensor product $\mathcal{T}M_{rd} = S^2(\mathcal{S}_{rd})$ which in turn specifies canonically a line subbundle $S^2(\Lambda^2\mathcal{S}^*_{rd}) \subset S^2(\emptyset\, M_{rd})$ which may be identified with the equivalence class C of conformally related complex metrics. Thus, if M is superconformal, then M_{rd} is conformal in the usual sense.

A scale on M induces a spinor metric ε_{rd} on $\mathcal{S}_{rd} \to M_{rd}$ and hence a particular complex metric g_{rd} in the conformal class C. It is clear that g_{rd} restricted to the subset of ρ-fixed points is real Lorentzian (resp. Euclidean) if M is Lorentzian (resp. Euclidean).

2. Superconformal affine connections. Let $D : T_1M \otimes T_1M \to T_1M$ be a *spinor* covariant differential satisfying, by definition, the Leibnitz rule $D(fX \otimes gY) = f(Xg)Y + fg(-1)^{\tilde{g}\tilde{X}} D(X \otimes Y)$ for any sections X, Y of T_1M and any functions f, g on M. For any section X of T_1M, define the covariant operator D_X as the composition $T_1M \longrightarrow X \otimes T_1M \xrightarrow{D} T_1M$. It is easy to check that the map

$$\begin{array}{rccl} j: & \Lambda^2(T_1M) & \longrightarrow & TM \\ & X \otimes Y & \longrightarrow & [i(X), i(Y)] - D_XY - D_YX \end{array}$$

is an $\mathcal{O}_M$-linear morphism which, therefore, splits extension (10) to give

$$TM = i(\Pi S) \oplus j\left(S^2(S)\right). \tag{11}$$

To simplify further arguments (and to explain the notations adopted in the vast physics literature on supergravity theories) we introduce a structure frame $e_A = (e_\alpha, e_{\mu\nu})$ associated to any chosen spinor derivative D as follows [1]: e_α denotes a local frame of T_1M, while $e_{\mu\nu} := j(e_\mu \otimes e_\nu)$. Then, according to (11), any tangent vector field V on M can be represented in the structure frame e_A as the set of functions $V^A = (V^\alpha, V^{\beta\gamma})$.

Let $\nabla : TM \otimes TM \to TM$ be a linear connection, i.e. a covariant operator satisfying the rule

$$\nabla(fX \otimes gY) = f(Xg)Y + fg(-1)^{\tilde{g}\tilde{X}} \nabla(X \otimes Y)$$

for any sections X, Y of TM and any functions f, g on M. As before, for any section X of TM we define the covariant operator ∇_X as the composition $TM \longrightarrow X \otimes TM \xrightarrow{\nabla} TM$.

If D is a spinor derivative, then the associated decomposition (11) of TM determines in turn the associated decompositions of the bundles of formal curvature and torsion tensors. It is convenient to describe the latter decompositions in a structure frame e_A as follows. With the identification $V \simeq V^A$, one can define the components of the torsion T and the curvature R tensors of a linear connection ∇ by the formula

$$[\nabla_A, \nabla_B] V^C = -T_{A,B}{}^E \nabla_E V^C + R_{A,B,E}{}^C V^E$$

where $\nabla_A := \nabla_{e_A}$.

Proposition-definition 18 *For any given spinor derivative $D : T_1M \otimes T_1M \to T_1M$, there is a unique linear connection $\nabla : TM \otimes TM \to TM$, called a connection consistent with the superconformal structure, such that the restriction of ∇ to $T_1M \otimes T_1M$ coincides with D and, in any structure frame associated to D, the torsion tensor T of ∇ satisfies the conditions*

$$T_{\alpha,\beta}{}^\gamma = 0, \quad T_{\alpha,\mu\nu}{}^\gamma = 0, \quad T_{\alpha,\beta}{}^{\mu\nu} = \delta^\mu_{(\alpha}\delta^\nu_{\beta)}. \tag{12}$$

---

[1] Small Greek letters $\alpha, \beta, \mu, \ldots$ run over the set $\{1,2\}$, while capital Latin indices $A, B, C, \ldots$ stand for the pairs of indices of the type $(\alpha, \mu\nu)$.

A sketch of proof. In a coordinate chart (x^m, θ^μ), $m = 1, 2, 3$, the local frame e_α is expressed as $e_\alpha = E^\mu_\alpha \partial_\mu + E^m_\alpha \partial_m$. Defining the coefficients of D in the frame e_α,

$$D(e_\alpha \otimes e_\beta) = -\omega_{\alpha\beta}{}^\gamma e_\gamma \tag{13}$$

we find $e_{\mu\nu} = [e_\mu, e_\nu] + 2\omega_{(\mu\nu)}{}^\gamma e_\gamma$. Defining the coefficients $\Gamma_{AB}{}^C$ of ∇ in e_A as in (13), one easily checks that (12) is a system of linear equations for $\Gamma_{AB}{}^C$ which has a unique solution in terms of $\omega_{\alpha\beta}{}^\gamma$ and the vielbeins E^μ_α, E^m_α. In particular, $\Gamma_{\alpha\beta}{}^\gamma = \omega_{\alpha\beta}{}^\gamma$. □

3. Levi-Civita connection. Let ε be a scale on M. A linear connection $\nabla : \mathcal{T}M \otimes \mathcal{T}M \to \mathcal{T}M$ is said to be *consistent with the scale* if $\nabla\varepsilon = 0$.

Proposition-definition 19 *Let M be a superconformal $(3|2)$-fold with a fixed scale ε. Then there is a unique linear connection ∇ on M, called the Levi-Civita connection, which is consistent with both superconformal structure and the scale and whose torsion-tensor satisfies the condition*

$$T_{\alpha,\mu\nu}{}^{\gamma\delta} = 0. \tag{14}$$

A sketch of proof. In a local structure frame the scale ε can be described as a symplectic metric

$$\varepsilon_{\mu\nu} = \Phi\eta_{\mu\nu},$$

where $\Phi = \Phi(x, \theta)$ is some even function on M, and $\nu_{11} = \nu_{22} = 0$, $\nu_{12} = -\nu_{21} = 1$. If $\omega_{\alpha\beta}{}^\gamma$ are the coefficients of D, then, as is not hard to check, equations (14) form a system of linear equations for the symmetrized part $\omega_{(\alpha\beta)\gamma}$ of $\omega_{\alpha\beta\gamma} := \varepsilon_{\gamma\delta}\omega_{\alpha\beta}{}^\delta$ which has a unique solution expressing $\omega_{(\alpha\beta)\gamma}$ in terms of the vielbeins E^μ_α and E^m_α. The requirement $D_\alpha \varepsilon_{\mu\nu} = 0$ fixes the antisymmetric part $\omega_{\alpha[\beta\gamma]} = \varepsilon_{\beta\gamma} D_\alpha \Phi$. This is sufficient to determine the full spinor connection

$$\omega_{\alpha\beta\gamma} = \omega_{(\alpha\beta)\gamma} - \omega_{(\alpha\gamma)\beta} + \omega_{(\beta\gamma)\alpha} - \omega_{\gamma[\beta\alpha]} + \omega_{\beta[\alpha\gamma]} - -\omega_{\alpha[\gamma\beta]}$$

in terms of E^μ_α, E^m_α and Φ. □

A standard analysis of the second Bianchi identity for the Levi-Civita connection implies the following decomposition of the non-vanishing components of the torsion and curvature tensors in terms of only two irreducible superfields R and $W_{\alpha\beta\gamma} = W_{(\alpha\beta\gamma)}$

$$\begin{aligned}
R_{\alpha,\beta,\gamma}{}^\delta &= -(\varepsilon_{\alpha\gamma}\delta^\delta_\beta + \varepsilon_{\beta\gamma}\delta^\delta_\alpha) R \\
R_{\alpha,\mu\nu,\beta}{}^\gamma &= -\nabla_{(\alpha} R_{\mu,\nu),\beta}{}^\gamma + \varepsilon^{\gamma\delta}(\varepsilon_{\alpha\mu} W_{\nu\beta\delta} + \varepsilon_{\alpha\nu} W_{\mu\beta\delta}) \\
T_{\alpha\beta,\mu\nu}{}^{\gamma\delta} &= -R_{\alpha,\beta,\mu\nu}{}^{\gamma\delta} = -2R_{\alpha,\beta,(\mu}{}^{(\gamma}\delta^{\delta)}_{\nu)} \\
T_{\alpha\beta,\mu\nu}{}^\gamma &= R_{\alpha,\mu\nu,\beta}{}^\gamma + R_{\beta,\mu\nu,\alpha}{}^\gamma \\
R_{\alpha\beta,\mu\nu,\gamma}{}^\delta &= \nabla_\gamma T_{\alpha\beta,\mu\nu}{}^\delta
\end{aligned}$$

which are related to each other as follows

$$\varepsilon^{\alpha\beta} D_\alpha W_{\beta\mu\nu} = \frac{1}{3}\nabla_{\mu\nu} R.$$

4. The action and component structure of $N = 1$, $D = 3$ supergravity. Let M be a superconformal (3|2)-fold with a fixed scale ε. The functional of local superfields $\Phi(x,\theta)$, $E^\mu_\alpha(x,\theta)$ and $E^m_\alpha(x,\theta)$

$$I = \int_M (R - \lambda)\varepsilon^{-2}$$

is called the action of $N = 1$, $D = 3$ Einstein supergravity with cosmological constant λ. It is invariant under the pseudo-group of general coordinate transformations

$$(x,\theta) \to (x' = f(x,\theta), \theta' = g(x,\theta)) \tag{15}$$

as well as changes of the frame

$$e_\alpha \to e'_\alpha = G^\beta_\alpha(x,\theta) e_\alpha. \tag{16}$$

Using the latter, we may always set $E^\mu_\alpha = \delta^\mu_\alpha$. Then the former transformations permit us to set all 'θ'-components of the remaining superfields $\Phi(x,\theta)$ and $E^m_\alpha(x,\theta)$ to zero except the following ones:

$$\begin{aligned} \Phi &= 1 + \eta_{\mu\nu}\theta^\mu\theta^\nu A(x) \\ E^m_\alpha &= \theta^\beta e^m_{\alpha\beta}(x) + \eta_{\mu\nu}\theta^\mu\theta^\nu \psi^m_\alpha. \end{aligned}$$

Thus, in the Wess-Zumino gauge the local structure of the scaled superconformal (3|2)-fold is described by the supermultiplet $(e^m_{\alpha\beta}, \psi^m_\alpha, A)$ consisting of a 3-dimensional metric tensor $g^{mn} = \eta^{\mu_1\nu_1}\eta^{\mu_2,\nu_2} e^m_{\mu_1\mu_2} e^n_{\nu_1\nu_2}$, Rarita-Schwinger spin-$\frac{3}{2}$ field ψ^m_α and the so-called auxiliary scalar field $A(x)$. The dynamics of these fields is described by the Euler-Lagrange equations of the functional I which, in invariant form, are

$$R = \lambda, \qquad W_{\alpha\beta\gamma} = 0.$$

5. Einstein-Weyl supergeometry. A *Weyl superspace* is, by definition, a superconformal (3|2)-fold equipped with a linear connection $\mathcal{D} : TM \otimes TM \to TM$ which is consistent with the superconformal structure and satisfies condition (14). The latter is called a *Weyl connection.*

In a local structure frame $(e_\alpha, e_{\mu\nu})$, a Weyl connection is represented by covariant differentials $\mathcal{D}_\alpha$ and $\mathcal{D}_{\mu\nu}$. If $\varepsilon_{\mu\nu} = \Phi\eta_{\mu\nu}$ is some scale on M, then

$$\mathcal{D}_\alpha \varepsilon_{\mu\nu} = \mathcal{A}_\alpha \varepsilon_{\mu\nu}$$

for some odd superfield $\mathcal{A}_\alpha$. Under a change of scale,

$$\Phi \to \hat{\Phi} = \Omega\Phi, \quad \mathcal{A}_\alpha \to \hat{\mathcal{A}}_\alpha = \mathcal{A}_\alpha - e_\alpha(\ln \Omega), \tag{17}$$

for some even non-vanishing function Ω. Note that the 1-form $\mathcal{A} = \mathcal{A}_\alpha e^{*\alpha}$ measures the difference between the Weyl connection $\mathcal{D}$ and the Levi-Civita connection ∇ of the chosen scale. Thus $\mathcal{D}$ can be thought of as the pair $(\mathcal{A}, \nabla)$ modulo changes of scale (17).

A standard analysis of the second Bianchi identity for the Weyl connection implies the following decomposition of the non-vanishing components of the torsion and curvature tensors in terms of two superfields $\mathcal{R}$ and $\mathcal{W}_{\alpha\beta\gamma}$

$$\begin{aligned}
R_{\alpha,\beta,\gamma}{}^{\delta} &= -(\varepsilon_{\alpha\gamma}\delta^{\delta}_{\beta} + \varepsilon_{\beta\gamma}\delta^{\delta}_{\alpha})\mathcal{R} \\
R_{\alpha,\mu\nu,\beta}{}^{\gamma} &= -\mathcal{D}_{(\alpha}R_{\mu,\nu),\beta}{}^{\gamma} + \varepsilon^{\gamma\delta}(\varepsilon_{\alpha\mu}\mathcal{W}_{\nu\beta\delta} + \varepsilon_{\alpha\nu}\mathcal{W}_{\mu\beta\delta}) \\
T_{\alpha\beta,\mu\nu}{}^{\gamma\delta} &= -R_{\alpha,\beta,\mu\nu}{}^{\gamma\delta} = -2R_{\alpha,\beta,(\mu}{}^{(\gamma}\delta^{\delta)}_{\nu)} \\
T_{\alpha\beta,\mu\nu}{}^{\gamma} &= R_{\alpha,\mu\nu,\beta}{}^{\gamma} + R_{\beta,\mu\nu,\alpha}{}^{\gamma} \\
R_{\alpha\beta,\mu\nu,\gamma}{}^{\delta} &= \mathcal{D}_\gamma T_{\alpha\beta,\mu\nu}{}^{\delta}
\end{aligned}$$

where the superfield $\mathcal{W}_{\alpha\beta\gamma}$, in contrast to the Levi-Civita connection case, is *not* irreducible — it only satisfies the condition $\mathcal{W}_{\alpha\beta\gamma} = \mathcal{W}_{(\alpha\beta)\gamma}$. If R and $W_{\alpha\beta\gamma}$ are the basic curvature superfields of the Levi-Civita connection ∇ of some fixed scale $\varepsilon_{\mu\nu}$ on M, then, as is not hard to check,

$$\begin{aligned}
\mathcal{R} &= R + \varepsilon^{\mu\nu}(\nabla_\mu A_\nu + 2A_\mu A_\nu) \\
\mathcal{W}_{\alpha\beta\gamma} &= W_{\alpha\beta\gamma} - \mathcal{A}_{(\alpha}\nabla_\beta\mathcal{A}_{\gamma)} + \nabla_{(\alpha}\nabla_\beta\mathcal{A}_{\gamma)} + \frac{1}{3}(\varepsilon_{\alpha\gamma}F_\beta + \varepsilon_{\beta\gamma}F_\alpha)
\end{aligned}$$

where

$$F_\alpha = \varepsilon^{\mu\nu}(\nabla_\mu\nabla_\nu\mathcal{A}_\alpha + \nabla_\mu\nabla_\alpha\mathcal{A}_\nu + \nabla_{\alpha\mu}\mathcal{A}_\nu).$$

A Weyl superspace M is called *Einstein-Weyl* if

$$\mathcal{W}_{(\alpha\beta\gamma)} = 0. \tag{18}$$

This is a system of non-linear differential equations for local superfields Φ, E^μ_α, E^m_α and A_α which is invariant under the pseudo-group of general coordinate transformations (15), changes of the local frame (16) and conformal rescalings (17). This gauge freedom can be used to set to zero all 'θ'-components of the local superfields, except for the following ones

$$\begin{aligned}
\Phi &= 1 + \eta_{\mu\nu}\theta^\mu\theta^\nu A(x) & E^\mu_\alpha &= \delta^\mu_\alpha \\
E^m_\alpha &= \theta^\beta e^m_{(\alpha\beta)}(x) + \eta_{\mu\nu}\theta^\mu\theta^\nu\psi^m_\alpha & A_\alpha &= \theta^\gamma B_{(\alpha\gamma)} + \eta_{\mu\nu}\theta^\mu\theta^\nu\psi_\alpha.
\end{aligned}$$

Thus in the Wess-Zumino gauge the local structure of a Weyl superspace is completely determined by the supermultiplet $(e^m_{\alpha\beta}, \psi^m_\alpha, A)$ of $N = 1$, $D = 3$ supergravity plus the Weyl 1-form $B_{\alpha\beta}$ and its spin-$\frac{1}{2}$ superpartner ψ_α.

6. Twistor transform of Einstein-Weyl superspaces. Let M be a superconformal (3|2)-fold. There is a natural (2|1)-conic structure

$$\begin{array}{ccc} i: \ P_M(\mathcal{S}) & \longrightarrow & G_M(2|1;\mathcal{T}M) \\ \pi_1 \downarrow & & \downarrow \pi_2 \\ M & = & M \end{array}$$

which is uniquely specified by the extension

$$0 \longrightarrow \Pi\mathcal{O}_F(-1) \longrightarrow i^*(\mathcal{I}) \longrightarrow \pi_1^*(\mathcal{S}) \otimes \mathcal{O}_F(-1) \longrightarrow 0,$$

where $\mathcal{I}$ is the tautological vector bundle on $G_M(2|1;\mathcal{T}M)$ and $\mathcal{O}_F(-1)$ is the tautological line bundle on $F = P_M(\mathcal{S})$. Using the arguments very similar to the ones of Chapter 1 §7, one can show that a linear connection $\nabla : \mathcal{T}M \otimes \mathcal{T}M \to \mathcal{T}M$, which is consistent with the superconformal structure, induces a conic connection on F which is integrable if and only if (M, ∇) is Einstein-Weyl. In the latter case we may suppose that M is sufficiently "smallßo that the quotient of the associated foliation of F is a supermanifold. Then we get a double fibration

$$Z \xleftarrow{\mu} F \xrightarrow{\pi_1} M,$$

associated with an Einstein-Weyl supermanifold M. The quotient space Z is a complex (2|1)-fold called a *minitwistor* superspace. By construction, it contains a (3|2)-parameter family of embedded rational curves $\mathbb{C}P^{1|0}$ (the images under μ of the fibres of $\pi_1 : F \to M$). It is not hard to check (cf. Chapter 2, §2 or §7) that the normal bundle of each curve fits into an exact sequence

$$0 \longrightarrow \Pi\mathcal{O}(1) \longrightarrow N \longrightarrow \mathcal{O}(2) \longrightarrow 0. \tag{19}$$

With the tools developed in §2 and §7 of Chapter 2 at hand, one may apply Vaintrob's supersymmetrization of the Kodaira theorem to show that, given a projective line $\mathbb{C}P^{1|0}$ embedded into a complex (3|2)-fold Z with normal bundle (19), the associated Kodaira moduli superspace M comes equipped with the induced Einstein-Weyl structure. This establishes a 1-1 correspondence

$$\left\{ \begin{array}{c} \text{embedding data, } \mathbb{C}P^{1|0} \hookrightarrow Z \\ \text{with normal bundle (19)} \end{array} \right\} \Longleftrightarrow \left\{ \begin{array}{c} \text{solutions of supersymmetric} \\ \text{Einstein-Weyl equations (18)} \end{array} \right\}$$

§ 2. Quaternionic Supermanifolds

1. Almost quaternionic structures. An *almost quaternionic supermanifold* M is, by definition, a complex $(4k|2k+2)$-fold, $k \geq 2$, together with the following structures (cf. Chapter 5, §7):

(a) Two integrable complex distributions $T_l M$, $T_r M \subset TM$ of ranks $0|2$ and $0|2k$ respectively whose sum in TM is direct. It is assumed that the Frobenious form

$$\begin{array}{rcl} \phi : \ T_l M \otimes T_r M & \longrightarrow & T_0 M := TM/(T_l M + T_r M) \\ X \otimes Y & \longrightarrow & [X,Y] \bmod (T_l M + T_r M) \end{array}$$

is an isomorphism.

(b) A real structure ρ on M of type $(-1,1,1)$ which has a $4k$-dimensional real manifold of fixed points in M_{rd} and which leaves both $T_l M$ and $T_r M$ invariant.

Being of purely odd rank, the distributions $T_l M$ and $T_r M$, respectively, are integrable to fibrations whose quotient spaces M_r and M_l, respectively, are supermanifolds with $\dim M_r = 4k|2k$ and $\dim M_l = 4k|2$. Thus the data (a) gives rise to the diagram

$$M_l \xleftarrow{\pi_l} M \xrightarrow{\pi_r} M_r.$$

Let $\rho' : \mathbb{C}^{2k+2|1} \to \mathbb{C}^{2k+2|1}$ be an antilinear automorphism given in natural coordinates as follows[2]:

$$(z^\alpha, z^{\dot\alpha}, \zeta)^{\rho'} = (\eta^{\alpha\beta}\bar z^\beta, \eta^{\dot\alpha\dot\beta}\bar z^{\dot\beta}, \bar\zeta),$$

where $\eta^{\alpha\beta}$ and $\eta^{\dot\alpha\dot\beta}$ are skew-symmetric matrices in the standard representation and the bar denotes complex conjugation. Then it is not hard to check (cf. Chapter 4, §3) that the flag superspace $F(2|0,2|1;\mathbb{C}^{2k+2|1})$ together with the real structure ρ which is induced from ρ', is an almost quaternionic supermanifold with the distributions $T_l M$ and $T_r M$ defined by the diagram

$$G(2|0;\mathbb{C}^{2k+2|1}) \xleftarrow{\pi_l} F(2|0,2|1;\mathbb{C}^{2k+2|1}) \xrightarrow{\pi_r} G(2|1;\mathbb{C}^{2k+2|1}).$$

Note that the real $4k$-fold of fixed points in the manifold M_{rd} underlying $F(2|0,2|1;\mathbb{C}^{2k+2|1})$ is the projective quaternionic space $\mathbb{H}P^k$.

2. Description by superfields. Let us choose a local coordinate system (x_l^m, θ_l^μ) in M_l and $(x_r^m, \theta_r^{\dot\mu})$ in M_r in such a way that $(x_l^m)_{rd}$ are equal to $(x_r^m)_{rd}$ and are real when restricted to $M_0 \subset M_{rd}$. Then the functions $(x^m := \frac{1}{2}(x_l^m + x_r^m), \theta_l^\mu, \theta_r^{\dot\mu})$ form a local coordinate system in M. As in Chapter 5, §7, define the nilpotent functions

$$H^m(x,\theta_l,\theta_r) = \frac{1}{2}(x_l^m - x_r^m)$$

and check that the vector fields

$$\nabla_\alpha = \frac{\partial}{\partial\theta_l^\alpha} + (A^{-1})_m^n \frac{\partial H^m}{\partial\theta_l^\alpha}\frac{\partial}{\partial x^n}, \quad \nabla_{\dot\alpha} = \frac{\partial}{\partial\theta_r^{\dot\alpha}} + (B^{-1})_m^n \frac{\partial H^m}{\partial\theta_r^{\dot\alpha}}\frac{\partial}{\partial x^n},$$

---

[2] Here and throughout §2, undotted Greek indices assume vales 1,2, while dotted ones $1,\ldots,2k$. Small Latin indices run over $1,\ldots,4k$.

where

$$A_n^m = \delta_n^m - \frac{\partial H^m}{\partial x^n}, \quad B_n^m = \delta_n^m + \frac{\partial H^m}{\partial x^n},$$

form a local frame of distributions $\mathcal{T}_l M$ and $\mathcal{T}_r M$ respectively. Therefore, the functions $H^m(x, \theta_l, \theta_r)$ encode the full local information about an almost quaternionic structure on M. They are defined up to coordinate transformations by

$$\begin{pmatrix} x_l \\ \theta_l \end{pmatrix} \to \begin{pmatrix} \tilde{x}_l = f_l(x_l, \theta_l) \\ \tilde{\theta}_l = g_l(x_l, \theta_l) \end{pmatrix}, \quad \begin{pmatrix} x_r \\ \theta_r \end{pmatrix} \to \begin{pmatrix} \tilde{x}_r = f_r(x_r, \theta_r) \\ \tilde{\theta}_r = g_r(x_r, \theta_r) \end{pmatrix},$$

which can be used to further specialize the coordinate system $(x^m, \theta_l^\mu, \theta_r^{\dot{\mu}})$ in a such way that the odd coordinate expansion of the superfield $H^m(x, \theta_l, \theta_r)$ acquires the form

$$\begin{aligned} H^m &= \theta_l^\mu \theta_r^{\dot{\nu}} e^m_{\mu\dot{\nu}}(x) + \theta_l^\mu \left(\sum_{i=2}^{2k} \theta_r^{\dot{\nu}_1} \dots \theta_r^{\dot{\nu}_i} \psi^m_{\mu, \dot{\nu}_1 \dots \dot{\nu}_i}(x) \right) \\ &\quad + \theta_l^1 \theta_r^2 \left(\sum_{i=1}^{2k} \theta_r^{\dot{\nu}_1} \dots \theta_r^{\dot{\nu}_i} \chi^m_{\mu, \dot{\nu}_1 \dots \dot{\nu}_i(x)} \right). \end{aligned}$$

Thus an almost quaternionic $(4k|2k+2)$-fold M can be viewed locally as a domain in $\mathbb{R}^{4k}$ equipped with an almost quaternionic structure given by fields $e^m_{\mu\dot{\nu}}(x)$ plus a collection of tensor fields $(\psi_{\mu, \dot{\nu}_1 \dots \dot{\nu}_i}(x), \chi^m_{\mu, \dot{\nu}_1 \dots \dot{\nu}_i}(x))$ which are antisymmetric over the dotted subscripts.

3. Structures induced on the underlying manifold. Let M_0 be a real $4k$-dimensional manifold of ρ-fixed points in M_{rd}. Its complexified tangent bundle factors as a tensor product $\mathbb{C} \otimes \mathcal{T}M_0 = \mathcal{S} \otimes \tilde{\mathcal{S}}$ of complex vector bundles $\mathcal{S} = (\Pi\mathcal{T}M_l)_{rd}$ and $\tilde{\mathcal{S}} = (\Pi\mathcal{T}M_r)_{rd}$ of ranks 2 and $2k$ respectively. The map ρ_* restricted to M_0 induces antilinear fibre preserving involutions $j_l : \mathcal{S} \to \mathcal{S}$ and $j_r : \tilde{\mathcal{S}} \to \tilde{\mathcal{S}}$ with $j_l^2 = j_r^2 = -1$. The subgroup of GL$(4k, \mathbb{C})$ which commutes with $j_l \otimes j_r$ is GL$(k, \mathbb{H}) \cdot$ Sp(1). Therefore, M_0 comes equipped with an induced GL$(k, \mathbb{H}) \cdot$Sp(1)-structure, which means that M_0 is almost quaternionic in the usual sense.

Conversely, let M_0 be an almost quaternionic $4k$-dimensional manifold. Then $\mathbb{C} \otimes \mathcal{T}M_0$ factors locally as a tensor product of some complex vector bundles $\mathcal{S}$ and $\tilde{\mathcal{S}}$ of rank 2 and $2k$ respectively. The obstruction to such a *global* factorization is a cohomology class $\epsilon \in \mathrm{H}^1(M_0, \mathbb{Z}_2)$. In other words, ϵ is an obstruction to the lifting of the given GL$(k, \mathbb{H}) \cdot$ Sp(1)-structure to a GL$(k, \mathbb{H}) \times$ Sp(1)-structure on M_0.

Fact 20 *There is a canonical functor from the category of analytic almost quaternionic $4k$-dimensional manifolds with $\epsilon = 0$ to the category of $(4k|2k+2)$-dimensional almost quaternionic supermanifolds.*

Proof. Let M be a complexification of M_0. We shall associate to M a $(4k|2k+2)$-dimensional almost quaternionic supermanifold $\mathcal{M}$.

Since $\epsilon = 0$, there is a factorization

$$\phi : \Omega^1 M \xrightarrow{\simeq} \mathcal{S}^* \otimes \tilde{\mathcal{S}}^*,$$

for some holomorphic vector bundles $\mathcal{S}$ and $\tilde{\mathcal{S}}$ of rank 2 and $2k$ respectively. Define the globally decomposable supermanifolds

$$N = \left(M, S_{\mathcal{O}_M}(\Pi\mathcal{S} + \Pi\tilde{\mathcal{S}})\right), \quad \mathcal{M}_l = (M, S_{\mathcal{O}_M}(\Pi\mathcal{S})), \quad \mathcal{M}_r = \left(M, S_{\mathcal{O}_M}(\Pi\tilde{\mathcal{S}})\right).$$

There is a chain of natural embeddings $M \overset{i}{\hookrightarrow} N \overset{j}{\hookrightarrow} \mathcal{M}_l \times \mathcal{M}_r$. Denote the ideal sheaves of M and N in $\mathcal{M}_l \times \mathcal{M}_r$ by I and J respectively. Since $\mathcal{O}_M = \mathcal{O}_{\mathcal{M}_l \times \mathcal{M}_r}/I$ and $J \subset I$, we have

$$0 \longrightarrow I/J^2 \longrightarrow \mathcal{O}_{\mathcal{M}_l \times \mathcal{M}_r}/J^2 \longrightarrow \mathcal{O}_M \longrightarrow 0,$$

where, as one may easily check, $I/J^2 = (j \circ i)_*(\Omega^1 M) + S_{\mathcal{O}_M}(\Pi\mathcal{S} + \Pi\tilde{\mathcal{S}})$. Next, we use the given almost quaternionic structure ϕ on M to "deformthe inclusion $(j \circ i)_* : \Omega^1 M \to I/J^2$ into the following morphism

$$\begin{array}{rccl} k: & \Omega^1 M & \longrightarrow & I/J^2 \\ & \omega & \longrightarrow & (j \circ i)_*\omega \ - \ <\phi(\omega), e_\mu \otimes e_{\dot{\nu}}> \theta^\mu \theta^{\dot{\nu}}, \end{array}$$

where $\{e_\mu\}$ and $\{e_{\dot{\nu}}\}$ are any local frames of $\mathcal{S}$ and $\tilde{\mathcal{S}}$ respectively, and $\{\theta^\mu\}$ and $\{\theta^{\dot{\nu}}\}$ are the associated dual frames of $\mathcal{S}^*$ and $\tilde{\mathcal{S}}^*$ respectively.

The final step in the construction is to define a $(4k|2k+2)$-fold $\mathcal{M} = (M, \mathcal{O}_{\mathcal{M}})$ by the exact sequence

$$0 \longrightarrow \Omega^1 M \xrightarrow{k} \mathcal{O}_{\mathcal{M}_l \times \mathcal{M}_r}/J^2 \longrightarrow \mathcal{O}_{\mathcal{M}} \longrightarrow 0,$$

and check that $\mathcal{M}$ is almost quaternionic as defined in subsection 1. In particular, in the Wes-Zumino gauge one has $H^m(x, \theta_l, \theta_r) = \phi^m_{\mu\dot{\nu}}(x)\theta_l^\mu \theta_r^{\dot{\nu}}$, where $\phi^m_{\mu\dot{\nu}}(x)$ are the coordinate components of ϕ. □

4. Quaternionic supermanifolds. Let $\mathcal{M}$ be an almost quaternionic $(4k|2k+2)$-fold. There is a natural $(2k|2k+1)$-conic structure

$$\begin{array}{rccc} i: & P_{\mathcal{M}}(\mathcal{S}) & \longrightarrow & G_{\mathcal{M}}(2k|2k+1; \mathcal{T}\mathcal{M}) \\ & \pi_1 \downarrow & & \downarrow \pi_2 \\ & \mathcal{M} & = & \mathcal{M} \end{array}$$

which is uniquely specified by the extension

$$0 \longrightarrow \Pi\mathcal{O}_{\mathcal{F}}(-1) + \pi_1^*(\tilde{\mathcal{S}}) \longrightarrow i^*(\mathcal{I}) \longrightarrow \pi_1^*(\tilde{\mathcal{S}}) \otimes \mathcal{O}_{\mathcal{F}}(-1) \longrightarrow 0,$$

where $\mathcal{I}$ is the tautological vector bundle on $G_M(2k|2k+1;TM)$ and $\mathcal{O}_F(-1)$ the tautological line bundle on $F := P_M(\mathcal{S})$.

The supermanifold M is called *quaternionic* if the conic structure F admi an integrable conic connection. In this case F is foliated, and we may assume th the associated quotient space Z is a supermanifold (this imposes no restrictions (the local structure of M but excludes certain global pathologies). Thus, if M quaternionic, there is a double fibration

$$Z \xleftarrow{\mu} F \xrightarrow{\pi_1} M,$$

where Z is a complex $(2k+1|1)$-fold called the *twistor* superspace of M. It com equipped with a $(4k|4k+2)$-parameter family of embedded rational curves $\mathbb{C}P^1$ (the images under μ of the fibres of $\pi_1 : F \to M$). This fact makes it very easy show that the underlying $4k$-manifold is quaternionic in the usual sense.

Notes

The information about contact and symplectic structures, compact complex hom geneous manifolds, holonomy groups, G-structures and Spencer cohomology grou used in Chapter A is standard – the appropriate references are [*1], [*5], [*13], [*2 and [*85]. The proof of Theorem 2 in §1 is a slight modification of arguments us in [*61]. The main references for §2 are [*39], [*62]–[*65]. The exposition of follows [*61] except Lemma 10 which is taken from [*41]. In Example 14, t Donaldson-Friedman theorem is stated in the form given in [*37].

Further details on the material presented in Part B can be found in [*57 [*60] and [*93]. In particular, supersymmetrizations of quaternionic Kähler a hyper-Kähler structures are given in [*59].

Among the important themes which have been not been discussed in the A dendum are the twistor theory of harmonic maps [*14], [*15], [*20], [*22], [*8 twistor approach to integrable systems [*18], [*55], [*56], [*92], Einstein equatio [*33], [*34], [*69], and several other systems of field equations [*23], [*24], [*3 [*91]; twistor analysis of symmetric manifolds [*4], [*10], [*16], [*17], [*38]; mer morphic structure of twistor spaces [*44], [*46], [*75], [*79]–[*81]; G-structures twistor type [*2] and conic structures [*28].

For the most complete exposition of SUSY-curves, flag superspaces and oth topics in holomorphic supergeometry we refer to the book [*53] and the pape [*51], [*52], [*54]; see also the papers [*30] and [*90] on the supersymmetrizati of the Kodaira-Spencer deformation theory, and [*47] on a super-analogue of t Kodaira embedding theorem.

Bibliography

Arnold, V.I.

*1. Mathematical Methods of Classical Mechanics, Springer, Berlin, 1978

Alekseevskii, D.V., Graev, M.M.

*2. Twistors and G-structures, Izvestya Math. 40, No.1 (1993), 1-31

Bailey, T.N., Eastwood, M.G.

*3. Complex paraconformal manifolds – their differential geometry and twistor theory, Forum Math. 3 (1991), 61–103

Baston, R.J.

*4. Almost hermitian symmetric manifolds I & II, Duke Math. J. 63 (1991), 81–112, 113–138

Baston, R.J., Eastwood, M.G.

*5. The Penrose transform: its interaction with representation theory, Oxford University Press, Oxford, 1989

Baston, R.J., Mason, L.

*6. Conformal gravity, the Einstein equations, and spaces of complex null geodesics, Class. Quantum Grav. 4 (1987), 815–826

Berger, M.

*7. Sur les groupes d'holonomie des variétés á connexion affine et des variétés Riemanniennes, Bull. Soc. Math. France 83 (1955), 279–330

Boyer, C.P.

*8. Conformal duality and compact complex surfaces, Math. Ann. 274 (1986), 517–526

*9. A note on hyper-Hermitian four-manifolds, Proc. Amer. Math. Soc. 102 (1988), 157–164

Bryant, R.

*10. Lie groups and twistor spaces, Duke Math. J. 52 (1985), 223–261

*11. Metrics with exceptional holonomy, Ann. of Math. 126 (1987), 525–576

*12. Two exotic holonomies in dimension four, path geometries, and twistor theory, Proc. Symposia in Pure Mathematics 83 (1991), 33–88

*13. Classical, exceptional, and exotic holonomies: a status report, preprint 1995

Burstall, F.E.

*14. A twistor description of harmonic maps of a 2-sphere into a Grassmanian, Math. Ann. 274 (1986), 61–74

*15. Twistor methods for harmonic maps, Springer Lect. Notes in Math. 1263 (1987), 55–96

Burstall, F., Gutt S., Rawnsley, J.

*16. Twistor spaces for Riemannian symmetric spaces, Math. Ann. 295 (1993), 729–743

Burstall, F.E., Rawnsley, J.H.

*17. Twistor theory for Riemannian symmetric spaces, Springer Lect. Notes in Math. 1424 (1990), 2–84

Carey, A., Hannabus, K., Mason, L.J., Singer, M.A.

*18. The Landau Lifschitz equations, elliptic curves and the Ward transform, Commun. Math. Phys. 154 (1993), 25–47

Chi, Q.-S., Schwachhöfer, L.J.

*19. Exotic holonomy on moduli spaces of rational curves, MPI preprint (1994) (to appear in Diff. Geo. and Appl.)

Davidov, J., Sergeev, A.G.

*20. Twistor spaces and harmonic maps, Russian Math. Surveys 48, No.3 (1993), 1–91

Donaldson, S., Friedman, R.

*21. Connected sums of self-dual manifolds and deformations of singular spaces, Nonlinearity 2 (1989), 191-239

Eastwood, M.G.

*22. Complexification, twistor theory, and harmonic maps from a Riemann surface, Bull. Amer. Math. Soc. 11 (1984), 317-328

*23. The generalized Penrose-Ward transform, Math. Proc. Camb. Phil. Soc. 97 (1985), 165-187

*24. The Penrose transform for curved ambitwistor space, Quart. Jour. Math. 39 (1988), 427–441

Gauduchon, P.

*25. Structure de Weyl-Einstein, Espaces de Twisteurs et Variétés de Type $S^1 \times S^3$, preprint.

*26. Structures de Weyl et théorèmes d'annulation sur une variété conforme autoduale, Ann. Sc. Norm. Sup. Pisa XVIII (1991), 563–629

*27. Self-dual Riemannian manifolds (according to Taubes, C.H., and others), Asterisque 216 (1993), 151–186

Goncharov, A.B.

*28. Generalized conformal structures on manifolds, Selecta Math. Sovietica 6 (1987), 307–340

Guillemin, V.

*29. The integrability problem for G-structures, Trans. Amer. Math. Soc. 116 (1965), 544–560

Flenner, H., Sundararaman, D.

*30. Analytic geometry of complex superspaces, Trans. Amer. Math. Soc. 330 (1992), 1–40

Hano, J., Ozeki, H.

*31. On the holonomy groups of linear connections, Nagoya Math. J. 10 (1956), 97–100

Hitchin, N.

*32. Complex manifolds and Einstein's equations. In: Twistor geometry and non-linear systems, Springer Lect. Notes Math. , vol. 970 (1982), 73–99

*33. Twistor spaces, Einstein metrics and isomonodromic deformations, J. Diff. Geom. 42 (1995), 30–112

Hitchin, N., Karlhede A., Lindström, U., Roček, M.

*34. Hyperkähler metrics and supersymmetry, Commun. Math. Phys. 108 (1987), 535–589

Jones, P.E., Tod, K.P.

*35. Minitwistor spaces and Einstein-Weyl geometry, Class. Quantum Grav. 2 (1985), 565–577

Kapranov, M.M., Manin, Yu.I.

*36. The twistor transformation and algebraic-geometric constructions of solutions of the field theory equations, Russian Math. Surveys, 41 (1986), 33–61

Kim, J., Pontecorvo, M.

*37. A new method of constructing scalar-flat Kähler surfaces, J. Diff. Geom. 41 (1995), 449–477

Kobak, P.Z.

*38. Birational correspondences between twistor spaces, Bull. London Math. Soc. 26 (1994), 186–190

Kodaira, K.

*39. A theorem of completeness of characteristic systems for analytic families of compact submanifolds of complex manifolds, Ann. Math. 75 (1962), 146–162.

LeBrun, C.

*40. Quaternionic-Kähler manifolds and conformal geometry, Math. Ann. 284 (1989), 353–376

*41. Thickenings and conformal gravity, Commun. Math. Phys. 139 (1991), 1–43

*42. On complete quaternionic-Kähler manifolds, Duke Math. J. 63 (1991), 723-743

*43. Explicit self-dual metrics on $\mathbb{C}P^2\#\ldots\#\mathbb{C}P^2$, J. Diff. Geom. 34 (1991), 223–253

*44. Twistors, Kähler manifolds, and bimeromorphic geometry I, J. Amer. Math. Soc. 5 (1992), 289–316

*45. Anti-self-dual Riemannian 4-manifolds, Marcel Dekker Lect. Notes in Pure and Applied Math. 169 (1995), 81–94

LeBrun, C., Poon, Y.S.

*46. Twistors, Kähler manifolds, and bimeromorphic geometry II, J. Amer. Math. Soc. 5 (1992), 317–325

LeBrun, C., Poon, Y.S., Wells, R.O., Jr.

*47. Projective embeddings of complex supermanifolds, Commun. Math. Phys. 126 (1990), 433–452

LeBrun, C., Salamon, S.

*48. Strong rigidity of positive quaternion-Kähler manifolds, Invent. Math. 118 (1994), 109–132

LeBrun, C., Singer, M.

*49. Existence and deformation theory for scalar-flat Kähler metrics on compact complex surfaces, Invent. Math. 112 (1993), 273–313

Madsen, A.B., Pedersen, H., Poon, Y.S., Swann, A.

*50. Compact Einstein-Weyl manifolds with large symmetry group, Odense preprint 1995

Manin, Yu.I.

*51. New dimensions in geometry, Springer Lect. Notes in Math. 1111 (1985), 59-99

*52. Neveu-Schwarz sheaves and differential equations for Mumford superforms, J. Geom. Phys. 5 (1988), 161–181

*53. Topics in Noncommutative Geometry, Princeton University Press, Princeton 1991

Manin, Yu.I., Voronov, A.

*54. Supercell decompositions of flag superspaces, Itogi Nauki i Tehniki 32 (1988), 27–70

Mason, L.J., Singer, M.A.

*55. The twistor theory of equations of KdV type I, Commun. Math. Phys. 166 (1994), 191–218

Mason, L.J., Sparling, G.A.J.

*56. Twistor correspondences for the soliton hierarchies, J. Geom Phys. 8 (1992), 24: 271

Merkulov, S.A.

*57. Twistor transform of Einstein-Weyl superspaces, Class. Quantum Grav. 8 (1991 2149–2162

*58. Superconformal geometry in three dimensions, J. Math. Phys. 33 (1992), 735–75'

*59. Quaternionic, quaternionic Kähler, and hyper-Kähler supermanifolds, Lett. Mat] Phys. 25 (1992), 7–16

*60. Simple supergravity, supersymmetric non-linear gravitons and supertwistor theor Class. Quantum Grav. 9 (1992), 2369–2393

*61. Moduli of compact Legendre submanifolds of complex contact manifolds, Mat] Research Lett. 1 (1994), 717–727; Existence and geometry of Legendre modu spaces, Math. Zeit. (1996)

*62. Geometry of relative deformations I, Marcel Dekker Lect. Notes in Pure ar Applied Math. 169 (1995), 199–237

*63. Moduli spaces of compact submanifolds of complex fibred manifolds, Math. Pro Cambridge Phil. Soc. 118 (1995), 71–91

Merkulov, S.A., Pedersen, H.

*64. Geometry of relative deformations II, Marcel Dekker Lect. Notes in Pure ar Applied Math. 169 (1995), 239–261

*65. Projective structures on moduli spaces of compact complex hypersurfaces, Pro Amer. Math. Soc. (1996)

Nitta, T., Takeuchi, M.

*66. Contact structures on twistor spaces, J. Math. Soc. Japan 39 (1987), 139–162

Onishchik, A.L.

*67. Inclusion relations between transitive compact transformations groups, Trudi Mos Mate. Obshchestva 11 (1962), 199-242 (Russian)

Pedersen, H.

*68. Einstein-Weyl spaces and $(1, n)$-curves in the quadric surface, Ann. Global An Geom. 4 (1986), 89–120

*69. Einstein metrics, spinning top motions and monopoles, Math. Ann. 274 (198(35–59

Pedersen, H., Poon, Y.S.

*70. Twistorial construction of quaternionic manifolds, In Proc. VIth Int. Coll. on Di Geom. , Cursos y Congresos, Univ. Santiago de Compostela, 61 (1989), 207–21ٍ

*71. Self-duality and differentiable structures on the connected sums of complex pr jective planes, Proc. Amer. Math. Soc. 121 (1994), 859–864

Pedersen, H., Swann, A.

*72. Riemannian submersions, four-manifolds and Einstein-Weyl geometry, Proc. Lo don Math. Soc. 66 (1993), 381–399

Pedersen, H., Tod, K.P.

*73. 3-dimensional Einstein-Weyl geometry, Adv. Math. 97 (1993), 74–109

Pontecorvo, M.

*74. Hermitian surfaces and a twistor space of algebraic dimension 2, Proc. Amer. Math. Soc. 113 (1991), 177–186

*75. Algebraic dimension of twistor spaces and scalar curvature of anti-self-dual metrics, Math. Ann. 291 (1991), 113–122

*76. On twistor spaces of anti-self-dual hermitian surfaces, Trans. Amer. Math. Soc. 331 (1992), 653–661

*77. Complex structures on quaternionic manifolds, Diff. Geom. and Its Appl. 4 (1994), 163–177

Poon, Y.S.

*78. Compact self-dual manifolds with positive scalar curvature, J. Diff. Geom. 24 (1986), 97–132

*79. Algebraic dimension of twistor spaces, Math. Ann. 282 (1988), 621–627

*80. Twistor spaces with meromorphic functions, Proc. Amer. Math. Soc. 111 (1991), 331–338

*81. On the algebraic structure of twistor spaces, J. Diff. Geom. 36 (1992), 451–491

Rawnsley, J.H.

*82. F-structures, F-twistor spaces and harmonic maps, Springer Lect. Notes in Math. 1164 (1985), 85–159

Salamon, S.M.

*83. Quaternionic-Kähler manifolds, Invent. Math. 67 (1982), 143–171

*84. Differential geometry of quaternionic manifolds, Ann. Sci. Éc. Norm. Sup. 19 (1986), 31–55

*85. Riemannian Geometry and Holonomy Groups, Pitman Research Notes Math. 201, Longman (1989)

Singer, I.M., Sternberg S.

*86. The infinite groups of Lie and Cartan I. J. Analyse Math. 15 (1965), 1–114

Schwachhöfer, L.

*87. Connections with exotic holonomy. Trans. Amer. Math. Soc. 345 (1994), 293–322

Taubes, C.H.

*88. The existence of anti-self-dual conformal structures, J. Diff. Geom. 36 (1992), 163–253

Tod, K.P.

*89. Compact 3-dimensional Einstein-Weyl structures, J. London Math. Soc. 45 (1992), 341–351

Vaintrob, A.Yu.

*90. Deformations of complex superspaces and coherent sheaves on them, Itogi Nauki i Teckhniki, Seriya Sovremennye Problemi Matematiki, 32 (1988), 125–211

Ward, R.S.,

*91. Classical solutions of the chiral model, unitons, and holomorphic vector bundles, Commun. Math. Phys. 128 (1990), 319–332

*92. Integrable systems in twistor theory, In Twistors in Mathematics and Physics", LMS Lect. Note Series 156 (1990), 246–259

Zupnik, B.M., Pak, D.G.

*93. Superfield formulations of simple three dimensional gauge and conformal supergravity theories, Teor. Mat. Fiz. 77 (1988), 97–105

INDEX

Grundlehren der mathematischen Wissenschaften

A Series of Comprehensive Studies in Mathematics

A Selection

210. Gihman/Skorohod: The Theory of Stochastic Processes I
211. Comfort/Negrepontis: The Theory of Ultrafilters
212. Switzer: Algebraic Topology – Homotopy and Homology
215. Schaefer: Banach Lattices and Positive Operators
217. Stenström: Rings of Quotients
218. Gihman/Skorohod: The Theory of Stochastic Procrsses II
219. Duvaut/Lions: Inequalities in Mechanics and Physics
220. Kirillov: Elements of the Theory of Representations
221. Mumford: Algebraic Geometry I: Complex Projective Varieties
222. Lang: Introduction to Modular Forms
223. Bergh/Löfström: Interpolation Spaces. An Introduction
224. Gilbarg/Trudinger: Elliptic Partial Differential Equations of Second Order
225. Schütte: Proof Theory
226. Karoubi: K-Theory. An Introduction
227. Grauert/Remmert: Theorie der Steinschen Räume
228. Segal/Kunze: Integrals and Operators
229. Hasse: Number Theory
230. Klingenberg: Lectures on Closed Geodesics
231. Lang: Elliptic Curves. Diophantine Analysis
232. Gihman/Skorohod: The Theory of Stochastic Processes III
233. Stroock/Varadhan: Multidimensional Diffusion Processes
234. Aigner: Combinatorial Theory
235. Dynkin/Yushkevich: Controlled Markov Processes
236. Grauert/Remmert: Theory of Stein Spaces
237. Köthe: Topological Vector Spaces II
238. Graham/McGehee: Essays in Commutative Harmonic Analysis
239. Elliott: Probabilistic Number Theory I
240. Elliott: Probabilistic Number Theory II
241. Rudin: Function Theory in the Unit Ball of C^n
242. Huppert/Blackburn: Finite Groups II
243. Huppert/Blackburn: Finite Groups III
244. Kubert/Lang: Modular Units
245. Cornfeld/Fomin/Sinai: Ergodic Theory
246. Naimark/Stern: Theory of Group Representations
247. Suzuki: Group Theory I
248. Suzuki: Group Theory II
249. Chung: Lectures from Markov Processes to Brownian Motion
250. Arnold: Geometrical Methods in the Theory of Ordinary Differential Equations
251. Chow/Hale: Methods of Bifurcation Theory
252. Aubin: Nonlinear Analysis on Manifolds. Monge-Ampère Equations
253. Dwork: Lectures on ρ-adic Differential Equations
254. Freitag: Siegelsche Modulfunktionen
255. Lang: Complex Multiplication
256. Hörmander: The Analysis of Linear Partial Differential Operators I
257. Hörmander: The Analysis of Linear Partial Differential Operators II
258. Smoller: Shock Waves and Reaction-Diffusion Equations
259. Duren: Univalent Functions
260. Freidlin/Wentzell: Random Perturbations of Dynamical Systems
261. Bosch/Güntzer/Remmert: Non Archimedian Analysis – A System Approach to Rigid Analytic Geometry

262. Doob: Classical Potential Theory and Its Probabilistic Counterpart
263. Krasnosel'skiĭ/Zabreĭko: Geometrical Methods of Nonlinear Analysis
264. Aubin/Cellina: Differential Inclusions
265. Grauert/Remmert: Coherent Analytic Sheaves
266. de Rham: Differentiable Manifolds
267. Arbarello/Cornalba/Griffiths/Harris: Geometry of Algebraic Curves, Vol. I
268. Arbarello/Cornalba/Griffiths/Harris: Geometry of Algebraic Curves, Vol. II
269. Schapira: Microdifferential Systems in the Complex Domain
270. Scharlau: Quadratic and Hermitian Forms
271. Ellis: Entropy, Large Deviations, and Statistical Mechanics
272. Elliott: Arithmetic Functions and Integer Products
273. Nikol'skiĭ: Treatise on the Shift Operator
274. Hörmander: The Analysis of Linear Partial Differential Operators III
275. Hörmander: The Analysis of Linear Partial Differential Operators IV
276. Liggett: Interacting Particle Systems
277. Fulton/Lang: Riemann-Roch Algebra
278. Barr/Wells: Toposes, Triples and Theories
279. Bishop/Bridges: Constructive Analysis
280. Neukirch: Class Field Theory
281. Chandrasekharan: Elliptic Functions
282. Lelong/Gruman: Entire Functions of Several Complex Variables
283. Kodaira: Complex Manifolds and Deformation of Complex Structures
284. Finn: Equilibrium Capillary Surfaces
285. Burago/Zalgaller: Geometric Inequalities
286. Andrianov: Quadratic Forms and Hecke Operators
287. Maskit: Kleinian Groups
288. Jacod/Shiryaev: Limit Theorems for Stochastic Processes
289. Manin: Gauge Field Theory and Complex Geometry
290. Conway/Sloane: Sphere Packings, Lattices and Groups
291. Hahn/O'Meara: The Classical Groups and K-Theory
292. Kashiwara/Schapira: Sheaves on Manifolds
293. Revuz/Yor: Continuous Martingales and Brownian Motion
294. Knus: Quadratic and Hermitian Forms over Rings
295. Dierkes/Hildebrandt/Küster/Wohlrab: Minimal Surfaces I
296. Dierkes/Hildebrandt/Küster/Wohlrab: Minimal Surfaces II
297. Pastur/Figotin: Spectra of Random and Almost-Periodic Operators
298. Berline/Getzler/Vergne: Heat Kernels and Dirac Operators
299. Pommerenke: Boundary Behaviour of Conformal Maps
300. Orlik/Terao: Arrangements of Hyperplanes
301. Loday: Cyclic Homology
302. Lange/Birkenhake: Complex Abelian Varieties
303. DeVore/Lorentz: Constructive Approximation
304. Lorentz/v. Golitschek/Makovoz: Constructive Approximation. Advanced Problems
305. Hiriart-Urruty/Lemaréchal: Convex Analysis and Minimization Algorithms I. Fundamentals
306. Hiriart-Urruty/Lemaréchal: Convex Analysis and Minimization Algorithms II. Advanced Theory and Bundle Methods
307. Schwarz: Quantum Field Theory and Topology
308. Schwarz: Topology for Physicists
309. Adem/Milgram: Cohomology of Finite Groups
310. Giaquinta/Hildebrandt: Calculus of Variations I: The Lagrangian Formalism
311. Giaquinta/Hildebrandt: Calculus of Variations II: The Hamiltonian Formalism
312. Chung/Zhao: From Brownian Motion to Schrödinger's Equation
313. Malliavin: Stochastic Analysis
314. Adams/Hedberg: Function Spaces and Potential Theory
315. Bürgisser/Clausen/Shokrollahi: Algebraic Complexity Theory
316. Saff/Totik: Logarithmic Potentials with External Fields

MIX
Papier aus verantwortungsvollen Quellen
Paper from responsible sources
FSC® C105338

If you have any concerns about our products,
you can contact us on
ProductSafety@springernature.com

In case Publisher is established outside the EU,
the EU authorized representative is:
Springer Nature Customer Service Center GmbH
Europaplatz 3, 69115 Heidelberg, Germany

Printed by Libri Plureos GmbH
in Hamburg, Germany